Combinatorics, Computation & Logic '99

Springer
*Singapore
Berlin
Heidelberg
New York
Barcelona
Budapest
Hong Kong
London
Milan
Paris
Tokyo*

Australian Computer Science Communications
Volume 21 Number 3

Combinatorics, Computation & Logic

99

PROCEEDINGS OF DMTCS'99 AND CATS'99
AUCKLAND, NEW ZEALAND, 18-21 JANUARY 1999

EDS.
C. S. Calude
M. J. Dinneen

Cris Calude and Michael J. Dinneen
Department of Computer Science
The University of Auckland
Private Bag 92019
Auckland
New Zealand

QA
164
.D62
1999

ISBN 981-4021-56-3

This work is subject to copyright. All rights are reserved, whether the whole or part of the material is concerned, specifically the rights of translation, reprinting, reuse of illustrations, recitation, broadcasting, reproduction on micro-films or in any other way, and storage in databanks or in any system now known or to be invented. Permission for use must always be obtained from the publisher in writing.

© Springer-Verlag Singapore Pte. Ltd. 1999
Printed in Singapore

The publisher makes no representation, express or implied, with regard to the accuracy of the information contained in this book and cannot accept any legal responsibility or liability for any errors or omissions that may be made.

Typesetting: Camera-ready by editors
SPIN 10712049   5 4 3 2 1 0

Preface

The second *Discrete Mathematics and Theoretical Computer Science conference* (DMTCS'99) was combined with the fifth *Computing: The Australasian Theory Symposium* (CATS'99). As part of the *Australasian Computer Science Week* (ACSW'99), January 18–21, 1999, these joint conferences were held in the beautiful New Zealand city of Auckland (City of Sails, Museums, Cafes, Restaurants, Polynesian and Maori Culture and more).

The five invited speakers of the conference were: R. Downey (Victoria University, NZ), J. Goguen (University of California at San Diego, USA), J. Pach (Hungarian Academy of Sciences, Hungary), A. Restivo (University of Palermo, Italy) and P.G. Walsh (University of Ottawa, Canada).

The Programme Committee consisting of R.J. Back (TUCS, Finland), M. Conder (University of Auckland, NZ), B. Cooper (University of Leeds, UK), M.J. Dinneen (University Auckland, NZ; Chair), R. Goldblatt (Victoria University, NZ) S. Goncharov (Novosibirsk University, Russia), J. Harland (RMIT, Australia), R.E. Hiromoto (UTSA, USA), H. Ishihara (JAIST, Japan), M. Ito (Kyoto Sangyo University, Japan), M. Li (University Waterloo, Canada), X. Lin (UNSW, Australia), R. Shore (Cornell University, USA), T. Tokuyama (IBM, Japan) and D. Wolfram (ANU, Australia) had a difficult task selecting 19 papers out of 53 submissions. The Commiittee thanks the following referees for their reports:

J. An	R.W. Doran	P. Hertling	R. Paterson
S. Arteniov	P. Eades	D. Holton	G. Păun
J.A. Bergstra	J.A. Ellis	S. Hong	F. Ruskey
H.L. Bodlaender	S. Ganguli	J. Hromkovič	A. Salomaa
C.P. Bonnington	J. Gibbons	H. Jürgensen	P. Sharp
D. Bridges	P.B. Gibbons	B. Khoussainov	J. Shepherd
C.S. Calude	G. Gimel'farb	M. Lipponen	Gh. Ştefănescu
E. Calude	O. Goldschmidt	S. Manoharan	C. Thomborson
V.E. Căzănescu	A. Gordon	R. Mathon	I. Tomescu
R. Coles	U. Guenther	D. Miller	Y. Toyama
C. Collberg	H. Guesgen	J. Miller	P.G. Walsh
S.B. Cooper	M.T. Hallett	M. Morton	H.T. Wareham
J.N. Crossley	P.R. Hafner	E. O'Brien	X.-F. Ye
J. Dassow	J. Hamer	H. Ono	S. Yu
P. Denny			

Finally, we want to acknowledge the dedication of the DMTCS'99+CATS'99 Conference Committee, which consisted of C.P. Bonnington (publicity), C.S. Calude (general chair), E. Calude (venue), R. Coles (records), P.B. Gibbons (accommodation), U. Guenther (registration) and B. Khoussainov (local chair). We thank our ACSW'99 contact members R.W. Doran (general chair) and P. Fenwick for organizing the 'big' Australasian conference. We also thank our Springer-Verlag, Singapore publishers R. Ali and G. Chee for producing the volume in time for the conference.

November 1998

C.S. Calude
M.J. Dinneen

Table of Contents

Invited papers

Rodney G. Downey and Michael R. Fellows
Parametric Complexity After (almost) Ten Years: Review and
Open Questions 1

Joseph A. Goguen
Hidden Algebra for Software Engineering 34

Filippo Mignosi and Antonio Restivo
On Negative Informations in Language Theory 60

János Pach
Crossroads in Flatland 73

P. Gary Walsh
Efficiency vs. Security in the Implementation of Public-Key
Cryptography 81

Contributed papers

Noriko H. Arai
No Feasible Monotone Interpolation for Cut-free Gentzen Type
Propositional Calculus with Permutation Inference 106

Michael D. Atkinson and Robert Beals
Permuting Mechanisms and Closed Classes of Permutations 117

Jan A. Bergstra and Alban Ponse
Process Algebra with Five-Valued Conditions 128

Vasco Brattka
A Stability Theorem for Recursive Analysis 144

**Alessandra Cherubini, Stefano Crespi–Reghizzi and
Pierluigi San Pietro**
Languages Based on Structural Local Testability 159

**Anders Strandløv Elkjær, Michael Höhle, Hans Hüttel and
Kasper Overgård**
Towards Automatic Bisimilarity Checking in the Spi Calculus 175

Manfred Göbel
Three Remarks on SAGBI Bases for Polynomial Invariants of
Permutation Groups 190

Raymond Greenlaw and Rossella Petreschi
Computing Prüfer Codes Efficiently in Parallel 202

Mohamed Hamada, Aart Middeldorp and Taro Suzuki
Completeness Results for a Lazy Conditional Narrowing Calculus 217

Toru Hasunuma
The Pagenumber of de Bruijn and Kautz Digraphs 232

Holger Hinrichsen, Hans Eveking and Gerd Ritter
Formal Synthesis for Pipeline Design 247

Costas S. Iliopoulos and Laurent Mouchard
An $O(n \log n)$ Algorithm for Computing all Maximal Quasiperiodicities
in Strings 262

Teodor Knapik and Hugues Calbrix
The Graphs of Finite Monadic Semi–Thue Systems Have a Decidable
Monadic Second–Order Theory 273

Lars Kristiansen
Low_n, $High_n$, and Intermediate Subrecursive Degrees 286

Ko-Wei Lih, Daphne D.-F. Liu and Xuding Zhu
Star-extremal Circulant Graphs 301

Laxmi Parida
On the Approximability of Physical Map Problems using Single Molecule
Methods 310

Arno Schönegge and David Kempe
On the Weakness of Conditional Equations in Algebraic Specification . 329

Naoki Shibata, Kozo Okano, Teruo Higashino and Kenichi Taniguchi
A Decision Algorithm for Prenex Normal Form Rational Presburger
Sentences Based on Combinatorial Geometry 344

Shao Chin Sung and Keisuke Tanaka
Lower Bounds on Negation-Limited Inverters 360

Parameterized Complexity After (Almost) Ten Years: Review and Open Questions

Rodney G. Downey[1] * and Michael R. Fellows[2] **

[1] School of Mathematics and Computing Sciences, P.O. Box 600, Victoria University, Wellington, New Zealand. rod.downey@vuw.ac.nz
[2] Department of Computer Science, University of Victoria, Victoria, B.C. V8W 3P6, Canada. mfellows@csr.uvic.ca

Abstract. We give a review of the development and some of the achievements of the theory of parameterized complexity in the last (nearly) ten years. We highlight what we see as some of the major open questions and programmes for future development.

Introduction

We begin our monograph [DF98] with the sentence:
"The idea for this book was conceived over the second bottle of Villa Maria's Cabernet Merlot '89, at the dinner of the Australasian Combinatorics Conference held at Palmerston North, New Zealand in May 1990, where the authors first met and discovered they had a number of interests in common."

It seems therefore most appropriate in this New Zealand conference to look back at the achievements of the area as well as looking at future directions for research. Even at this point we'd like to remark that the opportunities for future research are very exciting indeed!

So what is parameterized complexity? Classical complexity theory views questions as either decision problems "Does a graph have a red bandersnatch?" or, to a lesser extent, optimization questions, "What is the biggest red bandersnatch in the graph?" One of the "triumphs" of the last 30 years of complexity theory is the realization that most interesting combinatorial problems are essentially intractable, being NP-complete or worse.

One of the points stressed in Garey and Johnson [GJ79], is that a hardness result such as NP-completeness should be just the beginning of an attack on a problem. All it says is that the initial hope for an efficient exact general algorithm is in vain. The principal idea of parameterized complexity is to try to look more deeply at the *structure* of the input. In parameterized complexity we consider the input not just as a single entity, for example, a graph—but as a graph with

---

* Research supported by a grant from the United States/New Zealand Cooperative Science Foundation and by the New Zealand Marsden Fund for Basic Science.
** Research supported by grants from the National Science and Engineering Research Council of Canada.

an additional parameter. So the underlying languages are subsets of $\Sigma^* \times \Sigma^*$. The typical problem will be of the form below.
Input: $\langle x, k \rangle$
Parameter: k.
Question: Is $\langle x, k \rangle \in L$?

The point is that we are trying to look at the problem *by the slice*. That is, we ask what is the behaviour for a *fixed* k? There may be many different ways to slice a particular problem. The following three examples were in fact our original motivation and serve well to demonstrate that there is something of interest here.

VERTEX COVER
Instance: A graph $G = (V, E)$.
Parameter: A positive integer k.
Question: Does G have a vertex cover of size $\leq k$? (A vertex cover of a graph G is a collection of vertices V' of G such that for all edges $v_1 v_2$ of G either $v_1 \in V'$ or $v_2 \in V'$.)

DOMINATING SET
Input: A graph G.
Parameter: A positive integer k.
Question: Does G have a dominating set of size k? (A *dominating set* is a set $V' \subseteq V(G)$ where, for each $u \in V(G) - V'$ there is a $v \in V'$ such that $uv \in E(G)$.)

INDEPENDENT SET
Input: A graph G.
Parameter: A positive integer k.
Question: Does G have a set of k vertices $\{x_1, \ldots, x_k\}$ such that all $i \neq j$, x_i is not adjacent to x_j?

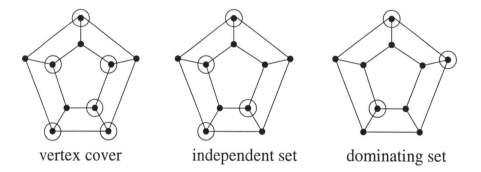

vertex cover   independent set   dominating set

Fig. 1. Three Vertex Set Problems.

All of these can be solved trivially by brute force, trying all possible k-subsets, and hence in time $\Omega(|G|^{k+1})$. For INDEPENDENT SET and DOMINATING SET, this is essentially the best known algorithm, while VERTEX COVER can be solved in time $O(|G|)$ for any fixed k. This fact has obvious implications for applications where the range of the parameter is small. Table 1 illustrates the difference between a running time of $\Omega(|G|^{k+1})$ (that is, where complete search is necessary), and one running in time $2^k|G|$. The latter has been achieved for several natural parameterized problems. (In fact, as we see below the constant 2^k can sometimes be significantly improved.)

	$n = 50$	$n = 100$	$n = 150$
$k = 2$	625	2,500	5,625
$k = 3$	15,625	125,000	421,875
$k = 5$	390,625	6,250,000	31,640,625
$k = 10$	1.9×10^{12}	9.8×10^{14}	3.7×10^{16}
$k = 20$	1.8×10^{26}	9.5×10^{31}	2.1×10^{35}

Table 1. The Ratio $\frac{n^{k+1}}{2^k n}$ for Various Values of n and k.

So what is the global issue? The point is that for *any* combinatorial problem we will deal with only a small finite fraction of the universe. We tend to forget that the original suggestions that asymptotic analysis was a reasonable way to measure complexity, and polynomial time was a reasonable measure of feasibility, were initially quite controversial. The reasons that the now central ideas of asymptotic polynomial time and NP-completeness have survived the test of time are basically two. Firstly, the methodologies associated with P-time and P-time reductions have proved to be readily amenable to mathematical analysis in most situations. In a word, P has good closure properties. It offers a readily negotiable mathematical currency. Secondly, and even more importantly, for "natural" problems the universe seems *kind* in the sense that if a "natural" computational problem is in P then usually we can find an algorithm having a polynomial running time with small degree and small constants. This last point, while seeming a religious, rather than mathematical, statement, seems the key driving force behind the universal use of P as a classification tool for "natural" problems.

The same kind of concrete considerations are behind the idea of *fixed parameter tractability*. Here a parameterized problem is called *fixed parameter tractable* (FPT) if, like VERTEX COVER above, there is an algorithm Φ solving the problem "efficiently by the slice". That is there is a constant c (independent of k and a function f such that Φ accepts $\langle x, k \rangle$ in time $f(k)|x|^c$. For VERTEX COVER, we can take $c = 1$.

If the constant $f(k)$ can be made small enough, then for a reasonable range of k, the algorithm is feasible for a very large instance size.

This consideration is extremely useful if we can get our hands on a good parameter with which to partition off the usual combinatorial explosion. Are such parameters usually avaiable? Are there any standard techniques to achieve parametric tractability? What should I do if I can't seem to attack the question in this way? Is the phenomenom widespread? Are there open questions? In general, why should I be interested in this area?

We attempt to answer these questions below.

Review

1 Why this is interesting: Applications often involve natural parameters

There are lots of areas of applications, and potential applications.

- Most computational problems involve several pieces of input, one or more of which may be a relevant parameter for various applications.
- For instance, VERTEX COVER can be often used to represent a *conflict graph* of some sort, and the parameter k in this situation models the cost of eliminating the conflicts. The FPT result for VERTEX COVER has many different applications through this natural generality.
- *Graph linear layout width metrics* are of interest in VLSI layout and routing problems and have important applications for width values of $k \leq 10$. Interval graphs of pathwidth $k \leq 10$ have applications in DNA sequence reconstruction problems and in other evolutionary problems such as language evolution and phylogenic tree reconstruction.
- *Logic and database problems* frequently are defined as having input consisting a formula (which may be small and relatively invariant), and some other structure (such as a database) which is typically quite large and changeable. Formula size, or other aspects of formula structure may be a relevant parameter.
- *Robotics.* The number of degrees of freedom in a robot motion-planning problem is commonly in the range $k \leq 10$.
- *VLSI.* The number of wiring layers in VLSI chip manufacture is typically bounded by $k \leq 30$.
- *Artificial Intelligence.* One of the central problems in Artificial Intelligence, STRIPS PLANNING, is known in general to PSPACE complete (Bylander [Byl94]) yet has parametric versions that are FPT (Downey-Fellows-Stege [DFS98]).
- *Network problems* may be naturally concerned with optimally locating a small number of facilities.
  These few examples are only suggestive and by no means exhaustive. There are myriad ways in which numbers that are small or moderately large (e.g., $k \leq 40$) arise naturally in problem specifications.

2 Why this is interesting: Parameters may be hidden and implicit

The above gave a sample of possible applications. There are many parameters that arise naturally in practice. Some of them are hidden, and some have been the arenas of intensive investigation in recent years. In this section we will look at this issue.

In classical complexity a decision problem is specified by two items of information:
(1) The input to the problem.
(2) The question to be answered.

In parameterized complexity there are three parts of a problem specification:
(1) The input to the problem.
(2) The aspects of the input that constitute the parameter.
(3) The question

- We have already seen *Graph Width Metrics*. These can often be *hidden* parameters in some computational problem. Examples include TREEWIDTH, CUTWIDTH, BANDWIDTH and PATHWIDTH. For example, it has been proposed that the syntactic structure of sentences of natural languages can be modeled by dependency graphs of pathwidth less than 10 (Kornai and Tuza [KT92]). CUTWIDTH and PATHWIDTH have applications for small parameter ranges in VLSI design [FL92]. TREEWIDTH and PATHWIDTH are important hidden parameters arising from the work of Robertson and Seymour which provide a generic methodology of solving generally hard problems on a wide class of graphs. (More on this later.)
- We have already seen *Implicit Parameters in Logic Programming, Compiler Design and Database Applications.* In databases, the database is often much larger than the queries. This is an obvious arena for this kind of analysis. Sometimes the parameter can be less obvious. For example, the problem of type inference arises in implementations of programming languages such as ML that are based on the polymorphic typed λ-calculus. In Henglein and Mairson [HM91] it is shown that the problem is complete for deterministic exponential time, yet it has been widely noted that in practice the problem is efficiently solved by an algorithm that is known to be exponential in the worst case. An explanation for this discrepancy comes from noting that the logic formulas that occur in natural programs tend to have small bounded depth of *let*'s. For a parameter k bounding this depth, the problem is linear fixed-parameter tractable. This observation can be articulated in the point below.
- *Possible explanation of unpredicted tractability.* Problems can be much easier in practice than one would expect from theoretical analyses. One example is in the point above. Another example concerns the register allocation problem for imperative programming languages. This is a problem that is NP-complete in general. It has been shown that for structured programs (with short-circuit evaluation) the control-flow graphs have treewidth less than 10, which allows linear FPT algorithms based on this hidden structural parameter to be developed in, for instance, Alstrup, Lauridsen and Thorup [ALT96]. Such considerations give a different possible explanation from those currently offered such as probabilistic behaviour.

- *Parameters Introduced by Engineering Practices.* Some proposals for implementations of public key cryptosystems have considered limiting the size or Hamming weight of keys in order to obtain faster processing times. A cautionary note is sounded by the result Fellows and Koblitz [FK93] that for every fixed k, with high probability it can be determined in time $f(k)n^3$ whether an n-bit positive integer has a prime divisor less than n^k.
- *Approximation Parameters in Computational Biology.* A parameter can be composed of any relevant aspects of a problem. In Jiang, Lawler, and Wang [JLW94] it is shown that the problem of computing a tree alignment for DNA sequences of length n for k species of cost within a factor of $1 + \epsilon$ of the cost of an optimal alignment can be accomplished in time $O((kn)^{c^{1/\epsilon}})$. It is natural to consider this problem with the parameter consisting of the pair $(k, 1/\epsilon)$ and to ask whether it is fixed-parameter tractable.
- *Problems considered* in practice *come from "people" situations and hence can often have very small parameters associated with them because of our "brain" limitations.*

3 Why this is interesting: A rich positive toolkit

So far we have seen that (i) parameterized complexity has lots of potential areas of application and (ii) there are many natural, often hidden, parameters. Another very important cog in the machinery is that *there is a very distinctive positive toolkit of FPT algorithm design methods, begging for further development.* We discuss some of these basic methods below. Due to space limitations we will keep the discussion fairly incomplete and refer the reader to the monograph [DF98] for further details.

- *Well-Quasi-Ordering.* The spectacular results of Robertson and Seymour give a method that in its most general form yields $O(n^3)$ fixed-parameter tractability results that are supremely impractical, primarily because they involve astronomical hidden constants. This is still useful, however, as a complexity classification tool, pointing the way to better FPT results based on a more detailed problem analysis.

    Briefly, a *quasi-order* on a set A is a relation $\preceq$ that is transitive and reflexive. $\preceq$ is a *well-quasi-order* (WQO) if every filter has a finite basis. That is, for any set $F \subseteq A$ such that $x \in F$ and $x \preceq y$ implies $y \in F$, there is a finite subset $F' \subseteq F$ such that $x \in F$ iff $\exists f \in F'(f \preceq x)$.

    The crucial point is that if, for a fixed f, the relation $f \preceq x$ can be decided in polynomial time, then membership in a filter F can be determined in polynomial time by the finiteness of F'.

    The best known example is Robertson and Seymour's proof that the minor relation on the class of finite graphs is a WQO. Here G is a minor of H iff G can be obtained from H be a finite sequence of edge contractions and vertex/edge deletions. Robertson and Seymour in the famous series of papers (e.g [RS85,RS96]) showed the WQO result and they also proved that for a fixed G the relation "G is a minor of H" can be decided in time $O(|H|^3)$. Thus, *any set of graphs closed under the minor relation has a cubic time recognition algorithm.* This allows is to immediately conclude that the following problems are FPT.

GRAPH GENUS
Instance: A graph $G = (V, E)$.
Parameter: A positive integer k.
Question: Does G have genus k? (That is, can G be embedded with no edges crossing on a surface with k handles?)

GRAPH LINKING NUMBER
Instance: A graph $G = (V, E)$.
Parameter: A positive integer k.
Question: Can G be embedded into 3-space such that the maximum size of a collection of topologically linked disjoint cycles is bounded by k?

- *Bounded Treewidth and Pathwidth.* These FPT methods are currently "approaching" practicality (see Bodlaender [Bod96]) with the best current parameter function being $f(k) = 2^{ck^2}$. Treewidth, for instance, measures how "treelike" a graph is. If a graph has bounded treewidth then many intractable problems become tractable by dynamic programming/automata techniques. In more detail, A *tree-decomposition* of a graph $G = (V, E)$ is a tree $\mathcal{T}$ together with a collection of subsets T_x (called *bags*) of V labelled by the vertices x of $\mathcal{T}$ such that $\cup_{x \in \mathcal{T}} T_x = V$ and 1. and 2. below hold:
1. For every edge uv of G there is some x such that $\{u, v\} \subseteq T_x$.
2. (Interpolation Property) If y is a vertex on the unique path in $\mathcal{T}$ from x to z then $T_x \cap T_z \subseteq T_y$.

The *width* of a tree decomposition is the maximum value of $|T_x| - 1$ taken over all the vertices x of the tree $\mathcal{T}$ of the decomposition. Figure 2 gives an example of a tree decomposition of width 2.

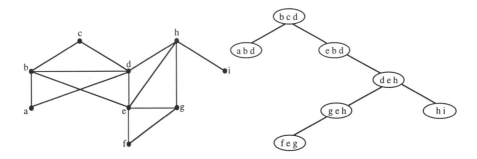

Fig. 2. Example of Tree Decomposition of Width 2.

Historically, authors often tried to get around classical intractability by looking at the problems at hand on special classes of graphs. Authors often discovered that intractable problems became tractable if the problems were restricted to say, "outerplanar" graphs. Such restriction is not purely an academic exercise since, in many practical situations, the graphs that arise do not in fact demonstrate the full pathology of the class of all graphs. [Consider, for instance TRAVELLING SALESMAN concerned with a given city. The graph will almost certainly be of small maximum clique size.] The table (from Van Leeuwin [vanL90]) below lists some families of graphs that have been studied and have treewidth.

Families of Graphs	Bound on Treewidth
Trees	1
Almost Trees (k)	$k+1$
Partial k-trees	k
Bandwidth k	k
Cutwidth k	k
Planar of Radius k	$3k$
Series Parrallel	2
Outerplanar	2
Halin	3
k-Outerplanar	$3k-1$
Chordal with Maximum Clique Size k	$k-1$
Undirected Path with Maximum Clique Size k	$k-1$
Directed Path with Maximum Clique Size k	$k-1$
Interval with Maximum Clique Size k	$k-1$
Proper Interval with Maximum Clique Size k	$k-1$
Circular Arc with Maximum Clique Size k	$2k-1$
Proper Circular Arc with Maximum Clique Size k	$2k-2$

Many of the algorithms for the classes above relied on heroic case analysis. We now know the key fact is that the bags of the tree decomposition give a kind of roadmap to the construction of the graph, and, more importantly, keep track of that part of the graph for which information must be kept in some algorithm. One can think of the algorithm as processing information in the bags down the branches of the tree decomposition. This picture can be formalized because we can associate with a graph of bounded treewidth a standard *parse tree*. The idea is that given some property such as HAMILTONICITY, we will build a kind of tree automaton. This will be used to accept or reject the relevant graph by feeding into the automaton the relevant parse tree of the graph. The culmination of this approach is Courcelle's Theorem [Co90] below. Recall the monadic second order logic is a two sorted logic with the usual logical connectives, vertex and edge variables, and variables for sets of vertices and sets of edges, together with incidence relations. Quantification is allowed over vertices and edges as well as over the sets of vertices and edges. Many properties are expressible in this logic. For instance, a graph G is Hamiltonian iff the edges of G can be partitioned into two sets *red* and *blue* such that each vertex has exactly two incident red edges, and the subgraph induced by the red edges spans G. It is routine to write this description out in monadic second order logic. This all leads to Courcelle's Theorem below.

Theorem 1 (Courcelle [Co90]). *Any property of graphs definable in monadic second order logic is recognizable for bounded treewidth, in the sense that we can build an automaton as above accepting graphs satisfying the property.*

Courcelle's Theorem gives a linear time FPT algorithm for such properties.
- *Color-Coding (Hashing).* This is a general method of devising FPT algorithms using k-perfect families of hash functions derived from sophisticated de-randomization techniques developed by Alon, Yuster and Zwick in [AYZ94] after earlier work of Dennenberg and Gurevich [DGS86]. The method works quite well for certain kinds of problems, such as finding small subgraphs in a graph (such as paths, cycles and matchings). The resulting algorithms offer significant practical parameter ranges and no sources of large hidden constants.

The main result of [AYZ94] proves that for every k and n there is a family $\mathcal{H}$ of functions $h : \{1, ..., n\} \to \{1, ..., k\}$ with the property that for any k-element set $S \subseteq \{1, ..., n\}$ there is some $h \in \mathcal{H}$ that is injective on S. The size of of the family $\mathcal{H}$ of hash functions is bounded by $|\mathcal{H}| = O(2^{ck} \log n)$ where c is a small constant independent of k and n.

For a simple example of how this method works, consider the problem of 3-DIMENSIONAL MATCHING, which is one of the six NP-complete problems singled out by Garey and Johnson as particularly useful starting points for NP-completeness reductions. This problem (essentially) takes as input a set T of triples $T \subseteq A \times B \times C$ and a positive integer k, and the question is whether there is a subset $S \subseteq T$ of k triples such that if $t = (a, b, c) \in S$ and $t' = (a', b', c') \in S$, then $a \neq a'$ and $b \neq b'$ and $c \neq c'$. In other words, the question is whether there is a set of k of the triples of T that are pairwise distinct in each component. Note that if we ask the same question about 2-tuples rather than 3-tuples, then the problem can be solved in polynomial time by computing a maximum matching in a graph.

To solve this question by the method of color-coding, note that if there is a solution set S of k triples, then this involves at most $3k$ elements in $A \cup B \cup C$. Let $n = |A \cup B \cup C|$, and let $\mathcal{H}$ be the set of perfect hash functions $h : \{1, ..., n\} \to \{1, ..., 3k\}$. If there is a solution set S, then at least one of the hash functions in $\mathcal{H}$ must assign each of the $3k$ elements of S a different "color". Conversely, suppose we choose a function $h \in \mathcal{H}$ and color the elements of $A \cup B \cup C$. After doing this, we can list all of the different color-triples that result. There can be at most $(3k)^3$ of these. In time that is just a function of k we can determine if there is a set of k of these that are pairwise "color-distinct" in each component. But this implies there is a set of k pre-images in T that are a solution. Checking each function in $\mathcal{H}$ in this way gives us an FPT algorithm for the problem.
- *Elementary Methods: search trees* and *reductions to a problem kernel.* An interesting aspect of these two approaches is that while they are simple

algorithmic strategies, they are in some sense "new ideas" that might not immediately come to mind, *because they involve costs that are exponential in the parameter.*

The idea behind search trees is that we might be able to bound the size of the nondeterminism in some search as a function of k. A simple minded approach using this idea for VERTEX COVER is to proceed as follows. Take any edge of G, say, xy. Now every vertex cover of G must include one of x or y. Begin a tree woth root λ and first nodes labelled respectively x and y. For the x branch remove all edges covered by x and consider the derived graph (and similarly for the y branch). Proceed inductively. The tree is binary and for a fixed k will have depth k. The time needed to see if this results in a vertex cover is $O(2^k|G|)$.

The idea of the method of problem kernel is to define some sort of reduction which takes an instance G of size n to one of size $f(k)$. In the VERTEX COVER case, for instance, we can remove vertices of high degree since they must be in any vertex cover. When one removes all such verteices, of which there can be at most k one should get a relatively small graph, since one can demonstrate that if a graph has only vertices of low degree then it can only have a k element vertex cover if the number of vertices is bounded by something like k^2. If the kernel obtained by the first reduction is small, then one checks for a vertex cover of the relevant size by examining all subsets (or by some other means). The reader should note that this gives an algorithm with the parameter function $f(k)$ as an *additive* constant. The two methodologies are compatible in the sense that often one uses first kernalization and then search trees.

Just these naive methods gives an algorithm with running time $O(|G|+2^k k^2)$ for VERTEX COVER. These simple strategies have been successfully applied to a wide variety of problems such as PLANAR DOMINATING SET, FEEDBACK VERTEX SET, FACE COVER NUMBER FOR PLANAR GRAPHS, (see [DF98]), MIMIMUM FILL-IN (Kaplan, Tarjan and Shamir [KST94]), for search trees, and k-LEAF SPANNING TREE, ([DF98]), various $\Pi_{i,j,k}$-GRAPH MODIFICATION PROBLEMS (Leizhen Cai [LeC95]), SET BASIS, UNIQUE HITTING SET ([DF98]), and various phylogenetic tree metric problems (Allen [Al98]).

4 Why this is interesting: The deal with the devil

The above discussion focuses on the fact that parameterized complexity describes a new paradigm for seeking tractability in computational problems. We call this *The Deal with the Devil of Intractability.* The basic idea is that we are trying to solve some important natural problem, and these (as we now know) are almost always NP-complete or PSPACE complete, or worse. What to do? Lets *actively search for a parameter to absorb the intractability* and make the problem feasible by the slice.

Another fundamental aspect of this line of attack we call *Lichenism*. By this we mean that initial efforts may yield an FPT algorithm that is useful only for a small range of the parameter k, but this can (we have seen for many problems) be subsequently improved, as we gradually expand the "viable margin" on the "rock of intractability" that we are dealing with (perhaps a planetary metaphor would speak more eloquently of this point of view).

Given the general abundance of "bad news" about computational tractability, an effort to slice the problem in various ways and apply FPT methods *can't hurt!* Using these techniques we have improved many known algorithms. For instance, using kernelization we forst improved VERTEX COVER from $2^k|G|$ to $|G| + 2^k k^2$. If kernelization can be applied to a problem (see the above described kernelization for VERTEX COVER for example) then surely dealing with the smaller problem that results can only make our life easier.

Another point worth mentioning is that many heuristic algorithms in the literature *already use* these techniques. The FPT notion "points the way" and systematizes much current practice in practical computing.

5 Intractability and its applications

Before we look towards the future, and compare this approach with other coping strategies, it is appropriate to also examine the phenomenom of parametric intractability.

There are two ingredients for any intractability theory.
- The identification of hard problems.
- Reductions to calibrate problems.

Reductions don't pose much of a problem. What is needed is a reduction that preserves the "slicewise" complexity of parameterized languages. The basic working definition used in concrete reductions is that there is a constant c, and three functions

- $k \mapsto k'$, an arbitary function,
- $k \mapsto f(k)$, another arbitrary function,
- $\langle x, k \rangle \mapsto \langle x', k' \rangle$, a function computable in time $f(k)|x|^c$, such that

$$\langle x, k \rangle \in L \text{ iff } \langle x', k' \rangle \in L'$$

There are other types of reductions but the one above suffices for most concrete applications. We call this type a *standard reduction*. Here is an example of a standard parameterized reduction. Consider the following two problems.
(k–)INDEPENDENT SET
Input: A graph G.
Parameter: A positive integer k.
Question: Does G have a set of k vectors $\{x_1, \ldots, x_k\}$ such that all $i \neq j$, x_i is not adjacent to x_j?
(k–)MAXIMAL IRREDUNDANT SET
Input: A graph G.
Parameter: A positive integer k.
Question: Does G have a set S of k vertices such that for each member x of

S there is a $y \in V(G)$ such that either $y = x$ or (x, y) is an edge, and y is not adjacent to nor a member of S, and furthermore S is maximal with this property?

We will show that INDEPENDENT SET $\leq$ MAXIMAL IRREDUNDANT SET via a standard reduction. We remark that H. Bodlaender and D. Kratsch have proven the apparently stronger result that WEIGHTED CNF SATISFIABILITY $\leq$ MAXIMAL IRREDUNDANT SET. (see below)

To prove our result, given a graph G and a k we construct a graph H and a k' so that G has a k element independent set iff H has a maximal irredundant set of size k'. H is constructed as follows. Construct k blocks consisting of $k+1$ columns each consisting of $n = |G|$ points. So column i in block j consists of n points $x(j, i, t)$ for $t = 1, ..., n$. The idea is that a block represents a choice for a vertex in the independent set. Now within a block join all vertices except those with the same last coordinate. (We refer to this as a *row*.) The first column is not connected to anything else and is called the *dummy column*. Now for each block i pick a column in each block $i' \neq i$ to identify with block i, and similarly identify a corresponding column i'' in block i. Do this in such a way that for each pair (i, j) there is exactly one column $c_i(i, j)$ identified with exactly one column $c_j(i, j)$ in block j, and these columns are used for no other identifications, for any other pair except (i, j). Let (q, r) be some such pair of columns. Now connect each $x(i, q, t)$ to $x(j, r, s)$ iff t and s are adjacent vertices in G.

Now let $k' = k(k + 1)$. First if G has an independent set of size k. Let $\{t_1, ...t_{k-1}\}$ be this independent set. The relevant irredundant set for H is obtained by taking the union of vertices of the t_i'th row of block i, for $i = 1, .., k$. They are each their own private neighbours as the corresponding set in G is independent, and it is maximal by the fact that the blocks are almost cliques. Conversely, take S a size k' maximal irredundant set in H. One can argue that it must be of the form above and hence corresponds to an independent set in G of size k. (There are 2 cases depending on whether some block B has more than k vertices of S. Remember that a maximal irredundant set must also be a dominating set.)

Actually, more instructive than the above example is the following *non-example*. Consider the two problems below.

WEIGHTED CNF SATISFIABILITY
Instance: A CNF formula X,
Parameter: A positive integer k.
Question: Does X have a satisfying assignment of weight k?

WEIGHTED q-CNF SATISFIABILITY
Instance: A q-CNF formula X, i.e., a CNF formula such that each clause has no more than q literals.
Parameter: A positive integer k.
Question: Does X have a satisfying assignment of weight k?

The *classical* reduction of Karp showing that CNF SATISFIABILITY is polynomial time reducible to 3SAT, (i.e. the unweighted versions of the above), works as follows. Let the given CNF formula be $X = c_1 \wedge c_2 \wedge ... \wedge c_n$, where the c_i

represent clauses. For those clauses with size ≥ 4, say, $c_i = \{y_1, ..., y_m\}$, add $m - 3$ new dummy variables, and get the clauses

$$\{y_1, y_2, z_1\}, \{\overline{z_1}, y_3, z_2\}, \{\overline{z_2}, y_4, z_3, \}, ...$$

Then the original clause is satisfiable iff there is some assigment of the new clauses that makes one of the original literals true. *The key thing is that this reduction does not preserve the Hamming weight of the truth assignments.* Thus, this is *not* a parameterized reduction.

We conjecture that there is *no* parameterized reduction between WEIGHTED CNF SATISFIABILITY and WEIGHTED q-CNF SATISFIABILITY. This is why we regard the Bodlaender-Kratsch result as being more significant than the one above.

However, all of this is somewhat irrelevant if all we are interested in is establishing fixed parameter intractability. What we need is a good measure of what we will regard as intractable. That is, what we need is an analog of Cook's Theorem. This leads us to the basic hardness class $W[1]$. Our analog of Cook's Theorem is based on the following problem.

SHORT NONDETERMINISTIC TURING MACHINE ACCEPTANCE
Instance: A nondeterministic Turing Machine M and a positive integer k.
Parameter: k.
Question: Does M have a computation path accepting the empty string in at most k steps?

If one believes the philosophical argument that Cook's Theorem provides compelling evidence that SAT is intractible, then one surely must believe the same for the parametric intractability of SHORT NONDETERMINISTIC TURING MACHINE ACCEPTANCE.

We now know that there are many natural problems of the same parameterized complexity as SHORT NONDETERMINISTIC TURING MACHINE ACCEPTANCE. These include parameterized versions of CLIQUE, INDEPENDENT SET, and a number of other well studied problems. We remark that given the sensitive nature of the reductions, we seem to get a hierarchy of intractable problems, which can be investigated in terms of the logical depth of formulae and circuits that describe the problems. (See [DF98] or [DF95a,DF95b] for details.)

$$FPT \subseteq W[1] \subseteq W[2] \subseteq ... \subseteq W[SAT] \subseteq W[P] \subset XP.$$

The following table gives the classification of some concrete problems based upon this hierarchy.

6 Insight into classical complexity

An interesting aspect of Table 2 is that it includes several problems that in their unparameterized form are considered unlikely to be NP-hard. An example is the problem of VC-DIMENSION, which is in DTIME($n^{\log n}$). If we believe that $W[1] \neq FPT$ then the $W[1]$-hardness result says that there is no polynomial time algorithm to solve this problem, which is interesting new information.

Class	Problem	Reference
$W[P]$	LINEAR INEQUALITIES	[ADF95]
	MINIMUM AXIOM SET	[DFKHW94]
	SHORT SATISFIABILITY	[ADF95]
	WEIGHTED CIRCUIT SATISFIABILITY	[ADF95]
$W[SAT]$	WEIGHTED SATISFIABILITY	[ADF95]
	⋮	
$W[t]$, for all t	LONGEST COMMON SUBSEQUENCE (k = no. of seqs.,$|\Sigma|$) (hard)	[BDFHW95]
	FEASIBLE REGISTER ASSIGNMENT (hard)	[BFH94]
	TRIANGULATING COLORED GRAPHS (hard)	[BFH94]
	BANDWIDTH (hard)	[BFH94]
	PROPER INTERVAL GRAPH COMPLETION (hard)	[BFH94]
	WEIGHTED t-NORMALIZED SATISFIABILITY	[DF95a]
	⋮	
$W[2]$	WEIGHTED $\{0,1\}$ INTEGER PROGRAMMING	[DF95a]
	DOMINATING SET	[DF95a]
	TOURNAMENT DOMINATING SET	[DF95c]
	UNIT LENGTH PRECEDENCE CONSTRAINED SCHEDULING (hard)	[BF95]
$W[1]$	SHORTEST COMMON SUPERSEQUENCE (k)(hard)	[FHK95]
	MAXIMUM LIKELIHOOD DECODING (hard)	[DFVW98]
	WEIGHT DISTRIBUTION IN LINEAR CODES (hard)	[DFVW98]
	NEAREST VECTOR IN INTEGER LATTICES (hard)	[DFVW98]
	SHORT PERMUTATION GROUP FACTORIZATION (hard)	[CCDF96]
	SHORT POST CORRESPONDENCE	[CCDF96]
	WEIGHTED q-CNF SATISFIABILITY	[DF95b]
	VAPNIK–CHERVONENKIS DIMENSION	[DEF93]
	LONGEST COMMON SUBSEQUENCE (k, m = length of subseq.)	[BDFW95]
	INDEPENDENT SET	[DF95b]
	SQUARE TILING	[CCDF96]
	MONOTONE DATA COMPLEXITY FOR RELATIONAL DATABASES	[DFT96]
	k-STEP DERIVATION FOR CONTEXT SENSITIVE GRAMMARS	[CCDF96]
	CLIQUE	[DF95b]
	SHORT NTM COMPUTATION	[CCDF96]
FPT	FEEDBACK VERTEX SET	[DF95c]
	GRAPH GENUS	[RS85]
	MINOR ORDER TEST	[RS85]
	TREEWIDTH	[Bod96]
	VERTEX COVER	[BFR98]

Table 2. A Sample of Parametric Complexity Classifications.

That is, *you can use parameterized complexity it to prove unparameterized problems to not be feasible even though they are probably not NP-complete or are unknown.*

There are also deep connections with approximation issues for classical optimization problems. Recall that an *NP optimization problem Q* is either a *minimization problem* or a *maximization problem* and is given as a 4-tuple (I_Q, S_Q, f_Q, opt_Q), where

(i) I_Q is the set of *input instances*. I_Q is recognizable in polynomial time.

(ii) $S_Q(x)$ is the set of *feasible solutions* for the input $x \in I_Q$ such that there is a polynomial p and a polynomial-time computable relation Π (p and Π only depend on Q) such that for all $x \in I_Q$, $S_Q(x)$ can be expressed as $S_Q(x) = \{y : |y| \leq p(|x|) \land \Pi(x, y)\}$;

(iii) $f_Q(x, y) \in N$ is the *objective function* for each $x \in I_Q$ and $y \in S_Q(x)$. The function f_Q is computable in polynomial time;

(iv) $opt_Q \in \{\max, \min\}$.

An *optimal solution* for an input instance $x \in I_Q$ is a feasible solution $y \in S_Q(x)$ such that $f_Q(x, y) = opt_Q\{f_Q(x, z) : z \in S_Q(x)\}$. We write

$$opt_Q(x) = opt_Q\{f_Q(x, z) : z \in S_Q(x)\}.$$

The recent work of Liming Cai, Chen and Bagzan has shown that $W[1]$ provides a vehicle for demonstrating that various problems do not have good approximation schemes of various types unless $W[1] = FPT$.

This is of more than passing interest. Aside from the original work in Garey and Johnson, there have been few advances which enable one to prove that an approximation problem has no good approximation algorithm, even assuming $P \neq NP$. The exception to this is the recent work of Arora *et al.* [ALMSS92] who demonstrated that many *NP* optimization problems *with a certain syntactic description* don't have good approximation algorithms unless $P \neq NP$. The Cai-Chen-Bagzan material demonstrates that, assuming $W[1] \neq FPT$, fixed-parameter intractability of an NP optimization problem implies non-approximability of the problem. This result applies to many problems *not* covered by the Arora el. al. results. Thus, the study of fixed-parameter complexity provides a new and potentially powerful approach to proving non-approximability of NP optimization problems. The connections are not fully explored as yet.

For a relation R from (e.g.) $\{\geq, \leq, =\}$ we can associate a natural decision problem.

DECISION PROBLEM ASSOCIATED WITH THE NP OPTIMIZATION PROBLEM $Q = (I_Q, S_Q, f_Q, opt_Q)$.
Input: $x \in I_Q$.
Parameter: A positive integer k.
Question: Does $R(opt_Q(x), k)$ hold?

We write Q_R for the decision problem associated with R. The complexity of the parameterized problems $Q_=, Q_\leq$, and $Q_\geq$ have been studied in the literature. For example, Leggett and Moore [LM81] showed that for many NP optimization problems Q, the problems $Q_\leq$ and $Q_\geq$ are in NP $\cup$ coNP, while the problem $Q_=$

is not in NP ∪ coNP unless NP = coNP. Using a simple elementary argument, Liming Cai and Chen [CC97] demonstrated that the problems $Q_=$, $Q_\leq$, and $Q_\geq$ have the same complexity *from the viewpoint of fixed-parameter tractability*. That is, *if any of $Q_=$, $Q_\leq$, and $Q_\geq$ are FPT they all are*. The following theorem was the first to demonstrate a connection between FPT and approximation.

Theorem 2 (Cai [Ca92], Liming Cai and Chen [CC97]). *If an NP optimization problem has a fully polynomial-time approximation scheme, then it is fixed-parameter tractable.*

Proof. Suppose that an NP optimization problem $Q = (I_Q, S_Q, f_Q, opt_Q)$ has a fully polynomial-time approximation scheme A. Then, the algorithm A takes as input both an instance $x \in I_Q$ of Q and an accuracy requirement $\epsilon > 0$, and then outputs a feasible solution in time $O(p(1/\epsilon, |x|))$ such that the relative error $\Re_A(x)$ is bounded by ϵ, where p is a two-variable polynomial.

First we assume that Q is a maximization problem. By the remarks above, we only need to show that the problem $Q_\leq$ is fixed-parameter tractable. Given an instance $\langle x, k \rangle$ for the parameterized problem $Q_\leq$, we run the algorithm A on input x and $1/2k$. In time $O(p(2k, |x|))$, the algorithm A produces a feasible solution $y \in S_Q(x)$ such that

$$\Re_A(x) = \frac{opt_Q(x) - f_Q(x,y)}{f_Q(x,y)} \leq \frac{1}{2k} < \frac{1}{k}$$

If $k \leq f_Q(x,y)$, then certainly $k \leq opt_Q(x)$ since Q is a maximization problem. On the other hand, if $k > f_Q(x,y)$, then $k - 1 \geq f_Q(x,y)$. Combining this with the inequality $(opt_Q(x) - f_Q(x,y))/f_Q(x,y) < 1/k$, we get immediately $k > opt_Q(x)$. Therefore, $k \leq opt_Q(x)$ if and only if $k \leq f_Q(x,y)$. Since the feasible solution y can be constructed by the algorithm A in time $O(p(2k, |x|))$, we conclude that the NP optimization problem Q is fixed-parameter tractable.

The case when Q is a minimization problem can be proved similarly by showing that the problem $Q_\geq$ is fixed-parameter tractable. □

The most important Corollary to this result is in its application in negative form.

Corollary 3 (Cai and Chen [CC97]). *If a NP optimization problem is fixed parameter intractable, then it has no fully polynomial time approximation scheme. In particular under the current working hypothesis that $W[1] \neq FPT$, the NP optimization problems that are $W[1]$-hard under the uniform reduction have no fully polynomial-time approximation scheme.*

Using similar arguments, Bagzan has extended the the Cai-Chen Theorem to a more general class of nonuniform PTAS's. There are a number of applications of Theorem 2. One not covered by the classical literature is that of the VC-DIMENSION, which has applications in computational learning theory and has been studied extensively in the literature. VC-DIMENSION is $W[1]$-hard, and so cannot have a fully polynomial time approximation scheme unless $W[1] = FPT$.

As we now know, approximation schemes also have intimate connections with syntactic descriptions of problems. The class MAXSNP was introduced by Papadimitriou and Yannakakis [PY91]. The class comes from consideration of the syntactic form of NP problems via Fagin's Theorem below. Fagin's Theorem is known in finite model theory as $NP = \Sigma_1^1$.

Theorem 4 (Fagin's Theorem, Fagin [Fag74]). *NP is the class of all problems that can be put in the form*

$$\{I : \exists S \in 2^{[n]^k} \forall x \in [n]^p \exists y \in [n]^q \phi(I, S, x, y)\}.$$

Here $I \subseteq [n]^m$ is the input relation, x and y are tuples ranging over $[n] = \{1, 2, ..., n\}$, and ϕ is quantifier free.

For example, SATISFIABILITY can be written as

$$\exists T \forall c \exists x[(P(c,x) \land x \in T) \lor (N(c,x) \land x \notin T)].$$

Papadimitriou and Yannakakis realized that some problems have a form that is *simpler* than the basic Fagin Theorem. We can write 3SAT with one less alternation of quantifiers by assuming that the input consists of four relations C_0, C_1, C_2, C_3 where C_j consists of all clauses with exactly j negative literals. Thus $(x_1, x_2, x_3) \in C_j$ means that there is a clause with $x_1, ..., x_j$ occurring negatively and $x_{j+1}, ..., x_3$ positively. 3SAT is then

$$\exists T \forall (x_1, x_2, x_3)[\land_{j=0}^3 ((x_1, x_2, x_3) \in C_j \to (\land_{i \leq j} x_i \notin T \land \land_{i > j} x_i \in T.))].$$

The problems such as 3SAT which can be expressed as $\exists S \forall x \psi(x, G, S)$ constitute a class called *simple* NP or SNP. For each $\Pi \in NP$, one can then define the *maximization version* $MAX\Pi$ as

$$\max_S |\{x : \exists y \psi(x, y, G, S)\}|.$$

The immediate relevance of all of the above to our studies comes from the following consequential definition of Papadimitriou and Yannakakis.

A maximization problem $Q = (I_Q, S_Q, f_Q, opt_Q)$ is in the class MAX SNP if its optimum $opt_Q(X)$, $X \in I_Q$, can be expressed as

$$opt_Q(X) = \max_S |\{v : \psi(v, X, S)\}|$$

where the input instance $X = (U, P^1, \cdots, P^b)$ is described by a finite structure over the finite universe U and P^i is a predicate of arity r_i for some integer $r_i \geq 1$, S is also a finite structure over the universe U, v is a vector of fixed arity of elements in U, and ψ is a quantifier-free formula.

Despite the fact that Papadimitriou and Yannakakis [PY91] proved that any problem in MAX SNP can be approximated to within some constant ratio, Arora et al. [ALMSS92] proved the beautiful result that *MAX SNP-complete (under L-reductions) optimization problems have no fully polynomial-time approximation scheme unless $P = NP$*. Because of this we cannot use Theorem 2 to demonstrate that MAXSNP problems are FPT. Nevertheless, Liming Cai and Chen were able to establish the fixed-parameter tractability of all problems in MAX SNP.

Theorem 5 (Cai and Chen [CC97]). *All maximization problems in the class MAX SNP are fixed-parameter tractable.*

The point of this result is that we can see that some problems are *FPT from syntactic descriptions alone*. Cai and Chen also showed that the important class of minimization problems MIN $F^+ \Pi_1(h)$ introduced by Kolaitis and Thakur [KoT95] are fixed-parameter tractable. We refer the reader to [CC97] or [DF98] for further details.

7 A comparison with other coping strategies

As we argued in Downey, Fellows, Stege [DFS98], there are two ways of viewing parametric tractability. The first is easiest to arrive at. In this view, FPT is as a kind of first aid that can sometimes be applied to problems that are NP-hard, PSPACE-hard or undecidable. That is, it can be viewed as a potential means of coping with *classical* intractability.

In [DF98], we argued that there is a more radical second way that one can view parameterized complexity. We can view it as a fundamentally richer and generally more productive *primary framework* for problem analysis and algorithm design, including the design of heuristic and approximation algorithms. We will amplify this point later.

In this section we will compare the parametric point of view with other current approaches of "coping with intractability." There is no doubt that there is a percieved gap between the work of theoretical computer scientists and the work of "practical" ones. It may indeed be a reflection of the societal mind set, but there have been a number of calls for reform especially among theorists ([HL92,Hart94,PGWRS96,AFGPR96]). In particular, there seems a desperate need for the development of general coping strategies for dealing with the ubiquitous nature of computational intractability, and the practical computer scientists don't seem to percieve the theory community as delivering the goods.

Of course, computer science theory has articulated a few general programs for systematically coping with the phenomenon of computational intractability. We list these basic approaches:

• The idea of focusing on average-case as opposed to worst-case analysis of problems.
• The idea of settling for approximate solutions to problems, and of looking for efficient approximation algorithms.
• The idea of using randomization in algorithms.
• The idea of harnessing quantum mechanics, or molecular chemistry, to create qualitatively more powerful computational mechanisms.

To this list of fundamentally mathematical strategies for coping with intractability, in [DF98], we argued the following should be added.

• The idea of devising *FPT* algorithms for parameterizations of a problem.
• The design of mathematically informed, but perhaps unanalyzable *heuristics*, that are empirically evaluated by their performance on sets of benchmark instances.

The actual state of the practical world of computing is that (with the exception of some areas) there is not much connection to work in theoretical computer science on algorithms and complexity. Overwhelmingly, heuristic algorithms are relied on to deal with the hard problems encountered in most applications. How do the above strategies compare?

Average-Case Analysis

In many applications practitioners would be happy with algorithms having good average-case performance, and this criterion is implicit in the common practice of evaluating heuristic algorithms on sets of benchmarks. The idea that average-case analysis is more realistic than worst-case analysis has been around since the beginnings of theoretical computer science, and its potential role as a method of coping with intractability is discussed by Garey and Johnson in their chapter on this subject in [GJ79].

As we noted in [DFS98], obtaining theorems about average-case complexity tends to be mathematically difficult, even for the simplest of algorithms and distributions. As a consequence, the methods are difficult to apply, and moreover, it is frequently unclear what constitutes a reasonable assumption about the distribution of problem instances. The completeness notion for average-case complexity introduced by Levin [Lev86] also seems to have limited applicability. Another drawback of the program is that it lacks a positive toolkit. If I want to design an algorithm with good average-case performance (however this is evaluated)—how do I do that?

The real strength of average case analysis as a means of coping is that for most applications of computing it is the *right idea* for how complexity should (usually) be measured, and it is what practitioners generally continue to *do* about complexity measurement in practice, though informally. Perhaps what is lacking is a general methodology relating benchmarks with average case behaviour.

Approximation

There has been enormous effort in the last decades towards the development of approximation schemes for combinatorial problems. Much of Garey and Johnson [GJ79] is devoted to explaining the basic ideas of *polynomial time approximation algorithms and schemes*. There were significant early successes for problems such as BIN PACKING and KNAPSACK. However, apart from a few similar results on problems mostly of this same general flavor, it now seems to be clear, on the basis of powerful new proof techniques [ALMSS92], that these results are *not* typical for *NP*-hard and otherwise intractable problems. It now seems to be the case that the majority of natural *NP*-hard optimization problems are also hard to approximate, under the usual assumptions (such as $P \neq NP$).

The great strength of polynomial time approximation as a program for coping with intractability is that for those problems to which it can be applied, it allows for the clever deployment of mathematics and can be a very effective cure for worst-case intractability. It is now also clear that the negative tools for showing non-approximability are mathematically deep and widely applicable.

Polynomial Time Randomization

Randomization is discussed in Chapter 6 of Garey and Johnson as a means of avoiding one of the weaknesses of average-case analysis as a coping strategy—the need to have some knowledge in advance of a realistic distribution of problem instances. An algorithm that flips coins as it works may be able to conform to whatever distribution it is given, and either produce an answer in polynomial-time that is correct with high probability (Monte Carlo randomization), or give an answer that is guaranteed to be correct after what is quite likely to be a polynomial amount of time (Las Vegas randomization).

These seemed at first to be potentially very powerful generalizations of polynomial time. Randomized Monte Carlo and Las Vegas algorithms are a workhorse of cryptography [SS77,GM84], and have important applications in computational geometry [Cl87], pattern matching, on-line algorithms and computer algebra (see [Karp86] and [MR95] for surveys), in part because they are often simple to program.

Despite these successes, it now seems that randomized polynomial time is better at delivering good algorithms for difficult problems that "probably" are in P anyway, than at providing a general means for dealing with intractable problems. There have recently been a number of important results replacing fast probabilistic algorithms with ordinary polynomial time algorithms through the use of sophisticated derandomization techniques.

The main weakness of randomization (in the sense of algorithms with performance guarantees) as a general program for coping with intractability is that it is unclear whether randomization actually provides real progress against problems that are truly hard. One of the strengths of the program is that it supports a rich positive toolkit of algorithm design and analysis techniques (see [MR95]).

New Forms of Computation: DNA and Quantum Mechanics

Although these programs have been launched with great fanfare and have received headline coverage in the newspapers, they so far offer much less of substance than the other items on this list in terms of a general program for coping with intractability. DNA computing essentially comes down to computation by molecular brute force. Combinatorial explosion quickly forces one to contemplate a very large test tube. It is still unclear whether quantum computers useful for any kind of computation can actually be built. The notion of quantum polynomial time QP is mathematically interesting, but so far appears to be applicable only to a few very special kinds of problems. Expert opinion appears to be gathering around conjectures that QP is not much of an extension of P.

Probably neither of these programs can really be considered to be a serious contender for systematically dealing with intractability in ways that will have real applications in the foreseeable future.

Parameterization

We can trace the idea of coping with intractability through parameterization to early discussions in Garey and Johnson [GJ79], particularly Chapter 4, where it is pointed out that parameters associated with different parts of the input to a problem can interact in a wide variety of ways in producing non-polynomial complexity, and that some forms of intractability might be preferable to others.

From the current point of view, we see that the main historical drawback of parameterization as a convincing program has been the lack of *industrial strength FPT algorithms*. Certainly one of the original engines for FPT was provided by the mathematically very deep results of Robertson and Seymour. The resulting algorithms involved constants far larger than the size of the universe.

The main strengths of parameterization as a program are that it does seem to be very generally applicable to hard problems throughout the classical hierarchy of intractability, and it supports a rich toolkit of both positive and negative techniques.

Heuristics

Since heuristic algorithms that work well in practice are now, and have always been, the workhorses of industrial computing, there is no question about the ultimate significance of this program for dealing with intractability. There has recently been a revival of interest in obtaining systematic empirical performance evaluations of these algorithms for hard problems [BGKRS95,Hoo95,JM93,JT96].

The main question is whether this can actually be considered a program of computer science theory in the sense intended for this list. It isn't fundamentally at present a *mathematical research program*, though theorists have sometimes contributed to the design of heuristic algorithms. Mathematical ideas are used, but there are frequently no theorems to prove! No theorems are called for—only empirical performance.

Mindless randomized heuristic algorithms such as simulated annealing are applicable to almost any problem and have become extremely popular. Although it is almost impossible to say anything about the performance of such algorithms from a mathematical point of view, these clearly have become the principle means by which practitioners do what they can about intractability.

Open Questions and Future Work

8 Industrial strength FPT

We see as a major research program the development of "industrial strength" FPT techniques. This can be illustrated by reviewing the history of results concerning the VERTEX COVER problem.

- 1986. Fellows-Langston used the Robertson-Seymour results to establish a complexity of $f(k)n^3$, where $f(k)$ was a tower of 2's $500k$ high. Furthermore the result was nonconstructive in that only the existence of an algorithm was established.
- 1987. Johnson constructively described an $f(k)n^3$ algorithm for VERTEX COVER based on a combination of tree-decomposition and finite-state dynamic programming techniques, with a much better $f(k)$.
- 1988. Fellows using the method of bounded search trees gave an algorithm running in time $2^k n$.

- 1989. S. Buss using a problem kernel approach gave an algorithm requiring time $O(kn + 2^k k^{2k+2})$.
- 1992. R. Balasubramanian, R. Downey, M. Fellows and V. Raman gave an algorithm with running time $O(kn + 2^k k^2)$ by combining the problem kernel and search tree approaches.
- 1993. Using maximum matching, Papadimitriou and Yannakakis gave an algorithm with running time $O(3^k n)$ and noted that this is P-time for the classical problem if k is $O(\log n)$.
- 1996. Balasubramanian, Fellows and Raman used kernelization and an improved search tree strategy to achieve $O(kn + (4/3)^k k^2)$.
- 1998. Downey, Fellows, Stege [DFS98] gave an $O(kn + 1.31951^k k^2)$ algorithm. The 1996 algorithm above actually gives a constant of 1.32472; this small improvement yields a 21% difference for $k = 60$.

Thus we would like to know whether the kind of progress illustrated above is particular to VERTEX COVER, and we believe the answer is "surely not." The following is the basic challenge.

The Challenge. *Develop methods to make the various FPT methods some of which are currently impractical all practical. Or show this is impossible in general and point out when it can be done.*

Looking at our general techniques for proving things to be FPT, we get the following picture.

- Good candidate: Bounded treewidth. Bring down Bodlaender's constant. Currently the constant in Bodlaender's bounded treewidth k recognition algorithm is 2^{32k^2}, which is impractical. However, experience with small treewidth seems to suggest that finding tree decompositions should be easier than this. Even an algorithm that quickly gave a treewidth $2k$ decomposition for a treewdith k graph would be very useful.
- Good candidate. More extensive use of Colour Coding and other hashing techniques to replace or improve other approaches. Color-Coding give constants around 2^k, which is not bad. Can they be further improved? Can we use Colour Coding to replace WQO methods? Can Color-Coding be used in the bounded treewidth situation?
- Good candidate. What about linear programming? Linear programming is a powerful engine for producing practical algorithms. It is a notable hole in the book [DF98], mainly because of the ignorance of the authors. Presumably the material from Grötschel, Lovasz, and Schrijver [GLS88] is relevant here.
- Bad candidate. WQO methods. Given the generality and complexity of the proofs of these techniques, it does not seem reasonable that they will ever be *in general* feasible. What would be nice is a real proof of this, perhaps using methods from logic. (e.g. [FRS87]). Also a delineation of when they might be feasible would be nice.
- Good candidate. Kernelization and Bounded Search Trees. It would be nice to explore these methodologies further and figure out more applications for them.

9 Connections with heuristics.

From our point of view the principal drawback of heuristics is that they are hand-crafted *ad hoc* techniques. Perhaps that is necessary. But perhaps not. What is interesting is that we have found many instances in the literature and in working with, e.g. computational biologists, where *existing heuristics are in fact FPT algorithms!* For instance, an important problem in computational biology is the STEINER PROBLEM FOR HYPERCUBES. It is not important precisely what this problem is, but after one of the authors described an FPT algorithm for this problem, the biologist replied simply
That's what I already do!
It is our belief that there is something much deeper here.

The Challenge. *Based on FPT techniques, develop general methodologies for building industrial strength heuristic algorithms for combinatorial problems.*

As an example of what such a general methodology might look like, note that many FPT algorithms work because of some finite combinatorial basis or bounded search tree. Why do we need to use all the basis? For instance, we know that a graph is not planar iff it has K_5 or $K_{3,3}$ as a minor. But in general, for almost all graphs, testing with $K_{3,3}$ suffices.

Can we work this sort of observation into a practical, general methodology? Use partial bases, partial obstruction sets, partial automata?

Similarly, because the kernelization phase simplifies and decreases the size of the problem instance, it is a reasonable first step for any general attack on the unparameterized problem. It would be interesting to know if any systematic connection can be made between the "optimum" size of a problem kernel (which requires a definition) and the *complexity threshold* for the problem [KS94].

10 Connections with classical complexity

We have already mentioned some connections in an earlier section. The general area we wish to point to is:

The Challenge. *Explore how the parametric and the classical complexity frameworks intertwine.*

Already in [CC97] and [DF98] we have some connections. Here are some particular suggestions for further exploration.

- What's the best we can do? Can VERTEX COVER be improved to $O(kn + (1 + \epsilon)^k)$ for arbitrarily small ϵ, or would there be some unexpected consequence?
- Can we use parameterized complexity to establish something about the complexity of concrete problems that are classically unresolved. For instance, can we prove, as we did for VC DIMENSION, that GRAPH ISOMORPHISM is unlikely to be in P?
- Can we reverse any of the approximation connections? Is there a class of problems that have approximation algorithms iff W[1]=FPT?

- Can we develop a parametric notion of approximability? For a concrete example, we can ask if there an fast algorithm for DOMINATING SET that either gives an answer "no dominating set of size k" or produces a dominating set of size $2k$.

11 Unknown classifications

There are a number of interesting and important parametric problems that have not been classified. Here are some. (We refer the reader to [DF98] for others.)

- GRAPH ISOMORPHISM
  Instance: Graphs G and H.
  Parameter: (Various, e.g. Max degree, max treewidth)
  Question: Is G isomorphic to H?
- (FPT?) TOPOLOGICAL CONTAINMENT
  Instance: A graph $G = (V, E)$.
  Parameter: A graph H.
  Question: Is H topologically contained in G?
- (FPT?) IMMERSION ORDER TEST
  Instance: A graph $G = (V, E)$ and a graph $H = (V', E')$.
  Parameter: A graph H.
  Question: Is $H \leq_{immersion} G$ where $\leq_{immersion}$ denotes the immersion ordering?
- (FPT?) DIRECTED FEEDBACK VERTEX SET
  Instance: A directed graph $D = (V, A)$.
  Parameter: A positive integer k.
  Question: Is there a set S of k vertices such that each directed cycle of G contains a member of S?
- (FPT?) PLANAR DIRECTED DISJOINT PATHS
  Input: A directed planar graph G and k pairs $\langle r_1, s_1 \rangle, ..., \langle r_k, s_k \rangle$ of vertices of G.
  Parameter: A positive integer k.
  Question: Does G have k vertex-disjoint paths $P_1, ..., P_k$ with P_i running from $r - i$ to s_i?
  Known to be solvable in $O(n^{f(k)})$ time by Schrijver.
- (FPT?) PLANAR MULTIWAY CUT
  Instance: A weighted planar graph $G = (V, E)$ with terminals $\{x_1, \ldots, x_k\}$.
  Parameter: A positive integer k.
  Question: Is there a set of edges of total weight $\leq k'$ whose removal disconnects each terminal from all the others?
  The general version of this problem is NP-complete by Dahlhaus *et al.* [DJPSY92]. Best known complexity is $O((4^k)^k n^{2k-1} \log n)$ by [DJPSY92] where it is asked if the problem is FPT.
- (W[1] Hard?) BOUNDED HAMMING WEIGHT DISCRETE LOGARITHM
  Instance: An n-bit prime, a generator g of F_p^*, an element $a \in F_p^*$, a positive

integer k.
Parameter: A positive integer k.
Question: Is there a positive integer x whose binary representation has at most k 1's (that is, x has a Hamming weight of k) such that $a = g$?

- (W[1] Hard?) CROSSING NUMBER
  Instance: A graph $G = (V, E)$.
  Parameter: k
  Question: Can G be embedded in the plane with at most k edges crossing?
- (W[1] Hard?) SHORT CHEAP TOUR
  Instance: A graph $G = (V, E)$, an edge weighting $w : E \to \mathbf{Z}$.
  Parameter: A positive integer k.
  Question: Is there a tour through at least k nodes of G of cost at most S?
- (W[1] Hard?) FIXED ALPHABET LONGEST COMMON SUBSEQUENCE (LCS)
  Instance: k sequences X_i over an alphabet Σ of fixed size (such as $|\Sigma| = 4$ for DNA sequences), and a positive integer m.
  Parameter: A positive integer k.
  Question: Is there a string $X \in \Sigma^*$ of length m that is a subsequence of each of the X_i.

The LCS problem for k sequences is known to be hard for $W[t]$ for all t when the alphabet is bounded by the parameter k by a very intricate reduction—but this is not the problem that is really of interest to biologists. A simple dynamic programming algorithm that is widely used in the packages available to molecular biologists runs in time $O(n^k)$, and is frequently applied with $k = 5$ (with the help of supercomputers). This is an almost ideal target for parameterized complexity analysis or algorithm design—yet it seems to be very difficult to resolve.

12 Structural issues and analogs of classical results

FPT and the parametric universe in general throw up a whole new world to explore. One natural thing to do is to look for inspiration towards the classical world, and then investigate the possibility of analogous structural results in the parametric world. Here are some examples.

One of the deep results of the theory is a *partial* analog of Ladner's theorem [Lad75] that the polynomial time degrees are dense [DF93]. Obtaining a full analog—if it is even true—remains a significant and very challenging open problem.

We currently don't even know a reasonable hypothesis implying that the $W[t]$-hierarchy is infinite. If the hierarchy collapses, does collapse propogate upwards? That is, if $W[1] = W[2]$ for instance, then does entail $W[2] = W[3]$? Or would collapse propagate downwards? Does the hierarch collapse under *randomized reductions*? What seems to be needed are uniform versions of Hastad's switching lemma or Razborov's techniques.

It is important for the reader interested in structural issues to realize that proving parameterized analogs of classical theorems is not at all routine, and that few analogs are currently known. Some would have very interesting concrete consequences. For example, if there is a parameterized analog of Toda's theorem, then the problem UNIQUE CLIQUE of determining whether a graph has a unique k-clique (for parameter k) is as hard as any problem at any level of the $W[t]$ hierarchy [DFR98a]. Currently, however, we lack even the basic probability amplification results that such an analog would presumably require.

What about oracle results? The ones from [DF93] are somehow unsatisfying. Perhaps there are better models.

What about highness/lowness? This might be relevant to GRAPH ISOMORPHISM.

Finally, what about the structure of FPT? For instance, define a parameterized reduction $\langle x, k \rangle \mapsto \langle x', k' \rangle$ to be (poly, poly) if the function $k \mapsto k'$ is also polynomial time. Then we could look at problems in FPT that only had FPT algorithms with the constant 2^k or 2^{2^k}. Are there complete problems etc? The reductions would be (poly,poly) ones.

A Final Word

The previous sections have explored the FPT world as a method of coping with classical infeasibility, and thus as a kind of fall-back after an initial pessimistic classification such as NP-completeness. In [DFS98], the authors finsihed with a much more radical view which we would like to summarize below. The view is that perhaps parametric analysis is a better *primary classification tool* than P for combinatorial problems.

The current approach to the analysis of concrete computational problems is dominated by two kinds of effort:

(1) The search for asymptotic worst-case polynomial-time algorithms.

(2) Alternatively, proofs of classical hardness results, particularly *NP*-hardness.

We expect that these will become substantially supplemented by:

(1') The design of *FPT* algorithms for various parameterizations of a given problem, and the development of associated heuristics.

(2') Alternatively, demonstrations of $W[1]$-hardness.

We think this will happen because we are inevitably forced towards something like an *ultrafinitist* [YV70] outlook concerning computational complexity because of the nature of the universe of interesting yet feasible computation. The main point of this outlook is that numbers in different ranges of magnitude should be treated in qualitatively different ways.

The pair of notions (1') and (2') are actually rather straightforward mutations of (1) and (2), and they inherit many of the properties that have made the framework provided by (1) and (2) so successful. We note the following in support of this position.

• The enrichment of the dialogue between practice and theory that parameterized complexity is based on always makes sense. It *always* makes sense to ask the users of algorithms, "Are there aspects of your problem that may typically belong to limited distributional ranges?"

- Fixed-parameter tractability is a more accurate notion of "the good". If you were concerned with inverting very large matrices and could identify a bounded structural parameter k for your application that allows this to be done in time $O(2^k n^2)$, then you might well prefer this *classically exponential-time* algorithm to the usual $O(n^3)$ polynomial-time algorithm.
- The "bad", $W[1]$-hardness, is based on a miniaturization of Cook's Theorem in a way that establishes a strong analogy between NP and $W[1]$. Proofs of $W[1]$-hardness are generally more challenging than NP-completeness, but it is obvious by now that this is a very applicable complexity measurement.
- Problems that are hard do not just go away. Parameterization allows for several kinds of sustained dialogue with a single problem, in ways that allow finer distinctions about the causes of intractability (and opportunities for practical algorithms, including systematically designed heuristics) to be made than the exploration of the "NP-completeness boundary" described in [GJ79].
- Polynomial time has thrived because it is a mathematically rich and productive notion allowing for a wide variety of algorithm design techniques. FPT seems to offer an even richer field of play, in part because it encompasses polynomial time as usually the best kind of FPT result. Beyond this, the FPT objective encompasses a rich and distinctive positive toolkit, including novel ways of defining and exploiting parameters.
- There is some evidence that not only are small polynomial exponents generally available when problems are FPT, but also that simple exponential parameter functions such as 2^k are frequently achievable, and that many of the problems in FPT admit kernelization algorithms that provide useful start-ups for *any* algorithmic attack on the problem.
- The complexity of approximation is handled more elegantly than in the classical theory, with $W[1]$-hardness immediately implying that there is no efficient PTAS. Moreover, FPT algorithm design techniques appear to be fruitful in the design of approximation algorithms.
- Parameterization is a very broad idea. It is possible to formulate and explore notions such as randomized FPT [FK93], parameterized parallel complexity [Ces96], parameterized learning complexity [DEF93], parameterized approximation [BFH97], parameterized cryptosystems based on $O(n^k)$ security, etc.

Finally, we feel that the methodology is widely applicable because of our very nature as people. The relevant parameter may not be obvious, but because of the inherent bounded depth of our minds, the natural situations that give rise to computational problems frequently exhibit distinct regularities and parameterizable constraints. It seems just a matter of looking for the correct parameter.

References

[ADF95] K. Abrahamson, R. Downey and M. Fellows, "Fixed Parameter Tractability and Completeness IV: On Completeness for $W[P]$ and $PSPACE$ Analogs," *Annals of Pure and Applied Logic* 73 (1995), 235–276.

[AF93] K. Abrahamson and M. Fellows, "Finite Automata, Bounded Treewidth and Wellquasiordering," In: *Graph Structure Theory*, American Mathematical Society, Contemporary Mathematics Series, vol. 147 (1993), 539–564.

[Al98] B. Allen, *Subtree Transfer Operations and their Induced Metrics on Evolutionary Trees*, MSc. Thesis, University of Canterbury, 1998.

[AFGPR96] E. Allender, J. Feigenbaum, J. Goldsmith, T. Pitassi and S. Rudich, "The Future of Computational Complexity Theory: Part II," *SIGACT News* 27 (1996), 3–7.

[ALT96] S. Alstrup, P.W. Lauridsen and M. Thorup. "Generalized Dominators for Structured Programs." *Proc. 3rd Static Analysis Symp.*, Springer Verlag Lecture Notes in Computer Science vol. 1145 (1996), 42–51.

[AMOV91] G. B. Agnew, R. C. Mullin, I. M. Onyszchuk and S. A. Vanstone, "An Implementation for a Fast Public-Key Cryptosystem," *J. Cryptology* 3 (1991), 63–79.

[ALMSS92] S. Arora, C. Lund, R. Motwani, M. Sudan and M. Szegedy, "Proof Verification and Intractability of Approximation Algorithms," *Proceedings of the IEEE Symposium on the Foundations of Computer Science* (1992).

[Ar96] S. Arora, "Polynomial Time Approximation Schemes for Euclidean TSP and Other Geometric Problems," In: *Proceedings of the 37th IEEE Symposium on Foundations of Computer Science*, 1996.

[AYZ94] N. Alon, R. Yuster and U. Zwick, "Color-Coding: A New Method for Finding Simple Paths, Cycles and Other Small Subgraphs Within Large Graphs," *Proc. Symp. Theory of Computing (STOC)*, ACM (1994), 326–335.

[Ba94] B. Baker, "Approximation Algorithms for NP-Complete Problems on Planar Graphs," *J.A.C.M.* 41 (1994), 153–180.

[Baz95] C. Bazgan, "Schémas d'approximation et complexité paramétrée," Rapport de stage de DEA d'Informatique à Orsay, 1995.

[BDFHW95] H. Bodlaender, R. Downey, M. Fellows, M. Hallett and H. T. Wareham, "Parameterized Complexity Analysis in Computational Biology," *Computer Applications in the Biosciences* 11 (1995), 49–57.

[BDFW95] H. Bodlaender, R. Downey, M. Fellows and H.T. Wareham, "The Parameterized Complexity of the Longest Common Subsequence Problem," *Theoretical Computer Science A* 147 (1995), 31-54.

[BF95] H. Bodlaender and M. Fellows, "On the Complexity of k-Processor Scheduling," *Operations Research Letters* 18 (1995), 93–98.

[BFH94] H. Bodlaender, M. R. Fellows and M. T. Hallett, "Beyond NP-completeness for Problems of Bounded Width: Hardness for the W Hierarchy," *Proc. ACM Symp. on Theory of Computing (STOC)* (1994), 449–458.

[BFH97] H. Bodlaender, M. Fellows, M. Hallett, "Parameterized Complexity and Parameterized Approximation for Some Problems About Trees," manuscript, 1997.

[BFR98] R. Balasubramanian, M. Fellows and V. Raman, "An Improved Fixed-Parameter Algorithm for Vertex Cover," *Information Processing Letters*, to appear.

[BFRS98] D. Bryant, M. Fellows, V. Raman and U. Stege, "On the Parameterized Complexity of MAST and 3-Hitting Sets," manuscript, 1998.

[BFW92] H. Bodlaender, M. Fellows and T. Warnow, "Two Strikes Against Perfect Phylogeny," in: *Proceedings of the 19th International Colloquium on Automata, Languages and Programming*, Springer-Verlag, Lecture Notes in Computer Science vol. 623 (1992), 273–283.

[BGKRS95] R. S. Barr, B. L. Golden, J. P. Kelly, M.G.C. Resende and W. R. Stewart, "Designing and Reporting on Computational Experiments with Heuristic Methods," *J. Heuristics* 1 (1995), 9–32.

[Bod96] H. Bodlaender, "A Linear Time Algorithm for Finding Tree Decompositions of Small Treewidth," *SIAM J. Comp.* 25 (1996), 1305–1317.

[Bry97] D. Bryant. Dissertation.

[Byl94] T. Bylander, "The Computational Complexity of Propositional STRIPS Planning," *Artificial Intelligence* 69 (1994), 165–204.

[LM81] E. Leggett and D. Moore, "Optimization problems and the polynomial time hierarchy," *Theoretical Computer Science*, Vol. 15 (1981), 279-289.

[LeC95] Leizhen Cai, "Fixed-parameter tractability of graph modification problems for hereditary properties," Tecnical Report, Department of Computer Science, The Chinese University of Hong Kong, Shatin, New Territories, Hong Kong, to appear in *Information Processing Letters*, 1995.

[Ca92] Liming Cai, "Fixed parameter tractability and approximation problems," Project Report, June 1992.

[CC97] L. Cai and J. Chen. "On Fixed-Parameter Tractability and Approximability of NP-Hard Optimization Problems," *J. Computer and Systems Sciences* 54 (1997), 465–474.

[CCDF96] L. Cai, J. Chen, R. G. Downey amd M. R. Fellows, "On the Parameterized Complexity of Short Computation and Factorization," *Arch. for Math. Logic* 36 (1997), 321–337.

[CCDF97] L. Cai, J. Chen, R. Downey and M. Fellows, "Advice Classes of Parameterized Tractability," *Annals of Pure and Applied Logic* 84 (1997), 119–138.

[CD94] K. Cattell and M. J. Dinneen, "A Characterization of Graphs with Vertex Cover up to Five," *Proceedings ORDAL'94*, Springer Verlag, Lecture Notes in Computer Science, vol. 831 (1994), 86–99.

[CDDFL98] K. Cattell, M. Dinneen, R. Downey and M. Fellows, "On Computing Graph Minor Obstruction Sets," *Theoretical Computer Science A*, to appear.

[Ces96] M. Cesati, "Structural Aspects of Parameterized Complexity," Ph.D. dissertation, University of Rome, 1995.

[CF96] M. Cesati and M. Fellows, "Sparse Parameterized Problems," *Annals of Pure and Applied Logic* 62 (1996).

[Cl87] K. L. Clarkson, "New Applications of Random Sampling in Computational Geometry," *Discrete and Computational Geometry* 2 (1987), 195–222.

[Co90] B. Courcelle, "Graph Rewriting: An Algebraic and Logical Approach," in: *Handbook of Theoretical Computer Science, vol. B*, J. van Leeuwen, ed., North Holland (1990), Chapter 5.

[CL95] B. Courcelle and J. Lagergren, "Equivalent Definitions of Recognizability for Sets of Graphs of Bounded Treewidth," *Mathematical Structure in Computer Science* 6 (1996), no. 2, 141–165.

[CT97] M. Cesati and L. Trevisan, "On the Efficiency of Polynomial Time Approximation Schemes," *Information Processing Letters* 64 (1997), 165–171.

[CW95] M. Cesati and H. T. Wareham, "Parameterized Complexity Analysis in Robot Motion Planning," *Proceedings 25th IEEE Intl. Conf. on Systems, Man and Cybernetics*.

[DJPSY92] E. Dahlhaus, D. Johnson, C. Papadimitriou, P. Seymour,and M. Yannakakis, "The complexity of multiway cuts," *STOC* (1992) 241–251

[DGS86] L. Dennenberg, Y. Gurevich and S. Shelah. Definability by constant-depth polynomial-size circuits. *Information and Control* 70 (1986), 216–240.

[DEF93] R. Downey, P. Evans and M. Fellows, "Parameterized Learning Complexity," *Proc. 6th ACM Workshop on Computational Learning Theory* (1993), 51–57.

[DF93] R. Downey and M. Fellows, "Fixed Parameter Tractability and Completeness III: Some Structural Aspects of the W-Hierarchy," in: K. Ambos-Spies, S. Homer and U. Schöning, editors, *Complexity Theory: Current Research*, Cambridge Univ. Press (1993), 166–191.

[DF95a] R. G. Downey and M. R. Fellows, "Fixed Parameter Tractability and Completeness I: Basic Theory," *SIAM Journal of Computing* 24 (1995), 873-921.

[DF95b] R. G. Downey and M. R. Fellows, "Fixed Parameter Tractability and Completeness II: Completeness for $W[1]$," *Theoretical Computer Science A* 141 (1995), 109-131.

[DF95c] R. G. Downey and M. R. Fellows, "Parametrized Computational Feasibility," in: *Feasible Mathematics II, P. Clote and J. Remmel (eds.)* Birkhauser, Boston (1995) 219-244.

[DF98] R. G. Downey and M. R. Fellows, *Parameterized Complexity*, Springer-Verlag, 1998.

[DFS98] R. Downey, M. Fellows and U. Stege, Parameterized Complexity: A Framework for Systematically Confronting Computational Intractability, (To appear) DIMACS series on Combinatorics in the 21st Century, AMS Publ.

[DFKHW94] R. G. Downey, M. Fellows, B. Kapron, M. Hallett, and H. T. Wareham. "The Parameterized Complexity of Some Problems in Logic and Linguistics," *Proceedings Symposium on Logical Foundations of Computer Science (LFCS)*, Springer-Verlag, Lecture Notes in Computer Science vol. 813 (1994), 89–100.

[DFR98a] R. G. Downey, M. R. Fellows and K. W. Regan, "Parameterized Circuit Complexity and the W Hierarchy," *Theoretical Computer Science A* 191 (1998), 91–115,

[DFR98b] R. G. Downey, M. Fellows and K. Regan. "Threshold Dominating Sets and an Improved Characterization of $W[2]$," *Theoretical Computer Science A*, to appear.

[DFT96] R. G. Downey, M. Fellows and U. Taylor, "The Parameterized Complexity of Relational Database Queries and an Improved Characterization of $W[1]$," in: *Combinatorics, Complexity and Logic: Proceedings of DMTCS'96*, Springer-Verlag (1997), 194–213.

[DFVW98] R. Downey, M. Fellows, A. Vardy and G. Whittle, "The Parameterized Complexity of Some Fundamental Problems in Coding Theory," *SIAM J. Computing*, to appear.

[DKL96] T. Dean, J. Kirman and S.-H. Lin, "Theory and Practice in Planning," Technical Report, Computer Science Department, Brown University, 1996.

[Fag74] R. Fagin, "Generalized first order spectra and polynomial time recognizable languages," in *Complexity of Computations*, (R. Karp, ed.) SIAM-AMS proceedings, Vol. 7 (1974), 43-73.

[Fel97] J. Felsenstein. Private communication, 1997.

[FHK95] M. Fellows, M. Hallett and D. Kirby, "The Parameterized Complexity of the Shortest Common Supersequence Problem," manuscript, 1985.

[FHW93] M. Fellows, M. Hallett and H. T. Wareham, "DNA Physical Mapping: Three Ways Difficult," in: *Algorithms – ESA'93: Proceedings of the First European Symposium on Algorithms*, Springer-Verlag, Lecture Notes in Computer Science vol. 726 (1993), 157–168.

[FK93] M. Fellows and N. Koblitz, "Fixed-Parameter Complexity and Cryptography," *Proceedings of the 10th Intl. Symp. on Applied Algebra, Algebraic Algorithms and*

Error-Correcting Codes, Springer-Verlag, Berlin, Lecture Notes in Computer Science vol. 673 (1993), 121–131.

[FKS98] M. Fellows, C. Korostensky and U. Stege, "Theory-Based Heuristics for Editing Gaps in Multiple Sequence Alignments," manuscript, 1998.

[FL87] M. Fellows and M. Langston, "Nonconstructive Proofs of Polynomial-Time Complexity," *Information Processing Letters* 26(1987/88), 157–162.

[FL88] M. Fellows and M. Langston, "Nonconstructive Tools for Proving Polynomial-Time Complexity," *Journal of the Association for Computing Machinery* 35 (1988) 727–739.

[FL89] M. R. Fellows and M. A. Langston, "An Analogue of the Myhill-Nerode Theorem and its Use in Computing Finite-Basis Characterizations," *Proceedings of the IEEE Symposium on the Foundations of Computer Science* (1989), 520–525.

[FL92] M. R. Fellows and M. A. Langston, "On well-partial-ordering theory and its applications to combinatorial problems in VLSI design," *SIAM J. of Discrete Math.* 5 (1992) 117-126.

[FL94] M. R. Fellows and M. A. Langston, "On Search, Decision and the Efficiency of Polynomial-Time Algorithms," *Journal of Computer and Systems Science* 49 (1994), 769–779.

[FN71] R. E. Fikes and N. J. Nilsson, "STRIPS: A New Approach to the Application of Theorem Proving to Problem Solving," *Artificial Intelligence* 2 (1971), 189–208.

[FPT95] M. Farach, T. Przytycka, and M. Thorup. "On the agreement of many trees" *Information Processing Letters* 55 (1995), 297–301.

[Fr97] J. Franco, J. Goldsmith, J. Schlipf, E. Speckenmeyer and R.P. Swaminathan, "An Algorithm for the Class of Pure Implicational Formulas,"

[FRS87] H. Friedman, N. Robertson, and Seymour, "The metamathematics of the graph minor theorem," *Logic and Combinatorics* (Arcata, Calif., 1985), 229–261, Contemp. Math., 65, Amer. Math. Soc., Providence, R.I., 1987.

[GJ79] M. Garey and D. Johnson. *Computers and Intractability: A Guide to the Theory of NP-completeness*. W.H. Freeman, San Francisco, 1979.

[GM75] G. R. Grimmet and C. H. J. McDiarmid, "On Colouring Random Graphs," *Math. Proc. Cambridge Philos. Soc.* 77 (1975), 313–324.

[GLS88] M. Grötschel, L. Lovasz and A. Schrijver, *Geometric Algorithms and Combinatorial Optimization*, Springer–Verlag, (1988).

[GM84] S. Goldwasser and S. Micali, "Probabilistic Encryption," *J. Computer and Systems Science* 28 (1984), 270–299.

[Gur89] Y. Gurevich, "The Challenger-Solver Game: Variations on the Theme of P=?NP," *Bulletin EATCS* 39 (1989), 112–121.

[Hal96] M. Hallett. "An Integrated Complexity Analysis of Some Problems in Computational Biology," Ph.D. dissertation, Department of Computer Science, University of Victoria, 1996.

[Hart94] J. Hartmanis, "About the Nature of Computer Science," *Bulletin EATCS* 53 (1994), 170–190.

[HL92] J. Hartmanis and H. Lin, editors, *Computing the Future: A Broader Agenda for Computer Science and Engineering*, National Academy Press, 1992.

[HM91] F. Henglein and H. G. Mairson, "The Complexity of Type Inference for Higher-Order Typed Lambda Calculi." In *Proc. Symp. on Principles of Programming Languages (POPL)* (1991), 119-130.

[Hoo95] J. N. Hooker, "Testing Heuristics: We Have It All Wrong," *J. Heuristics* 1 (1995), 33–42.

[GHS98] G. Gonnet, M. Hallett and U. Stege, "Vertex Cover Revisited: A Hybrid Algorithm of Theory and Heuristic," to appear.

[IK75] O. H. Ibarra and C. E. Kim, "Fast Approximation for the Knapsack and Sum of Subset Problems," *J. Assoc. Computing Machinery* 22 (1975), 463–468.
[JDUGG74] D. S. Johnson, A. Demers, J. D. Ullman, M. R. Garey and R. L. Graham, "Worst-Case Performance Bounds for Simple One-Dimensional Packing Algorithms," *SIAM J. Computing* 3 (1974), 299–325.
[Jo87] D. S. Johnson, "The NP-Completeness Column: An Ongoing Guide," *Journal of Algorithms* 1987.
[JM93] D. S. Johnson and C. McGeoch (eds.), *Network Flows and Matching: First DIMACS Implementation Challenge*, American Mathematical Society, 1993.
[JT96] D. S. Johnson and M. Trick (eds.), *Cliques, Coloring and Satisfiability: Second DIMACS Implementation Challenge*, American Mathematical Society, 1996.
[Karp72] R. M. Karp, "Reducibility Among Combinatorial Problems," in: *Complexity of Computer Applications*, Plenum Press, New York (1972), 85–103.
[Karp86] R. M. Karp, "Combinatorics, Complexity and Randomness," *Communications of the ACM* 29 (1986), 98–109.
[KS94] S. Kirkpatrick and B. Selman, "Critical Behavior in the Satisfiability of Boolean Formulae," *Science* 264 (1994), 1297–1301.
[KST94] H. Kaplan, R. Shamir and R. E. Tarjan, "Tractability of Parameterized Completion Problems on Chordal and Interval Graphs: Minimum Fill-In and DNA Physical Mapping," in: *Proc. 35th Annual Symposium on the Foundations of Computer Science (FOCS)*, IEEE Press (1994), 780–791.
[KoT95] P. Kolaitis and M. Thakur, "Approximation properties of NP minimization classes," *J. C. S. S.*, Vol. 50 (1995), 391-411.
[KT92] A. Kornai and Z. Tuza, "Narrowness, Pathwidth and Their Application in Natural Language Processing," *Discrete Applied Mathematics* 36 (1992), 87-92.
[JLW94] T. Jiang, E. Lawler, and L. Wang, "Aligning sequences via an evolutionary tree," in *Proceedings of the 26th Annual Symposium on the Theory of Computing*, ACM Press, 1994, 760-769.
[Lad75] R. Ladner, "On the structure of polynomial time reducibility," *J. Assoc. Comput. Mach.*, Vol. 22 (1975), 155-171.
[Len90] T. Lengauer, "VLSI Theory," in: *Handbook of Theoretical Computer Science vol. A*, Elsevier (1990), 835-868.
[Lev86] L. Levin, "Average Case Complete Problems," *SIAM J. Computing* 15 (1986), 285-286.
[Luks82] E. Luks, "Isomorphism of Graphs of Bounded Valence Can Be Tested in Polynomial Time," *J. Comput. and Systems Sci.* 25 (1982), 42-65.
[LP85] O. Lichtenstein and A. Pneuli. "Checking That Finite-State Concurrents Programs Satisfy Their Linear Specification." In: *Proceedings of the 12th ACM Symposium on Principles of Programming Languages* (1985), 97-107.
[MP94] S. Mahajan and J. G. Peters, "Regularity and Locality in k-Terminal Graphs," *Discrete Applied Mathematics* 54 (1994), 229-250.
[MR98] M. Mahajan and V. Raman, "Parameterizing Above the Guarantee: MaxSat and MaxCut," to appear in *J. Algorithms*.
[MR95] R. Motwani and P. Raghavan, *Randomized Algorithms*, Cambridge Univ. Press, 1995.
[MSL92] D. Mitchell, B. Selman and H. Levesque, "Hard and Easy Distributions of SAT Problems," in: *Proceedings AAAI-92*, AAAI (1992), 459-465.
[NP85] J. Nešetřil and S. Poljak, "On the Complexity of the Subgraph Problem," *Comm. Math. Univ. Carol.* 26 (1985), 415-419.

[PGWRS96] C. Papadimitriou, O. Goldreich, A. Wigderson, A. Razborov and M. Sipser, "The Future of Computational Complexity Theory: Part I," *SIGACT News* 27 (1996), 6–12.

[PY91] C. Papadimitriou and M. Yannakakis, "Optimization, approximation, and complexity classes," *Journal Comput. Sys. Sci* Vol. 43 (1991) 425–440.

[PY96] C. Papadimitriou and M. Yannakakis, "On Limited Nondeterminism and the Complexity of the VC Dimension," *J. Computer and Systems Sciences* 53 (1996), 161–170.

[RS85] N. Robertson and P. D. Seymour, "Graph Minors: A Survey," in *Surveys in Combinatorics*, I. Anderson, ed. (Cambridge University Press: Cambridge, 1985), 153-171.

[RS96] N. Robertson and P. D. Seymour, "Graph Minors XV. Giant Steps," *J. Comb. Theory Ser. B* 68 (1996), 112–148.

[Sha97] R. Shamir. Private communication, 1997.

[SOWH96] D. L. Swofford, G. J. Olsen, P. J. Waddell and D. M. Hillis, "Phylogenetic Inference," in: D. M. Hillis, C. Moritz and B. K. Mable (eds.) *Molecular Systematics*. Sinauer Associates, Inc. (1996), 407–514.

[SS77] R. Solovay and V. Strassen, "A Fast Monte Carlo Test for Primality," *SIAM J. Computing* (1977), 84–85.

[SW97] M. Steel and T. Warnow,

[Th97] M. Thorup, "Structured Programs Have Small Tree-Width and Good Register Allocation," *Proceedings 23rd International Workshop on Graph-Theoretic Concepts in Computer Science, WG'97*, R. Möhring (ed.), Springer Verlag, Lecture Notes in Computer Science vol. 1335 (1997), 318–332.

[To91] S. Toda, "PP is as Hard as the Polynomial Time Hierarchy," *SIAM J. Comput.* (1991), 865–877.

[vanL90] L. Van Leeuwin, *Handbook of Theoretical Computer Science, Vol. A*. North Holland, 1990.

[Wilf85] H. Wilf, "Some Examples of Combinatorial Averaging," *Amer. Math. Monthly* 92 (1985), 250–261.

[Yan95] M. Yannakakis. "Perspectives on Database Theory," *Proceedings of the IEEE Symposium on the Foundations of Computer Science* (1995), 224–246.

[YV70] A. S. Yessenin-Volpin, "The Ultraintuitionistic Criticism and the Antitraditional Program for Foundations of Mathematics," in: *Intuitionism and Proof Theory: Proceedings of the Summer Conference at Buffalo, N.Y., 1968*, A. Kino, J. Myhill and R. E. Vesley (eds.), North-Holland, 1970.

Hidden Algebra for Software Engineering

Joseph A. Goguen

Department of Computer Science & Engineering,
University of California at San Diego,
La Jolla CA 92093-0114 USA. goguen@cs.ucsd.edu

Abstract. This paper is an introduction to recent research on hidden algebra and its application to software engineering; it is intended to be informal and friendly, but still precise. We first review classical algebraic specification for traditional "Platonic" abstract data types like integers, vectors, matrices, and lists. Software engineering also needs changeable "abstract machines," recently called "objects," that can communicate concurrently with other objects through visible "attributes" and state-changing "methods." Hidden algebra is a new development in algebraic semantics designed to handle such systems. Equational theories are used in both cases, but the notion of satisfaction for hidden algebra is *behavioral*, in the sense that equations need only *appear* to be true under all possible experiments; this extra flexibility is needed to accommodate the clever implementations that software engineers often use to conserve space and/or time. The most important results in hidden algebra are powerful *hidden coinduction* principles for proving behavioral properties. This paper also includes some comparison with the closely related area called coalgebra, and some bits of history.

1 Introduction

Algebra in its modern abstract sense seems to have begun with Emmy Noether around 1927 [76]. Today this area builds on equational theories for monoids, groups, rings, etc. Garrett Birkhoff pioneered the general study of equational theories under what we now call "loose semantics," giving in particular rules of deduction and a completeness theorem [6]; Alfred Tarski also made significant early contributions to this area, which is now called "universal algebra" (or sometimes "general algebra"). ADJ[1] pioneered "initial semantics" and its application to abstract data types (hereafter abbreviated ADTs) in computer science [26, 46, 45], which has since blossomed into an area of research called "algebraic semantics" (also "algebraic specification"), with hundreds of workers, its own conferences and journals, and even its own professional society (called AMAST). Hidden algebra is one of the most recent developments in this lively area.

---
[1] This name was used for the set {Goguen, Thatcher, Wagner, Wright}; see [29] for historical details.

1.1 Notes on the state of the software arts

Software development is very difficult. To understand it better, we can distinguish among designing, coding, and verifying (i.e., proving properties of) programs. Most of the literature addresses code verification, but this is very difficult in practice, and empirical studies have shown that little of the cost of software arises from errors in coding: most comes from errors in design and requirements [7]. Moreover, many programs are written in obscure and/or obsolete languages, with complex ugly semantics (like Cobol, Jovial, and Mumps), are very poorly documented, are indispensable to some enterprise, and are very large, often several million lines, sometimes more. Therefore it is usually an enormous effort to verify real code, and it isn't usually worth the trouble. I like to call this the *semantic swamp*; it is a place to avoid.

Moreover, programs in everyday use usually evolve, because computers, operating systems, tax laws, user requirements, etc. are all changing rapidly. Therefore the effort of verifying yesterday's version is wasted, because even small code modifications can require large proof modifications—proof is a discontinuous function of truth.

This suggests that we should focus on *design* and *specification*. But even this is difficult, because the properties that people really want, such as security, deadlock freedom, liveness, ease of use, and ease of maintenance, are complex, not always formalizable, and even when they are formalizable, may involve subtle interactions among remote parts of systems. However, this is an area where mathematics can make a contribution.

It is well known that most of the effort in programming goes into debugging and maintaining (i.e., improving and updating) programs [7]. Therefore anything that can be done to ease these processes has enormous economic potential. One step in this direction is to "encapsulate data representations"; this means to make the actual structure of data invisible, and to provide access to it only via a given set of operations which retrieve and modify the hidden data structure. Then the implementing code can be changed without having any effect on other code that uses it. On the other hand, if client code relies on properties of the representation, it can be extremely hard to track down all the consequences of modifying a given data structure (say, changing a doubly linked list to an array), because the client code may be scattered all over the program, without any clear identifying marks. This is why the so-called year 2,000 (Y2K) problem is so difficult.

An encapsulated data structure with its accompanying operations is called an *abstract data type*. The crucial advance was to recognize that operations should be associated with data representations; this is exactly the same insight that advanced algebra from mere *sets* to *algebras*, which are sets *with* their associated operations. In software engineering this insight seems to have been due to David Parnas [71, 70], and in algebra to Emmy Noether [76]. (Parallel developments in software engineering and abstract algebra are a theme of this paper.)

It turns out that although abstraction as isomorphism is enough for algebras representing data values (numbers, vectors, etc.), other important problems in software engineering need the more general notion of *behavioral abstraction*, where two models are considered abstractly the same if they exhibit the same behavior. The usual many sorted algebra is not rich enough for this: we have to add structure to distinguish sorts used for data values from sorts used for states, and we need a more general, behavioral, notion of satisfaction, as discussed in Section 3.

In line with the above general discussion of software methodology, we want to prove properties of specifications, not properties of code. Often the most important property of a specification is that it *refines* another specification, in the sense that any model (i.e., any code realizing) the second is also a model of the first. Methodologically speaking, a refinement embodies a set of closely related design decisions for realizing one set of behaviors from another[2]. In line with the discussion of the previous paragraph, we will often want to prove *behavioral* refinements. Behavioral refinement is much more general than ordinary refinement, and many of the clever implementation techniques that so often occur in practice require this extra generality.

The need for improved software development methods is very great. Large complex software systems fail much more often than seems to be generally recognized. One highly visible example is the 1996 cancellation by the US Federal Aviation Agency of an 8 billion dollar contract with IBM to build the next generation air traffic control system for the entire US [72]. This is perhaps the largest default in history, but there are many more examples, including the 1995 cancellation by the US Department of Defense of a 2 billion dollar contract with IBM to provide modern information systems to replace myriads of obsolete, incompatible systems. Other highly publicized failures include IBM software for delivering real time sports data to the media at the 1996 Olympic Games in Atlanta, the 2 billion dollar loss of the European Ariane 5 satellite, and the failure of the software for the United Airlines baggage delivery system at Denver International Airport, delaying its 1994 opening by one and a half years [25], and losing 1.1 million dollars per day.

What these examples have in common is that they were hard to hide. Anyone who has worked in the software industry has seen numerous examples of projects that were over time, over cost, or failed to meet crucial requirements, and hence were cancelled, curtailed, diverted, replaced, or released anyway, sometimes with dire consequences, and frequently with loud declarations of success, even though the system was never used, and may well have been unusable. For obvious reasons, the organizations involved usually try to hide their failures, but experience suggests that half or more of large complex systems fail in some significant way,

---

[2] Empirical studies show that real software development projects involve many false starts, redesigns, prototypes, patches, etc. [11]. Although an idealized view of a project as a sequence of refinements is a useful way to organize and document work retrospectively, it is important not to confuse this with the actual development process.

and that the frightening list in the previous paragraph is just the tip of an enormous iceberg. Today attention has shifted to the Y2K problem. The cost of necessary software repairs has been estimated to lie in the multi-trillion dollar range. It remains to be seen what the effects will be when that fateful night arrives.

1.2 Overview of this paper

Section 2 briefly summarizes some basics of classical algebraic specification theory, following [30] and [32]. Then Section 3 explains why software engineering also needs objects, and why their behavioral properties are important, followed by a brief overview of hidden algebra that follows [38]. The notation of OBJ3 [37, 48] is used in some examples. Section 4 briefly discusses some other related work.

Although I'm far from the only one working on these kinds of problem, this survey is focused on the work with which I'm most familiar, from UCSD and from the Japanese CafeOBJ project. The UCSD group has put much material on the web, including several hidden algebraic proofs with tutorial background information, remote proof execution, and Java applets to illustrate the main ideas; see http://www.cs.ucsd.edu/groups/tatami and the papers [35, 34, 43], which are available (along with many others) from my website:

http://www.cs.ucsd.edu/users/goguen

2 Algebraic specification of "Platonic" abstract data types

In the early 1970s, the ADT concept was understood due to the pioneering work of David Parnas [70, 71], but there was no precise semantics, so it was impossible to verify the correctness of an implementation for an ADT, or even to formulate what correctness might mean. The first to try an algebraic solution to this problem was Steve Zilles [77], or at least, hearing about his work via Jack Dennis led to my giving it a try. The initial algebra theory of abstract data types provided the first rigorous formalization of ADTs; this approach works especially well for things like integers, which do not change over time[3]; it also works for ADTs with state, but not as well. John Guttag [49] developed a different approach, which while not quite rigorous, was actually more suitable for handling states. This section gives precise definitions for the initial algebra approach to ADTs, and some of their basic properties, especially that an ADT is uniquely determined by its specification as an initial algebra, that abstract data types are indeed abstract, and that initial algebra semantics can capture all computable data types.

---

[3] These are the sort of eternal "ideas" with which Plato was concerned, as opposed to the later view of Aristotle, which was more concerned with change.

2.1 Signature, algebra and homomorphism

We begin with syntax, which raises more issues than you might suspect at first. The classical one sorted case of Birkhoff, Tarski etc. is not adequate for computer science applications, which involve many different "types" of data. Benabou [1] seems to have been the first to consider many sorted universal algebra, which he did in an elegant category theoretic setting; it has since been developed by many others in more conventional settings; see [40] for more historical information. Goguen [26] introduced *overloaded* operation symbols for many sorted algebra. This extension is important for applications like dynamic binding in object oriented programming, and it is now quite common in the computer science literature. The approach[4] is based on many sorted sets: Given a set S, whose elements are called **sorts**, then an S-**sorted set** A is a family of sets A_s, one for each $s \in S$; the notation $A = \{A_s \mid s \in S\}$ will be used for such sets. Recall that S^* denotes the set of strings over S; we will let "[]" denote the empty string. We are now ready for the main concepts.

Definition 1. An S-**sorted signature** Σ is an $(S^* \times S)$-sorted set $\{\Sigma_{w,s} \mid \langle w, s \rangle \in S^* \times S\}$. The elements of $\Sigma_{w,s}$ are called **operation symbols** of **arity** w, **sort** s, and **rank** $\langle w, s \rangle$; in particular, $\sigma \in \Sigma_{[],s}$ is a **constant symbol**. Σ is a **ground signature** iff $\Sigma_{[],s} \cap \Sigma_{[],s'} = \emptyset$ whenever $s \neq s'$ and $\Sigma_{w,s} = \emptyset$ unless $w = []$. □

By convention, $|\Sigma| = \bigcup_{w,s} \Sigma_{w,s}$ and $\Sigma' \subseteq \Sigma$ means $\Sigma'_{w,s} \subseteq \Sigma_{w,s}$ for each w, s. Similarly, **union** is defined by $(\Sigma \cup \Sigma')_{w,s} = \Sigma_{w,s} \cup \Sigma'_{w,s}$. A common special case is union with a ground signature X, for which we use the notation $\Sigma(X) = \Sigma \cup X$.

Definition 2. A Σ-**algebra** A consists of an S-sorted set also denoted A, plus an **interpretation** of Σ in A, which is a family of arrows $i_{s_1...s_n,s} \colon \Sigma_{s_1...s_n,s} \to [A^{s_1...s_n} \to A_s]$ for each rank $\langle s_1...s_n, s \rangle \in S^* \times S$, which interpret the operation symbols in Σ as actual operations on A. For constant symbols, the interpretation is given by $i_{[],s} \colon \Sigma_{[],s} \to A_s$. Usually we write just σ for $i_{w,s}(\sigma)$, but if we need to make the dependence on A explicit, we may write σ_A. A_s is called the **carrier** of A of sort s.

Given Σ-algebras A, A', a Σ-**homomorphism** $h \colon A \to A'$ is an S-sorted arrow $h \colon A \to A'$ such that $h_s(\sigma_A(a_1, ..., a_n)) = \sigma_{A'}(h_{s_1}(s_1), ..., h_{s_n}(a_n))$ for each $\sigma \in \Sigma_{s_1...s_n,s}$ and $a_i \in A_{s_i}$ for $i = 1, ..., n$, and such that $h_s(c_A) = c_{A'}$ for each constant symbol $c \in \Sigma_{[],s}$. □

An increased emphasis on homomorphisms characterizes the modern era of general algebra; this is related to the central role that morphisms play in category theory. Given homomorphisms $f \colon A \to B$ and $g \colon B \to C$, we denote their composition $A \to C$ by $f; g$. Also we denote the identity homomorphism on an algebra A by 1_A.

---

[4] This was developed for my course Information Science 329, Algebraic Foundations of Computer Science, first taught in 1969 at the University of Chicago.

2.2 Term, equation and specification

Given an S-sorted signature Σ, the S-sorted set T_Σ of (**ground**) Σ-**terms** is the smallest set of lists of symbols that contains the constants (i.e., $\Sigma_{[],s} \subseteq T_{\Sigma,s}$), and such that given $\sigma \in \Sigma_{s_1...s_n,s}$ and $t_i \in T_{\Sigma,s_i}$ then $\sigma(t_1 \ldots t_n) \in T_{\Sigma,s}$. We view T_Σ as a Σ-algebra by interpreting $\sigma \in \Sigma_{[],s}$ as just σ, and $\sigma \in \Sigma_{s_1...s_n,s}$ as the operation sending $t_1, \ldots, t_n$ to the list $\sigma(t_1 \ldots t_n)$. Then T_Σ is called the Σ-**term algebra**. Note that because of overloading, terms do not always have a unique parse. The following is the key property of this algebra:

Theorem 3. *(Initiality) Given a signature Σ with no overloaded constants[5] and a Σ-algebra M, there is a unique Σ-homomorphism $T_\Sigma \to M$.* □

This is proved in many places, for example [45]. The Σ-term algebra T_Σ serves as a standard model for a specification with no equations. For example, if Σ is the signature for the natural numbers with just zero and successor, then T_Σ is the natural numbers in Peano notation. If Σ consists of the operation symbols 0, s, + and *, then T_Σ consists of all *expressions* formed using these symbols (with the right arities); these are simple numerical expressions. In order to get the natural numbers with addition and multiplication from this, we need to impose some equations.

Definition 4. A Σ-**equation** consists of a ground signature X of **variable symbols** (disjoint from Σ) plus two $\Sigma(X)$-terms of the same sort $s \in S$; we may write such an equation abstractly in the form $(\forall X)\ t = t'$ and concretely in the form $(\forall x, y, z)\ t = t'$ when $|X| = \{x, y, z\}$ and the sorts of x, y, z can be inferred from their uses in t and in t'. A **specification** is a pair (Σ, E), consisting of a signature Σ and a set E of Σ-equations. □

Conditional equations can be defined in a similar way, but we omit this here. Given Σ and a ground signature X disjoint from Σ, we can form the $\Sigma(X)$-algebra $T_{\Sigma(X)}$ and then view it as a Σ-algebra by forgetting the names of the new constants in X; let us denote this Σ-algebra by $T_\Sigma(X)$. It has the following universal **freeness** property:

Proposition 5. *Given a Σ-algebra A and an interpretation $a\colon X \to A$, there is a unique Σ-homomorphism $\bar{a}\colon T_\Sigma(X) \to A$ extending a, in the sense that $\bar{a}_s(x) = a_s(x)$ for each $x \in X_s$ and $s \in S$.* □

Definition 6. A Σ-algebra A **satisfies** a Σ-equation $(\forall X)\ t = t'$ iff for any $a\colon X \to A$ we have $\bar{a}(t) = \bar{a}(t')$ in A, written $A \models_\Sigma (\forall X)\ t = t'$. A Σ-algebra A satisfies a set E of Σ-equations iff it satisfies each one, written $A \models_\Sigma E$. We may also say that A is a P-algebra, and write $A \models P$ where $P = (\Sigma, E)$. The class of all algebras that satisfy P is called the **variety** defined by P. Given sets E and E' of Σ-equations, let $E \models E'$ mean $A \models E'$ for all E-models A. □

---

[5] Actually, every signature Σ has an initial Σ-algebra, but when Σ has overloaded constants, terms must be annotated by their sort; we will use the same notation T_Σ for this case.

The following simple result is much used in equational theorem proving, but is rarely stated explicitly. Its proof is very simple because it uses the *semantics* of satisfaction rather than some particular rules of deduction, and because it exploits the *initiality* of the term algebra. We have found this typical of proofs in this area; commutative diagrams and other universal properties also help give elegant conceptual proofs.

Fact 7. (Lemma of Constants) *Given a signature Σ, a ground signature X disjoint from Σ, a set E of Σ-equations, and $t, t' \in T_{\Sigma(X)}$, then $E \models_\Sigma (\forall X)\, t = t'$ iff $E \models_{\Sigma \cup X} (\forall \emptyset)\, t = t'$.*

Proof. Each condition is equivalent to the condition that $\bar{a}(t) = \bar{a}(t')$ for every $\Sigma(X)$-algebra A satisfying E and every $a\colon X \to A$. □

Theorem 8. *$T_{\Sigma, E} = T_\Sigma / \equiv_E$ is an initial (Σ, E)-algebra, where $\equiv_E$ is the Σ-congruence relation generated by the ground instances of equations in E.* □

The proof can be found in many places, e.g., [45].

Usually we want proofs about software to be independent of how the data types involved happen to be represented; for example, we are usually not interested in properties of the decimal or binary representations of the natural numbers, but instead are interested in abstract properties of the abstract natural numbers. The following result shows that satisfaction of an equation by an algebra is an "abstract" property, in the sense that it is independent of how the algebra happens to be represented. This result implies that exactly the same equations are true of any one initial P-algebra as any other.

Proposition 9. *Given a specification $P = (\Sigma, E)$, any two initial P-algebras are Σ-isomorphic; in fact, if A and A' are two initial P-algebras, then the unique Σ-homomorphisms $A \to A'$ and $A' \to A$ are both isomorphisms, and indeed, are inverse to each other. Moreover, given isomorphic Σ-algebras A and A', and given a Σ-equation e, then $A \models e$ iff $A' \models e$.*

Proof. Let $f\colon A \to A'$ and $g\colon A' \to A$ be the two homomorphisms guaranteed by initiality. Now notice that the unique homomorphisms $A \to A$ and $A' \to A'$ must each be identities, namely 1_A and $1_{A'}$. Therefore $f;g\colon A \to A$ is 1_A, and $g;f\colon A' \to A'$ is $1_{A'}$. The proof of the second assertion is omitted here (see [30]). □

The word "abstract" in the phrase "abstract algebra" means "uniquely defined up to isomorphism"; for example, an "abstract group" is an isomorphism class of groups, indicating that we are not interested in properties of any particular representation, but only in properties that hold for all representations; e.g., see [57]. Because Theorem 9 implies that all the initial models of a specification $P = (\Sigma, E)$ are abstractly the same in precisely this sense, the word "abstract" in "abstract data type" has *exactly* the same meaning. This is not a mere pun, but a significant fact about software engineering.

Another fact suggesting we are on the right track is that any computable abstract data type has an equational specification; moreover, this specification tends to be reasonably simple and intuitive in practice. The following result from [65] slightly generalizes the original version due to Bergstra and Tucker [2]. (M is **reachable** iff the unique Σ-homomorphism $T_\Sigma \to M$ is surjective):

Theorem 10. *(Adequacy of Initiality) Given any computable reachable Σ-algebra M with Σ finite, there is a finite specification $P = (\Sigma', A')$ such that $\Sigma \subseteq \Sigma'$, such that Σ' has the same sorts as Σ, and such that M is Σ-isomorphic to T_P viewed as a Σ-algebra.* □

We do not here define the concept of a "computable algebra", but it corresponds to what one would intuitively expect: all carrier sets are decidable and all operations are total computable functions; see [65]. This result tells us that all of the data types that are of interest in computer science can be defined using initiality, although sometimes it may be necessary to add some auxiliary functions. All of this motivates the following fundamental conceptualization, which goes back to 1975 [46, 45]:

Definition 11. The **abstract data type** (abbreviated **ADT**) defined by a specification P is the class of all initial P-algebras. □

The importance of initiality for computer science developed gradually. The term "initial algebra semantics" and its first applications appeared in [26]; these included formulating (Knuthian) attribute semantics as a homomorphism from an initial many sorted syntactic algebra generated by a context free grammar, into a semantic algebra. In his PhD thesis [58], William Lawvere used the characterization of the natural numbers as an initial algebra in the axiom of infinity for his category theoretical axiomatization of sets. A key step in formalizing ADTs this way was my realization that Lawvere's use of initiality could be extended to characterize other data types abstractly; see [46], and for a more complete and rigorous exposition, see [45]. More on initiality can be found in [47] and [65]; the latter especially develops connections with induction and computability. See [29] for more historical information about this period, and [37, 32] for more recent results, examples and references. The adequacy of initiality was studied by Bergstra and Tucker in an important series of papers, of which [2] is most relevant for present purposes.

Real software has many features that are difficult or impossible to treat with ordinary many sorted algebra. These include the raising and handling of exceptions, overloaded operators, subtypes, inheritance, coercions, and multiple representations. Order sorted algebra (abbreviated **OSA**) is an attempt to extend many sorted algebra to address these problems. Introduced in [28], this reached fruition in joint work with Meseguer [41, 66]; OSA provides a partial ordering relation on sorts, interpreted semantically as subset inclusion among model carriers. Meseguer and I proved [66] that there are simple ADTs with no adequate many sorted equational specification, because (what we call) the constructor-selector problem can't be solved in this setting. Order sorted initial algebras capture all *partial* recursive functions (and algebras), not just the total ones, as in the many sorted case. OSA is only slightly more difficult than many sorted algebra, and essentially all results generalize without much fuss; in particular, there are complete rules of deduction and initial models. Because OSA is strongly typed, many terms that intuitively should be well formed because they evaluate to well formed terms, are actually ill formed; [41] introduced *retracts* to handle this problem. There are now many different variants and extensions of OSA, too numerous to mention here, although Meseguer's membership equational logic [64] should not be omitted.

2.3 OBJ

Because it is all too easy to write incorrect specifications, it is important to have a tool that can not only check syntax but also execute test cases, and verify properties; such a tool is also very helpful in teaching. Around 1974, I conceived the OBJ language for this purpose, using order sorted algebra[6] with overloaded mixfix syntax, and with term rewriting as its operational semantics [27]; the goal was to make specifications as readable and testable as possible. The final OBJ3 [48] version[7] of OBJ resulted from the efforts of many people, including José Meseguer, Kokichi Futatsugi, Jean-Pierre Jouannaud, Claude and Hélène Kirchner, David Plaisted, and Joseph Tardo [44, 42, 22, 48]; this system provided loose and initial semantics, rewriting modulo equations, generic modules, order sorted algebra with retracts, and user definable builtins[8]; OBJ2 was heavily used in designing OBJ3 [56], and I think greatly speeded up this effort, by facilitating team communication and documenting interfaces. Many other languages have followed OBJ's lead, including: ACT ONE [21], which was used in the LOTOS hardware description language; the CafeOBJ system [23, 20], an industrial strength version of the OBJ language being built by a large Japanese national project, that includes hidden algebra to handle behavioral properties of objects with states; the CASL system [15], being developed by a diverse European group called CoFI (for Common Framework Initiative), which aspires to be a Common Algebraic Specification Language, although it does not support order sorted algebra or hidden algebra; and Maude [13], which efficiently extends OBJ3 with rewriting logic [64], membership logic [64], and reflection.

OBJ has notation for expressing both initial and loose specifications, using overloaded order sorted syntax [48, 37, 32]. OBJ modules that are to be interpreted loosely begin with the keyword **theory** (or **th**) and close with the keyword **endth**. Between these two keywords come declarations for sorts and operations, plus (as discussed later) variables and equations. For example, the following OBJ3 code specifies the theory of automata:

---

[6] Actually, a precursor called error algebra, motivated by the importance of error handling in real systems.

[7] "OBJ" refers to the general design, while "OBJ3" refers to a specific implementation.

[8] These were originally intended for providing builtin data structures like numbers, but were later used in implementing complex systems on top of OBJ, since they allow access to the underlying Lisp system [48].

```
th AUTOM is
  sorts Input State Output .
  op s0 : -> State .
  op f : Input State -> State .
  op g : State -> Output .
endth
```

Any number of sorts can be declared following sorts (or equivalently, sort), and operations are declared with their arity between the : and the ->, with their sort following the ->.

The keyword pair obj...endo indicates that initial semantics is intended. For example, the Peano natural numbers are given by

```
obj NATP is
  sort Nat .
  op 0 : -> Nat .
  op s_ : Nat -> Nat .
endo
```

which uses "mixfix" syntax for the successor operation symbol: in the expression before the colon, the underbar character is a place holder, showing where the operation's arguments should go; hence successor has prefix syntax here.

The following implements a simple flag object by representing its state as a natural number. The keyword pr indicates an importation, in this case of EVENAT, which is the natural numbers enriched with a Boolean valued function even, which is true iff its argument is even; the expression after EVENAT renames the sort Nat of EVENAT to be Flag. The next line introduces three prefix operations all at once, each with the same source and target; these are methods for changing state. The line that begins with var declares a variable and its sort, and the next three lines are equations that define the three methods and the Boolean valued attribute up?.

```
obj NFLAG is
  pr EVENAT *(sort Nat to Flag) .
  ops (up_)(dn_)(rev_) : Flag -> Flag .
  op up?_ : Flag -> Bool .
  var F : Flag .
  eq up F = 2 * F .
  eq dn F = 2 * F + 1 .
  eq rev F = F + 1 .
  eq up? F = even F .
endo
```

The next section gives an abstract specification for this object.

All the OBJ3 code in this paper is executable, and (once suitable definitions for EVENAT and one other module are added) executing it actually proves the simple result about flags discussed below; the OBJ output is given in Appendix A.

3 Hidden algebra

Initial semantics works very well for data structures like integers, lists, booleans, vectors and matrices, but is more awkward for situations that involve a state, i.e., an internal representation that is changed by commands and never viewed directly, but only through external "attributes." For example, it is usually more appropriate to view stacks as machines with an encapsulated (invisible) internal state, having "top" as an attribute. Although initial models exist for any reasonable specification of stacks, real stacks are more likely to be implemented by a model that is not initial, such as a pointer plus an array. This implies that a new notion of implementation is needed, different from the simple notion of initial model. Moreover, in considering (for example) stacks of integers, the sorts for stacks and for integers must be treated differently, since the latter are still modeled initially as data. Although these issues have been successfully addressed in an initial framework (e.g., [45]), it is really better to take a different viewpoint.

Hidden algebra explicitly distinguishes between "visible" sorts for data and "hidden" sorts for states. It makes sense to declare a fixed collection of shared data values, bundled together in a single algebra, because the components of a system must use the same representations for the data that they share, or else they cannot communicate[9].

Definition 12. Let D be a fixed **data algebra**, with Ψ its signature and V its sort set, such that each D_v with $v \in V$ is non-empty and for each $d \in D_v$ there is some $\psi \in \Psi_{[],v}$ such that ψ is interpreted as d in D; we call V the **visible sorts**. For convenience, we assume $D_v \subseteq \Psi_{[],v}$ for each $v \in V$. □

The above concerns semantics; but the prudent verifier needs an effective specification for data values to support proofs, and it is especially convenient to use initial algebra semantics for this purpose, because it supports proofs by induction. We now generalize the notion of signature:

Definition 13. A **hidden signature (over a data algebra** (V, Ψ, D)) is a pair (H, Σ), where H is a set of **hidden sorts** disjoint from V, Σ is an $S = (H \cup V)$-sorted signature with $\Psi \subseteq \Sigma$, such that

(S1) each $\sigma \in \Sigma_{w,s}$ with $w \in V^*$ and $s \in V$ lies in $\Psi_{w,s}$, and
(S2) for each $\sigma \in \Sigma_{w,s}$ at most one hidden sort occurs in w.

We may abbreviate (H, Σ) to just Σ. If $w \in S^*$ contains a hidden sort, then $\sigma \in \Sigma_{w,s}$ is called a **method** if $s \in H$, and an **attribute** if $s \in V$. If $w \in V^*$ and $s \in H$, then $\sigma \in \Sigma_{w,s}$ is called a **(generalized) hidden constant**.

A **hidden** (or **behavioral**) **theory** (or **specification**) is a triple (H, Σ, E), where (H, Σ) is a hidden signature and E is a set of Σ-equations that does not include any Ψ-equations; we may write (Σ, E) or just E for short. □

---
[9] In practice, there may be multiple representations for data with translations among them, and representations may change during development, so this is a simplifying assumption; however, it can easily be relaxed.

Condition (S1) expresses data encapsulation, that Σ cannot add any new operations on data items. Condition (S2) says that methods and attributes act singly on (the states of) objects. Every operation in a hidden signature is either a method, an attribute, or else a constant[10]. Equations about data (Ψ-equations) are not allowed in specifications; any such equation needed as a lemma should be proved and asserted separately, rather than being included in a specification. The following example should help to clarify this definition; it is the simplest possible example where something beyond pure equational reasoning and induction is needed.

Example 14. Below is a hidden specification for flag objects, where intuitively a flag can be either up or down, with methods to put it up, to put it down, and to reverse it:

```
th FLAG is sort Flag .
  pr DATA .
  ops (up_) (dn_) (rev_) : Flag -> Flag .
  op up?_ : Flag -> Bool .
  var F : Flag .
  eq up? up F = true .
  eq up? dn F = false .
  eq up? rev F = not up? F .
endth
```

Here FLAG is the name of the module and Flag is the name of the class of flag objects. The methods and attribute are as in the previous NFLAG example, which is actually a model of the above specification, in a sense made precise below. □

If Σ is the signature of FLAG, then Ψ is a subsignature of Σ, and so a model of FLAG should be a Σ-algebra whose restriction to Ψ is D, providing functions for all the methods and attributes in Σ, and behaving as if it satisfies the given equations. Elements of such models are possible states for Flag objects. This motivates the following:

Definition 15. Given a hidden signature (H, Σ), a **hidden Σ-algebra** A is a (many sorted) Σ-algebra A such that $A|_\Psi = D$. □

We next define behavioral satisfaction of an equation, an idea introduced by Reichel [73]. Intuitively, the two terms of an equation 'look the same' under every 'experiment' consisting of some methods followed by an 'observation,' i.e., an attribute. More formally, such an experiment is given by a *context*, which is a term of visible sort having one free variable of hidden sort:

Definition 16. Given a hidden signature (H, Σ) and a hidden sort h, then a Σ-**context** of sort h is a visible sorted Σ-term having a single occurrence of a

---

[10] In our most recent work [75], this assumption is weakened by allowing operations with more than one hidden sort; this is important for examples like sets, where union and intersection have two hidden arguments.

new variable symbol z of sort h. A context is **appropriate** for a term t iff the sort of t matches that of z. Write $c[t]$ for the result of substituting t for z in the context c.

A hidden Σ-algebra A **behaviorally satisfies** a Σ-equation $(\forall X)\ t = t'$ iff for each appropriate Σ-context c, A satisfies the equation $(\forall X)\ c[t] = c[t']$; then we write $A \models_\Sigma (\forall X)\ t = t'$.

A **model** of a hidden theory $P = (H, \Sigma, E)$ is a hidden Σ-algebra A that behaviorally satisfies each equation in E. Such a model is also called a (Σ, E)-**algebra**, or a P-algebra, and then we write $A \models P$ or $A \models_\Sigma E$. Also we write $E' \models_\Sigma E$ iff $A \models_\Sigma E'$ implies $A \models_\Sigma E$ for each hidden Σ-algebra A. □

Example 17. Let's look at a simple Boolean cell C as a hidden algebra. Here, $C_{\text{Flag}} = C_{\text{Bool}} = \{true, false\}$, up $F = true$, dn $F = false$, up? $F = F$, and rev $F = not\ F$.

A more complex implementation H keeps complete histories of interactions, so that the action of a method is merely to concatenate its name to the front of a list of method names. Then $H_{\text{Flag}} = \{up, dn, rev\}^*$, the lists over $\{up, dn, rev\}$, while $H_{\text{Bool}} = \{true, false\}$, up $F = up\frown F$, dn $F = dn\frown F$, rev $F = rev\frown F$, while up? $up\frown F = true$, up? $dn\frown F = false$, and up? $rev\frown F = not$ up? F, where $\frown$ is the concatenation operation. Note that C and H are *not* isomorphic. □

For visible equations, there is no difference between ordinary satisfaction and behavioral satisfaction. But these concepts can be very different for hidden equations. For example,

    rev rev F = F

is strictly satisfied by the Boolean cell model C, but it is *not* satisfied by the history model H, nor by the model NFLAG, for which the left side has value F+2. However, the equation is *behaviorally* satisfied by all these models. This illustrates why behavioral satisfaction is so often more appropriate for computer science applications.

Previously we gave a semantic definition of an abstract data type as an isomorphism class of initial algebras for some specification; equivalently, by Theorem 10, we could define it to be an isomorphism class of computable algebras, or in the order sorted case, of partial computable algebras. The hidden analog of this defines an **abstract object** (or **machine**) to be a class of all hidden algebras that satisfy some hidden specification (in practice, it is often desirable to restrict attention to reachable models).

3.1 Coinduction

The first effective algebraic proof technique for behavioral properties was context induction, introduced by Rolf Hennicker [51] and developed further in joint work with Michel Bidoit [5, 4]. Their research programme is similar to ours in several ways, but is more concerned with semantics than with proofs. Unfortunately, context induction can be awkward to apply in practice, as first noticed in

[24]. We proposed hidden coinduction as a way to avoid this awkwardness. The technique resembles one introduced by Robin Milner [67], but is more general. Peter Padawitz is also developing similar notions [68, 69].

Induction is a standard technique for proving properties of initial (or more generally, reachable) algebras of a theory, and principles of induction can be justified from the fact that an initial algebra has no proper subalgebras [32, 65]. Final (terminal) algebras play an analogous role[11] in justifying reasoning about behavioral properties with hidden coinduction. Before describing the final algebra, note that its use is not precisely dual to that of the initial algebra for abstract data types. The semantics of a hidden specification is not the final algebra, but rather is the variety of all hidden algebras that satisfy the spec; in fact, final algebras do not even exist in general. However, their existence for certain signatures, with no equations, plays an important technical role.

Given a hidden signature Σ without generalized hidden constants (recall these are hidden operations with no hidden arguments), the hidden carriers of the final Σ-algebra F_Σ are given by the following "magical formula," for h a hidden sort:

$$F_{\Sigma,h} = \prod_{v \in V} [C_\Sigma[z_h]_v \to D_v] ,$$

the product of the sets of functions taking contexts to data values (of appropriate sort). Elements of F_Σ can be thought of as 'abstract states' represented as functions on contexts, returning the data values resulting from evaluating a state in a context. This also appears in the way F_Σ interprets attributes: let $\sigma \in \Sigma_{hw,v}$ be an attribute, let $p \in F_{\Sigma,h}$ and let $d \in D_w$; then we define $F_{\Sigma,\sigma}(p,d) = p_v(\sigma(z_h,d))$; i.e., p_v is a function taking contexts in $C_\Sigma[z_h]_v$ to data values in D_v, so applying it to the context $\sigma(z_h,d)$ gives the data value resulting from that experiment. Methods are interpreted similarly; see [38] for details.

Definition 18. Given a hidden signature Σ, a hidden subsignature $\Phi \subseteq \Sigma$, and a hidden Σ-algebra A, then **behavioral Φ-equivalence** on A, denoted $\equiv_\Phi$, is defined as follows, for $a, a' \in A_s$:

(E1) $a \equiv_{\Phi,s} a'$ iff $a = a'$

when $s \in V$, and

(E2) $a \equiv_{\Phi,s} a'$ iff $A_c(a) = A_c(a')$ for all $v \in V$ and all $c \in C_\Phi[z]_v$

when $s \in H$, where z is of sort s and A_c denotes the function interpreting the context c as an operation on A, that is, $A_c(a) = \theta_a^*(c)$, where θ_a is defined by $\theta_a(z) = a$ and θ_a^* denotes the free extension of θ_a.

When $\Phi = \Sigma$, we call $\equiv_\Phi$ the **behavioral equivalence** and denote it $\equiv$.

For $\Phi \subseteq \Sigma$, a **hidden Φ-congruence** on a hidden Σ-algebra A is a Φ-congruence $\simeq$ which is the identity on visible sorts, i.e., such that $a \simeq_v a'$ iff $a = a'$ for all $v \in V$ and $a, a' \in A_v = D_v$. We call a hidden Σ-congruence just a **hidden congruence**. □

---

[11] Though no longer so in our latest work [75].

The key property is the following:

Theorem 19. *If Σ is a hidden signature, Φ is a hidden subsignature of Σ, and A is a hidden Σ-algebra, then behavioral Φ-equivalence is the* largest *behavioral Φ-congruence on A.* □

There is beautiful abstract proof of this result for those who know a little category theory in [38]; it uses the existence of final algebras when there are no hidden constants, which is also shown in [38]. The congruence relation can be seen as a generalization of the so called Nerode equivalence in the classical theory of abstract machines, e.g., see [65].

Theorem 19 implies that if $a \simeq a'$ under some hidden congruence $\simeq$, then a and a' are behaviorally equivalent. This justifies a variety of techniques for proving behavioral equivalence (see also [36]). In this context, a relation may be called a **candidate relation** before it is proved to be a hidden congruence. Probably the most common case is $\Phi = \Sigma$, but the generalization to smaller Φ is useful, for example in verifying behavioral refinements.

Example 20. Let A be any model of the FLAG theory in Example 14, and for $f, f' \in A_{\texttt{Flag}}$, define $f \simeq f'$ iff up? f = up? f' (and $d \simeq d'$ iff $d = d'$ for data values d, d'). Then we can use the equations of FLAG to show that $f \simeq f'$ implies up $f \simeq$ up f' and dn $f \simeq$ dn f' and rev $f \simeq$ rev f', and of course up? $f \simeq$ up? f'. Hence $\simeq$ is a hidden congruence on A.

Therefore we can show $A \models (\forall F : \texttt{Flag})$ rev rev F = F just by showing $A \models (\forall F : \texttt{Flag})$ up? rev rev F = up? F. This follows by ordinary equational reasoning, since up? rev rev F = not(not(up? F)). Therefore the equation is behaviorally satisfied by any FLAG-algebra A.

It is easy to do this proof mechanically using OBJ3, since all the computations are just ordinary equational reasoning. We set up the proof by opening FLAG and adding the necessary assumptions; here R represents the candidate relation $\simeq$:

```
openr FLAG .
  op _R_ : Flag Flag -> Bool .
  var F1 F2 : Flag .
  eq F1 R F2 = (up? F1 == up? F2) .
  ops f1 f2 : -> Flag .
close
```

The new constants f1, f2 are introduced to stand for universally quantified variables, following the lemma of constants, and == is OBJ3's builtin equality test[12]. We now show that R is a hidden congruence:

```
open . eq up? f1 = up? f2 .
  red (up f1) R (up f2) . ***> should be: true
  red (dn f1) R (dn f2) . ***> should be: true
  red (rev f1) R (rev f2) . ***> should be: true
close
```

---

[12] This operation reduces its two arguments, and then checks whether the results are identical. It is known that this gives equality under certain general conditions, which do hold here [37].

where red is a command that tells OBJ to "reduce" the subsequent term, i.e., to apply equations as left-to-right rewrite rules, until a term is obtained where no rule applies.

Finally, we show that all FLAG-algebras behaviorally satisfy the equation with:

red (rev rev f1) R f1 .

All the above code runs in OBJ3, and gives true for each reduction, provided the following lemma about the Booleans is added somewhere,

eq not not B = B .

where B is a Boolean variable. I think this proof is about as simple as could be hoped for. □

Just as there is a rich lore about doing inductive proofs, so more and more lore is accumulating about doing coinductive proofs. For example, the third reduction in the example above is unnecessary; however, it is more trouble to justify its elimination than it is to ask OBJ to do it; see [38].

3.2 Nondeterminism

Because nondeterminism is very problematic for ordinary algebraic specification, it is perhaps surprising that it is already an inherent facet of hidden algebra. We first illustrate this with the following very simple example:

```
th C is pr DATA .
  op c : -> Nat .
endth
```

Here c has some natural number value in every model, and every number can occur; each model chooses exactly one. However, there can also be arbitrary junk in models, so it makes sense to restrict to reachable models; then the choice of a value for c completely characterizes a model. I like to describe this by saying that each model is a "possible world" in which some fixed choice has been made for each nondeterministic possibility.

It is also easy to restrict the choice of a value for c, by adding an equation like one of the following:

```
eq c => 1 = true .
eq 2 => c = true .
eq odd(c) = true .
eq prime(c) = true .
eq c == 1 or c == 2 = true .
```

The last equation suggests a rather cute way to specify nondeterministic choice in hidden algebra:

```
th CH is pr DATA .
  op _|_ : Nat Nat -> Nat .
  vars N M : Nat .
  eq N | M == N or N | M == M = true .
endth
```

Here again, models are "possible worlds," where some choice of one of N, M is made for each pair N, M. It is not hard to prove that this choice function is idempotent, i.e., satisfies the equation

```
eq N | N = N .
```

However, the commutative and associative properties fail for some models (the reader is invited to find the appropriate models) and hence for the theory.

Neither example of nondeterminism above involves state, which is the most characteristic feature of hidden algebra, so we really should give an example of nondeterminism with a hidden sort. For some reason, vending machines are very popular for illustrating various aspects of systems, especially nondeterminism and concurrency. The spec below describes perhaps the simplest vending machine that is not entirely trivial: when you put a coin in, it nondeterministically gives you either coffee or tea, represented say by true and false, respectively; and then it goes into a new state where it is prepared to do the same again. In this spec, init is the initial state, in(init) is the state after one coin, in(in(init)) is the state after two coins, etc., while out(init) is what you get after the first coin, out(in(init)) after the second, etc.

```
th VCT is sort St .
  pr DATA .
  op in : St -> St .
  op out : St -> Bool .
endth
```

As before, it is easy to restrict behavior by adding equations like

```
cq out(in(in(S))) = not out(S) if out(S) = out(in(S)) .
```

which says that you cannot get the same substance three times in a row. (It is interesting to notice that this equation will guarantee fairness.)

For examples like this, it is also interesting to look at the final algebra F, for the signature without the constant init: according to the "magic formula," it consists (up to isomorphism) of all Boolean sequences – i.e., it is the algebra of (what are called) *traces*; in fact, contexts are the natural generalization of traces to a non-monadic world. Since there is a unique (hidden) homomorphism $M \to F$ for any model M of VCT, the image of init under this map characterizes the behavior of M. This simple and elegant situation holds for nondeterministic concurrent systems in general. (More information about nondeterminism and final models can be found in [38].)

The approach to nondeterminism in hidden algebra is quite different from that which is traditional in automaton theory: in hidden algebra, each possible behavior appears in a different possible world, whereas a nondeterministic automaton includes all choices in a single model. The possible worlds approach corresponds to real computers, which are always deterministic, and must simulate nondeterminism, e.g., using pseudo-random numbers. Chip makers don't make nondeterministic Turing machines or automata; if they could then $P = NP$ wouldn't be a problem!

3.3 Proving behavioral refinement

The simplest view of behavioral refinement assumes a specification (Σ, E) and an implementation A, and asks if $A \models_\Sigma E$; the use of behavioral satisfaction is significant here, because it allows us to treat many subtle implementation tricks that only 'act as if' correct, e.g., data structure overwriting, abstract machine interpretation, and much more.

Unfortunately, trying to prove $A \models_\Sigma E$ directly dumps us into the semantic swamp mentioned in the introduction. To rise above this, we work with a specification E' for A, rather than an actual model[13]. This not only makes the proof far easier, but it also has the advantage that the proof will apply to any other model A' that (behaviorally) satisfies E'. Hence, what we prove is $E' \models E$; in semantic terms, this means that any A (behaviorally) satisfying E' also (behaviorally) satisfies E; and very significantly, it also means that we can use hidden coinduction to do the proof. The method is just to prove that each equation in E' is a behavioral consequence of E, i.e., a behavioral property of every model (implementation) of E. More details and some examples are given in [38], including the proof that a pointer with an array gives a behavioral refinement of the stack spec.

3.4 The object paradigm

Objects have local states with visible local "attributes" and "methods" to change state. Objects also come in "classes," which can "inherit" from other classes, and objects can communicate concurrently and nondeterministically with other objects in the same system. This paradigm has become dominant in many important application areas. We have already seen how to handle most of these features with hidden algebra. Aspects of concurrency and inheritance are treated in [33, 38]. A full treatment of inheritance requires the use of order sorted algebra for subclasses [41].

---

[13] Some may object that this maneuver isolates us from the actual code used to define operations in A, preventing us from verifying that code. However, we contend that this isolation is actually an *advantage*, since only about 5% of the difficulty of software development lies in the code itself [7], with much more of the difficulty in specification and design; our approach addresses these directly, without assuming the heavy burden of a messy programming language semantics. But of course we can use algebraic semantics to verify code if we wish, as extensively illustrated in [37]. Thus we have achieved a significant separation of concerns.

4 Summary and related work

This paper has presented hidden algebra as a natural next step in the evolution of algebraic specification, that can handle the main features of the object paradigm. Of course, no one would invent a method like coinduction for examples as simple as our flag example; this was chosen just to bring out the basic ideas clearly. Much more complex examples have been done, including correctness proofs for an optimizing compiler and for a novel communication protocol [50, 34]. Hidden algebra first appeared in [31], and was subsequently elaborated in papers including [33, 10]; an important precursor was work by Goguen and Meseguer on what they called "abstract machines" [39]. The rapidly growing literature on hidden algebra includes [17, 38, 59, 12, 52, 18, 60]. Coinduction seems to give proofs that are about as simple as possible, but more experience is needed before this can be said with complete certainty. The closely related area of coalgebra also uses coinduction, and also has a rapidly growing literature, including [74, 53–55]. However, it seems that coalgebra has difficulty with nondeterminism, concurrency, and operators with multiple state arguments.

It can be argued that algebraic specification is now entering a golden age, in which new techniques are bringing old goals to fruition in unexpected ways, and are also opening new horizons from which exciting new goals seem reachable. We have discussed hidden algebra and its cousin coalgebra. Another important new development is rewriting logic [61, 62], a weakening of equational logic that provides an ideal operational semantics for rapidly implementing many term rewriting algorithms [14], as well as for describing and comparing various kinds of concurrency [62]; rewriting logic has been efficiently implemented in Maude [63, 13]. The CafeOBJ system [20, 23] provides industrial strength implementations of rewriting logic, as well as of ordinary order sorted equational logic, hidden sorted equational logic, *and* all their combinations [19, 20]! The designs for both Maude and CafeOBJ are heavily indebted to that of OBJ, and indeed can be considered versions of OBJ. There is also exciting new work in term rewriting [16] (which is the basis for implementing systems like OBJ3, Maude and CafeOBJ), for example in France around Jean-Pierre Jouannoud and Adel Bouhoula, on induction and termination proofs [9], including the SPIKE and CiME systems [8], and some new work on proving behavioral properties with a similar technology [3]. The issues discussed in this paper seem to be of increasing importance for computer science, and I think we can look forward to continuing progress.

References

1. Jean Benabou. Structures algébriques dans les catégories. *Cahiers de Topologie et Géometrie Différentiel*, 10:1–126, 1968.
2. Jan Bergstra and John Tucker. Characterization of computable data types by means of a finite equational specification method. In Jaco de Bakker and Jan van Leeuwen, editors, *Automata, Languages and Programming, Seventh Colloquium*, pages 76–90. Springer, 1980. Lecture Notes in Computer Science, Volume 81.

3. Narjes Berregeb, Adel Bouhoula, and Michaël Rusinowitch. Observational proofs with critical contexts. In *Fundamental Approaches to Software Engineering*, volume 1382 of *Lecture Notes in Computer Science*, pages 38–53. Springer, 1998.
4. Michael Bidoit and Rolf Hennicker. Behavioral theories and the proof of behavioral properties. *Theoretical Computer Science*, 165:3–55, 1996.
5. Michel Bidoit, Rolf Hennicker, and Martin Wirsing. Behavioural and abstractor specifications. *Science of Computer Programming*, 25(2–3), 1995.
6. Garrett Birkhoff. On the structure of abstract algebras. *Proceedings of the Cambridge Philosophical Society*, 31:433–454, 1935.
7. Barry Boehm. *Software Engineering Economics*. Prentice-Hall, 1981.
8. Adel Bouhoula. Automated theorem proving by test set induction. *Journal of Symbolic Computation*, 23(1):47–77, 1997.
9. Adel Bouhoula and Jean-Pierre Jouannaud. Automata-driven automated induction. In *Proceedings, 12th Symposium on Logic in Computer Science*, pages 14–25. IEEE, 1997.
10. Rod Burstall and Răzvan Diaconescu. Hiding and behaviour: an institutional approach. In Andrew William Roscoe, editor, *A Classical Mind: Essays in Honour of C.A.R. Hoare*, pages 75–92. Prentice-Hall, 1994. Also Technical Report ECS-LFCS-8892-253, Laboratory for Foundations of Computer Science, University of Edinburgh, 1992.
11. Graham Button and Wes Sharrock. Occasioned practises in the work of implementing development methodologies. In Marina Jirotka and Joseph Goguen, editors, *Requirements Engineering: Social and Technical Issues*, pages 217–240. Academic, 1994.
12. Corina Cîrstea. A semantical study of the object paradigm. Transfer thesis, Oxford University Computing Laboratory, 1996.
13. Manuel Clavel, Steven Eker, Patrick Lincoln, and José Meseguer. Principles of Maude. In José Meseguer, editor, *Proceedings, First International Workshop on Rewriting Logic and its Applications*. Elsevier Science, 1996. Volume 4, *Electronic Notes in Theoretical Computer Science*.
14. Manuel Clavel, Steven Eker, and José Meseguer. Current design and implementation of the Cafe prover and Knuth-Bendix tools, 1997. Presented at CafeOBJ Workshop, Kanazawa, October 1997.
15. CoFI. CASL summary, 1998. http://www.brics.dk/Projects/CoFI/.
16. Nachum Dershowitz and Jean-Pierre Jouannaud. Rewriting systems. In *Handbook of Theoretical Computer Science, Volume B*, pages 243–309. North-Holland, 1990.
17. Răzvan Diaconescu. Foundations of behavioural specification in rewriting logic. In José Meseguer, editor, *Proceedings, First International Workshop on Rewriting Logic and its Applications*. Elsevier Science, 1996. Volume 4, *Electronic Notes in Theoretical Computer Science*.
18. Răzvan Diaconescu. Behavioural coherence in object-oriented algebraic specification. Technical Report IS-RR-98-0017F, Japan Advanced Institute for Science and Technology, June 1998. Submitted for publication.
19. Răzvan Diaconescu and Kokichi Futatsugi. Logical semantics for CafeOBJ. Technical Report IS-RR-96-0024S, Japan Advanced Institute for Science and Technology, 1996.
20. Răzvan Diaconescu and Kokichi Futatsugi. *CafeOBJ Report: The Language, Proof Techniques, and Methodologies for Object-Oriented Algebraic Specification*. World Scientific, 1998. AMAST Series in Computing, volume 6.

21. Hartmut Ehrig, Werner Fey, and Horst Hansen. ACT ONE: An algebraic specification language with two levels of semantics. Technical Report 83–03, Technical University of Berlin, Fachbereich Informatik, 1983.
22. Kokichi Futatsugi, Joseph Goguen, Jean-Pierre Jouannaud, and José Meseguer. Principles of OBJ2. In Brian Reid, editor, *Proceedings, Twelfth ACM Symposium on Principles of Programming Languages*, pages 52–66. Association for Computing Machinery, 1985.
23. Kokichi Futatsugi and Ataru Nakagawa. An overview of Cafe specification environment. In *Proceedings, ICFEM'97*. University of Hiroshima, 1997.
24. Marie-Claude Gaudel and Igor Privara. Context induction: an exercise. Technical Report 687, LRI, Université de Paris-Sud, 1991.
25. W. Wyat Gibbs. Software's chronic crisis. *Scientific American*, pages 72–81, September 1994.
26. Joseph Goguen. Semantics of computation. In Ernest Manes, editor, *Proceedings, First International Symposium on Category Theory Applied to Computation and Control*, pages 151–163. Springer, 1975. (San Fransisco, February 1974.) Lecture Notes in Computer Science, Volume 25.
27. Joseph Goguen. Abstract errors for abstract data types. In Eric Neuhold, editor, *Proceedings, First IFIP Working Conference on Formal Description of Programming Concepts*, pages 21.1–21.32. MIT, 1977. Also in *Formal Description of Programming Concepts*, Peter Neuhold, Ed., North-Holland, pages 491–522, 1979.
28. Joseph Goguen. Order sorted algebra. Technical Report 14, UCLA Computer Science Department, 1978. Semantics and Theory of Computation Series.
29. Joseph Goguen. Memories of ADJ. *Bulletin of the European Association for Theoretical Computer Science*, 36:96–102, October 1989. Guest column in the 'Algebraic Specification Column.' Also in *Current Trends in Theoretical Computer Science: Essays and Tutorials*, World Scientific, 1993, pages 76–81.
30. Joseph Goguen. Proving and rewriting. In Hélène Kirchner and Wolfgang Wechler, editors, *Proceedings, Second International Conference on Algebraic and Logic Programming*, pages 1–24. Springer, 1990. Lecture Notes in Computer Science, Volume 463.
31. Joseph Goguen. Types as theories. In George Michael Reed, Andrew William Roscoe, and Ralph F. Wachter, editors, *Topology and Category Theory in Computer Science*, pages 357–390. Oxford, 1991. Proceedings of a Conference held at Oxford, June 1989.
32. Joseph Goguen. *Theorem Proving and Algebra*. MIT, to appear.
33. Joseph Goguen and Răzvan Diaconescu. Towards an algebraic semantics for the object paradigm. In Hartmut Ehrig and Fernando Orejas, editors, *Proceedings, Tenth Workshop on Abstract Data Types*, pages 1–29. Springer, 1994. Lecture Notes in Computer Science, Volume 785.
34. Joseph Goguen, Kai Lin, Akira Mori, Grigore Roşu, and Akiyoshi Sato. Distributed cooperative formal methods tools. In Michael Lowry, editor, *Proceedings, Automated Software Engineering*, pages 55–62. IEEE, 1997.
35. Joseph Goguen, Kai Lin, Akira Mori, Grigore Roşu, and Akiyoshi Sato. Tools for distributed cooperative design and validation. In *Proceedings, CafeOBJ Symposium*. Japan Advanced Institute for Science and Technology, 1998. Nomuzu, Japan, April 1998.
36. Joseph Goguen and Grant Malcolm. Proof of correctness of object representation. In Andrew William Roscoe, editor, *A Classical Mind: Essays in Honour of C.A.R. Hoare*, pages 119–142. Prentice-Hall, 1994.

37. Joseph Goguen and Grant Malcolm. *Algebraic Semantics of Imperative Programs*. MIT, 1996.
38. Joseph Goguen and Grant Malcolm. A hidden agenda. Technical Report CS97-538, UCSD, Dept. Computer Science & Eng., May 1997. To appear in *Theoretical Computer Science*. Early abstract in *Proc., Conf. Intelligent Systems: A Semiotic Perspective, Vol. I*, ed. J. Albus, A. Meystel and R. Quintero, Nat. Inst. Science & Technology (Gaithersberg MD, 20–23 October 1996), pages 159–167.
39. Joseph Goguen and José Meseguer. Universal realization, persistent interconnection and implementation of abstract modules. In M. Nielsen and E.M. Schmidt, editors, *Proceedings, 9th International Conference on Automata, Languages and Programming*, pages 265–281. Springer, 1982. Lecture Notes in Computer Science, Volume 140.
40. Joseph Goguen and José Meseguer. Completeness of many-sorted equational logic. *Houston Journal of Mathematics*, 11(3):307–334, 1985. Preliminary versions have appeared in: *SIGPLAN Notices*, July 1981, Volume 16, Number 7, pages 24–37; SRI Computer Science Lab, Report CSL-135, May 1982; and Report CSLI-84-15, Center for the Study of Language and Information, Stanford University, September 1984.
41. Joseph Goguen and José Meseguer. Order-sorted algebra I: Equational deduction for multiple inheritance, overloading, exceptions and partial operations. *Theoretical Computer Science*, 105(2):217–273, 1992. Drafts exist from as early as 1985.
42. Joseph Goguen, José Meseguer, and David Plaisted. Programming with parameterized abstract objects in OBJ. In Domenico Ferrari, Mario Bolognani, and Joseph Goguen, editors, *Theory and Practice of Software Technology*, pages 163–193. North-Holland, 1983.
43. Joseph Goguen, Akira Mori, and Kai Lin. Algebraic semiotics, ProofWebs and distributed cooperative proving. In Yves Bartot, editor, *Proceedings, User Interfaces for Theorem Provers*, pages 25–34. INRIA, 1997. (Sophia Antipolis, 1–2 September 1997).
44. Joseph Goguen and Joseph Tardo. An introduction to OBJ: A language for writing and testing software specifications. In Marvin Zelkowitz, editor, *Specification of Reliable Software*, pages 170–189. IEEE, 1979. Reprinted in *Software Specification Techniques*, Nehan Gehani and Andrew McGettrick, editors, Addison Wesley, 1985, pages 391–420.
45. Joseph Goguen, James Thatcher, and Eric Wagner. An initial algebra approach to the specification, correctness and implementation of abstract data types. In Raymond Yeh, editor, *Current Trends in Programming Methodology, IV*, pages 80–149. Prentice-Hall, 1978.
46. Joseph Goguen, James Thatcher, Eric Wagner, and Jesse Wright. Abstract data types as initial algebras and the correctness of data representations. In Alan Klinger, editor, *Computer Graphics, Pattern Recognition and Data Structure*, pages 89–93. IEEE, 1975.
47. Joseph Goguen, James Thatcher, Eric Wagner, and Jesse Wright. Initial algebra semantics and continuous algebras. *Journal of the Association for Computing Machinery*, 24(1):68–95, January 1977. An early version is "Initial Algebra Semantics", by Joseph Goguen and James Thatcher, IBM T.J. Watson Research Center, Report RC 4865, May 1974.
48. Joseph Goguen, Timothy Winkler, José Meseguer, Kokichi Futatsugi, and Jean-Pierre Jouannaud. Introducing OBJ. In Joseph Goguen and Grant Malcolm, editors, *Algebraic Specification with OBJ: An Introduction with Case Studies*. World

Scientific, to appear. Also Technical Report SRI-CSL-88-9, August 1988, SRI International.
49. John Guttag. *The Specification and Application to Programming of Abstract Data Types.* PhD thesis, University of Toronto, 1975. Computer Science Department, Report CSRG-59.
50. Lutz Hamel. *Behavioural Verification and Implementation of an Optimizing Compiler for OBJ3.* PhD thesis, Oxford University Computing Lab, 1996.
51. Rolf Hennicker. Context induction: a proof principle for behavioural abstractions. *Formal Aspects of Computing*, 3(4):326–345, 1991.
52. Shusaku Iida, Michihiro Matsumoto, Răzvan Diaconescu, Kokichi Futatsugi, and Dorel Lucanu. Concurrent object composition in CafeOBJ. Technical Report IS-RR-96-0024S, Japan Advanced Institute for Science and Technology, 1997.
53. Bart Jacobs. Objects and classes, coalgebraically. In B. Freitag, Cliff Jones, C. Lengauer, and H.-J. Schek, editors, *Object-Orientation with Parallelism and Persistence*, pages 83–103. Kluwer, 1996.
54. Bart Jacobs. Invariants, bisimulations and the correctness of coalgebraic refinements. In M. Johnson, editor, *Algebraic Methodology and Software Technology*, pages 276–291. Springer, 1997. Lecture Notes in Computer Science, Volume 1349.
55. Bart Jacobs and Jan Rutten. A tutorial on (co)algebras and (co)induction. *Bulletin of the European Association for Theoretical Computer Science*, 62:222–259, June 1997.
56. Claude Kirchner, Hélène Kirchner, and Aristide Mégrelis. OBJ for OBJ. In Joseph Goguen and Grant Malcolm, editors, *Algebraic Specification with OBJ: An Introduction with Case Studies*. Academic, to appear.
57. Saunders Mac Lane and Garrett Birkhoff. *Algebra.* Macmillan, 1967.
58. F. William Lawvere. An elementary theory of the category of sets. *Proceedings, National Academy of Sciences, U.S.A.*, 52:1506–1511, 1964.
59. Grant Malcolm. Behavioural equivalence, bisimilarity, and minimal realisation. In Magne Haveraaen, Olaf Owe, and Ole-Johan Dahl, editors, *Recent Trends in Data Type Specifications: 11th Workshop on Specification of Abstract Data Types*, pages 359–378. Springer Lecture Notes in Computer Science, Volume 1130, 1996. (Oslo Norway, September 1995).
60. Michihiro Matsumoto and Kokichi Futatsugi. Test set coinduction: Toward automated verification of behavioural properties. In *Proceedings of the Second International Workshop on Rewriting Logic and its Applications*, Electronic Notes in Theoretical Computer Science. Elsevier Science, to appear 1998.
61. José Meseguer. Conditional rewriting logic: Deduction, models and concurrency. In Stéphane Kaplan and Misuhiro Okada, editors, *Conditional and Typed Rewriting Systems*, pages 64–91. Springer, 1991. Lecture Notes in Computer Science, Volume 516.
62. José Meseguer. Conditional rewriting as a unified model of concurrency. *Theoretical Computer Science*, 96(1):73–155, 1992.
63. José Meseguer. A logical theory of concurrent objects and its realization in the Maude language. In Gul Agha, Peter Wegner, and Aki Yonezawa, editors, *Research Directions in Object-Based Concurrency*. MIT, 1993.
64. José Meseguer. Membership algebra as a logical framework for equational specification, 1997. Draft manuscript. Computer Science Lab, SRI International.
65. José Meseguer and Joseph Goguen. Initiality, induction and computability. In Maurice Nivat and John Reynolds, editors, *Algebraic Methods in Semantics*, pages 459–541. Cambridge, 1985.

66. José Meseguer and Joseph Goguen. Order-sorted algebra solves the constructor selector, multiple representation and coercion problems. *Information and Computation*, 103(1):114–158, March 1993. Revision of a paper presented at LICS 1987.
67. Robin Milner and Mads Tofte. Co-induction in relational semantics. *Theoretical Computer Science*, 87(1):209–220, 1991.
68. Peter Padawitz. Towards the one-tiered design of data types and transition systems. In *Proceedings, WADT'97*, pages 365–380. Springer, 1998. Lecture Notes in Computer Science, Volume 1376.
69. Peter Padawitz. Swinging types = functions + relations + transition systems, 1999. Submitted to *Theoretical Computer Science*.
70. David Parnas. Information distribution aspects of design methodology. *Information Processing '72*, 71:339–344, 1972. Proceedings of 1972 IFIP Congress.
71. David Parnas. On the criteria to be used in decomposing systems into modules. *Communications of the Association for Computing Machinery*, 15:1053–1058, 1972.
72. Tekla Perry. In search of the future of air traffic control. *IEEE Spectrum*, 34(8):18–35, August 1997.
73. Horst Reichel. Behavioural validity of conditional equations in abstract data types. In *Contributions to General Algebra 3*. Teubner, 1985. Proceedings of the Vienna Conference, June 21-24, 1984.
74. Horst Reichel. An approach to object semantics based on terminal co-algebras. *Mathematical Structures in Computer Science*, 5:129–152, 1995.
75. Grigore Roşu and Joseph Goguen. Hidden congruent deduction. In Gernot Salzer, editor, *Proceedings, 1998 Workshop on First Order Theorem Proving*. Johannes Kepler Univ. Linz, 1998.
76. Bartel van der Waerden. *A History of Algebra*. Springer, 1985.
77. Steven Zilles. Abstract specification of data types. Technical Report 119, Computation Structures Group, Massachusetts Institute of Technology, 1974.

A OBJ3 output

Below is the output that OBJ3 produces when it executes this paper (there is a little program that extracts the executable code from the paper, and passes it to OBJ3 for execution; the source file for this paper has two "invisible" OBJ3 modules, EVENAT and DATA, which are needed to make others work):

```
             \|||||||||||||||||/
             --- Welcome to OBJ3 ---
             /|||||||||||||||||\
  OBJ3 version 2.04oxford built: 1994 Feb 28 Mon 15:07:40
      Copyright 1988,1989,1991 SRI International
            1998 Nov 8 Sun 10:21:48

==========================================
th AUTOM
==========================================
obj NATP
==========================================
***> This is the invisible EVENAT module:
```

```
========================================
obj EVENAT
========================================
obj NFLAG
========================================
***> This is the invisible DATA module:
========================================
obj DATA
========================================
th FLAG
========================================
***> prove rev rev F = F :
========================================
openr FLAG
========================================
op _ R _ : Flag Flag -> Bool .
========================================
var F1 F2 : Flag .
========================================
eq F1 R F2 = ( up? F1 == up? F2 ) .
========================================
ops f1 f2 : -> Flag .
========================================
close
========================================
open
========================================
eq up? f1 = up? f2 .
========================================
reduce in FLAG : up f1 R up f2
rewrites: 4
result Bool: true
========================================
***> should be: true
========================================
reduce in FLAG : dn f1 R dn f2
rewrites: 4
result Bool: true
========================================
***> should be: true
red (rev f1) R (rev f2) .
========================================
***> should be: true
close
========================================
```

```
reduce in FLAG : rev (rev f1) R f1
rewrites: 7
result Bool: true
==========================================
th C
==========================================
th CH
==========================================
th VCT
OBJ> Bye.
```

The true results above indicate that OBJ3 has in fact done the computations that constitute the proof.

On Negative Informations in Language Theory

Filippo Mignosi and Antonio Restivo

University of Palermo, Dipartimento di Matematica ed Applicazioni
Via Archirafi 34. 90123 Palermo, Italy.
{mignosi,restivo}@altair.math.unipa.it

Abstract. In this paper we survey some recent results that show how negative informations, described in terms of *minimal forbidden words*, are useful in language theory and in related areas.

1 Introduction

In many problems concerning language theory and related areas, such as combinatorics on words, string processing, symbolic dynamics, etc., one considers words that occur as factors of some word in a given language. For instance, in the study and in the classification of infinite words, one defines the *complexity* of an infinite word α as the function that counts, for any natural number n, the numbers of words of length n that occur as factors in α. As another example, many *text compression* algorithms make use of *dictionaries*, i.e. particular sets of words that occur as factors in the text to be compressed.

In this paper we present some problems in language theory that, on the contrary, take advantage by some *negative* informations about the language, i.e. by considering words that do not occur as factors of words in the language L, and that we call *forbidden* (for L).

In a more formal way, a word w is *forbidden* for L if $w \notin F(L)$, where $F(L)$ denotes the set of factors of words in L. An important point about this notion is that we can introduce a condition of minimality: a word w is a *minimal forbidden word* for L if w is forbidden and all proper factors of w belong to $F(L)$. We denote by $MF(L)$ the language of minimal forbidden words for L.

From an algebraic point of view the complement of $F(L)$ in the monoid A^* is an ideal of A^* and the set $MF(L)$ of the minimal forbidden words is its (unique) base. Such concept of minimal forbidden word synthetizes effectively some negative information about a language and plays an important role in several applications.

It turns out that the combinatorial properties of $MF(L)$ helps to investigate the structure of the language L. Consider, for instance, the case of locally testable factorial languages (cf [22]): they are characterized by the fact that the corresponding languages of minimal forbidden words are finite. In the context of symbolic dynamics they correspond to systems of finite type. Another example is given by a language L which is the set of factors of an infinite word (or of a set of infinite words): in this case, as we show in Section 2, the elements of $MF(L)$

are closely related to the *bispecial* factors (cf. [16, 17, 10]) of the infinite word. Minimal forbidden words have been also considered in the study of complexity in the framework of a hierarchical modeling of physical systems (cf. [4, 5]).

In this paper we survey some recent results obtained in [7, 13, 14], in which the notion of minimal forbidden word plays a central role. In particular, in Section 3 we focus on the transformations between a factorial language L and the corresponding antifactorial language $MF(L)$. We further report the construction of an automaton accepting L, that is built from the language $M = MF(L)$.

Section 4 considers the special case of a language $L(v)$ that is the set of factors of a single word v. We show that the construction of Section 3 produces the *minimal* automaton recognizing $L(v)$. As a corollary we obtain a non-trivial upper bound on the number of minimal forbidden words of a single word.

In Section 5 we introduce a topological invariant of symbolic dynamical systems, based on minimal forbidden words and independent of the entropy. As an application we derive, for any Sturmian word s, a characterization of the language $\mathcal{F}(s)$ of factors of s, based on the counting of elements of $\mathcal{F}(s)$ and on the counting of minimal forbidden words of s.

2 Minimal forbidden words

For any notation not explicitely defined in this paper we refer to [21] and to [18].

Let A be a finite alphabet and A^* be the set of finite words drawn from the alphabet A, the empty word ϵ included. Let $L \subseteq A^*$ be a *factorial language*, *i.e.* a language satisfying: $\forall u, v \in A^*$ $uv \in L \Longrightarrow u, v \in L$. The complement language $L^c = A^* \setminus L$ is a (two-sided) ideal of A^*. Denote by $MF(L)$ the base of this ideal, we have $L^c = A^* MF(L) A^*$.

The set $MF(L)$ is called the set of *minimal forbidden words* for L. A word $v \in A^*$ is forbidden for the factorial language L if $v \notin L$, which is equivalent to say that v occurs in no word of L. In addition, v is minimal if it has no proper factor that is forbidden.

One can note that the set $MF(L)$ uniquely characterizes L, just because

$$L = A^* \setminus A^* MF(L) A^*. \qquad (1)$$

The following simple observation provides a basic characterization of minimal forbidden words.

Remark 1. A word $v = a_1 a_2 \cdots a_n$ belongs to $MF(L)$ iff the two conditions hold:

- v is forbidden, (*i.e.*, $v \notin L$),
- both $a_1 a_2 \cdots a_{n-1} \in L$ and $a_2 a_3 \cdots a_n \in L$ (the prefix and the suffix of v of length $n-1$ belong to L).

The remark translates into the equality:

$$MF(L) = AL \cap LA \cap (A^* \setminus L). \qquad (2)$$

As a consequence of both equalities (1) and (2) we get the following proposition.

Proposition 2. *For a factorial language L, languages L and $MF(L)$ are simultaneously rational, that is, $L \in Rat(A^*)$ iff $MF(L) \in Rat(A^*)$.*

The set $MF(L)$ is an *anti-factorial language* or a *factor code*, which means that it satisfies: $\forall u, v \in MF(L)$ $u \neq v \Longrightarrow u$ is not a factor of v, property that comes from the minimality of words of $MF(L)$.

We introduce a few more definitions.

Definition 3. A word $v \in A^*$ *avoids the set M*, $M \subseteq A^*$, if no word of M is a factor of v, (i.e., if $v \notin A^*MA^*$). A language L *avoids M* if every words of L avoid M.

From the definition of $MF(L)$, it readily comes that L is the largest (according to the subset relation) factorial language that avoids $MF(L)$. This shows that for any anti-factorial language M there exists a unique factorial language $L(M)$ for which $M = MF(L)$. The next remark summarizes the relation between factorial and anti-factorial languages.

Remark 4. There is a one-to-one correspondence between factorial and anti-factorial languages. If L and M are factorial and anti-factorial languages respectively, both equalities hold: $MF(L(M)) = M$ and $L(MF(L)) = L$.

A word $v \in L$ is *special on the left* with respect to B, where $B \subset A$ and $Card(B) \geq 2$, if for any $b \in B$, bv belongs to L. Anagously we define words *special on the right*. Given $B, C \subset A$ such that $Card(B) \geq 2$ and $Card(C) \geq 2$, we say that a word $v \in L$ is *bispecial* with respect to (B,C) if it is special on the left with respect to B and special on the right with respect to C.

In the case of a two letters alphabet A, special and bispecial words have been extensively studied (cf. [9, 16, 17, 10]). Remark that, since $Card(A) = 2$, there is no need to specify the sets B and C (both must be equal to A). In this case let us denote by $BS(L)$ the set of bispecial elements of L.

Example 5. Let K be the set of factors of the Fibonacci infinite word **f** (cf. [9]). Then
$$BS(K) = \{v|\ v \text{ is a palindrome prefix of } \mathbf{f}\}$$
$$= \{\epsilon, a, aba, abaaba, abaababaaba, abaababaabaababaaba, \cdots\}.$$
$$MF(K) = \{w|w = bvb,\ v \text{ is the } n\text{-th palindrome prefix of } \mathbf{f},\ n \text{ is even}\}$$
$$\cup \{w|w = ava,\ v \text{ is the } n\text{-th palindrome prefix of } \mathbf{f},\ n \text{ is odd}\}$$
$$= \{bb, aaa, babab, aabaabaa, babaababaabab, aabaababaabaababaabaa, \cdots\}.$$

As can be seen by previous example, the sets $MF(K)$ and $BS(K)$ are "similar". This fact is motivated by the following two easy propositions, that relate bispecial and minimal forbidden words.

Proposition 6. *If $u \in A^*$ is bispecial with respect to (B,C) and $buc \notin L$ for some $b \in B$ and $c \in C$, then $buc \in MF(L)$.*

The converse of previous proposition holds true under the supplementary hypothesis that L is *extensible*. Recall that a language L is extensible if, for any $v \in L$ there exist $x, y \in A$ such that $xv \in L$ and $vy \in L$.

Proposition 7. *Let L be a factorial extensible language. If $w = buc \in MF(L)$, $b, c \in A$, then there exist $B, C \subset A$, Card(B)≥ 2, Card(C)≥ 2, $b \in B$ and $c \in C$ such that u is bispecial with respect to (B, C).*

3 From minimal forbidden words to the automaton

In this section we consider a factorial language L and we focus on the transformations between L and $MF(L)$. In particular we report (cf. [11, 13, 14]) the construction of an automaton accepting L that is built from the language $M = MF(L)$. This construction is useful in several applications.

Let M be a finite anti-factorial language. We define a tree-like finite automaton $\mathcal{A}(M)$ associated with M as described below. The automaton is deterministic and complete, and, as shown later, the automaton accepts the language $L(M)$.

The automaton $\mathcal{A}(M)$ is the tuple (Q, A, i, T, F) where

- the set Q of states is $\{w \mid w$ is a prefix of a word in $M\}$,
- A is the current alphabet,
- the initial state i is the empty word ϵ,
- the set T of terminal states is $Q \setminus M$.

States of $\mathcal{A}(M)$ that are words of M are sink states. The set F of transitions is partitioned into the three (pairwise disjoint) sets F_1, F_2, and F_3 defined by:

- $F_1 = \{(u, a, ua) \mid ua \in Q, a \in A\}$ (forward edges or tree edges),
- $F_2 = \{(u, a, v) \mid u \in Q \setminus M, a \in A, ua \notin Q, v$ longest suffix of ua in $Q\}$ (backward edges),
- $F_3 = \{(u, a, u) \mid u \in M, a \in A\}$ (loops on sink states).

The transition function defined by the set F of arcs of $\mathcal{A}(M)$ is noted δ.

Remark 8. One can easily prove from definitions that
1. if $q \in Q \setminus (M \cup \{\epsilon\})$, all transitions arriving on state q are labeled by the same letter $a \in A$,
2. from any state $q \in Q$ we can reach a sink state, *i.e.*, q can be extended to a word of M.

Denoting by $Lang(\mathcal{A})$ the language accepted by an automaton $\mathcal{A}$, we get the main theorem of the section.

Theorem 9. *For any anti-factorial language M, $Lang(\mathcal{A}(M)) = L(M)$.*

Let us recall that a set of words M is called *unavoidable* if the language that avoids M is finite. As a consequence of previous theorem, one obtains the following result of C. Choffrut and K. Culik (cf. [11]).

Theorem 10. *An anti-factorial set of words M is unavoidable if and only if automaton $\mathcal{A}(M)$ has no cicles.*

The above definition of $\mathcal{A}(M)$ turns into the algorithm below, called *L-automaton*, that builds the automaton from a finite anti-factorial set of words. The input is the trie $\mathcal{T}$ that represents M. It is a tree-like automaton accepting the set M and, as such, it is noted (Q, A, i, T, δ'). The procedure can be adapted to test whether $\mathcal{T}$ represents an anti-factorial set, or even to generate the trie of the anti-factorial language associated with a set of words.

The design of the algorithm remains to adapt the construction of a pattern matching machine (see [2] or [12]). The algorithm uses a function f called a *failure function* and defined on states of $\mathcal{T}$ as follows. States of the trie $\mathcal{T}$ are identified with the prefixes of words in M. For a state au ($a \in A$, $u \in A^*$), $f(au)$ is $\delta'(i, u)$, quantity that may happen to be u itself. Note that $f(i)$ is undefined, which justifies a specific treatment of the initial state in the algorithm.

```
L-automaton (trie T = (Q, A, i, T, δ'))
 1. for each a ∈ A
 2.     if δ'(i, a) defined
 3.         set δ(i, a) = δ'(i, a);
 4.         set f(δ(i, a)) = i;
 5.     else
 6.         set δ(i, a) = i;
 7. for each state p ∈ Q \ {i} in width-first search and each a ∈ A
 8.     if δ'(p, a) defined
 9.         set δ(p, a) = δ'(p, a);
10.         set f(δ(p, a)) = δ(f(p), a);
11.     else if p ∉ T
12.         set δ(p, a) = δ(f(p), a);
13.     else
14.         set δ(p, a) = p;
15. return (Q, A, i, Q \ T, δ);
```

Example 11. Figure 1 displays the trie that accepts $M = \{\text{aa}, \text{bbaa}, \text{bbb}\}$. It is an anti-factorial language. The automaton produced from the trie by algorithm *L-automaton* is shown in Figure 2. It accepts the prefixes of $(\text{ab} \cup \text{b})(\text{ab})^*\text{ba}$ that are all the words avoiding M.

Theorem 12. *Let $\mathcal{T}$ be the trie of an anti-factorial language M. Algorithm L-automaton builds a complete deterministic automaton accepting $L(M)$.*

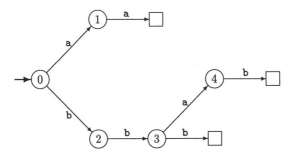

Fig. 1. Trie of the factor code {aa, bbaa, bbb} on the alphabet {a, b}. Squares represent terminal states.

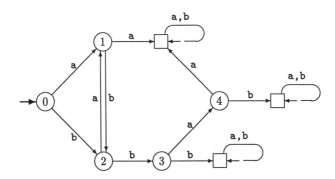

Fig. 2. Automaton accepting the words that avoid the set {aa, bbaa, bbb}. Squares represent non-terminal states (sink states).

4 Minimal forbidden words of a single word

In this section we consider the particular situation of a language that is the set of factors of a single word v (cf. [13, 14]).

The construction of its *factor automaton*, the minimal deterministic automaton accepting the factors of v (see [12]) is known to be rather intricate. It is remarkable that the transformation given in the previous section yields exactly the factor automaton of v when the input if the set M of minimal forbidden words of v. We also give an algorithm that realizes the converse transformation, building the trie of M from the factor automaton of v. A corollary of the algorithm is a non-trivial upper bound on the number of minimal forbidden words of a word.

Let us remark that the minimality of the automaton seems to be exceptional because, for example, the same construction applied to the set $\{aa, ab\}$ does not provide a minimal automaton.

The reverse construction that produces the trie of minimal forbidden words from the factor automaton is also described.

We consider a fixed word $v \in A^*$ and denote by $\mathcal{F}(v)$ be the language of factors of v.

We first remark that the language $MF(\mathcal{F}(v))$ is finite. Indeed factors of v, of lengths less than $|v| + 1$, avoid all words of length exactly $|v| + 1$. Therefore, every minimal forbidden word of $\mathcal{F}(v)$ has length at most $|v| + 1$.

The next statement gives a complete characterization of the automaton buit from the language $MF(\mathcal{F}(v))$ as the factor automaton of v.

Theorem 13. *For any $v \in A^*$, the automaton obtained from $\mathcal{A}(MF(\mathcal{F}(v)))$ by removing its sink states is the minimal deterministic finite automaton accepting the language $\mathcal{F}(v)$ of factors of v.*

Let $\sim_v$ denotes the Nerode equivalence of the language $\mathcal{F}(v)$: for $x, y \in A^*$, $x \sim_v y$ if and only if for any $z \in A^*$, $xz \in \mathcal{F}(v) \iff yz \in \mathcal{F}(v)$. The zero class correspond to the unique class of all words of A^* that does not belong to $\mathcal{F}(v)$.

Since the element of minimal length in a Nerode class correspond to the shortest path from the intial state to the state corresponding to the class, one derive the following corollary.

Corollary 14. *The elements of minimal length of the non-zero classes of the Nerode equivalence $\sim_v$ correspond to the proper prefixes of minimal forbidden words of $\mathcal{F}(v)$.*

We end this section by reporting an algorithm that builds, starting from the factor automaton of v and its suffix function, the trie accepting the language $MF(\mathcal{F}(v))$.

> **MF-trie** (factor automaton $\mathcal{A} = (Q, A, i, T, \delta)$ and its suffix function s)
> 1. **for** each state $p \in Q$ in width-first search from i and each $a \in A$
> 2.   **if** $\delta(p, a)$ undefined **and** ($p = i$ **or** $\delta(s(p), a)$ defined
> 3.   set $\delta'(p, a) = $ new sink;
> 4.   **else if** $\delta(p, a) = q$ **and** q not already treated
> 5.   set $\delta'(p, a) = q$;
> 6. **return** $(Q, A, i, \{sinks\}, \delta')$;

The input of algorithm *MF-trie* is the factor automaton of word v. It includes the failure function defined on the states of the automaton and called s. This function is a by-product of efficient algorithms that build the factor automaton (see [12]). It is defined as follows. Let $u \in A^+$ and $p = \delta(i, u)$. Then, $s(p) = \delta(i, u')$ where u' is the longest suffix of u for which $\delta(i, u) \neq \delta(i, u')$. It can be shown that the definition of $s(p)$ does not depend on the choice of u.

Example 15. Consider the word $v = $ abbab on the alphabet $\{a, b, c\}$. Its factor automaton is displayed in Figure 3. The failure function s defined on states has values: $s(1) = s(5) = 0$, $s(2) = s(3) = 5$, $s(4) = 1$, $s(6) = 2$. Algorithm *MF-trie* produces the trie of Figure 4 that represents the set of five words $\{$aa, aba, babb, bbb, c$\}$.

Theorem 16. *Let $\mathcal{A}$ be the factor automaton of a word $v \in A^*$. (It accepts the language $\mathcal{F}(v)$.) Algorithm MF-trie builds the tree-like deterministic automaton accepting $MF(\mathcal{F}(v))$ the set of minimal forbidden words of $\mathcal{F}(v)$.*

Since it is known that the factor automaton of a word v has at most $2|v| - 2$ states (cf. [12]), one obtains a non-trivial upper-bound on the number of minimal forbidden words of a word.

Corollary 17. *A word $v \in A^*$ has no more than $2(|v| - 2)(|A_v| - 1) + |A|$ minimal forbidden words if $|v| \geq 3$, where A_v is the set of letters occurring in v. The bound becomes $|A| + 1$ if $|v| < 3$.*

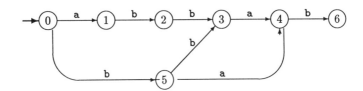

Fig. 3. Factor automaton of abbab; all states are terminal.

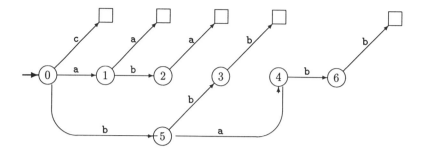

Fig. 4. Trie of minimal forbidden words of $\mathcal{F}(\mathsf{abbab})$ on the alphabet $\{\mathsf{a},\mathsf{b},\mathsf{c}\}$. Squares represent terminal states.

5 Minimal forbidden words in symbolic dynamics

In this section we show that the notion of minimal forbidden word plays an important role in symbolic dynamics.

Symbolic dynamics is a field born at the beginning of the years 20 with the work in topology of Marston Morse (cf. [23]). Later the theory was developed as a branch of ergodic theory. There are deep connections between the theory of automata and formal languages and symbolic dynamics (cf. [24, 8] and references therein). Several results from symbolic dynamics have a natural interpretation in terms of formal languages and conversely.

We start this section with a short introduction to the concepts of symbolic dynamics. Basic definitions and notations are from [24] and [8]. For any other notations not explicitely defined here we refer to [6].

Let A^Z be the set of bi-infinite sequence of letters of an alphabet A. An element of A^Z, $\mathbf{x} = (x_n)_{n \in Z}$, $x_i \in A$ is called an *infinite word*.

Let us endow A with the discrete topology and A^Z with the product topology. Then A^Z is a compact space.

The *shift* σ is a function defined on A^Z. It associates to $\mathbf{x}$ the element $\mathbf{y} = \sigma(\mathbf{x})$ defined by the rule: $y_n = x_{n+1}$ for any integer n.

A *symbolic dynamical system* $\mathcal{S}$ is a closed subset of A^Z that is invariant by the shift function; i.e. $\sigma(\mathcal{S}) = \mathcal{S}$.

Given a system $\mathcal{S}$, the set $L(\mathcal{S})$ of all factors of all infinite words in $\mathcal{S}$ is a factorial and extensible language. Conversely it is possible to associate to a factorial and extensible language L a system $\mathcal{S}(L)$ composed by all the infinite words $\mathbf{x}$ such that all the factors of $\mathbf{x}$ belong to L. It is easy to see that the system associated to $L(\mathcal{S})$ is again $\mathcal{S}$ and that $L(\mathcal{S}(L))$ is again L. Hence any system $\mathcal{S}$ is uniquely specified by the associated language $L(\mathcal{S})$. Let us recall that a system $\mathcal{S}$ is called *sofic* if $L(\mathcal{S})$ is a rational language.

Given a language $L \subset A^*$, the *complexity function* $f_L(n)$ of L is defined as

$$f_L(n) = Card\{v \in L \text{ such that } |v| = n\}.$$

A measure of the complexity of a language L is given by its *topological entropy* H_L, defined as follows:

$$H_L = \limsup_{n \to \infty} \log_2(\sqrt[n]{f_L(n)}).$$

It is known that for a factorial extensible language previous lim sup can be substituted by lim.

The entropy H_S of a system S is the topological entropy of the corresponding language $L(S)$.

A *morphism* between two systems S and T is a map $\phi\colon S \longrightarrow T$ which is continuous and commutes with the shift, i.e. such that $\sigma \circ \phi = \phi \circ \sigma$.

An *isomorphism* is a bijective morphism. A property of a system S that is preserved under isomorphism is said to be a *topological invariant* of S. A long standing open problem of the theory (cf. [27, 20, 24, 8]) is to decide whether two given sofic systems are isomorphic.

A property of a system S that is preserved under isomorphism is said to be a *topological invariant* of S. Several topological invariants have been found, like topological entropy and zeta functions. None of them is characteristic; for instance there exist systems S and T that are not isomorphic but that have same topological entropy and same zeta function.

Let g, f be two functions from N to N. We say that f and g are *linearly equivalent*, and we write $f \simeq g$, if there exist a constant K such that

1) For any $n \geq K$, $f(n) \leq K \sum_{i=-K}^{K} g(n+i)$.
2) For any $n \geq K$, $g(n) \leq K \sum_{i=-K}^{K} f(n+i)$.

It is easy to verify that $\simeq$ is an equivalence relation.

Consider two systems S and T, S over the alphabet A, and T over the alphabet B. Let F_S and F_T, be the functions that count, respectively, the number of minimal forbidden words of S and of T.

We can now state the main result of this section (cf. [7]).

Theorem 18. *If two symbolic dynamical systems S and T are isomorphic then the two functions F_S and F_T are linearly equivalent.*

In other word, this result states that the equivalence class of F_S is a topologial invariant of S.

Let us first remark that, as a particular case of Theorem 18, we obtain the following well known result of symbolic dynamics concerning systems of finite type. Recall that a system S is of finite type if $MF(L(S))$ is finite.

Corollary 19. *Let S be a system of finite type and let T be a system isomorphic to S. Then T is also of finite type.*

It is well known that the entropy of a system is a topological invariant. Theorem 18 allows us to state that the entropy of minimal forbidden words is an invariant too.

Corollary 20. *Let S be a dynamical system. $H_{MF(L(S))}$ is a topological invariant.*

Theorem 18 and Corollary 20 provide useful tools to prove that two systems are not isomorphic. In particular, if S is a sofic system, i.e. $L(S)$ is a rational language, by Proposition 2, $MF(L(S))$ is also a rational language and its entropy can be easily computed.

However there exist sofic systems that have different zeta functions (and hence are non-isomorphic) but that have linearly equivalent growths of minimal forbidden words. Moreover there exist sofic systems that have equal zeta functions and that have linearly equivalent growths of minimal forbidden words but that are non-isomorphic.

Let us now consider the case of dynamical systems associated to Sturmian words. Recall that a Sturmian word can be also defined by considering the intersections with a squared-lattice of a semi-line having a slope which is an irrational number $\alpha > 0$ (cf. [16]). A vertical intersection is denoted by the letter a, a horizontal intersection by b and the intersection with a corner by ab or ba. It is possible to prove that the language of factors of a Sturmian word defined in this way (and hence the associated system S_α) depends only on the slope of the line.

Let S_α, S_β be the dynamical systems associated to two Sturmian words such that $\alpha \neq \beta$ and $\alpha \neq \frac{1}{\beta}$. It is possible to verify that the entropy of both systems and the entropy of their minimal forbidden words are zero. However in what follows, as a consequence of Theorem 18, we prove in a purely combinatorial way that that S_α, S_β are not isomorphic; this also shows that Theorem 18 is stronger than Corollary 20.

Theorem 21. *If $\alpha \neq \beta$ and $\alpha \neq \frac{1}{\beta}$, then S_α is not isomorphic to S_β.*

Remark that if $\alpha = \frac{1}{\beta}$ then $\mathbf{x}_\beta$ is obtained from $\mathbf{x}_\alpha$ by exchanging letter a with letter b. Hence S_α and S_β are trivially isomorphic.

Let $\mathbf{x}$ be a Sturmian word and let $L(\mathbf{x})$ be the set of finite factors of $\mathbf{x}$. Let $f_\mathbf{x}(n)$ and $F_\mathbf{x}(n)$ denote respectively the complexity function of $L(\mathbf{x})$ and of $MF(L(\mathbf{x}))$.

As a consequence of previous theorem we obtain the following corollary.

Corollary 22. *The language $L(\mathbf{x})$ is uniquely specified (up to the automorphism exchanging the two letters a and b) by the two functions $f_\mathbf{x}$ and $F_\mathbf{x}$.*

Finally let us briefly discuss some consequences of the main result reported in this section in the framework of formal languages. The fact that the growth of the function $F_L(n)$, that counts the number of minimal forbidden words of L of length n, is invariant under some natural transformation suggests that this function is a "good" tool to investigate the structure of the language L. The equivalence relation $\simeq$ introduced in this section between functions from N to N induces an equivalence relation between languages as follows: $L_1 \simeq L_2$ if and only if $F_{L_1} \simeq F_{L_2}$. This suggests a new classification for (factorial) formal languages.

6 Concluding remarks

In learning theory negative informations are commonly used in order to obtain fast inference algorithms (cf. for instance [1]).

In formal language theory the importance of this "negative" point of view has been, to our knowledge, up to now underestimated.

The results presented in this paper shows that negative informations, described in terms of minimal forbidden words, are useful in some problems arising in language theory.

Let us conclude this paper by mentioning a related work that introduces a new text compression scheme based on forbidden words (cf. [15]). Contrary to other methods which make use, as a main tool, of dictionaries, i.e. particular sets of words occurring as factors in the text (cf. for instance [26]), the method of [15] takes advantage from words that do not occur as factor in the text, i.e. that are forbidden. Such sets of words are called there *antidictionaries*.

References

1. Learning regular sets from queries and couterexamples. Information and Control **75** (1987), 87-106.
2. A. V. Aho and M. J. Corasick. Efficient string matching: an aid to bibliographic search, *Comm. ACM* **18:6** (1975) 333-340.
3. R. Adler, D. Coppersmith, M. Hassner. *Algorithms for Sliding-Block Codes*, IEEE Trans. Inf. Theory, **IT-29** (1983), 5-22.
4. R. Badii. *Quantitave Characterization of Complexity and Predictability*, Physics Letters A **160** (1991), 372-377.
5. R. Badii. *Complexity and Unpredictable Scaling of Hierarchical Structures*, in "Chaotic Dynamics: Theory and Practice", T. Bountis Ed., Plenum Press, New York, 1992.
6. M. P. Béal. *Codage Symbolique*, Masson, 1993.
7. M.-P. Béal, F. Mignosi, and A. Restivo. Minimal Forbidden Words and Symbolic Dynamics. in (*STACS'96*, C. Puech and R. Reischuk, eds., LNCS 1046, Springer, 1996) 555–566.
8. M. P. Béal, D. Perrin. *Symbolic Dynamics and Finite Automata*, Handbook of Formal Languages, G. Rozenberg, A. Salomaa eds., Vol. 2 Ch. 10, Springer Verlag, 1997.
9. J. Berstel. *Fibonacci Words - a Survey*, in "The Book of L", G. Rozenberg, A. Salomaa eds., Springer Verlag 1986.
10. J. Cassaigne. *Complexité et Facteurs Spéciaux*, Actes des Journées Montoises, 1994.
11. C. Choffrut, K. Culik II. On the Extendibility of Unavoidable Sets. Discrete Applied Math. **9** (1984), 125-137.
12. M. Crochemore, C. Hancart. Automata for matching patterns, in *Handbook of Formal Languages*, G. Rozenberg, A. Salomaa, eds.", Springer-Verlag", 1997, Volume 2, *Linear Modeling: Background and Application*, Chapter 9, 399–462.
13. M. Crochemore, F. Mignosi and A. Restivo. Minimal Forbidden Words and Factor Automata. in *MFCS'98*, Lubos Brim, ed., LNCS, **1450**, 665-673. Springer, 1998).

14. M. Crochemore, F. Mignosi and A. Restivo. *Automata and Forbidden Words.* Information Processing Letters **67** (1998) 111-117.
15. M. Crochemore, F. Mignosi, A. Restivo and S. Salemi. Text Compression Using Antidictionaries. Tech. Rept. IGM-98-10, Institut Gaspard Monge, 1998. DCA home page at URL http://www-igm.univ-mlv.fr/~mac/DCA.html
16. A. de Luca, F. Mignosi. *Some Combinatorial Properties of Sturmian Words,* Theor. Comp. Science, **136** (1994), 361-385.
17. A. de Luca, L. Mione. *On Bispecial Factors of the Thue-Morse Word,* Inf. Proc. Lett., **49** (1994), 179-183.
18. S. Eilenberg. *Automata, Languages, Machines,* Vol. A, Academic Press, 1974.
19. G. A. Hedlund. *Endomorphisms and Autorphisms of the Shift Dynamical System,* Math. System Theory, **3** (1969), 320-375.
20. K. H. Kim, F. W. Roush. *Williams Conjecture is False for Reducible Matrices,* J. Amer. Math. Soc. **5** (1992), 213-125.
21. M. Lothaire. *Combinatorics on Words,* Addison-Wesley, Reading, MA 1983.
22. R. McNaughton, S. Papert. *Counter-Free Automata,* M.I.T. Press, MA 1970.
23. M. Morse. *Recurrent geodesics on a surface of negative curvature,* Trans. Amer. Math. Soc. **22** (1921), 84-110.
24. D. Perrin. *Symbolic Dynamics and Finite Automata,* invited lecture in Proc. MFCS'95, Lect. Notes in Comp. Sci., **969**.
25. G. Rauzy, *Mots infinis en arithmétique* in "Automata on Infinite words", M. Nivat and D. Perrin eds., Lecture Notes in Comp. Science, **192**, Springer, Berlin 1984.
26. J. A. Storer. *Data Compression: Methods and Theory.* Computer Science Press, Rockville, MD, 1988.
27. R. Williams. *Classification of shifts of finite type,* Annals of Math., **98** (1973), 120-153. Errata, Annals of Math., **99** (1974), 380-381.

Crossroads in Flatland

János Pach *

Mathematical Institute of the Hungarian Academy of Sciences, H-1364 Budapest, Pf. 127, Hungary, and City College, New York. USA. pach@math-inst.hu

Drawing is one of the most ancient human activities. Our ancestors drew their pictures (pictographs or, simply, "*graphs*") on walls of caves, nowadays we use mostly computer screens for this purpose. From the mathematical point of view, there is not much difference: both surfaces are "flat," they are topologically equivalent.

1 Crossings—the brick factory problem

Every graph consists of *vertices* and *edges*. The vertex set of a graph G is a finite set $V(G)$, and its edge set, $E(G)$, is a collection of unordered pairs from $V(G)$. By a *drawing* of G, we mean a representation of G in the plane such that each vertex is represented by a distinct point and each edge by a simple (non-selfintersecting) continuous arc connecting the corresponding two points. If it is clear whether we talk about an "abstract" graph G or its planar representation, these points and arcs will also be called vertices and edges, respectively. For simplicity, we assume that in a drawing (a) no edge passes through any vertex other than its endpoints, (b) no two edges touch each other (i.e., if two edges have a common interior point, then at this point they properly cross each other), and (c) no three edges cross at the same point.

Every graph has many different drawings. If G can be drawn in such a way that no two edges cross each other, then G is *planar*. According to an observation of István Fáry [11], if G is planar then it has a drawing, in which every edge is represented by a straight-line segment.

Not every graph is planar. It is well known that K_5, the *complete graph* with 5 vertices, and $K_{3,3}$, the *complete bipartite graph* with 3 vertices in its classes are not planar. According to Kuratowski's famous theorem, a graph is planar if and only if it has no subgraph which can be obtained from K_5 or from $K_{3,3}$ by subdividing some (or all) of its edges with distinct new vertices. In the next section, we give a completely different representation of planar graphs (see Theorem 3).

If G is not planar then it cannot be drawn in the plane without crossing. Paul Turán [38] raised the following problem: find a drawing of G, for which the number of crossings is minimum. This number is called the *crossing number* of G and is denoted by $\mathrm{CR}(G)$. More precisely, Turán's (still unsolved) original

---

* Supported by the National Science Foundation (USA) and the National Fund for Scientific Research (Hungary).

problem was to determine $\mathrm{CR}(K_{n,m})$, for every $n, m \geq 3$. According to an assertion of Zarankiewicz, which was down-graded from theorem to conjecture [14], we have

$$\mathrm{CR}(K_{n,m}) = \left\lfloor \frac{m}{2} \right\rfloor \cdot \left\lfloor \frac{m-1}{2} \right\rfloor \cdot \left\lfloor \frac{n}{2} \right\rfloor \cdot \left\lfloor \frac{n-1}{2} \right\rfloor,$$

but we do not even know the limits

$$\lim_{n\to\infty} \frac{\mathrm{CR}(K_{n,n})}{n^4}, \quad \lim_{n\to\infty} \frac{\mathrm{CR}(K_n)}{n^4}$$

(cf. [34, 20]).

Turán used to refer to this question as the "brick factory problem," because it occurred to him at a factory yard, where, as forced labour during World War II, he moved waggons filled with bricks from kilns to storage places. According to his recollections, it was not a very tough job, except that they had to push much harder at the crossings. Had this been the only "practical application" of crossing numbers, much fewer people would have tried to estimate $\mathrm{CR}(G)$ during the past quarter of a century. In the early eighties, it turned out that the chip area required for the realization (VLSI layout) of an electrical circuit is closely related to the crossing number of the underlying graph [22]. This discovery gave an impetus to research in the subject.

2 Thrackles—Conway's conjecture

A drawing of a graph is called a *thrackle*, if any two edges which do not share an endpoint cross precisely once, and if two edges share an endpoint then they have no other point in common.

It is easy to verify that e.g. C_4, a cycle of length 4, cannot be drawn as a thrackle, but any other cycle can [41]. If a graph cannot be drawn as a thrackle, then the same is true for all graphs that contain it as a subgraph. Thus, a thrackle does not contain a cycle of length 4, and, according to an old theorem of Erdős in extremal graph theory, the number of its edges cannot exceed $n^{3/2}$, where n denotes the number of its vertices.

The following old conjecture states much more.

Conjecture (J. Conway). *Every thrackle has at most as many edges as vertices.*

The first upper bound on the number of edges of a thrackle, which is linear in n, was found in [23].

Theorem 1 ([23]). *Every thrackle has at most twice as many edges as vertices.*

Thrackle and planar graph are, in a certain sense, opposite notions: in the former any two edges intersect, in the latter there is no crossing pair of edges. Yet the next theorem shows how similar these concepts are.

A drawing of a graph is said to be a *generalized thrackle* if every pair of its edges intersect an odd number of times. Here the common endpoint of two edges also counts as a point of intersection. Clearly, every thrackle is a generalized thrackle, but not the other way around. For example, a cycle of length 4 can be drawn as a generalized thrackle, but not as a thrackle.

Theorem 2 ([23]). *A bipartite graph can be drawn as a thrackle if and only if it is planar.*

According to an old observation of Erdős, every graph has a bipartite subgraph which contains at least half of its edges. Clearly, every planar bipartite graph of $n \geq 3$ vertices has at most $2n - 4$ edges. Hence, Theorem 2 immediately implies that every thrackle with $n \geq 3$ vertices has at most $2(2n - 4) = 4n - 8$ edges. This statement is slightly weaker than Theorem 1.

In a drawing of a graph, a triple of internally disjoint paths $(P_1(u,v), P_2(u,v), P_3(u,v))$ between the same pair of vertices (u,v) is called a *trifurcation*. (The three paths cannot have any vertices in common, other than u and v, but they can cross at points different from their vertices.) A trifurcation $(P_1(u,v), P_2(u,v), P_3(u,v))$ is said to be a *converter* if the cyclic order of the initial pieces of P_1, P_2, and P_3 around u is opposite to the cyclic order of their final pieces around v.

Theorem 3 ([23]). *A graph is planar if and only if it has a drawing, in which every trifurcation is a converter.*

The second half of the theorem is trivial: if a graph is planar, then it can be drawn without crossing, and, clearly, every trifurcation in this drawing is a converter. The first half of the statement can be proved using Kuratowski's theorem.

Recently, G. Cairns and Y. Nikolayevsky [7] has improved the factor *two* in Theorem 1 to *one and a half*.

3 Different crossing numbers?

As is illustrated by Theorem 3, the investigation of crossings in graphs often requires parity arguments. This phenomenon can be partially explained by the 'banal' fact that if we start out from the interior of a simple (non-selfintersecting) closed curve in the plane, then we find ourselves inside or outside of the curve depending on whether we crossed it an even or an odd number of times.

Next we define three variants of the notion of crossing number.

(1) The *rectilinear crossing number*, LIN-CR(G), of a graph G is the minimum number of crossings in a drawing of G, in which every edge is represented by a straight-line segment.

(2) The *pairwise crossing number* of G, PAIR-CR(G), is the minimum number of crossing pairs of edges over all drawings of G. (Here the edges can be represented by arbitrary continuous curves, so that two edges may cross more than once, but every pair of edges can contribute to PAIR-CR(G) by at most one.)

(3) The *odd-crossing number* of G, ODD-CR(G), is the minimum number of those pairs of edges which cross an odd number of times, over all drawings of G.

It readily follows from the definitions that

$$\text{ODD-CR}(G) \leq \text{PAIR-CR}(G) \leq \text{CR}(G) \leq \text{LIN-CR}(G).$$

Bienstock and Dean [6] exhibited a series of graphs with crossing number 4, whose rectilinear crossing numbers are arbitrary large. However, we cannot rule out the possibility that

$$\text{ODD-CR}(G) = \text{PAIR-CR}(G) = \text{CR}(G),$$

for every graph G.

The determination of the odd-crossing number can be rephrased as a purely combinatorial problem, thus the possible coincidence of the above three crossing numbers would offer a spark of hope that there exists an efficient approximation algorithm for computing their value.

According to a remarkable theorem of Hanani (alias Chojnacki) [8] and William Tutte [39], if a graph G can be drawn in the plane so that any pair of its edges cross an even number of times, then it can also be drawn without any crossing. In other words, $\text{ODD-CR}(G) = 0$ implies that $\text{CR}(G) = 0$. Note that in this case, by the observation of Fáry mentioned in Section 2, we also have that $\text{LIN-CR}(G) = 0$.

The main difficulty in this problem is that a graph has so many essentially different drawings that the computation of any of the above crossing numbers, for a graph of only 15 vertices, appears to be a hopelessly difficult task even for a very fast computer [10].

Theorem 4 ([12, 31]). *The computation of the crossing number, the pairwise crossing number, and the odd-crossing number are NP-complete problems.*

All we can show is that the three parameters in Theorem 4, $\text{CR}(G)$, $\text{PAIR-CR}(G)$, and $\text{ODD-CR}(G)$, are not completely unrelated.

Theorem 5 ([31]). *For any graph G, we have*

$$\text{CR}(G) \leq 2(\text{ODD-CR}(G))^2.$$

The proof of the last statement is based on the following sharpening of the Hanani–Tutte Theorem.

Theorem 6 ([31]). *An arbitrary drawing of any graph in the plane can be redrawn in such a way that no edge, which originally crossed every other edge an even number of times, would participate in any crossing.*

In [28], we apply the original form of the Hanani–Tutte Theorem to answer a question raised in robotics [19].

4 Straight-line drawings

For "straight-line thrackles," Conway's conjecture discussed in Section 2 had been settled by H. Hopf–E. Pannwitz [15] and (independently) by Paul Erdős much before the problem was raised.

If every edge of a graph is drawn by a straight-line segment, then we call the drawing a *geometric graph* [24–26]. Two geometric graphs are considered isomorphic (identical), if and only if there is a rigid motion of the plane which takes one into the other.

Hopf–Pannwitz–Erdős Theorem. *If any two edges of a geometric graph intersect (in an endpoint or an internal point), then it can have at most as many edges as vertices.*

The systematic study of extremal problems for geometric graphs was initiated by S. Avital–H. Hanani [4], Erdős, Micha Perles, and Yaakov Kupitz [21]. In particular, they asked the following question: what is the maximum number of edges of a geometric graph of n vertices, which does not have k pairwise disjoint edges? (Here, by "disjoint" we mean that they cannot cross and cannot even share an endpoint.) Denote this maximum by $e_k(n)$.

Using this notation, the above theorem says that $e_2(n) = n$, for every $n > 2$. Noga Alon and Erdős [2] proved that $e_3(n) \leq 6n$. Since then, this bound was reduced by a factor of two [13]. It had been an open problem for a long time to decide whether $e_k(n)$ is linear in n for every fixed $k > 3$.

Theorem 7 ([32]). *For every k and every n, we have $e_k(n) \leq (k-1)^4 n$.*

This bound was improved successively by Géza Tóth–Pavel Valtr [37], and by Tóth to $e_k(n) \leq 100k^2 n$. It is very likely that the dependence of $e_k(n)$ on k is also (roughly) linear.

Analogously, one can try to determine the maximum number of edges of a geometric graph with n vertices, which does not have k pairwise crossing edges. Denote this maximum by $f_k(n)$. It follows from Euler's Polyhedral Formula that, for $n > 2$, every planar graph with n vertices has at most $3n - 6$ edges. Equivalently, we have $f_2(n) = 3n - 6$.

Theorem 8 ([1]). $f_3(n) = O(n)$.

Theorem 9 ([27]). *For a fixed $k > 3$, we have $f_k(n) = O(n \log^{2k-6} n)$.*

Recently, Valtr [40] has shown that $f_k(n) = O(n \log n)$, for any $k > 3$, but it can be conjectured that $f_k(n) = O(n)$. Moreover, it cannot be ruled out that there exists a constant c such that $f_k(n) \leq ckn$, for every k and n. However, we cannot even decide whether every complete geometric graph with n vertices contains at least (a positive) constant times n pairwise crossing edges. The strongest result in this direction is the following

Theorem 10 ([3]). *Every complete geometric graph with n vertices contains at least $\lfloor \sqrt{n/12} \rfloor$ pairwise crossing edges.*

In a recent series of papers [16–18], we established some *Ramsey-type* results for geometric graphs, closely related to the subject of this section. In [9], we generalized the above results for *geometric hypergraphs* (systems of simplices).

5 An application in computer graphics

It is a pleasure for the mathematician to see his research generate some interest outside his narrow field of studies. It is a source of even greater satisfaction if his results can be applied in other disciplines or, at some special and rare occasions, in practice.

During the past twenty years, combinatorial geometers have been fortunate enough to experience this feeling quite often. Automated production lines revolutionized *robotics*, and started an avalanche of questions whose solution required new combinatorial geometric tools [35]. *Computer graphics*, whose group of users encompasses virtually everybody from engineers to film-makers, has had a similar effect on our subject [5].

Finally, I would like to sketch a mathematical result which has applications in computer graphics. Most graphics packages available on the market contain some (so-called *warping* or *morphing*) program suitable for deforming figures or pictures. Originally, these programs were written for making commercials and animated movies, but today they are widely used.

An important step in programs of this type is to fix a few basic points of the original picture (say, the vertices of the straight-line drawing of a planar graph), and then to choose new locations for these points. We would like to re-draw the graph without creating any crossing. In general, we cannot now insist that the edges be represented by segments, because such a drawing may not exist. Our goal is to produce a drawing with polygonal edges, whose total number of segments is small. The complexity and the running time of the program is proportional to this number.

Theorem 11 ([33]). *Every planar graph with n vertices can be re-drawn in such a way that the new positions of the vertices are arbitrarily prescribed, and each edge is represented by a polygonal path consisting of at most $24n$ segments. There is an $O(n^2)$-time algorithm for constructing such a drawing.*

The next result shows that Theorem 11 cannot be substantially improved.

Theorem 12 ([33]). *For every n, there exist a planar graph G_n with n vertices and an assignment of new locations for the vertices such that in any polygonal drawing of G_n there are at least $n/100$ edges composed of at least $n/100$ segments.*

The proof of this theorem is based on a result discovered by Leighton [22] (and slightly generalized in [27]), which turned out to play a crucial role in the solution of many other extremal and algorithmic problems related to graph embeddings.

The *bisection width* of a graph is the minimum number of edges whose removal splits the graph into two pieces such that there are no edges running between them and the larger piece has at most twice as many vertices as the smaller.

Theorem 13 ([22, 27]). *Let G be a graph of n vertices whose degrees are $d_1, d_2, \ldots, d_n$. Then the bisection width of G is at most*

$$1.58\sqrt{16\operatorname{CR}(G) + \sum_{i=1}^{n} d_i^2}.$$

References

1. P. Agarwal, B. Aronov, J. Pach, R. Pollack, and M. Sharir: Quasi-planar graphs have a linear number of edges, *Combinatorica* **17** (1997), 1-9.
2. N. Alon and P. Erdős: Disjoint edges in geometric graphs, *Discrete and Computational Geometry* **4** (1989), 287-290.
3. B. Aronov, P. Erdős, W. Goddard, D. J. Kleitman, M. Klugerman, J. Pach, and L. J. Schulman: Crossing families, *Combinatorica* **14** (1994), 127-134.
4. S. Avital and H. Hanani: Graphs, *Gilyonot Lematematika* **3** (1966), 2-8 (in Hebrew).
5. M. de Berg, M. van Kreveld, M. Overmars, and O. Schwarzkopf: *Computational Geometry – Algorithms and Applications*, Springer-Verlag, Berlin, Heidelberg, 1997.
6. D. Bienstock and N. Dean: Bounds for rectilinear crossing numbers, *Journal of Graph Theory* **17** (1993), 333-348.
7. G. Cairns and Y. Nikolayevsky: Bounds for generalized thrackles, *Discrete and Computational Geometry*, to appear.
8. Ch. Chojnacki (A. Hanani): Über wesentlich unplättbare Kurven im dreidimensionalen Raume, *Fund. Math.* **23** (1934), 135-142.
9. T. K. Dey and J. Pach: Extremal problems for geometric hypergraphs, *Discrete and Computational Geometry* **19** (1998), 473-484.
10. P. Erdős and R. K. Guy: Crossing number problems, *American Mathematical Monthly* **80** (1973), 52-58.
11. I. Fáry: On straight line representation of planar graphs, *Acta Univ. Szeged. Sect. Sci. Math.* **11** (1948), 229-233.
12. M. R. Garey and D. S. Johnson: Crossing number is NP-complete, *SIAM Journal of Algebraic and Disccrete Methods* **4** (1983), 312-316.
13. W. Goddard, M. Katchalski, and D. J. Kleitman: Forcing disjoint segments in the plane, *European Journal of Combinatorics* **17** (1996), 391-395.
14. R. K. Guy: The decline and fall of Zarankiewicz's theorem, in: *Proof Techniques in Graph Theory*, Academic Press, New York, 1969, 63-69.
15. H. Hopf and E. Pannwitz: Aufgabe No. 167, *Jahresbericht der Deutschen Mathematiker-Vereinigung* **43** (1934), 114.
16. G. Károlyi, J. Pach, and G. Tóth: Ramsey-type results for geometric graphs. I, *Discrete and Computational Geometry* **18** (1997), 247-255.
17. G. Károlyi, J. Pach, G. Tardos, and G. Tóth: An algorithm for finding many disjoint monochromatic edges in a complete 2-colored geometric graph, in: Intuitive Geometry (I. Bárány and K. Böröczky, eds.) Bolyai Society Mathematical Studies **6**, Budapest, 1997, 367-372.
18. G. Károlyi, J. Pach, G. Tóth, and P. Valtr: Ramsey-type results for geometric graphs. II, *Discrete and Computational Geometry* **20** (1998), 375-388.

19. K. Kedem, R. Livne, J. Pach, and M. Sharir: On the union of Jordan regions and collision–free translational motion amidst polygonal obstacles, *Discrete and Computational Geometry* **1** (1986), 59–71.
20. D. J. Kleitman: The crossing number of $K_{5,n}$, *Journal of Combinatorial Theory* **9** (1970), 315–323.
21. Y. Kupitz: Extremal problems in combinatorial geometry, *Aarhus University Lecture Notes Series* **53**, Aarhus University, Denmark, 1979.
22. T. Leighton: *Complexity Issues in VLSI, Foundations of Computing Series*, MIT Press, Cambridge, MA, 1983.
23. L. Lovász, J. Pach, and M. Szegedy: On Conway's thrackle conjecture, *Discrete and Computational Geometry* **18** (1997), 369–376.
24. J. Pach: Notes on geometric graph theory, in: Discrete and Computational Geometry (J.E. Goodman et al., eds.), DIMACS Series, Vol 6, Amer. Math. Soc., Providence, 1991, 273–285.
25. J. Pach: Geometric graphs and geometric hypergraphs, *Graph Theory Notes of New York* **31** (1996), 39–43.
26. J. Pach and P.K. Agarwal: *Combinatorial Geometry*, J. Wiley and Sons, New York, 1995.
27. J. Pach, F. Shahrokhi, and M. Szegedy: Applications of the crossing number, *Algorithmica* **16** (1996), 111–117.
28. J. Pach and M. Sharir: On the boundary of the union of planar convex sets, *Discrete and Computational Geometry* (1998), to appear.
29. J. Pach, J. Spencer, and G. Tóth: New bounds for crossing numbers, in preparation.
30. J. Pach and G. Tóth: Graphs drawn with few crossings per edge, *Combinatorica* **17** (1997), 427–439.
31. J. Pach and G. Tóth: Which crossing number is it, anyway?, in: FOCS'98, to appear. Also in: *Journal of Combinatorial Theory, Ser. B*.
32. J. Pach and J. Törőcsik: Some geometric applications of Dilworth's theorem, *Discrete and Computational Geometry* **12** (1994), 1–7.
33. J. Pach and R. Wenger: Embedding planar graphs with fixed vertex locations, in: Graph Drawing '98 (Sue Whitesides, ed.), Lecture Notes in Computer Science, Springer-Verlag, Berlin, 1999, to appear.
34. R. B. Richter and C. Thomassen: Relations between crossing numbers of complete and complete bipartite graphs, *American Mathematical Monthly*, February 1997, 131–137.
35. M. Sharir: Motion planning, in: *Handbook of Discrete and Computational Geometry*, (J. E. Goodman and J. O'Rourke, Eds.), CRC Press, Boca Raton, Florida, 1997, 733–754.
36. G. Tóth: Geometric graphs with few disjoint edges II, in preparation.
37. G. Tóth and P Valtr: Geometric graphs with few disjoint edges, in: Proc. 14th Annual Symp. on Computational Geometry, ACM Press, 1998, 184–191. Also in: *Discrete and Computational Geometry*, to appear.
38. P. Turán: A note of welcome, *Journal of Graph Theory* **1** (1977), 7–9.
39. W. T. Tutte: Toward a theory of crossing numbers, *Journal of Combinatorial Theory* **8** (1970), 45–53.
40. P. Valtr: On geometric graphs with no k pairwise parallel edges, *Discrete and Computational Geometry* **19** (1998), 461–469.
41. D. R. Woodall: Thrackles and deadlock, in: *Combinatorial Mathematics and Its Applications* (D.J.A. Welsh, ed.), Academic Press, London, 1969, 335–348.

Efficiency vs. Security in the Implementation of Public-Key Cryptography

P. Gary Walsh

Department of Mathematics, University of Ottawa, 585 King Edward, Ottawa, Ontario, Canada K1N-6N5. gwalsh@mathstat.uottawa.ca

1 Introduction

In 1976, Diffie and Hellman [19] discovered the concept of public-key cryptography. This was independently found by Merkle [46], and in 1970, in an unpublished paper [20], by J.H. Ellis of CESG. The basic idea of public-key cryptography is that keys could come in pairs, a public encryption key and a secret decryption key. The applications of this concept include key distribution for symmetric (single key) cryptography, confidentiality, and through the use of digital signatures, message authentication, data integrity, and non-repudiation. In particular, these applications lend themselves to such services as secure on-line transactions and communications, third party authentication services, and a host of smart card applications (see [45] for more details). The usefulness of these applications is evidenced by the enormous growth in the number of implementations of public-key cryptography worldwide. Most current implementations of public-key cryptography use either some variation of the RSA scheme [62], whose security relies on the difficulty of factoring large composite integers of a specific form, or a discrete logarithm based scheme, whose security relies on the difficulty of computing the discrete logarithm of an element in a finite abelian group. Mathematical research on these two computational problems has increased dramatically in the past twenty years, with many surprising successes. As a result of this research, key lengths in current public-key implementations are hundreds of bits long. Since encryption and decryption in such cryptosystems usually involve an exponentiation of some kind, the time and computational cost for an encryption or a decryption can often be prohibitive for certain applications. Consequently, implementors of such cryptosystems must discover methods for improving efficiency. Unfortunately, there have been many instances in which such an efficiency improvement has provided the cryptanalyst with just enough information to weaken the security provided by the cryptosystem.

The purpose of this paper is to examine several classic examples of how public-key cryptosystems have been improved from the point of view of efficiency, and how this has later led to cryptanalytic attacks, leading to anywhere from a small reduction in the presumed level of security, to a total loss of security. Also, for the reader interested in the topic of the Mathematics of Public-Key Cryptography and the Cryptanalysis of public-key cryptosystems, we provide pointers to relevant papers on the subject, and to some of the more pertinent recent developments.

We will be primarily interested in discussing public-key cryptosystems that are, or have been, implemented for use. In Section 2, we discuss the RSA cryptosystem. In Section 3, we discuss the cryptanalysis of cryptosystems based on the difficulty of computing discrete logarithms in the group of nonzero elements in a finite field. In Section 4, we discuss the cryptanalysis of cryptosystems based on the difficulty of computing discrete logarithms in the group of points of an elliptic curve over a finite field. There are a host of other public key cryptosystems have been suggested, which we will not discuss. The reader should refer to Chapter 8 of [45] for a description of these.

The Subexponential Function. We will make reference to a function that is ever-present in the analysis of running times of algorithms to factor integers and compute discrete logarithms. Let A be an algorithm whose inputs are either elements of a finite field $GF(q)$ or an integer q. The running time of A is said to be *subexponential* if there are constants $c > 0$ and $0 \leq \alpha < 1$, such that the expected time of A is

$$L_q[\alpha, c] = O(\exp((c + o(1))(\ln q)^\alpha (\ln \ln q)^{1-\alpha}),$$

where $O()$ denotes a function dominated by that inside the brackets for large enough q, and $o(1)$ denotes a function tending to 0 as q goes to infinity.

An important observation is that if $\alpha = 0$ (resp. $\alpha = 1$), then $L_q[\alpha, c]$ represents a *polynomial* (resp. *exponential*) running time. Thus, the above subexponential function interpolates between polynomial and exponential time.

2 The RSA cryptosystem

2.1 Description

In 1977, Adleman, Rivest, and Shamir [62] published a practical way to achieve public-key cryptography. A similar scheme was discovered by C. Cocks of CESG in 1973, and described in an unpublished manuscript [12]. The RSA scheme uses integer arithmetic modulo a composite integer, the security of which relies on the difficulty of factoring the given integer. In its most basic form, the RSA cryptosystem can be described as follows. Each user of a system obtains or generates two large primes p and q, which are kept secret, and publishes the integer $N = pq$, commonly referred to as the user's *RSA modulus*. The user then generates or obtains a public encryption key e satisfying $1 < e < \phi(N) = (p-1)(q-1)$ and $\gcd(e, \phi(N)) = 1$, and computes the secret decryption key $d = e^{-1} \pmod{\phi(N)}$. In most implementations, the function $\lambda(N) = lcm(p-1, q-1)$ is used in place of $\phi(N)$, but we forego this minor difference in favour of a clearer presentation. Note that $\phi(N)$ must be kept secret by the user since with knowledge of N and $\phi(N)$, an attacker can recover the primes p and q.

Once the above parameters are in place, secure communications can be achieved as follows. If another communicator, say B, wishes to send a message M to the above user, say A, then B follows the following steps. B first uses some standard method for associating a numeric value m to M, with $1 < m < N$ and

$\gcd(m, N) = 1$. B then looks up the public encryption key e and encrypts m by computing the ciphertext
$$c = m^e \ (mod \ N).$$
Upon receiving c, A can easily recover the original message M by computing
$$m = c^d \ (mod \ N),$$
and using the standard method for obtaining the message M from m. Euler's generalization of Fermat's "little" theorem, which states that for any integer N and a coprime to N,
$$a^{\phi(N)} \equiv 1 \ (mod \ N),$$
forms the mathematical basis for the RSA cryptosystem, since from it one can deduce that
$$(m^e)^d \equiv m \ (mod \ N)$$
for any pair (e, d) satisfying $ed \equiv 1 \ (mod \ \phi(N))$.

We remark that, in the case of small encryption exponent RSA, if a single message M is sent to a multiple number of recipients, then the corresponding values of m for each recipient will need to differ by the use of padding or appending M with some pseudorandom bit string. This requirement comes from an attack by J. Hastad [27]. This attack has since been considerably generalized by Coppersmith, the details of which will be discussed in detail in section 2.4.

The security of the RSA cryptosystem comes from the assumption that computing M from the knowledge of c and N is as difficult as computing the factors p, q of N, and the fact that, as far as we know, computing such factors is computationally infeasible if N is sufficiently large, and if p and q are generated in an appropriate manner. We note that there is some recent evidence that in certain situations, breaking RSA is theoretically easier than factoring the modulus. The interested reader should see [5] and [6] for details on this matter.

2.2 Factoring integers

The history of factoring integers, and of primality testing, is long and certainly very interesting, both from the point of view of the theoretical mathematics that has been used, but also from the perspective that these problems have greatly contributed to the development of computational devices, as they continue to be a cornerstone problem in the development of computers today. The reader is referred to any of [7, 30, 37] for more information on the factoring problem. For a brief historical summary, see the notes at the end of Chapter 3.6 of [45].

There are many classes of integers for which there are fairly efficient methods for factoring, but in the case that an integer is constructed properly, factoring it remains a computationally infeasible problem. Large integers which are the product of two randomly generated primes of the same bitlength, satisfying certain arithmetic properties, which we forego for the moment, are exceedingly difficult to factor. These integers are sometimes referred to as *cryptographic integers*. The number field sieve is currently the fastest algorithm for factoring cryptographic integers of at least 120 digits in length. Thus, for the purpose of breaking RSA, with modulus at least 512 bits long, one would likely have to resort to an implementation of the number field sieve. This algorithm, in its most primitive form, was first described by Pollard to factor special integers of the form $r^e - s$, with r and s very small (see details in [37]), but has since been made practical for factoring general integers [8]. This algorithm has expected running time $L_p[\frac{1}{3}, c]$ with $c = (64/9)^{1/3} \approx 1.923$, although variations have been described which may enable c to be lowered somewhat. The interested reader should consult the volume [45] for more details about the number field sieve factoring algorithm. The current record factorization is 130 decimal digit RSA challenge number [60], and it is believed that 512-bit RSA moduli does not provide adequate medium or long term security at present [63]. In [56], Odlyzko gives a detailed analysis of the time and cost to factor integers of size 768 through 2048 bits based on the number field sieve, and it is recommended that no less than a 1024-bit modulus be used for a medium level of security, and 2048-bit modulus be used for any kind of long term security.

2.3 Small secret exponents

In an effort to make RSA more efficient, one may decide to use small secret exponents d in order to speed up decryption. On the surface, this seems like a perfectly good thing to do for several reasons. Firstly, the corresponding public encryption exponent would appear to be reasonably random. Secondly, provided that d contained sufficient many bits, guessing it would be at least as hard as factoring the modulus N.

We will describe an attack discovered by Mike Wiener [84] on RSA in the case that small decryption exponents are used. The mathematical basis for the attack is Diophantine Approximation, in particular, continued fraction expansions of real numbers, although we will only apply them to rational numbers. For more on the subject of Diophantine Approximation, see [70].

Definition. Let α be a real number. The *simple continued fraction expansion* of α is

$$\alpha = a_0 + \cfrac{1}{a_1 + \cfrac{1}{a_2 + \cfrac{1}{a_3 + \cdots}}} = <a_0; a_1, a_2, \ldots>,$$

where a_0 is defined to be the integer part of α, a_1 is defined to be the integer part of $\frac{1}{(\alpha - a_0)}$ provided $\alpha \neq a_0$, and so on.

The integers $a_0, a_1, \ldots$ are called the *partial quotients* of the expansion. They are all positive integers except possibly a_0, since it is defined as the truncation of α. For our purposes α will always be a rational number. It is well known that in this case, the continued fraction expansion has only finitely many terms, and so there exists a positive integer $l = l(\alpha)$ such that $\alpha = <a_0, a_1, \ldots, a_l>$.

Definition. Let α be a rational number, and $l = l(\alpha)$ as defined above. For $0 \leq k \leq l$ define the *k-th convergent* to α to be the fraction $p_k/q_k = <a_0, a_1, \ldots, a_k>$, where $gcd(p_k, q_k) = 1$.

It is well known that convergents satisfy many interesting and important properties. We will not cover this subject exhaustively, but rather point out the required facts. It is easy to see that the k-th convergent can be computed in time which is polynomial in the number of bits of α (and linear in k). Moreover, the numerators and denominators grow exponentially with k. The following result consists of two statements which are very well known. The first was proved by Lagrange, the second by Legendre. The reader may consult [70] or any standard number theory textbook covering the topic of continued fractions to see the proofs.

Lemma 1.
1. Each convergent p_k/q_k to α satisfies $|\alpha - p_k/q_k| < 1/q_k^2$.
2. If a rational number p/q satisfies $|\alpha - p/q| < 1/(2q^2)$, then p/q is one of the convergents to α.

Wiener's Attack. The idea of the attack can be summarized in the following theorem.

Theorem 2. *(Wiener, 1990) Assume that p and q are r-bit primes, $N = pq$, and that (e, d) is an RSA public/private key pair with $d < (N/18)^{1/4}$. Then k/d is a convergent in the continued fraction expansion of e/N.*

From the equation
$$ed \equiv 1 \ (mod \ \phi(N)),$$
there is an integer k such that
$$ed = 1 + k\phi(N).$$
Dividing by dN results in

(1) $$\frac{e}{N} = \frac{1}{dN} + \frac{k}{d}\frac{\phi(N)}{N}.$$

The prime factors p and q of N are generated using some random input, and so being of the same bit length, they would both be somewhat close to $\sqrt{N}$, and it is no loss of generality to assume that $2\sqrt{N} < p + q < \frac{3}{\sqrt{2}}\sqrt{N}$. From the fact that $\phi(N) = pq - (p+q) + 1$, and $k < d$, equation (1) gives

(2) $$\left|\frac{e}{N} - \frac{k}{d}\right| < \frac{c}{\sqrt{N}},$$

for some positive $c < \frac{3}{\sqrt{2}}$. Provided that $d < (N/18)^{1/4}$), we deduce from (2) that
$$\left|\frac{e}{N} - \frac{k}{d}\right| < \frac{1}{2d^2},$$
and so by the second part of Lemma 1, we find that k/d is a convergent to e/N.

Thus, if small enough secret exponents are used in RSA, then an attacker can compute the secret decryption exponent d simply by computing a few convergents of the publicly known number e/N. The idea of using small decryption exponents therefore results in a total loss of security. We remark that despite Wiener's claim that his method could not be extended to full length secret keys d, the method has been extended to such keys in [11]. This observation is only of theoretical interest since the essential point of Wiener's result is to show that the improved efficiency of using small secret keys results in a total loss of security. Pinch [58] has extended Wiener's attack to other RSA-type cryptosystems. Boneh et. al. [5] have recently improved Wiener's result by showing that RSA is weak if $d < N^{.293}$.

2.4 Small public keys

In general, RSA is implemented with small encryption exponents, the most common values being $3, 17$, and $65537 = 2^{16} + 1$. The use of small encryption exponents is an obvious way to increase the efficiency of an implementation of RSA. Moreover, the fact that it is the encryption exponent that is small can be a substantial advantage in the case that a message is being sent to many recipients, and if an RSA signature is required for each message.

It is unknown, yet widely agreed, that RSA with small encryption exponents is as secure as RSA with full-length public keys. On the other hand, there are a few ways that the scheme can break down in this scenario. If one is to use small public keys, then it is worthwhile being a little bit careful about the implementation and for what purpose RSA is being used.

There many attacks on RSA with small public exponent. For an exhaustive list of these, the reader is referred to the notes in Ch. of [45]. Perhaps the most devastating and most interesting mathematically is that devised by Don Coppersmith, and presented at Eurocrypt '96 [15]. Using computational methods from the Geometry of Numbers, specifically lattice basis reduction methods (see [38]), the following theorem was proved.

Theorem 3. *Let N denote a composite integer and let $p(x)$ denote a polynomial with integer coefficients of degree k. If there is an integer solution x_0 to the congruence*

$$p(x_0) \equiv 0 \ (mod \ N),$$

with $|x_0| < N^{1/k}$, then x_0 can be computed in time which is polynomial in k and $\log(N)$.

The method given in the proof is sufficiently complicated that we refer the reader directly to Coppersmith's paper for the details. The applications of this theorem include cryptanalysis of stereotyped messages, but more importantly, the cryptanalysis of padded messages, which we describe below.

It was remarked in Section 2.1 that because of an attack by Hastad, one must pad messages before RSA encryption in the case that there are multiple recipients. Coppersmith's theorem applies to this scenario as follows. Suppose that a message m is padded twice, with t and t', say. For simplicity replace t by $t - t'$, so that the two padded messages are now m and $m + t$. Assume that an RSA implementation with public encryption exponent $e = 3$ is encrypting these two messages. Then the ciphertexts being transmitted will be

$$c = m^3 \ (mod \ N),$$

and

$$c' = (m + t)^3 = m^3 + 3m^2t + 3mt^2 + t^3 \ (mod \ N).$$

Regarding c and c' as constants, m and t as variables, m can be eliminated by computing the resultant of the two polynomials

$$R(t) = Res_m(c - m^3, c' - (m + t)^3),$$

and from the above congruences, we have further that

$$R(t) \equiv 0 \ (mod \ N).$$

By Coppersmith's theorem, if $|t| < N^{1/9}$, then t can be computed, from which the original message m can be recovered.

As a result of this attack, there must be considerable care taken when one is padding a message before encryption. For 1024-bit RSA, padding with at least 256-bits or more is recommended, or alternatively, spreading the padding throughout the message is recommended. If one really wants to be safe, avoid $e = 3$ altogether, and use $e = 2^{16} + 1$.

A remark on full length public keys. Based on Coppersmith's attack, one may be sufficiently motivated to use full length public encryption keys. Of course this results in an enormous loss of efficiency, but may also lead to a total loss of security if one is not careful. In many instances the user may not actually be in control of the public key parameters being chosen, and the user, not being aware, may be given a potentially weak public key e. For example, for a given modulus $N = pq$, a substantial portion of the bit string of one of the primes, say p, may have been imbedded somewhere in the bitstring of e. There are obviously many other simple ways that a given e could be weak, and so it is important in this case that e be provably random in the sense of being the output of a hash function, or the like. This may be a moot point, since to this author's knowledge, every implementation of RSA uses small public encryption keys.

2.5 Prescribed bits

In [81], Vanstone and Zuccherato describe a variant of the RSA cryptosystem in which some of the high order bits of the prime factors of the modulus are set in advance. The motivation for this scheme comes from the recent advances on the problem of integer factorization. For security purposes, it is desirable that an RSA modulus be at least 1024 bits in length. Unfortunately, the resulting time and cost of prime generation and modular exponention is prohibitive for certain applications of the RSA scheme.

In their RSA variant, Vanstone and Zuccherato propose that the 512-bit prime factors p and q be of the form

$$p = 2^{384} \cdot f_1 + a_1, \quad q = 2^{384} \cdot f_2 + a_2,$$

where f_1 and f_2 are publicly known 128-bit numbers, and a_1 and a_2 are secret 248-bit numbers. We remark that the above scheme allows an attacker to know the high order 264 bits of the prime factors of the RSA modulus. The authors proceed to give heuristic arguments for the security of the above scheme, which we will forego for the following reason. At Eurocrypt '96, Don Coppersmith presented an outstanding theorem, which formed the basis for a devastating attack on the Vanstone-Zuccherato proposal. As with Theorem 2, Coppersmith's method involves an ingenius use of lattice basis reduction to find small integer solutions to equations of the form $p(x, y) = 0$, with p being a polynomial with integral coefficients. The method then is applied to the polynomial

$$p(x, y) = (p_0 + x)(q_0 + y) - pq,$$

where p_0 and q_0 are known estimates for the primes p and q respectively, and pq is the known RSA modulus. It is shown that if $|x|$ and $|y|$ are sufficiently small, roughly $N^{1/4}$, then there is a polynomial time algorithm to compute them. The details are quite elaborate, so we will refer the interested reader directly to Coppersmith's paper [14].

Coppersmith's main result is as follows.

Theorem 4. *Given an integer $N = pq$, and an estimate x for p satisfying*

$$|x - p| \leq N^{1/4},$$

then p and q can be computed in time which is polynomial in $\log(N)$.

In the scheme of Vanstone and Zuccherato, the high order 264 bits of the primes are publicly known, and so an attacker can very easily produce a quantity x satisfying the hypotheses of Theorem 3, and hence the scheme is entirely insecure.

Previous to Coppersmith's theorem is a result of Rivest and Shamir [61], in which they require an estimate x satisfying $|x - p| \leq N^{1/6}$. The proof of this result is based on Lenstra's theorem on integer programming [35], although the same result can be shown by simply using the theory of continued fractions. Also, a different attack on the Vanstone-Zuccherato scheme is described in [34], showing that the scheme provided at most 45 bits of security. Finally, in a recent paper [6], it is shown how Coppersmith's result can be used to prove that if a quarter of the low order bits of d are exposed, then the remaining unknown bits of d can be computed in time which is polynomial in $\log N$ and e. This is very useful from the point of view of timing attacks on RSA (see [32]).

Public-key cryptography 89

3 Discrete logarithms in $GF(2^n)$

3.1 Introduction

Let G denote an abelian group, and let $\langle g \rangle$ denote the cyclic subgroup generated by $g \in G$. The *discrete logarithm problem* in $\langle g \rangle$ is the problem of computing the exponent a to which g is raised so that $h = g^a$ for a given $h \in \langle g \rangle$. This computational problem is not only of interest from a mathematical perspective, but has significant importance for the security of several public-key cryptosystems. The best known of such cryptosystems is that of Diffie and Hellman [19], in which two users A and B who wish to share a common key for a symmetric cryptographic algorithm, such as DES, choose random secret integers a and b respectively, transmit g^a and g^b over an insecure line to each other, and proceed to compute the common value $g^{ab} = (g^a)^b = (g^b)^a$, which is modified in some standard way to produce a DES key which they will use for secure communications. M.J. Williamson of CESG had earlier discovered this scheme in 1974, the details of which are now available in [86]. Other discrete logarithm based schemes providing key distribution and methods for digital signatures have been proposed. The interested reader should consult Chapter 8 of [45] for a description of these.

Unlike symmetric cryptography, in which a k-bit key should provide k bits of security, discrete logarithm based cryptosystems can offer at most $k/2$ bits of security, and most current implementations offer considerably less than this. The value $k/2$ arises from at least two methods for computing discrete logarithms. The first is Shanks' baby step-giant step algorithm, and the second is Pollard's rho algorithm (see Chapter 3.6 of [45]). It is worth noting that the rho algorithm has recently been shown to be parallelizable by Van Oorschot and Wiener [80], and for a general group G, this is the current best known attack on discrete logarithm based cryptosystems. One can obtained improved performance of the rho algorithm by using the ideas presented by Teske in [79].

Another important contribution is that of Pohlig-Hellman [59], who showed how one can distribute a discrete logarithm computation into smaller such computations in the factor groups of G. Therefore, for cryptographic purposes, it is best to have the order of the group a prime, or at least divisible by a large prime.

In this section we will be concentrating on the computation of discrete logarithms in the multiplicative group of nonzero elements in a finite field. In this case, there is extra structure that the cryptanalyst can use. As a result, subexponential *factor base* methods have been discovered. A detailed history of this is given in the notes of Chapter 3.6 of [45]. Currently, the algorithm with the lowest expected running time for computing discrete logarithms in $GF(p)$ is that of Schirokauer [68], with an expected running time of $L_p[\frac{1}{3}, c]$, with $c = (64/9)^{1/3} \approx 1.923$. Note that this is roughly the same running time as the number field sieve factoring algorithm. The reader may also wish to consult and compare with the methods described in [33, 25, 16]. In the case of $GF(2^n)$, Coppersmith [13] has described an algorithm which has heuristic expected running time $L_{2^n}[\frac{1}{3}, c]$ for some $c < 1.587$. We will describe this algorithm in some detail in the section to follow. For fields not of these two forms, Adleman [1] has described the *function field sieve*, which has expected running time similar to that of Coppersmith's, and applies to fields of the form $GF(p^m)$ with $\log p \leq m^{g(m)}$, and where $g(m)$ is any function satisfying $0 < g(m) < .98$, and $\lim_{m \to \infty} g(m) = 0$. It is worth noting that Coppersmith's algorithm can be viewed as a special case of the function field sieve (see [69]). Although there is no known algorithm with similar running time for all finite fields, one should probably base their security estimates on the fact there is such an algorithm. This is somewhat of a moot point since almost all implementations of finite fields for cryptographic purpose use either $GF(2^n)$ for some n, or $GF(p)$ for large prime p. For an extensive and recent survey on this entire topic, we refer the reader to [69].

3.2 Coppersmith's algorithm

In this section we present an overview of the main phase of Coppersmith's algorithm, which computes the discrete logarithms of all irreducible polynomials in $GF(2^n)$ of bounded degree. In the case that one is implementing this algorithm, there are many variations, and many fine details that one must consider. We will not discuss these matters at all, but only describe the salient features of Coppersmith's algorithm. The interested reader should consult Odlyzko's extended analysis in [55] of Coppersmith's algorithm for the finer details. For more on the general theory of finite fields we refer to [40] and [44].

Let $P(x) \in GF(2)[x]$ denote a primitive irreducible polynomial of degree n. We identify the field $GF(2^n)$ as $GF(2)[x]/(P(x))$, and hence its elements are simply polynomials of degree at most $n - 1$. For Coppersmith's algorithm one must select $P(x)$ in such a way that

$$P(x) = x^n + P_1(x),$$

where $P_1(x)$ has very small degree. Heuristics (see [Appendix A in Odl]), show that $P(x)$ can be chosen so that $P_1(x)$ has degree bounded by $\log_2(n)$.

Several parameters must be defined at this point. Let b denote a bound for the degree of all irreducible polynomials, whose discrete logarithms we will compute. This is referred to as the *smoothness bound*. Let FB, the factor base, denote the set of all irreducible polynomials over $GF(2)$ of degree at most b. The integer b will be somewhat close to $n^{1/3}(\ln n)^{2/3}$, although its precise value will depend on the implementation.

Let k denote an integer so that 2^k is close to $n^{1/3}(\ln n)^{-1/3}$, and define

$$h = \lfloor n2^{-k} \rfloor + 1.$$

These definitions have been determined to produce a most efficient version of the algorithm. Finally, let B denote an integer close to b, although its precise value may vary from b somewhat, depending on the implementation.

With all of these parameters set, we are now in position to describe a method for generating certain linear relations among the logarithms of the elements in FB. Once enough such relations are found, linear algebra is then used to solve the linear system.

Let $A(x)$ and $B(x)$ be coprime polynomials over $GF(2)$ of degree at most B. For any such pair, define

$$C(x) = A(x)x^h + B(x),$$

and

(1) $$D(x) = C(x)^{2^k} \ (mod \ P(x)), \ deg D(x) < n.$$

From the fact that we are working over a field of characteristic 2, we find that

$$D(x) = A(x^{2^k})x^{h2^k} + B(x^{2^k}) \ (mod \ P(x)),$$

and from the definition of h, it follows that

$$D(x) = A(x^{2^k})x^{h2^k - n}P_1(x) + B(x^{2^k}) \ (mod \ P(x)).$$

Note that $deg C \leq h + B$ and $deg D \leq (B+1)2^k + \log_2 n$, and by our choices for the size of these parameters, $C(x)$ and $D(x)$ are polynomials of degree no larger than about $n^{2/3}$.

From Equation (1) we have that

(2) $$\log D(x) \equiv 2^k \log C(x) \ (mod \ 2^n - 1),$$

where log represents the discrete logarithm in the cyclic group of nonzero elements in $GF(2^n)$. If $C(x)$ and $D(x)$ factor into irreducibles in FB, then upon performing this factorization, Equation (2) yields a linear relation $(mod \ 2^n - 1)$ of the discrete logarithms of elements in FB. Note that there are efficient methods for factoring polynomials over finite fields (see Ch.2 of [44]), so that this computation is not overly costly.

To complete Coppersmith's algorithm, many pairs $A(x), B(x)$ must be tested as above until enough linear relations among the discrete logarithms of the elements in FB are obtained. Once enough relations are found, roughly the same number as the number of elements in FB, then Gaussian elimination of some form can be used to solve the linear system.

The most important feature of Coppersmith's algorithm is the ability to keep the polynomials $C(x)$ and $D(x)$ of low degree as possible, which as stated above is on the order of $n^{2/3}$. This fact drastically increases the probability that a given pair $(A(x), B(x))$ leads to a linear relation among the logarithms of elements in the factor base FB, and is precisely where Coppersmith's algorithm far supercedes previous algorithms.

3.3 Efficient implementation of finite field arithmetic

For a fixed level of security, Coppersmith's algorithm forces one to implement fields $GF(2^n)$ with n very large. In the course of performing a Diffie-Hellman key exchange, say, the cost of exponentiating becomes rather costly and slow. In order to provide a practical cryptosystem, an implementor must find ways to speed up these computations. One way to do this is through the use of normal bases, and optimal normal bases, which we describe below.

Definition. For a prime power q, a *normal* basis for $GF(q^n)$ over $GF(q)$ is a basis of the form $\{\alpha, \alpha^q, \ldots, \alpha^{q^{n-1}}\}$.

We will sometimes refer to a normal basis $\{\alpha_0, \alpha_1, \ldots, \alpha_{n-1}\}$, which is defined by $\alpha_i = \alpha^{q^i}$, thus coinciding with our definition above.

It is known that normal bases always exist (see Chapter 4 of [44]), and in fact Ore [57] gave a sharp estimate on the number of normal elements, by which we mean elements α generating a normal basis. As importantly, it is not difficult to compute a normal basis. In fact, choosing random α until $\{\alpha, \alpha^q, \ldots, \alpha^{q^{n-1}}\}$ are linearly independent is a probabilistic polynomial-time algorithm, although if one is a little smarter about it, one can reduce the work factor (see section 4.5 of [44]).

For our study, we are concerned with the case that $q = 2$. The big win in using an normal basis in finite fields of characteristic two comes from the following trivial observation.

Proposition 5. *Let* $\{\alpha, \alpha^2, \ldots, \alpha^{2^{n-1}}\}$ *denote a normal basis for* $GF(2^n)$ *over* $GF(2)$. *For* $x \in GF(2^n)$, *with*

$$x = a_0\alpha + a_1\alpha^2 + \cdots + a_{n-1}\alpha^{2^{n-1}}, \quad a_i \in GF(2),$$

then

$$x^2 = a_{n-1}\alpha + a_0\alpha^2 + \cdots + a_{n-2}\alpha^{2^{n-1}}.$$

In other words, squaring an element in the field reduces to rotating the bits in the word representing the field element. As exponentiation in the field is usually accomplished by a sequence of squarings and multiplying the resulting squares together, this provides an enormous efficiency upon implementation. It remains to find an efficient method for multiplication.

Definition. Let $\{\alpha_0, \alpha_1, \ldots, \alpha_{n-1}\}$ with $\alpha_i = \alpha^{q^i}$ denote a normal basis B for $GF(q^n)$ over $GF(q)$. The *complexity* c_B of B is number of nonzero entries in the matrix $T = T(B) = (t_{ij})$ defined by

$$\alpha_0\alpha_i = \sum_{j=0}^{n-1} t_{ij}\alpha_j, \quad 0 \le i \le n-1, \ t_{ij} \in GF(q).$$

The matrix T provides a measure for the complexity of multiplication in the field $GF(q^n)$ with respect to a given basis B. In order to simplify multiplication, it is desirable to have c_B as small as possible. The following important result from [54] provides some information on c_B.

Proposition 6. *For any normal basis B of $GF(q^n)$ over $GF(q)$, $c_B \geq 2n - 1$.*

Definition. A normal basis B of $GF(q^n)$ over $GF(q)$ is *optimal* if $c_B = 2n - 1$.

In the case that optimal normal bases exist, one can obtain some remarkable speedups of exponentiation in finite fields, particularly in the case of characteristic two. The interested reader should consult [64] for a detailed architecture design for an implementation of $GF(2^{593})$.

Unfortunately, unlike normal bases, optimal normal bases may not exist. We describe two ways to construct an optimal normal basis via theorems by Mullin et. al. in [54]. It is noteworthy that these two constructions are actually the only ways to produce optimal normal bases, as proved by Gao and Lenstra in [24].

Theorem 7. *Suppose that $n+1$ is prime and q is a generator of the multiplicative group of nonzero elements in $\mathbf{Z}_{n+1}$. Then the n nonunit $(n + 1)$th roots of unity in $GF(q^n)$ form an optimal normal basis of $GF(q^n)$ over $GF(q)$.*

Theorem 8. *Let $2n + 1$ be a prime and assume that either 2 generates the multiplicative group of nonzero elements in $\mathbf{Z}_{2n+1}$, or that $2n + 1 \equiv 3 \pmod 4$ and 2 generates the multiplicative group of nonzero squares in $\mathbf{Z}_{2n+1}$. Then $\alpha = \gamma + \gamma^{-1}$ generates an optimal normal basis of $GF(2^n)$ over $GF(2)$, where γ is a primitive $(2n + 1)$th root of unity in an algebraic closure of $GF(2)$.*

A list of fields with an optimal normal basis over $GF(2)$ is given in Chapter 5 of [44]. We will see in the next section that not only do optimal normal bases provide an enormous efficiency improvement, but also provide additional structure to the field $GF(2^n)$ that will enable us to greatly improve Coppersmith's discrete logarithm computation.

3.4 Semaev's improvement to Coppersmith's algorithm

In [73], Semaev describes how one can use the structure of a field containing an optimal normal basis to systematically produce the linear relations in Coppersmith's algorithm, similar to those in Equation (1). Moreover, in many cases the resulting polynomials $C(x)$ and $D(x)$ that we are hoping to be smooth (factor into irreducibles in the factor base FB), have significantly smaller degree than in the relations generated by Coppersmith's method, thereby resulting in a much greater probability of smoothness.

In fact, Semaev describes three such algorithms for the different possible scenarios given in Theorems 4 and 5. We will describe the first case arising from Theorem 5. In this case, we wish to compute logarithms in $GF(2^n)$ where, by hypotheses, $2n + 1$ is prime and the order of 2 modulo $(2n + 1)$ is $2n$. Under these conditions, there is a $(2n+1)$th root of unity γ in an algebraic closure of $GF(2)$ with the property that $\alpha = \gamma + \gamma^{-1}$ generates an optimal normal basis of $GF(2^n)$ over $GF(2)$. It is not hard to see that $\gamma \in GF(2^{2n})$.

Definition. For $i = 0, \ldots, n-1$ define $\alpha_i = \gamma^i + \gamma^{-i}$.

Given i in the range above, let s be an integer such that $2^s \equiv i \pmod{2n+1}$. Such an integer s exists be our assumption on the order of 2 modulo $(2n+1)$. It follows that
$$\alpha^{2^s} = \gamma^{2^s} + \gamma^{-2^s} = \gamma^i + \gamma^{-i} = \alpha_i,$$
therefore the optimal normal basis from Theorem 5 is precisely $\{\alpha_0, \alpha_1, \ldots, \alpha_{n-1}\}$.

As $\alpha = \alpha_0$ satisfies the property that $GF(2^n) = GF(2)(\alpha)$, there are polynomials $p_i(x)$ ($i \geq 0$), of degree at most $n-1$, with the property that
$$\alpha_i = p_i(\alpha).$$
What is interesting, and most important to Semaev's method, is that

Proposition 9. *The polynomials $p_i(x)$ have degree i.*

This is easily proved by induction on i.

Let $P(x)$ denote the minimal polynomial of α over $GF(2)$. From the construction we have so far, the following result is an immediate consequence.

Proposition 10. *Let h, k, s be integers for which $h2^s \equiv \pm k \pmod{2n+1}$, then*
$$p_h(x)^{2^s} \equiv p_k(x) \pmod{P(x)}.$$

The polynomials $p_i(x)$ satisfy several interesting properties. The following result is Lemma 1 of [73], and can be proved by induction.

Proposition 11. *For $1 \leq i \leq n$,*
$$p_{n-i}(x) \equiv (p_i(x) - p_{i-1}(x) + \cdots + (-1)^{i-1}p_1(x) + (-1)^i)p_n(x) \pmod{P(x)}.$$

Define $q_0(x) = 1$ and for $1 \leq i \leq n$, define
$$q_i(x) = p_i(x) - p_{i-1}(x) + \cdots + (-1)^{i-1}p_1(x) + (-1)^i,$$
so that, by Proposition 5,
$$p_{n-i}(x) \equiv q_i(x)p_n(x) \pmod{P(x)}.$$
Note that $q_i(x)$ has degree i for each $i = 0, 1, \ldots, n-1$.

Construction of $C(x)$ and $D(x)$

At this point, Semaev defines certain parameters which are determined by an analysis of the running time of the entire algorithm, and provide an optimal running time.

Let s be an integer such that $c = p^s \pmod{2n+1}$ satisfies $1 \leq c \leq n$. In practice, s is chosen so that c is very small. Let d be an integer satisfying

Public-key cryptography   95

$\frac{n}{c+1} \leq d \leq \frac{n+1}{c-1}$ and $1 \leq d \leq n$. Let m be an integer satisfying $m = \left\lfloor \frac{(c+1)d-n}{c} \right\rfloor$.
For $i \geq 0$, let $n_i = i2^s \pmod{(2n+1)}$. Then, with t an integer satisfying $1 \leq t \leq (2n+1)$ and $p^t \equiv n \pmod{(2n+1)}$, define

$$C(x) = \sum_{i=d-m}^{d} c_i p_i(x)$$

and

$$D(x) = x^{2^t} \sum_{i=d-m}^{d} c_i q_{k_i}(x),$$

where the $c_i \in GF(2)$ and $k_i = n - n_i$ or $k_i = n - n_i - 1$, depending on whether $n_i \leq n$ or $n_i > n$ respectively.

With everything as above we have the following fundamental result, which allows us to systematically produce many equations like Equation (1) which will, on average, have a much better chance of resulting in a linear relation, like Equation (2), than those produced by Coppersmith's method. This is Theorem 1 from [73].

Theorem 12. *For any vector* $(c_{d-m}, \ldots, c_d) \in GF(2)^{m+1}$,

$$C(x)^{2^s} \equiv D(x) \pmod{P(x)}.$$

Moreover, $deg(C(x)) \leq d$ *and* $deg(\frac{D(x)}{x^{2^t}}) \leq d$.

A Comparison for $GF(2^{593})$

To give an indication of how much is gained through all of this, we examine the particular field $GF(2^{593})$, which has an optimal normal basis of the appropriate type, and has been implemented in hardware for cryptographic purposes, as described in [64].

We will give a rough estimate for the work to compute the database of logarithms of small degree polynomials. This will be measured in the number of smoothness tests. A smoothness test determines whether a given polynomial $w(x)$ factors over the factor base. This can be achieved by computing

$$w'(x) \prod_{i=\lceil b/2 \rceil}^{b} (x^{2^i} + x) \pmod{w(x)},$$

and checking whether the resulting polynomial is zero or not.

From Coppersmith's algorithm described above, the smoothness bound for FB is $b = 28$, k is chosen so that 2^k is about 4, i.e. $k = 2$, and $h = 149$. This results in $degC(x) \leq 177$ and $degD(x) \leq 122$. From the analysis in Appendix A of [55], for a given pair $C(x), D(x)$, the probability that they are both smooth with respect to FB (i.e. factor into irreducibles of degree at most 28) is roughly

$$\left(\frac{177}{28}\right)^{\left(\frac{-177}{28}\right)} \left(\frac{122}{28}\right)^{\left(\frac{-122}{28}\right)},$$

while the number of elements in FB is roughly $2^{29}/28$. Thus, in order for the number of linear relations in the logarithms of elements in FB be greater than the number of elements in FB, the number of pairs $(C(x), D(x))$ needed to test is about 2^{41}.

For Semaev's algorithm, after some consideration, we choose the parameters $s = 1111$, $c = 5$, $d = 141$, $m = 36$, find that the corresponding values k_i are bounded by 111. For a given pair $C(x), D(x)$, the probability that they are both smooth with respect to FB is roughly

$$\left(\frac{141}{28}\right)^{\left(\frac{-141}{28}\right)} \left(\frac{122}{28}\right)^{\left(\frac{-122}{28}\right)},$$

and it turns out in this case that only 2^{36} pairs $C(x), D(x)$ need to be tested until we should have enough linear relations to solve the system. Thus, for $GF(2^{593})$, with the parameters given above, we find that Semaev's algorithm reduces the security by roughly 5 bits, or in other words, is an attack which will break the system 32 times faster. One can argue that Semaev's algorithm has the advantage over Coppersmith's in that there is a little more freedom in choosing the power of the Frobenius automorphism $x \to x^2$ of the field $GF(2^n)$. Coppersmith's algorithm allows only for very small powers of this map.

Two remarks are in order. We have only considered the cost of collecting enough linear relations of the logarithms of the low degree irreducible polynomials. We have neglected the cost of the linear algebra, and the cost of computing individual logarithms. The latter is relatively insignicant, but the former is not. Depending on the smoothness bound, the linear algebra can be prohibitively costly. In particular, the value $b = 28$ in the above analysis may not be optimal. If this value were reduced, it would lower the probability of smoothness, and hence more $(C(x), D(x))$ pairs would need to be tested. On the other hand, reducing the smoothness bound b would reduce the size of the linear system needed to be solved. In any implementation, one must balance these two seemingly disjoint problems.

Furthermore, in any implementation, there are a host of variations of this particular algorithm. The interested reader should consult [55] for more details. We have considered Coppersmith's and Semaev's algorithms in their most basic form for a comparison.

4 Elliptic curve cryptography

As we have seen, the discrete-logarithm based public key algorithms are, at least in theory, not particular to the group being implemented. The only requirements are that exponentiation in the group should be efficient, implementible in software or hardware, and that the discrete logarithm problem in the group should be computationally infeasible. It was suggested by Miller [47], and analyzed further by Koblitz in [28], how one could use the group of points on an elliptic curve defined over a finite field. This type of group satisfies the above implementation requirements, and so the remaining question is whether such a group would be suitable from the point of view of security. The immediate answer to this question is: potentially yes. The security shortcoming of the multiplicative group of invertible elements in a finite field is that there is an obvious choice for a factor base, which has led to subexponential attacks which have running time on the order of $L_c(1/3; N)$. In the case of elliptic curves, there is no obvious choice for a factor base, and as a result, there is still no known general subexponential algorithm for computing discrete logarithms in the group of points on an elliptic curve over a finite field.

In this section we will briefly overview the current state-of-the-art on elliptic curve cryptography, maintaining the ever-present theme of efficiency at the cost of security in mind. We will be primarily concerned with the case that the finite field over which the curve is defined is of characteristic two. For an excellent reference on this entire topic, the reader should the book by Menezes [41]. For a more general presentation of the theory of elliptic curves, the reader is advised to look at the now-standard text of Silverman [74].

4.1 Definitions and basic results

Let $K = GF(q)$ denote the finite field with q elements, where $q = p^r$ for some prime p and integer $r \geq 1$. The algebraic closure of K is $\cup_{m \geq 1} GF(p^m)$, and denoted by $\overline{K}$. The *projective plane* $P^2(K)$ over K is the set of equivalence classes of the relation $\sim$ on $K^3 \setminus \{(0,0,0)\}$, where $(x_1, y_1, z_1) \sim (x_2, y_2, z_2)$ if and only if $(x_1, y_1, z_1) = k(x_2, y_2, z_2)$ for some nonzero $k \in K$. The equivalence class containing (x, y, z) is denoted by $(x : y : z)$. Let $a_1, a_2, a_3, a_4, a_6 \in K$, and let

$$F(X, Y, Z) = Y^2 Z + a_1 XYZ + a_3 YZ^2 - X^3 - a_2 X^2 Z - a_4 XZ^2 - a_6 Z^3$$

be a homogeneous polynomial of degree 3. This is referred to as a *Weierstrass equation*. F is said to be *smooth* or *non-singular*, if for any triple $(x_1 : y_1 : z_1) \in P^2(\overline{K})$, with $F(x_1, y_1, z_1) = 0$, one of the partials $\frac{\partial F}{\partial X}$, $\frac{\partial F}{\partial Y}$, $\frac{\partial F}{\partial Z}$, does not vanish at (x_1, y_1, z_1). Notice that the point $(0 : 1 : 0)$, satisfies $F(0, 1, 0) = 0$.

Definition. Let

(3) $\qquad f(x, y) = F(x, y, 1) = y^2 + a_1 xy + a_3 y - x^3 - a_2 x^2 - a_4 x - a_6,$

then the *elliptic curve* E, determined by a_1, a_2, a_3, a_4, a_6, is the set of points $(x_1, y_1) \in \overline{K}^2$ satisfying $f(x_1, y_1) = 0$ together with the *point at infinity* $\mathcal{O}$.

The point $\mathcal{O}$ corresponds to the projective point $(0 : 1 : 0)$. Equation (3) is referred to as the *affine Weierstrass equation* of the curve E. Since $a_1, a_2, a_3, a_4, a_6 \in K$, the curve E is said to be *defined over* K. The set $E(K) = (K^2 \cap E) \cup \mathcal{O}$ is the subgroup of K-*rational* points of E.

It is well known that there is a binary operation $+$ on E so that E is an abelian group with $\mathcal{O}$ as the identity element. For a proof of this fact the reader may wish to consult [74]. For our purpose, it will suffice to describe the group operation in the case that K is of characteristic 2, which we do below.

Group Law when $char(K) = 2$

In the case that K is of characteristic 2, Equation (3) can be simplified, and it follows that elliptic curves are of one of two forms. The first is the class of curves with so-called j-invariant equal to 0

$$E_1 : y^2 + a_3 y = x^3 + a_4 x + a_6,$$

where $a_3, a_4, a_6 \in K$, $a_3 \neq 0$, together with $\mathcal{O}$. These are referred to as *super-singular* curves. The second is the class of curves with nonzero j-invariant

$$E_2 : y^2 + xy = x^3 + a_2 x^2 + a_6,$$

where $a_2, a_6 \in K$, $a_6 \neq 0$, together with $\mathcal{O}$. We refer to these as *ordinary* curves. (We do not pursue the j-invariant here.) The group law is then given as follows: For any $P \in E$, $P + \mathcal{O}$ and $\mathcal{O} + P$ are defined to be P, $-\mathcal{O}$ is defined to be $\mathcal{O}$, and whatever $-P$ is defined as, $-P + P$ and $P + (-P)$ are defined to be $\mathcal{O}$. The operation $+$ for all other points is given as follows:

Supersingular Case: Let $P = (x_1, y_1) \in E_1$; then $-P = (x_1, y_1 + a_3)$. If $Q = (x_2, y_2) \in E_1$ and $Q \neq -P$, then $P + Q = (x_3, y_3)$, where

$$x_3 = \begin{cases} (\frac{y_1+y_2}{x_1+x_2})^2 + x_1 + x_2 & (P \neq Q) \\ \frac{x_1^4 + a_4^2}{a_3^2} & (P = Q) \end{cases}$$

$$y_3 = \begin{cases} (\frac{y_1+y_2}{x_1+x_2})(x_1 + x_3) + y_1 + a_2 & (P \neq Q) \\ (\frac{x_1^2+a_4}{a_3})(x_1 + x_3) + y_1 + a_3 & (P = Q) \end{cases}$$

Ordinary Case: Let $P = (x_1, y_1) \in E_1$; then $-P = (x_1, y_1 + x_1)$. If $Q = (x_2, y_2) \in E_1$ and $Q \neq -P$, then $P + Q = (x_3, y_3)$, where

$$x_3 = \begin{cases} (\frac{y_1+y_2}{x_1+x_2})^2 + \frac{y_1+y_2}{x_1+x_2} + x_1 + x_2 + a_2 & (P \neq Q) \\ (\frac{y_1}{x_1})^2 + \frac{y_1}{x_1} + x_1^2 + x_1 + a_2 & (P = Q) \end{cases}$$

$$y_3 = \begin{cases} (\frac{y_1+y_2}{x_1+x_2})(x_1 + x_3) + y_1 + x_3 & (P \neq Q) \\ x_1^2 + (x_1 + \frac{y_1}{x_1})x_3 + x_3 & (P = Q) \end{cases}$$

It is important for many reasons to know the exact order of the group. Counting the number of points on an elliptic curve is of fundamental importance for the implementation of elliptic curve cryptosystems, obviously due to the attack of Pohlig and Hellman [59]. Some information in this direction is the following fundamental theorem due to Hasse.

Theorem 13. *Let K be a field with q elements, then the number of points in $E(K)$ is $q + 1 - t$, where t satisfies $|t| \leq 2\sqrt{q}$.*

The integer t is called the *trace* of $E(K)$. In the case that $q = p$ is prime, it follows from the work of Deuring [17] (see also [4]) that as $E(K)$ runs over all of the elliptic curves defined over K, the value of t is nearly uniformly distributed in the interval of length $2\sqrt{p}$, centered at $p+1$. This is an important property in order to find elliptic curves with prime, or near-prime, order for cryptographic purposes. We remark that for certain crpytographic applications one does require that the order be a prime number.

For reference purposes we record the following simple result.

Proposition 14. *$E(K)$ is supersingular if and only if the characteristic of K divides the trace of $E(K)$.*

4.2 Supersingular curves

Exponentiating in an elliptic curve group can be achieved by the usual squaring and collection of squares method that is most common. In order to achieve maximum efficiency, the two problems of speeding up a square and of speeding up a multiplication must be addressed. In the context of elliptic curves, squaring is replaced by *doubling*, while a multiplication is replaced by the *addition* of two points. In other words, the group is simply regarded additively. We will briefly discuss how one can improve the efficiency in the case that K is a finite field of characteristic 2.

We see, from the group law given in the previous section, that in both the supersingular and ordinary cases an inversion of a field element will need to be accomplished, and for those who are aware, this can be the most costly operation. Upon further inspection, it is apparent that one can avoid this altogether for the case of doubling a point provided that $E(K)$ is supersingular, and provided that one chooses $a_3 = 1$. This was the suggestion of Menezes and Vanstone in [42]. An analysis of the efficiency of an implementation of the particular curve

$$y^2 + y = x^3$$

over the fields $GF(2^{191})$ and $GF(2^{251})$ was described in [42]. This analysis, together with the fact that no subexponential attacks were available to compute discrete elliptic logarithms, showed that, by comparison to the implementation of $GF(2^{593})$, elliptic curves would provide an enormous improvement of speed and space for the same level of security.

In 1991, the Waterloo group, consisting of Alfred Menezes and Scott Vanstone, together with Tatsuaki Okamoto, proved the most remarkable advance on the elliptic curve discrete logarithm problem to date (see [43]).

Theorem 15. *(Menezes, Okamoto, Vanstone) Let K denote the finite field with q elements. If $E(K)$ is a supersingular curve, then there is a reduction of the elliptic curve logarithm problem in $E(K)$ to the discrete logarithm in $GF(q^k)$, where $k \in \{1, 2, 3, 4, 6\}$, and the reduction is probabilistic polynomial time in $\ln q$.*

Thus, E is supersingular if the number of points on E divides $q^k - 1$ for some small integer k. In the course of constructing an elliptic curve suitable for cryptographic purposes, one must not only determine that the order N of the curve is prime, but also that $q^k \not\equiv 1 \pmod{N}$ for any small value of k. The proof of Theorem 8 uses divisor theory of elliptic curves, and the effective construction of the Weil pairing, given by Victor Miller in [48] (see Chapter 5 of [41] for the details).

Aside from its intrinsic theoretical interest, Theorem 8 has substantial consequences to the security of elliptic curve cryptosystems in the case that the curve is supersingular. In particular, because of the subexponential factor base attack of Coppersmith, it follows that subexponential attacks exist to compute discrete elliptic logarithms for supersingular curves. It is worth noting that this *Weil-pairing* attack applies to any curve whose order divides $q^r - 1$ for some small integer r. In the case of supersingular curves, this is one of the most classical examples of how an implementation efficiency, as described above, has later been superceded by a significant loss in security.

As for supersingular curves, there may still be a chance they could be used for cryptographic purposes. Koblitz has recently suggested that the speed one obtains with supersingular curves may outweigh the loss of security created by the existence of a subexponential attack in the case that the parameter k in Theorem 8 is at least 4.

4.3 Recent advances and conclusions

In recent years the subject of elliptic curve cryptography has been of considerable interest, both from the point of view of implementation, and that of cryptanalysis. On the topic of implementation, several authors have considered the possibility of implementing curves over fields of characteristic 2 with composite degree [26, 18]. This particular method of speeding up an elliptic curve implementation has the shortcoming that the Pollard ρ-method can be improved by a factor of $\sqrt{2m}$, where the curves are defined over a subfield of index m (see [24, 84]). There are a host of papers which attempt to provide efficient implementation in the case of ordinary curves ([3, 10, 31, 49–52]). An optimized method for inversion in the ground field is given in [71]. Solinas has provided an even newer approach in [77]. A detailed description of an implementation of elliptic curves over $GF(2^{155})$ is described in [2].

Another important aspect of implementing elliptic curve cryptosystems is the problem of counting the number of points on a curve in order to obtain a curve whose order is either prime, or divisible by a large prime with a small cofactor. This is of critical importance in the case that the field is at least 200-bits in size. Lercier and Morain are regarded as the leaders in this area, and have publicly available software to perform this task. Their algorithms are optimized versions of the Schoof-Elkies-Atkin method. We refer the reader to the recent paper [39] for an update and a useful bibliography.

On the topic of cryptanalysis, there have been some interesting recent results, but nothing on the order of magnitude of Theorem 8, and certainly nothing giving an indication the randomly chosen ordinary elliptic curves over $GF(2^n)$ or $GF(p)$ can be attacked in subexponential time. Smart [76] and Satoh and Araki [66] have independently showed that the class of anomalous curves over $GF(p)$ (curves of order p with p an odd prime) can be attacked in polynomial time. Schaeffer [67], based on a result of Frey and Ruck [22], has shown a similar result in the case that the order of the group is $p+1$. Rück and Semaev [72] have independently found a polynomial time algorithm for discrete logarithms in the p-subgroup of a curve defined over a finite field of characteristic p. This was later generalized by Voloch in [82] and [83]. These results have little bearing on any implementation which chooses coefficients, and hence curves, at random, since the probability of choosing a curve with order that close to p is insignificant. It is therefore recommended that an elliptic curve implemented for cryptography have the property that each of its coefficients be provably random, perhaps derived in some standard way from the value of hash function. Some very interesting results of Stein [78] and Zuccherato [87] show that the elliptic curve discrete logarithm problem is equivalent to the logarithm problem in the infrastructure of a quadratic function field. This equivalence may provide significant credence to the security provided by elliptic curve cryptography, since it is widely believed that the infrastructure discrete logarithm problem is computationally infeasible. In a very recent paper [75], Silverman describes his so-called Xedni-Calculus algorithm, which is a modified version of Index Calculus possibly suitable for computing elliptic logarithms on curves over $GF(p)$. At this point there is only experimental data showing no successes of the attack, but from a security point of view, it would be extremely desirable to quantify, even heuristically, the chances of this attack succeeding. Finally, Frey [21] has described how the action of the galois group $Gal(GF(p^k)/GF(p))$ can be used to devise a method for computing discrete logarithms on curves over $GF(p^k)$, in the case that the order of this galois group has small factors. Whether this leads to any sort of practical attack on elliptic curve cryptosystems is somewhat questionable at this point in time. In [53], Müller and Paulus give a detailed description of the above attacks as well as a method for efficient generation of cryptographically strong elliptic curves.

Elliptic Logarithm Challenges

Certicom Corporation has recently published several challenge curves over fields of varying size and characteristic, along with points P and Q on the curve for which it is known that an integer k exists satisfying $Q = kP$. Some of the smaller size challenges have been solved (see [9]), but several of the larger challenges remain, which carry with them a substantial monetary prize. The interested reader can consult the Certicom web page [9] for all of the details.

In conclusion, the group of points on an elliptic curve over a finite field can provide a very efficient method for implementations of public-key cryptography. Provided that the coefficients of the curve are verifiably random, say the values of a hash function, and that the curve is ordinary with prime order, one can be fairly certain that an efficient secure public-key algorithm has been implemented. This comment applies to any finite field of the appropriate size, since at present, the only known algorithms for computing discrete logarithms in such a group runs in time which is roughly the square-root of the size of the field.

Acknowledgements. The author would like to express gratitude to Dan Boneh, Cris Calude, Andrew Odlyzko, and Joe Silverman for their suggestions on the presentation of this paper.

References

1. L.M. ADLEMAN, *The function field sieve*, Proceedings of ANTS I, (LNCS **877**), 108-121, 1994.
2. G.B. AGNEW, R.C. MULLIN, S.A. VANSTONE, *An implementation of elliptic curve cryptosystems over $F_{2^{155}}$*, IEEE Journal on Selected Areas of Communications, **11** no.5 (1993), 804-813.
3. T. BETH AND F. SCHAEFER, *Nonsupersingular elliptic curves for public-key cryptosystems*, Advances in Cryptology: EUROCRYPT '91 (1991), 316-327.
4. B.J. BIRCH, *How the number of points of an elliptic curve over a fixed prime field varies*, J. London Math. Soc. **43** (1968), 57-60.
5. D. BONEH, G. DURFEE, NEW RESULTS ON THE CRYPTANALYSIS OF LOW EXPONENT RSA, preprint 1998, (http://www.cs.stanford.edu/ dabo/).
6. D. BONEH, G. DURFEE, AND Y. FRANKEL, *An attack on RSA given a small fraction of the private key bits*, To Appear in the proceedings of Asiacrypt '98 (LNCS).
7. D.M. BRESSOUD, *Factorization and Primality Testing*, UTM Springer-Verlag, New York, 1989.
8. J. BUHLER, H.W. LENSTRA, JR., AND C. POMERANCE, *Factoring integers with the number field sieve*, in Volume 1554 of Lecture Notes in Math. (1993), 50-94.
9. THE CERTICOM ECC CHALLENGE, http://www.certicom.com/chal/, 1997.
10. J. CHAO, K. TANADA, S. TSUJII, *Design of elliptic curves with controllable lower bound of extension degree for reduction attacks*, Advances in Cryptology: CRYPTO'94, 1994.
11. C.Y. CHEN, C.C. CHANG, AND W.P. YANG, *Cryptanalysis of the secret exponent of the RSA scheme*, J. Information Science and Engineering, **12** (1996), 277-290.
12. C.C. COCKS, *A note on non-secret encryption*, CESG Report 20, 1973 (available from http://www.cesg.gov.uk/pkc.htm).
13. D. COPPERSMITH, *Fast evaluation of logarithms in fields of characteristic two*, IEEE Transactions on Information Theory, **30** (1984), 587-594.
14. D. COPPERSMITH, *Finding a small root of a bivariate integer equation; factoring with the high bits known*, Advances in Cryptology: EUROCRYPT '96 (LNCS 1070), 178-189, 1996.
15. D. COPPERSMITH, M. FRANKLIN, J. PATARIN, AND M. REITER, *Low exponent RSA with related messages*, Advances in Cryptology: EUROCRYPT '96 (LNCS 1070), 1-9, 1996.
16. D. COPPERSMITH, A.M. ODLYZKO, AND R. SCHROEPPEL, *Discrete logarithms in GF(p)*, Algorithmica **1** (1986), 1-15.
17. M. DEURING, *Die Typen der Multiplikatorenringe elliptischer Funktionenkörper*, Abh. Math. Sem. Hansischen Univ. **14** (1941), 197-272.
18. E. DE WIN, A. BOSSELAERS, S. VANDENBERGHE, P. DE GERSEM, AND J.VANDEWALLE, *A fast software implementation for arithmetic operations in $GF(2^n)$*, Advances in Cryptology: ASIACRYPT'96 (LNCS **1163**), 1996, 65-76.

19. W. DIFFIE AND M. HELLMAN, *New directions in cryptography*, IEEE Transactions on Information Theory **22** (1976), 644-654.
20. J.H. ELLIS, *The possibility of secure non-secret digital encryption*, CESG Report, 1970 (available from http://www.cesg.gov.uk/pkc.htm).
21. G. FREY, *A lecture on the elliptic curve discrete logarithm problem*, Elliptic Curve Workshop, Waterloo, Ontario, November 1997.
22. G. FREY AND H-G. RÜCK, *A remark concerning m-divisibility and the discrete logarithm in the divisor class group of curves*, Math. Comp. **62** (1994), 865-874.
23. R. GALLANT, R. LAMBERT, AND S.A. VANSTONE, *Improving the parallelized Pollard lambda search on binary anomalous curves*, preprint, 1998.
24. S. GAO AND H.W. LENSTRA, JR., *Optimal normal bases*, Designs, Codes, and Cryptography, 1992.
25. D. GORDON, *Discrete logarithms in $GF(p)$ using the number field sieve*, SIAM J. Discrete Math. **6** (1993), 124-138.
26. J. GUAJARDO AND C. PAAR, *Efficient algorithms for elliptic curve cryptosystems*, Advances in Cryptology: CRYPTO'97, (LNCS **1294**), 1997, 342-356.
27. J. HASTAD, *Solving simultaneous modular equations of low degree*, Siam J. Computing **17** (1988), 336-341.
28. N. KOBLITZ, *Elliptic curve cryptosystems*, Math. Comp. **48** (1987), 203-209.
29. N. KOBLITZ, *Hyperelliptic curve cryptosystems*, J. Cryptology **1** (1989), 139-150.
30. N. KOBLITZ, *A course in Number Theory and Cryptography*, Springer-Verlag, New York, 2nd Ed., 1994.
31. N. KOBLITZ, *CM-curves with good cryptographic properties*, Advances in Cryptology: CRYPTO'91, (LNCS **576**), 1992, 279-287.
32. P.C. KOCHER, *Timing attacks on implementations of Diffie-Hellman, RSA, DSS, and other systems*, Advances in Cryptology: CRYPTO'96, 1996, 104-113.
33. B.A. LAMACCHIA AND A.M. ODLYZKO, *Computation of discrete logarithms in prime fields*, Designs, Codes, and Cryptography **1** no.1 (1991), 47-62.
34. P.J. LEE AND C.H. LIM, *Use of RSA moduli with prespecified bits*, Electronic Letters **31** no. 10 (1995), 785-786.
35. H.W. LENSTRA, JR., *Integer programming in a fixed number of variables*, Report 81-03, Mathematisch Institut, Universitat ban Amsterdam (1981).
36. H.W. LENSTRA, JR., *Factoring integers with elliptic curves*, Annals of Mathematics **126** (1987), 649-673.
37. A.K. LENSTRA AND H.W. LENSTRA, JR., *Development of the Number Field Sieve*, (LNM **1554**), Springer-Verlag, New York, 1993.
38. A.K. LENSTRA, H.W. LENSTRA, JR., AND L. LOVASZ, *Factoring polynomials with rational coefficients*, Math. Annalen. **261** (1982), 515-534.
39. R. LERCIER AND F. MORAIN, *Counting the number of points on elliptic curves over finite fields: strategies and performances*, Advances in Cryptology: EUROCRYPT'95 (LNCS **921**), 1995, 79-94.
40. R. LIDL AND H. NIEDERREITER, *Finite Fields*, Cambridge University Press, Cambridge, 1984.
41. A. MENEZES, *Elliptic Curve Public Key Cryptosystems*, Kluwer Academic Publishers, Boston, 1993.
42. A. MENEZES AND S. VANSTONE, *The implementation of elliptic curve cryptosystems*, Advances in Cryptology: AUSCRYPT'90, (LNCS **453**), 1990, 2-13.
43. A. MENEZES, T. OKAMOTO, AND S. VANSTONE, *Reducing elliptic curve logarithms to logarithms in a finite field*, Proc. STOC'91, (1991), 80-89.
44. A. MENEZES, I. BLAKE, X. GAO, R. MULLIN, AND S. VANSTONE, *Applications of Finite Fields*, Kluwer Academic Publishers, Boston, 1993.

45. A. MENEZES, P.C. VAN OORSCHOT, AND S. VANSTONE, *Handbook of Applied Cryptography*, CRC Press, New York, 1997.
46. R.C. MERKLE, *Secure communications over insecure channels*, Communications of the ACM **21** no.4 (1978), 294-299.
47. V. MILLER, *Use of elliptic curves in Cryptography*, Advances in Cryptology: CRYPTO'85 (LNCS **218**), 1986, 417-426.
48. V. MILLER, *Short programs for functions on curves*, unpublished manuscript, 1986.
49. A. MIYAJI, *Fast elliptic curve cryptosystems*, Tech. Report of IEICE, COMP93-25, 1993.
50. A. MIYAJI, *On secure and fast elliptic curve cryptosystems over F_p*, IEICE Trans. Fundamentals **E77-A** no.4 (1994), 630-635.
51. A. MIYAJI, *Elliptic curve cryptosystems immune to any reduction into the discrete logarithm problem*, IEICE Trans. Fundamentals **E76-A** no.1 (1993), 50-54.
52. A. MIYAJI, *Elliptic curves suitable for cryptosystems*, IEICE Trans. Fundamentals **E77-A** no.1 (1994), 98-105.
53. V. MÜLLER AND S. PAULUS, *On the generation of cryptographically strong elliptic curves*, preprint, 1997 (available from http://www.informatik.th-darmstadt.de/TI/Mitarbeiter/sachar.html).
54. R. MULLIN, I. ONYSZCHUK, S. VANSTONE, AND R. WILSON, *Optimal normal bases in $GF(p^n)$*, Discrete Applied Math. **22** (1988/89), 149-161.
55. A.M. ODLYZKO, *Discrete logarithms and their cryptographic significance*, Advances in Cryptology: EUROCRYPT'84, 1985, 224-314.
56. A.M. ODLYZKO, *The future of integer factorization*, CryptoBytes, **1** no.2 (1995), 5-12.
57. O. ORE, *Contributions to the theory of finite fields*, Trans. Amer. Math. Soc. **36** (1934), 285-287.
58. R. PINCH, *Extending the Wiener attack to any RSA-type cryptosystem*, Electronic Letters, **31** (1995), 1736-1738.
59. S.C. POHLIG AND M.E. HELLMAN, *An improved algorithm for computing logarithms over $GF(p)$ and its cryptographic significance*, IEEE Transactions on Information Theory **24** no.1 (1978), 106-110.
60. RSA CHALLENGE INFORMATION, *http://www.rsa.com/rsalabs/html/factoring.html*,
61. R. RIVEST AND A. SHAMIR, *Efficient factoring based on partial information*, Advances in Cryptology: EUROCRYPT'85 (LNCS **219**), 1986, 31-34.
62. R. RIVEST, A. SHAMIR, AND L. ADLEMAN, *A method for obtaining digital signatures and public-key algorithms*, Communications of the ACM **21** no.2 (1978), 120-126.
63. M. ROBSHAW, *Security Estimates for RSA-512*, RSA Preprint, 1995.
64. T. ROSATI, *A high speed data encryption processor for public-key cryptography*, Proceedings of the IEEE Custom Integrated Circuits Conference, San Diego, 1989, 12.3.1-12.3.5.
65. H-G. RÜCK, *On the discrete logarithm in the divisor class group of curves*, To appear in Mathematics of Computation.
66. T. SATOH, AND K. ARAKI, *Fermat quotients and the polynomial time discrete logarithm algorithm for anomalous elliptic curves*, preprint, 1997.
67. E. SCHAEFER, Personal Communication, 1996.
68. O. SCHIROKAUER, *discrete logarithms and local units*, Phil. Trans. Royal Soc. London, Series A **345** (1993), 409-423.
69. O. SCHIROKAUER, D. WEBER, T. DENNY, *Discrete Logarithms: The Effectiveness of the Index Calculus Method*, Proceedings of ANTS II, (LNCS **1122**) 1996, 337-361.

70. W.M. SCHMIDT, *Diophantine Approximation*, Lecture Notes in Math. **785**, Springer-Verlag, 1980.
71. R. SCHROEPPEL, H. ORMAN, S. O'MALLEY, AND O. SPATSCHECK, *Fast key exchange with elliptic curve systems*, Advances in Cryptology: CRYPTO'95 (LNCS **963**), 1995, 43-56.
72. I.A. SEMAEV, *Evaluation of discrete logarithms in a group of p-torsion points of an elliptic curve in characteristic p*, Math. Comp. **67** (1998), 353-356.
73. I.A. SEMAEV, *An algorithm for evaluation of discrete logarithms in some nonprime fields*, To appear in Math. Comp.
74. J.H. SILVERMAN, *The Arithmetic of Elliptic Curves*, Graduate Texts in Mathematics **106**, Springer-Verlag, New York, 1986.
75. J.H. SILVERMAN, *Xedni Calculus*, preprint 1998.
76. N. SMART, *Announcement of an attack on the ECDLP for anomalous elliptic curves*, preprint, 1997.
77. J.A. SOLINAS, *An improved algorithm for arithmetic on a family of elliptic curves*, Advances in Cryptology: CRYPTO'97 (LNCS **1294**), 1997, 357-371.
78. A. STEIN, *Equivalences between elliptic curves and real quadratic congruence function fields*, J. Théorie des Nombres de Bordeaux **9** (1997), 75-95.
79. E. TESKE, *Speeding up Pollard's rho method for computing discrete logarithms*, Proceedings of ANTS III (LNCS **1423**), 1998, 541-554.
80. P. VAN OORSCHOT AND M. WEINER, *Parallel collision search with application to hash functions and discrete logarithms*, 2nd ACM Conference on Computer and Communications Security, ACM Press, 1994, 210-218.
81. S.A. VANSTONE AND R.J. ZUCCHERATO, *Short RSA keys and their generation*, J. Cryptology **8** (1995), 101-114.
82. J. VOLOCH, *Relating the Smart-Satoh-Araki and Semaev approaches to the discrete logarithm problem on anomalous elliptic curves*, preprint, 1998.
83. J. VOLOCH, *The discrete logarithm problem on elliptic curves and descents*, preprint, 1998.
84. M.J. WEINER, *Cryptanalysis of short RSA exponents*, IEEE Transactions on Information Theory **36** (1990), 553-558.
85. M.J. WEINER AND R.J. ZUCCHERATO, *Faster attacks on elliptic curve cryptosystems*, Preprint, 1998.
86. M.J. WILLIAMSON, *Non-secret encryption using a finite field*, CESG Report 21, 1974 (available from http://www.cesg.gov.uk/pkc.htm).
87. R.J. ZUCCHERATO, *The equivalence between elliptic curve and quadratic function field discrete logarithm problems*, Proceedings of ANTS III (LNCS **1423**), 1998, 621-638.

No Feasible Monotone Interpolation for Cut-free Gentzen Type Propositional Calculus with Permutation Inference

Noriko H. Arai

Department of Computer Science, Hiroshima City University 151 Ozuka, Asaminami-ku, Hiroshima 731-31 Japan. arai@cs.hiroshima-cu.ac.jp

Abstract. The feasible monotone interpolation method has been one of the main tools to prove the exponential lowerbounds for relatively weak propositional systems. In [2], we introduced a simple combinatorial reasoning system, GCNF+permutation, as a candidate for an automatizable, though powerful, propositional calculus. We show that the monotone interpolation method is not applicable to prove the superpolynomial lower bounds for GCNF+permutation. At the same time, we show that Cutting Planes, Hilbert's Nullstellensatz and the polynomial calculus do not p-simulate GCNF+permutation.

1 Introduction

In Arai (1996) [2], we introduced a new system for propositional calculus, which gives a natural framework for combinatorial reasoning using "without loss of generality" argument and brute force induction. This system, GCNF+permutation, is a fusion of the cut-free sequent calculus and the substitution Frege, and it inherits two virtues from its origin. It is a well-known result by Gentzen that cut-free sequent calculus proofs enjoy the *subformula property*, which Frege and LK with cuts fail to have. GCNF+permutation inherits this nice property from cut-free sequent calculus, hence it is likely for machines to find the shortest proof by simply breaking down a given formula in the "bottom-up" manner.

The substitution Frege system is known to be a quite powerful system: no tautology is suggested as a candidate to show this system is not polynomially bounded. GCNF+permutation inherits the efficiency from the substitution Frege. Amazingly, GCNF+permutation is strong enough to prove pigeonhole principle, mod k principles, Bondy's theorem and many other combinatorial theorems in polynomial time [3]; these are notorious tautologies which have been used to show the exponential lower bounds for analytic tableaux, resolution, Hilbert's Nullstellensatz, and polynomial calculus. There is no sequence of tautologies which requires superpolynomial size proofs in GCNF+permutation found so far.

It is a well-known result in classical logic that when $A(p,q) \supset B(q,r)$ is a tautologies with the occurrences of variables fully indicated, there exists a formula C called *interpolant* such that the variables in C are from q's, and both

$A \supset C$ and $C \supset B$ are tautologies. The question whether or not the interpolant is obtainable in polynomial time algorithm is answered by Mundici somewhat negatively [10]: interpolation functions are not always computable in polynomial-time unless $P = NP \cap co-NP$. Nevertheless, it is possible to find such a procedure or to bound the (circuit) size of the interpolants polynomially for particular propositional systems. In some cases, one can pick monotone circuits as interpolants: the resolution and the cutting plane system are among of those which enjoy such property. This fact is used to show that these propositional systems do not have polynomial-size proofs for a sequence of tautologies, called k-$Test(n)$, which expressing the positive and the negative test for k-clique problem [6], [11], [12].

Below we will use k-$Test(n)$ show the opposite result. GCNF+permutation is powerful enough to polynomially prove k-$Test(n)$, hence it does not enjoy feasible monotone interpolation. At the same time, k-$Test(n)$ witnesses the fact that the cutting plane system does not polynomially simulate GCNF+permutation.

The paper is organized as follows: Section 2 contains definitions of the system of GCNF+permutation. Section 3 reviews the results we acheived so far. Section 3 contains polynomial-size proofs of GCNF+permutation for k-$Test(n)$. Section 4 contains some conjectures and open problems.

2 The system of GCNF+permutation

The system $GCNF$ is a subsystem of cut-free Gentzen type propositional calculus, called cut-free LK. GCNF is designed exclusively for the conjunctive normal form formulas. One can understand that GCNF is a generalization of analytic tableaux written as directed acyclic graphs.

A *literal* is a propositional variable p or a conjugate $\bar{p}$. A *clause* is a finite set of literals, where the meaning of the clause is the disjunction of the literals in the clause. A finite set of clauses is called a *cedent*. For simplicity, we express as if cedents as a sequence of clauses. In the rest of our argument, literals are denoted by l's, clauses by C's and cedents by capital Greek letters.

$GCNF$ *refutation* is a sequence of cedents in which every sequent is an *initial sequent* of the form, $\{p\}, \{\bar{p}\}$ or derived from previous cedents by one of following *inference rules*.

$$\text{structural inference} \qquad \frac{\Gamma}{\Gamma \cup \Delta}$$

$$\text{logical inference} \qquad \frac{\Gamma, C_1, \ldots, C_k \quad \Pi, \{l\}}{\Gamma \cup \Pi, C_1 \cup \{l\}, \ldots, C_k \cup \{l\}} \ (l)$$

l is an arbitrary literal, which is called the *auxiliary literal* of this inference. The clauses $C_1 \cup \{l\}, \ldots, C_k \cup \{l\}$ are called the *principal formulas*.

Now we introduce a new inference rule, *permutation*, to GCNF.

$$\text{permutation} \quad \frac{\Gamma(p_1,\ldots,p_m)}{\Gamma(\pi(p_1)/p_1,\ldots,\pi(p_m)/p_m)}\,\pi$$

π is a permutation on $\{p_1,\ldots,p_m\}$ and $\Gamma(\pi(p_1)/p_1,\ldots,\pi(p_m)/p_m)$ is the result of replacing every occurrence of p_i $(1\le i\le m)$ in $\Gamma(p_1,\ldots,p_m)$ by $\pi(p_i)$.

When P is a sequence of symbols (such as formulas and proofs). The *size* of P is the number of all the symbols used in P, that is denoted by $size(P)$. Next we define a scale to measure the efficiency of a proof system. We say that a propositional system S_1 *polynomially simulates* (*p-simulates*) another propositional system S_2 if there is a polynomial-time algorithm which, given an S_2-proof of a formula A, produces an S_1-proof of A.

Proposition 1. *GCNF is a sound and complete propositional system.*

When GCNF is written in tree form, it polynomially equivalent to the tree resolution system.

Proposition 2. *1. Let P be a tree GCNF refutation of $C_1,\ldots,C_n$. Then, there exists a tree resolution refutation R of $C_1,\ldots,C_n$ with*

$$size(R)\le size(P).$$

2. Let R be a tree resolution refutation of $C_1,\ldots,C_n$. Then, there exists a GCNF refutation of $C_1,\ldots,C_n$ with

$$size(P)\le (size(R))^2.$$

When a clause C consists of a single literal l, we express C by l instead of $\{l\}$ for the sake of simplicity.

3 GCNF+permutation is powerful

In the previous section, we defined the systems of GCNF and GCNF+permutation, and show that GCNF is polynomially equivalent to the system of resolution when they are written in tree form. In this section, we review the results we achieved so far on the system of GCNF+permutation written in DAG, and show that how the permutation inference rule makes GCNF powerful.

Definition. (Pigeonhole principle)
The *pigeonhole principle* states that for each n, if $f:\{0,\ldots,n\}\to\{0,\ldots,n-1\}$ then f is not one-to-one.

For each i and j with $0\le i\le n$ and $0\le j\le n-1$ we will have the variable $p_{i,j}$ which 'means' $f(i)=j$.

$$PHP_n: \quad \bigwedge_{0\le i\le n}\bigvee_{0\le j\le n-1} p_{i,j},\ \bigwedge_{0\le i<m\le n}\bigwedge_{0\le j\le n-1}\{\overline{p}_{i,j},\overline{p}_{m,j}\}$$

$\bigvee_{0\leq i\leq n} p_i$ is an abbreviation for the clause $\{p_0,\ldots,p_n\}$. $\bigwedge_{0\leq i\leq n} C_i$ is an abbreviation for the cedent $C_0,\ldots,C_n$.
The number of all the literals contained in PHP_n is $n^3 + 2n^2 + n$.

Definition. (k-equipartition)
The k-equipartition states that if an integer n is not evenly divisible by k, then there is no partition of $\{1,\ldots,n\}$ into disjoint sets of size k.

Let $J_n^k = \{(j_1,\ldots,j_k) : 1 \leq j_1 < \ldots j_k \leq n\}$ For $j \in J_n^k$, we write $i \in j$ to mean that there exists $1 \leq l \leq k$ such that $i = j_l$. Suppose that $n \not\equiv 0 \pmod{k}$. We introduce new variables $x_{i,(j_1,\ldots,j_k)}$ for $1 \leq i, j_1,\ldots,j_k \leq n$ to mean that $(j_1,\ldots,j_k)$ is a partition of $\{1,\ldots,n\}$ and $i \in \{j_1,\ldots,j_k\}$.
k-$Eq(n)$ is defined as the following cedent;

$$k - Eq(n) : \bigwedge_{1\leq i\leq n} \bigvee_{\substack{j\in J_n^k \\ i\in j}} x_{i,j}, \quad \bigwedge_{\substack{j\in J_n^k \\ i_1,i_2 \in j \\ i_1 \neq i_2}} \{\overline{x}_{i_1,j}, x_{i_2,j}\}, \quad \bigwedge_{\substack{j_1,j_2 \in J_n^k \\ i\in j_1, i\in j_2, j_1 \neq j_2}} \{\overline{x}_{i,j_1}, \overline{x}_{i,j_2}\}$$

The number of all literals contained in k-$Eq(n)$ is

$$n\binom{n-1}{k-1} + 2\binom{n}{k}\binom{k}{2} + n\binom{n-1}{k-1}^2 - n\binom{n-1}{k-1}.$$

The first $\bigwedge$ of clauses expresses that "each i is contained in some partition whose size is k." The second $\bigwedge$ of clauses expresses that "if $(i_1,\ldots,i_k)$ is a partition containing i_1, then it is also a partition containing $i_2,\ldots$ and i_k." The last $\bigwedge$ of clauses means that "if $i_s = j_t$ for some $1 \leq s \leq k$ and $1 \leq t \leq k$ and if $(i_1,\ldots,i_k) \neq (j_1,\ldots,j_k)$, then either $(i_1,\ldots,i_k)$ or $(j_1,\ldots,j_k)$ is not a partition."

We show that GCNF+permutation has polynomial size refutations for PHP_n and k-$Eq(n)$.

Theorem 3. *There exists a GCNF+permutation refutation of PHP_n whose length is $O(n)$ and the size is $O(n^4)$.*

Proof. We prove PHP_n backwards and reduce it to PHP_{n-1}. Then, we show that the length of the proof of PHP_n is bounded by $O(n)$.

First, we decompose the clause $\{p_{n,0},\ldots,p_{n,n-1}\}$ by applying logical inferences backwards. As a result, we obtain the cedents Γ_j ($0 \leq j \leq n-1$)

$$p_{n,j}, \bigwedge_{0\leq i\leq n-1} \bigvee_{0\leq j\leq n-1} p_{i,j}, \bigwedge_{0\leq i<m\leq n} \bigwedge_{0\leq j\leq n-1} \{\overline{p}_{i,j}, \overline{p}_{m,j}\}.$$

Each Γ_j for $0 \leq j \leq n-2$ is obtainable by exchanging $p_{i,j}$ by $p_{i,n-1}$ for each $0 \leq i \leq n$. Hence, we only need to consider Γ_{n-1}. By applying a structural inference and a logical inference backwards, of which auxiliary literal is $\overline{p}_{n,n-1}$ to Γ_{n-1}, we obtain an initial cedent, $p_{n,n-1}, \overline{p}_{n,n-1}$, and a cedent of the form

$$\overline{p}_{0,n-1},\ldots,\overline{p}_{n-1,n-1}, \bigwedge_{0\leq i\leq n-1} \bigvee_{0\leq j\leq n-1} p_{i,j}, \bigwedge_{0\leq i<m\leq n-1} \bigwedge_{0\leq j\leq n-2} \{\overline{p}_{i,j}, \overline{p}_{m,j}\}.$$

By applying logical inferences backwards, of which auxiliary literals are $p_{0,n-1}, \ldots, p_{n-1,n-1}$, we obtain a cedent of the form

$$\bigwedge_{0\leq i\leq n-1} \bigvee_{0\leq j\leq n-2} p_{i,j}, \bigwedge_{0\leq i<m\leq n-1} \bigwedge_{1\leq j\leq n-2} \{\overline{p}_{i,j}, \overline{p}_{m,j}\},$$

which is PHP_{n-1}. By examing closely, we can conclude the length of the proof obtained above is bounded by a linear function of n. Since the size of the each line is bounded by $O(n^3)$, the size of the proof is bounded by $O(n^4)$. □

Theorem 4. *There exists a polynomial function p, independent from n, and a GCNF+permutation refutation of k-$Eq(n)$ whose size is bounded by $p(n)$.*

Proof. We prove k-$Eq(n)$ backwards and reduce it to k-$Eq(n-k)$. Then, we show that the length of the proof of k-$Eq(n)$ is bounded by $O(n^k)$.

First, we decompose the clause $\bigvee_{j\in J^k_{n+k}, n\in j} x_{n,j}$ by applying logical inferences backwards. Then, we obtain cedents, Γ_j, of the form

$$x_{n,j}, \bigwedge_{1\leq i\leq n-1} \bigvee_{j\in J^k_n, i\in j} x_{i,j}, \bigwedge_{\substack{j\in J^k_n \\ i_1,i_2\in j, i_1\neq i_2}} \{\overline{x}_{i_1,j}, x_{i_2,j}\}, \bigwedge_{\substack{j_1,j_2\in J^k_n \\ i\in j_1, i\in j_2, j_1\neq j_2}} \{\overline{x}_{i,j_1}, \overline{x}_{i,j_2}\},$$

for $j \in J^k_n$ and $n \in j$. Note that the number of such cedents are $\binom{n-1}{k-1}$. All of them are obtainable by applying permutation to Γ_{j_0},

$$x_{n,j_0}, \bigwedge_{1\leq i\leq n-1} \bigvee_{j\in J^k_n, i\in j} x_{i,j}, \bigwedge_{j\in J^k_n,\ i_1,i_2\in j,\ i_1\neq i_2} \{\overline{x}_{i_1,j}, x_{i_2,j}\}, \bigwedge_{\substack{j_1,j_2\in J^k_n \\ i\in j_1, i\in j_2 \\ j_1\neq j_2}} \{\overline{x}_{i,j_1}, \overline{x}_{i,j_2}\}$$

where j_0 is $(n-k+1,\ldots,n)$. Hence, we only need to consider Γ_{j_0}.

Now we apply a logical inference backwards of which auxiliary literal is $\overline{x}_{n,j_0}$. Then we obtain an initial cedent, $x_{n,j_0}, \overline{x}_{n,j_0}$, and a cedent of the form

$$\bigwedge_{n-k+1\leq i\leq n-1} x_{i,j_0}, \bigwedge_{1\leq i\leq n-1} \bigvee_{j\in J^k_n, i\in j} x_{i,j}, \bigwedge_{\substack{j\in J^k_n-\{j_0\} \\ i_1,i_2\in j, i_1\neq i_2}} \{\overline{x}_{i_1,j}, x_{i_2,j}\},$$

$$\bigwedge_{j\in J^k_n-\{j_0\}} \overline{x}_{n,j}, \bigwedge_{\substack{j_1,j_2\in J^k_n \\ i\in j_1, i\in j_2, j_1\neq j_2}} \{\overline{x}_{i,j_1}, \overline{x}_{i,j_2}\}.$$

By applying a structural inference and logical inferences backwards of which auxiliary literals are $\overline{x}_{n-k+1,j_0},\ldots,\overline{x}_{n-1,j_0}$, we obtain a cedent of the form

$$\bigwedge_{\substack{1\leq i\leq n-k \\ j\in J^k_n - J^k_{n-k}}} \overline{x}_{i,j}, \bigwedge_{1\leq i\leq n-k} \bigvee_{j\in J^k_n, i\in j} x_{i,j}, \bigwedge_{\substack{j\in J^k_{n-k} \\ i_1,i_2\in j, \\ i_1\neq i_2}} \{\overline{x}_{i_1,j}, x_{i_2,j}\}, \bigwedge_{\substack{j_1,j_2\in J^k_{n-k} \\ i\in j_1, i\in j_2 \\ j_1\neq j_2}} \{\overline{x}_{i,j_1}, \overline{x}_{i,j_2}\}.$$

By applying logical inferences backwards of which auxiliary literals are $x_{i,j}$ for every $j \in J_n^k - J_{n-k}^k$, we obtain

$$\bigwedge_{1 \le i \le n-k} \bigvee_{j \in J_{n-k}^k, i \in j} x_{i,j}, \quad \bigwedge_{\substack{j \in J_{n-k}^k \\ i_1, i_2 \in j, \ i_1 \ne i_2}} \{\overline{x}_{i_1,j}, x_{i_2,j}\}, \quad \bigwedge_{\substack{j_1, j_2 \in J_{n-k}^k \\ i \in j_1, i \in j_2, j_1 \ne j_2}} \{\overline{x}_{i,j_1}, \overline{x}_{i,j_2}\},$$

which is k-$Eq(n-k)$. By examing closely, we can conclude the length of the proof obtained above is bounded by $O(n^k)$. The size of every line in the proof is bounded by $O(n^{2k})$. Hence the size of the whole proof is bounded by a polynomial of n. □

Theorem 3 and theorem 4 witness the fact that none of resolution, bounded depth Frege, polynomial calculus, Hilbert's Nullstellensatz do not p-simulate the system of GCNF+permutation.

Proposition 5. *[9] There exists a constant c , c > 1 such that, for sufficiently large n, every resolution refutation of PHP_n contains at least c^n different cedents.*

Proposition 6. *[1] There exists a constant c , c > 1 so that, for sufficiently large n, every constant-depth Frege proof of $\neg(k\text{-}Eq(n))$ contains at least c^n different cedents.*

Note that any refutation of k-$Eq(n)$ in resolution with limited extension can be converted to a constant-depth Frege proof of $\neg(k\text{-}Eq(n))$ within a linear factor. Hence, it also gives an exponential lower bound for the system of resolution with limited extension. Likewise, it gives an exponential lower bound for the system of cut-free LK or cut-free LK with analytic cut (cuts of which cut-formulae must be subformulas of the end-sequent) written in DAG form.

Proposition 7. *[14] Polynomial calculus refutations of PHP_n must have degree at least $\lceil n/2 \rceil + 1$ over any field.*

Proposition 8. *[5] There is no Hilbert's Nullstellensatz refutation of k-$Eq(n)$ of constant degrees.*

Corollary 9. *The following systems of propositional calculus do not p-simulates GCNF+permutation.*

- *Resolution,*
- *Resolution with limited extension,*
- *Bounded depth Frege,*
- *Cut-free LK,*
- *Cut-free LK with analytic cut,*
- *Polynomial calculus of constant depth, and*
- *Hilbert's Nullstellensatz of constant depth.*

4 No feasible interpolation for GCNF+permutation

We say that a propositional system S admits *feasible interpolation* when there exists a polynomial function f satisfying the following property. When a formula $A(p,q) \supset B(q,r)$, with the variables fully indicated, has an S-proof P, there exists a formula $C(q)$, with variables fully indicated, such that

1. both $A \supset C$ and $C \supset B$ are valid, and
2. the DAG (circuit) size of C is bounded by $f(size(P))$.

Some proof systems, such as resolution and Cutting Planes, admit even a stronger version of feasible interpolation. When q occurs only positively either in A or in B, we can pick a monotone circuit C as an interpolant. This property is called *feasible monotone interpolation*.

We show that GCNF+permutation does not admit feasible monotone interpolation by using Razborov's theorem on the lowerbounds for monotone circuits size.

Define a cedent k-$Clique(n)$ by the set of the following clauses.

1. $\{q_{i,1}, \ldots, q_{i,n}\}$ for $1 \le i \le k$,
2. $\{\bar{q}_{i,m}, \bar{q}_{j,m}\}$ for $1 \le m \le n$ and $1 \le i < j \le k$, and
3. $\{\bar{q}_{i,m}, \bar{q}_{j,l}, p_{m,l}\}$ for $1 \le m < l \le n$ and $1 \le i,j \le k$.

The above clauses encode a graph which has n vertices and contains k-clique as follows. We enumerate all the vertices of the graph $\{1, \ldots, n\}$. The q's encode a function f from $\{1, \ldots, k\}$ to $\{1, \ldots, n\}$. The literal $q_{i,l}$ means that $f(i) = l$. (The intuitive meaning of $f(i) = l$ is that the vertex named i in the graph is actually the vertex named l in the k-clique.) The $p_{m,l}$ encode that there exists an edge between m and l. Hence, the first clause means that the function f is defined for all i ($i = 1, \ldots, k$). The second clause means that f is one-to-one. The third clause means that if there exists i, j such that $f(i) = m$ and $f(j) = l$, then there exists an edge between m and l. Note that k-$Clique(n)$ corresponds to the positive test graph in [13] and [7].

Now we define a cedent k'-$Color(n)$ by the set of the following clauses.

1. $\{r_{m,1}, \ldots, r_{m,k'}\}$ for $1 \le m \le n$,
2. $\{\bar{r}_{m,i}, \bar{r}_{m,j}\}$ for $1 \le m \le n$ and $1 \le i < j \le k'$, and
3. $\{\bar{r}_{m,i}, \bar{r}_{l,i}, \bar{p}_{m,l}\}$ for $1 \le m < l \le n$ and $1 \le i \le k'$.

The above clauses encode a graph which is a k'-partite graph as follows. The r's encode a coloring function g from $\{1, \ldots, n\}$ to $\{1, \ldots, k'\}$. The literal $r_{m,i}$ means that the vertex named m is colored by i. Hence, the first clause means that every vertex is colored. The second clause means that none of the vertices has more than one color. The third clause means that the coloring is proper: when the vertices m and l has the same color, then there is no edge between m and l. Note that k'-$Color(n)$ corresponds to the negative test graph.

We define $k\text{-}Test(n)$ by the cedent consists of all the clauses in $k\text{-}Clique(n)$ and $(k{-}1)\text{-}Color(n)$ together. The size of $k\text{-}Test(n)$ is $O(n^4)$. It is obvious that $k\text{-}Test(n)$ is unsatisfiable as follows. When all the clauses in $k\text{-}Clique(n)$ is true, that means the graph contains a k-clique. Any k-clique does not have a proper $(k{-}1)$ coloring. Than means at least one of the clauses in $(k{-}1)\text{-}Color(n)$ must be false.

Theorem 10. $k\text{-}Test(n)$ has a proof of length $O(n^5)$ and size $O(n^9)$ in GCNF + permutation.

Proof. We prove $k\text{-}Test(n)$ backwards and reduce it to propositional pigeonhole principle. Then, we show that the length of the proof of $k\text{-}Test(n)$ is bounded by $O(n^5)$. The cedent $k\text{-}Test(n)$ consists of the clauses listed below.

1. $\{q_{i,1},\ldots,q_{i,n}\}$ for $1 \leq i \leq k$,
2. $\{\bar{q}_{i,m}, \bar{q}_{j,m}\}$ for $1 \leq m \leq n$ and $1 \leq i < j \leq k$, and
3. $\{\bar{q}_{i,m}, \bar{q}_{j,l}, p_{m,l}\}$ for $1 \leq m < l \leq n$ and $1 \leq i, j \leq k$.
4. $\{r_{m,1},\ldots,r_{m,k-1}\}$ for $1 \leq m \leq n$,
5. $\{\bar{r}_{m,i}, \bar{r}_{m,j}\}$ for $1 \leq m \leq n$ and $1 \leq i < j \leq k-1$, and
6. $\{\bar{r}_{m,i}, \bar{r}_{l,i}, \bar{p}_{m,l}\}$ for $1 \leq m < l \leq n$ and $1 \leq i \leq k-1$.

First, we decompose the clause $\{q_{1,1},\ldots,q_{1,n}\}$ in $k\text{-}Test(n)$ by applying logical inferences backwards. Then, we obtain n-many cedents of the form $q_{1,m}, \Gamma_1$ ($1 \leq m \leq n$), where Γ_1 denote the cedent obtained from $k\text{-}Test(n)$ by deleting the clause $\{q_{1,1},\ldots,q_{1,n}\}$. Note that $q_{1,m}, \Gamma_1$ is obtainable from $q_{1,1}, \Gamma_1$ by exchanging $q_{1,m}$ by $q_{1,1}$. Hence, we only need to consider $q_{1,1}, \Gamma_1$.

Secondly, we decompose the clause $\{q_{2,1},\ldots,q_{2,n}\}$ in Γ_1 by applying logical inferences backwards. Then, we obtain n-many cedents of the form $q_{2,m}, q_{1,1}, \Gamma_2$ ($1 \leq m \leq n$), where Γ_2 denote the cedent obtained from Γ_1 by deleting the clause $\{q_{2,1},\ldots,q_{2,n}\}$. For $m = 1$, the cedent $q_{2,m}, q_{1,1}, \Gamma_2$ is reducible to the cedent $q_{2,1}, q_{1,1}, \{\bar{q}_{2,1}, \bar{q}_{1,1}\}$, which has a simple proof. For $m > 1$, $q_{2,m}, q_{1,1}, \Gamma_2$ is obtainable from $q_{2,2}, q_{1,1}, \Gamma_2$ by applying a permutation. Hence, we only need to consider $q_{2,2}, q_{1,1}, \Gamma_2$.

We keep go on until we obtain the cedent $q_{k,k},\ldots,q_{1,1}, \Gamma_k$ where Γ_k consists of the following clauses.

1. $\{q_{i,1},\ldots,q_{i,n}\}$ for $k+1 \leq i \leq k$,
2. $\{\bar{q}_{i,m}, \bar{q}_{j,m}\}$ for $1 \leq m \leq n$ and $1 \leq i < j \leq k$, and
3. $\{\bar{q}_{i,m}, \bar{q}_{j,l}, p_{m,l}\}$ for $1 \leq m < l \leq n$ and $1 \leq i, j \leq k$.
4. $\{r_{m,1},\ldots,r_{m,k-1}\}$ for $1 \leq m \leq n$,
5. $\{\bar{r}_{m,i}, \bar{r}_{m,j}\}$ for $1 \leq m \leq n$ and $1 \leq i < j \leq k-1$, and
6. $\{\bar{r}_{m,i}, \bar{r}_{l,i}, \bar{p}_{m,l}\}$ for $1 \leq m < l \leq n$ and $1 \leq i \leq k-1$.

The cedent $q_{k,k},\ldots,q_{1,1}, \Gamma_k$ intuitively means that the vertices $\{1,\ldots,k\}$ forms a clique, and it has a proper $(k{-}1)$-coloring. The length of the proof up to here is bounded by $O(n^2)$.

By applying logical inferences of which auxiliary literals are $\bar{q}_{i,i}$ ($1 \leq i \leq k$) and a structural inference backwards, we obtain the cedent Δ which consists of the following clauses.

1. $\{q_{i,1}, \ldots, q_{i,n}\}$ for $k+1 \leq i \leq k$,
2. $\{p_{m,l}\}$ for $1 \leq m < l \leq k$,
3. $\{r_{m,1}, \ldots, r_{m,k-1}\}$ for $1 \leq m \leq k$,
4. $\{\bar{r}_{m,i}, \bar{r}_{m,j}\}$ for $1 \leq m \leq k$ and $1 \leq i < j \leq k-1$, and
5. $\{\bar{r}_{m,i}, \bar{r}_{l,i}, \bar{p}_{m,l}\}$ for $1 \leq m < l \leq k$ and $1 \leq i \leq k-1$.

By applying logical inferences of which auxiliary literals are $\bar{p}_{m,l}$ ($1 \leq m < l \leq k$), we obtain a propositional pigeonhole principle PHP_k. In the previous section, we already showed that PHP_k has a GCNF+permutation proof of length $O(k)$ and size $O(k^4)$. The length of the proof combined together is bounded by $O(n^2)$. The size of every line is $O(n^4)$. Consequently the size of the proof given above is bounded by $O(n^6)$. □

Razborov showed that any small size monotone circuit either almost always outputs 0 for the positive test graph, or almost always outputs 1 for the negative test graph. That means there is no small-size monotone circuit $C(\boldsymbol{p}_{m,l})$ such that

$$C(\boldsymbol{p}_{m,l}) \text{ is false} \to k\text{-}Clique(n) \text{ is false, and}$$
$$C(\boldsymbol{p}_{m,l}) \text{ is true} \to (k\text{--}1)\text{-}Color(n) \text{ is false.}$$

Consequently, we have the following corollary.

Corollary 11. *GCNF+permutation does not admit feasible monotone interpolation.*

The cutting plane system CP is an extension of resolution, which has polynomial-size proofs for the pigeonhole principle, the $s-t$ connectivity, and the non-unique endnode principle [8]. No exponential lower bound was known for CP until recently: Pudlák proved an exponential lower bound by showing that CP admits feasible monotone interpolation property [11], [12]. It requires exponential size proofs in CP for $k\text{-}Test(n)$ (under an adequate translation). This technique has been applied for other propositional calculi, including polynomial calculus, Hilbert's Nullstellensatz, and generalized cutting plane system to show that they do not have polynomial-size proofs for $k\text{-}Test(n)$.

Corollary 12. *Any propositional calculus which admits feasible monotone interpolation does not p-simulate GCNF+permutation. More specifically, CP and generalized CP do not p-simulate GCNF+permutation.*

5 Open problems

Corollary 11 and Corollary 12 suggest that GCNF+permutation may be more efficient than resolution, cut-free LK and many algebraic systems, which are popular systems for automated theorem proving: there is no doubt that we can obtain a much faster theorem prover by implementing GCNF+permutation. We implemented GCNF in tree form [4]. Amazingly, even tree form GCNF (without permutation) already has the ability to produce a proof of PHP_n of size $O(n!)$

automatically [4], and it is as short as the best known DAG resolution refutation of PHP_n. Our next goal is to design an algorithm (deterministic or probablistic) to find suitable permutations without the exhaustive search.

There are many open problems on the relative efficiency of this system.

- Does GCNF+permutation p-simulates resolution?
- Does GCNF+permutation p-simulates CP?

Note that is a well-known open problem whether or not GCNF (or cut-free LK in DAG) p-simulates resolution. We conjecture that

- GCNF+permutation does not p-simulate Frege system.

In [3], we showed that Frege system p-simulates GCNF+renaming if and only if Frege system p-simulates extended Frege system. However, it may be possible that Frege p-simulates GCNF+permutation, since permutation rule is a very restricted form of renaming rule.

We conjecture that

- Frege+permutation does not p-simulate extended Frege.

It is also an open problem whether or not GCNF+permutation admits feasible nonmonotone interpolation.

References

1. M.Ajtai, "The complexity of the pigeonhole principle", *29th Annual Symposium on the Foundations of Computer Science* (1988), 346-55.
2. N.H. Arai, "Tractability of cut-free Gentzen type propositional calculus with permutation inference", *Theoretical Computer Science*, Vol.170 (1996), 129-144.
3. N.H. Arai, "Tractability of cut-free Gentzen type propositional calculus with permutation inference II", submitted for publication.
4. N.H. Arai and R. Masukawa, "An implementation of cut-free Gentzen type propositional calculus", manuscript.
5. P. Beame, R. Impagliazzo, J. Krajíček, T. Pitassi and P. Pudlák, "Lower bounds on Hilbert's Nullstellensatz and propositional proofs", *35th Annual Symposium on the Foundations of Computer Science* (1994), 794-806.
6. M. Bonet, T. Pitassi and R. Raz, "Lower bounds for cutting planes proofs with small coefficients", *Proc. ACM Symp. Theory of Computing* (1995) 575-584.
7. R. Boppana and M. Sipser, "The complexity of finite functions", in *Handbook of Theoretical Computer Science, Volume A: Algorithms and Complexity*, ed. J. van Leeuwen, MIT Press/Elsevier (1990) 757-804.
8. S. Buss and P. Clote,"Cutting planes, connectivity, and threshold logic", *Archive for Mathematical Logic*, Vol.35 (1996) 33-62.
9. A. Haken, "The intractability of resolution", *Theoretical Computer Science*, Vol.39 (1985), 297-308.
10. D. Mundici, "A lower bound for the complexity of Craig's interpolants in sentential logic", *Archiv fur Math. Logik*, Vol.23 (1983) 27-36.
11. P. Pudlák,"Lower bounds for resolution and cutting planes proofs and monotone computations", *J. Symbolic Logic*, Vol.62 (1997) 981-998.

12. P. Pudlák and J. Sgall, "Algebraic models of computation and interpolation for algebraic proof systems", submitted.
13. A. Razborov, "Lower bounds on the monotone complexity of some Boolean functions", *Dokl. Akad. Nauk. SSSR*, Vol.281 (1985) 798–801.
14. A. Razborov, "Lower bounds for the polynomial calculus", manuscript.

Permuting Mechanisms and Closed Classes of Permutations

Michael D. Atkinson[1] and Robert Beals[2]

[1] School of Mathematical and Computational Sciences, North Haugh,
St Andrews, Fife, KY16 9SS, UK. mda@dcs.st-and.ac.uk
[2] Department of Mathematics, University of Arizona, Tucson, AZ 85721, USA.
beals@math.arizona.edu

Abstract. The classes of permutations produced by certain abstract permuting mechanisms are considered. Under a natural condition on the mechanism these classes are closed under a partial order called involvement. Serial and parallel composition of the permuting mechanisms induce corresponding compositions of the associated closed classes of permutations. A significant question is whether a closed class can be characterised by a finite set of excluded permutations. The extent to which this property is preserved by serial and parallel composition is investigated. Closed classes in which the set of permutations of each degree form a group are characterised.

1 Introduction

A *permuting mechanism* M is a device that accepts any finite input sequence of objects (normally denoted by $1, 2, \ldots$) and produces a permutation of these objects. We let perm(M) denote the set of possible permutations that M might produce.

Examples.

1. A *riffle shuffler* divides the input sequence into two segments $1, 2, \ldots, m$ and $m+1, \ldots, n$ and then interleaves them in any way. Thus perm$(M) = \{1, 12, 21, 123, 132, 213, 231, 312, 1234, 1243, 1342, \ldots \text{etc.}\}$
2. A *stack* receives members of the input sequence and outputs them under a last-in-first-out discipline
3. A *transportation network* [2] is any finite directed graph with a node to represent the input queue and a node to represent the output queue. The other nodes can each hold one of the input objects and the objects are moved around the graph until they emerge at the output node.

All of these examples are *hereditary* in that a permutation effected on any subsequence of the input is a permutation that could be produced if the subsequence was the entire input sequence. For example, in any riffle shuffle of $1, 2, \ldots, n$ every subsequence of the input is permuted as though by a riffle shuffle.

Fig. 1. A permuting mechanism.

We say that two sequences $[\alpha_1, \ldots, \alpha_n]$ and $[\beta_1, \ldots, \beta_n]$ are *order isomorphic* if $\alpha_i < \alpha_j$ precisely when $\beta_i < \beta_j$. Then, for two permutations π, σ, we define $\pi \preceq \sigma$, and say that π is *involved* in σ if σ has a subsequence which is order isomorphic to π. A *closed* class X of permutations is a class with the property $\tau \in X$ and $\sigma \preceq \tau$ implies $\sigma \in X$. Clearly, if M is an hereditary permuting mechanism then $\text{perm}(M)$ is closed. On the other hand, given a closed class X, we can simply define a black box permuting mechanism M for which $\text{perm}(M) = X$ and then M will be hereditary.

The permuting mechanisms which arise in practice generally process inputs of all lengths and so give rise to infinite closed classes. The two simplest infinite closed classes are I, the set of all identity permutations $[1, 2, \ldots, n]$, and R, the set of all reversed identities $[n, n-1, \ldots, 1]$.

Lemma 1. *Every infinite closed class contains I or R.*

Proof. If the closed class X contains neither I nor R then, for some fixed r, s, X contains neither $[1, 2, \ldots, r]$ nor $[s, s-1, \ldots, 1]$. By the closure property the permutations of X will have neither an increasing subsequence of length r nor a decreasing subsequence of length s. A theorem of Erdös and Szekeres [4] then implies that all permutations in X have length at most $(r-1)(s-1)$.

Because closed classes are generally infinite it is useful to have some finite description of them. In certain cases this can be achieved by using a set of permutations called the basis. The *basis* of a closed class X is the set of permutations σ minimal with respect to the property $\sigma \notin X$. The basis of a closed class is important because of the following lemma, a direct consequence of the definition, which shows that a closed class is determined by its basis.

Lemma 2. *If X is a closed class with basis B then $\sigma \in X$ if and only if $\beta \not\preceq \sigma$ for all $\beta \in B$.*

Notice that the basis of a closed class necessarily consists of permutations incomparable with respect to $\preceq$. Also, in view of the lemma, every set of incomparable permutations is the basis of some closed class. In [6, 7] examples of infinite sets of incomparable permutations were given. It follows immediately that there are uncountably many closed classes of which but countably many have a finite basis. It is therefore a significant question to distinguish the finitely based closed classes from those which are not finitely based.

This paper is concerned with three constructions which correspond to combining permuting mechanisms in series and in parallel, and endowing them with feedback. In particular we study how they respect the finite basis property.

Fig. 2. Serial composition of permuting mechanisms.

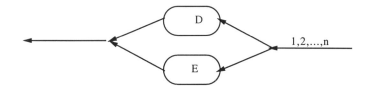

Fig. 3. Parallel composition of permuting mechanisms.

Let D, E be hereditary permuting mechanisms with associated closed classes X, Y. The serial composition of D with E is shown in Figure 2 and it is clear that the corresponding class of permutations is the serial composition of X and Y defined as $XY = \{\theta \mid \theta = \sigma\tau, \sigma \in X, \tau \in Y\}$

The parallel composition of two hereditary permuting mechanisms D, E with associated closed classes X, Y is shown in Figure 3. This mechanism begins by partitioning the input into two subsequences which are then fed into D and E whose outputs are then merged. We denote the set of permutations so obtained by $X \sqcup Y$, the parallel composition.

Intuitively it is clear that the serial and parallel compositions of hereditary permuting mechanisms is also hereditary. A formal proof of this is given in [1].

One might hope that the serial composition of two finitely based closed classes was also finitely based but this is false in general [1]. Indeed, very little is known about when a serial composition is finitely based; for example it is not even known whether the class corresponding to two stacks in series is finitely based. In Section 2 we give some special cases of finitely based serial compositions. We do not know whether the parallel composition of finitely based classes is always finitely based but it seems unlikely. In Section 3 we give some finitely based parallel compositions.

Suppose that M is a permuting mechanism and we 'enhance' it by allowing the output sequence to be channelled back into the input as many times as we like (see Figure 4) and call the new permuting mechanism M^*. Let $X = \text{perm}(M)$. Then M^* can produce all the permutations in X (by disregarding the feedback option), all the permutations of X^2 by feeding the output of M once more through M, and all other powers X^k by $k - 1$ feedbacks. Therefore $X^* = \text{perm}(M^*) = X \cup X^2 \cup X^3 \ldots$ and this is closed. Clearly $X^* \cap S_n$ is a group for all n. In Section 4 we give a complete characterisation of those closed classes Z with the property that $Z \cap S_n$ is a group for all n. We shall see that the condition is very restrictive.

We end this section with some notation that will be useful elsewhere in the paper.

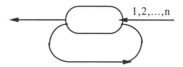

Fig. 4. A permuting mechanism with feedback.

If σ is a permutation of $\{1, 2, \ldots, n\}$ and $1 \leq i \leq n$ then we let $\sigma - i$ denote the permutation obtained from σ by deleting the ith symbol (image of i) and relabelling to obtain a permutation. Let $C(i, n)$ be the cycle $(i, i+1, \ldots, n)$. Then, by a slight abuse of notation in which we identify $\sigma - i$ with the permutation of degree n that fixes n and otherwise acts as $\sigma - i$, it is easy to see that $\sigma - i = C(i, n)\sigma C(i^\sigma, n)^{-1}$. Direct calculation then confirms that:

Lemma 3. $(\sigma - i)(\tau - i^\sigma) = \sigma\tau - i$.

2 Serial composition

Our first example of a finitely based serial composition class features the closed class C whose permutations of length n are $[1, 2, \ldots, n]$ and $[n, 1, 2, \ldots, n-1]$ (the identity permutation together with an n-cycle). C is a finitely based class with basis $\{[3, 2, 1], [2, 3, 1], [1, 3, 2], [2, 1, 3]\}$.

Theorem 4. *If D is any finitely based closed class then CD is finitely based.*

Proof. Let $\pi = p_1, p_2 \ldots, p_n$ be a permutation of degree n. Then $\pi \notin CD$ if and only if $p_1, p_2, \ldots, p_n \notin D$ and $p_2, p_3, \ldots, p_n, p_1 \notin D$. We choose a smallest such π (a basis element of CD). Then there exists a subsequence σ of $p_1, p_2, \ldots, p_n$ order isomorphic to a basis element of D and a subsequence τ of $p_2, p_3, \ldots, p_n, p_1$ also order isomorphic to a basis element of D.

If σ were contained in $p_2, \ldots, p_n$ then it would be contained in both $p_1\sigma$ and σp_1 and $p_1\sigma$ (a subsequence of π) would also not belong to CD. Then, by the minimality of π, we would have $\pi = p_1\sigma$. Since D is finitely based there are only a finite number of permutations π of this type. A similar argument applies if τ is contained in $p_2, \ldots, p_n$.

We may now assume that $\sigma = p_1\lambda$ and $\tau = \mu p_1$ for subsequences λ and μ of $p_2, \ldots, p_n$. Now let ω be the union of λ and μ. It follows that σ is a subsequence of $p_1\omega$ and that τ is a subsequence of ωp_1. Thus neither of $p_1\omega$ and ωp_1 are in D and, by the minimality of π again, $\pi = p_1\omega$ and so is of bounded length.

The class C has only two permutation of each length; it is a *bounded* class. One of the simplest unbounded closed classes is the *Coxeter* class in which the permutations of degree n are the identity permutation and the $n - 1$ transpositions $\xi_i = (i, i+1)$. This is a finitely based class and the basis is the set $\{[3, 2, 1], [3, 1, 2], [2, 3, 1], [2, 1, 4, 3]\}$.

Theorem 5. *Let X be the Coxeter class and Y any finitely based closed class. Then XY is finitely based.*

Proof. Let $\pi = p_1, p_2, \ldots, p_n$ be a permutation of degree n. Then $\pi \notin XY$ if and only if $\xi_i \pi \notin Y$ for all $i = 1, \ldots, n-1$. We choose π to be minimal with respect to this property, that is, π is a basis element of XY. Certainly $\pi \notin Y$ and so it has some subsequence $\beta = b_1 b_2 \ldots b_k$ which is order isomorphic to a basis element of Y; and, as Y is finitely based, k is bounded independently of n. Since β is a subsequence of π, π has a segment of the form $\beta_1 \gamma_1 \beta_2 \ldots \beta_{r-1} \gamma_{r-1} \beta_r$ where each β_i, γ_i is a non-empty segment of π, $\beta_1 \beta_2 \ldots \beta_r = \beta$, and $r \leq k$. For each $i = 1, 2, \ldots, r-1$ let g_i be an element of γ_i.

Consider any $b_j b_{j+1}$ contained in some β_i. The interchange of b_j and b_{j+1} in π results in a permutation that is not in Y and so there must be subsequences λ_j, μ_j of π such that $\lambda_j b_j b_{j+1} \mu_j$ is a subsequence of π and $\lambda_j b_{j+1} b_j \mu_j$ is order isomorphic to a basis element of Y (and so of length bounded independently of n).

Now define θ to be the union of all β_i, g_i and all λ_j, μ_j. This subsequence of π is not order isomorphic to any permutation of XY. To see this we consider the effect of interchanging any two adjacent symbols of θ. If the adjacent symbols are not adjacent in some β_i the result will have β as a subsequence. On the other hand, if the adjacent symbols are of the form b_j, b_{j+1} within some β_i, the result will have $\lambda_j b_{j+1} b_j \mu_j$ as a subsequence. In each case, the result has a subsequence order isomorphic to a basis element of Y. Thus θ is order isomorphic to a permutation which does not lie in XY and, by the minimality of π, we have $\pi = \theta$. However, θ has been defined as a union of a bounded number of sequences of bounded length; therefore there are only a finite number of possibilities for π.

3 Parallel composition

Parallel composition appears to be a slightly easier notion than serial composition. For example, it is commutative while serial composition is not. Moreover, while we have not been able to find a method for producing explicit elements not in XY from bases of X and Y we have the following argument for $X \sqcup Y$.

If σ and τ are permutations of m and n elements respectively define $\sigma \otimes \tau$ as a permutation $\rho = r_1, \ldots, r_{mn}$ of degree mn where each of the n segments $[r_{mi+1}, \ldots, r_{mi+m}], 0 \leq i < n$ is a rearrangement of $\{mi+1, mi+2, \ldots, mi+m\}$ order isomorphic to σ and one (and so every) sequence formed by choosing a representative from among each $[r_{mi+1}, \ldots, r_{mi+m}]$ is order isomorphic to τ. It is then straightforward to confirm:

Proposition 6. *Suppose that $\sigma \notin X$ and $\tau \notin Y$ then $\sigma \otimes \tau \notin X \sqcup Y$.*

Simple examples of the parallel composition are $I \sqcup I$ which is the set of all permutations with no decreasing subsequence of length 3, and $I \sqcup R$ which is the set of permutations which are the merge of an increasing sequence and a decreasing sequence; the latter class has recently been studied in [3,5].

Lemma 7. Let F_k denote the (finite) closed class of all permutations of length at most k and let X be any closed class. Then

1. $X \sqcup F_{m+n} = X \sqcup F_m \sqcup F_n$
2. If X is finitely based so is $X \sqcup F_n$ for all n.

Proof. The first part follows directly from the definitions. For the second part it is enough to prove that $X \sqcup F_1$ is finitely based since $X \sqcup F_n = (X \sqcup F_{n-1}) \sqcup F_1$. Let π be a basis element of $X \sqcup F_1$ and let $\mathcal{B}$ denote the set of all subsequences of π that are order isomorphic to basis elements of X. Since X is finitely based, all subsequences in $\mathcal{B}$ have length at most t for some fixed t.

Since $\pi \notin X \sqcup F_1$, $\pi - i \notin X$ for all $i = 1, \ldots, \text{length}(\pi)$, and so each $\pi - i$ contains some subsequence in $\mathcal{B}$. In particular, it follows that the intersection of all the sequences in $\mathcal{B}$ is empty since, if π_j were in this intersection, $\pi - j$ could not have a subsequence in $\mathcal{B}$.

It follows that we may choose subsequences $\beta_1, \beta_2, \ldots, \beta_t \in \mathcal{B}$ such that, for each s, $\cap_{i=1}^{s+1} \beta_i$ is properly contained in $\cap_{i=1}^{s} \beta_i$ and therefore $\cap_{i=1}^{t} \beta_i$ is empty. Put $\beta = \cup_{i=1}^{t} \beta_i$.

Then β clearly has the property that every $\beta - i$ has at least one of $\beta_1, \beta_2, \ldots, \beta_t$ as a subsequence. Therefore, if $\beta^\star$ denotes the permutation order isomorphic to β then $\beta^\star - i \notin X$ for all i and so $\beta^\star \notin X \sqcup F_1$. But $\beta^\star \preceq \pi$ and, as π is a basis element of $X \sqcup F_1$, $\beta^\star = \pi$ and so π has length at most t^2. Since there are only a finite number of such permutations π the result is proved.

4 Feedback

In this section we investigate when a closed class X can have the property that $X(n) = X \cap S_n$ is a group for all n. We call such a class a *group class*. Group classes arise out of permuting mechanisms with feedback. We shall see that the sequence of groups $X(n)$ eventually settles down to one of the following families of groups:

1. For some fixed k and ℓ, $X(n) = S_k \times S_\ell$, acting as S_k on $\{1, \ldots, k\}$, fixing each point in $\{k+1, \ldots, n-\ell\}$, and acting as S_ℓ on $\{n-\ell+1, \ldots, n\}$.
2. The cyclic groups: $X(n) = \langle [2, 3, 4, \ldots, n, 1] \rangle$.
3. The full symmetric group: $X(n) = S_n$.
4. The group generated by any of the above (with $k = \ell$ in example 1) together with the reversals R.

Our results also give information about the sort of groups $X(n)$ which can arise before convergence to one of the above types.

4.1 Intransitive groups

For each n we define k_n in terms of the orbit $1^{X(n)}$ of $X(n)$: $k_n + 1$ is the smallest point not in this orbit. In a similar way we define ℓ_n using the orbit of $X(n)$ containing n: $n - \ell_n$ is the largest point not in this orbit.

We first note that (k_n) is a non-increasing sequence. For, if $1 \leq j \leq k_n$, then $1^g = j$ for some $g \in X(n)$. Then, as $(k_n + 1)^g > k_n$, $1^{g-k_n} = j$. This shows that $k_{n-1} \geq k_n$. A similar argument shows that (ℓ_n) is non-increasing. Since the two sequences $(k_n), (\ell_n)$ are non-increasing and bounded below they have limits k, ℓ. Therefore we have:

Lemma 8. *There exist constants k, ℓ, N such that, for all $n \geq N$, $\{1, 2, \ldots, k\} \subseteq 1^{X(n)}$, $k + 1 \notin 1^{X(n)}$ and $\{n - \ell + 1, \ldots, n\} \subseteq n^{X(n)}$, $n - \ell \notin n^{X(n)}$.*

Lemma 9. *Let $t - 1, t$ be in different orbits of $X(n-1)$ and let $g \in X(n)$. Then either*

1. *$t^g = t$ and g preserves $\{1, \ldots, t-1\}$ and $\{t+1, \ldots, n\}$ or*
2. *$t^g = n - t + 1$, $(n - t + 1)^g = t$ and g interchanges $\{1, \ldots, t-1\}$ and $\{n - t + 2, \ldots, n\}$.*

Proof. Let $i, j \in \{1, \ldots, n\}$ be such that $i < t < j$. Then, for any $g \in X(n)$, $(t-1)^{g-i} \neq t^{g-j}$. This can only occur if $i^g < t^g < j^g$ or $j^g < t^g < i^g$. In other words, the triple i, t, j is monotonically mapped by any $g \in X(n)$. Since the choices of i and j were independent, it follows that any $g \in X(n)$ either preserves the sets $\{1, \ldots, t-1\}$ and $\{t+1, \ldots, n\}$ or maps $\{1, \ldots, t-1\}$ onto $\{t^g + 1, \ldots, n\}$. In the first case, $t^g = t$ and in the second case $t^g = n - t + 1$. However, in the second case we can consider $h = g^2$ and conclude that $t^h = t$ and $\{1, \ldots, t-1\}^h = \{1, \ldots, t-1\}$ which completes the proof.

Lemma 10. *There exists a constant $M > N$ such that either*

1. *For all $n \geq M$, $1^{X(n)} = \{1, \ldots, k\}$ and $n^{X(n)} = \{n - \ell + 1, \ldots, n\}$ or*
2. *$k = \ell$ and for all $n \geq M$ $1^{X(n)} = \{1, \ldots, k, n - k + 1, \ldots, n\}$.*

Proof. Let $n \geq N$ and let $g \in X(n)$. By Lemma 8 and Lemma 9 we have either

(i) $(k+1)^g = k+1$ and $\{1, \ldots, k\}^g = \{1, \ldots, k\}$ or
(ii) $(k+1)^g = n - k$ and $\{1, \ldots, k\}^g = \{n - k + 1, \ldots, n\}$

Suppose that (i) holds for all $g \in X(n)$ (so that $1^{X(n)} = \{1, \ldots, k\}$). Then k and $k+1$ are in different orbits of $X(n)$ and by Lemma 9 again we have, for all $h \in X(n+1)$, either $(k+1)^h = k+1$ or $(k+1)^h = n + 1 - (k+1) + 1$ and $(n+1)^h < n + 1 - k$. But the last alternative is impossible for it would imply that the element $h - (n+1) \in X(n)$ maps $k+1$ to $n - k$, a contradiction.

This proves that either $\{1, \ldots, k\}$ is the entire orbit of 1 under $X(n)$ for all n from some point on, or that this set is never the entire orbit for $n \geq N$. In the former case we can consider $n - \ell$ and $n - \ell + 1$, which are not in the same orbit of $X(n)$, and apply Lemma 9. This gives the first alternative of the lemma. In the latter case, the element g satisfying (ii) above always exists and the second alternative must hold.

Lemma 11. *If the first case of Lemma 10 holds then there exists K such that, for all $n > K$, $X(n)$ fixes all points in $k+1, \ldots, n-\ell$. If the second case holds then there exists K such that for all $n > K$ and all $g \in X(n)$ either g fixes every point in $k+1, \ldots, n-k$ or g maps every such point s to $n-s+1$.*

Proof. Suppose that the first alternative of Lemma 10 holds. Suppose that $n \geq M$. Then k and $k+1$ are in different orbits of $X(n)$ as are $n-\ell$ and $n-\ell+1$. Hence, by Lemma 9, both $k+1$ and $n-\ell+1$ are fixed in $X(n+1)$. But then $k, k+1, k+2$ are in different orbits of $X(n+1)$ and so $k+1$ and $k+2$ are fixed in $X(n+2)$; similarly, $n-\ell+1$ and $n-\ell+2$ are fixed in $X(n+2)$. Continuing this argument we see that, whenever $n > 2M$, all of the points in $k+1, \ldots, n-\ell$ are fixed in $X(n)$.

If the second alternative of Lemma 10 holds the same argument shows that if $g \in X(n)$ and $k+1 \leq s \leq n-k$ then, provided $n > 2M$, $s^g = s$ or $s^g = n-s+1$. However, Lemma 9 proves that $s^g = s$ precisely when $\{1, 2, \ldots, k\}$ is preserved by g. Thus $s^g = s$ for all $k+1 \leq s \leq n-k$ or $s^g = n-s+1$ for all such s.

Theorem 12. *Let X be a group class in which not every $X(n)$ is transitive. Then one of the following holds:*

1. *$X(n) = S_k \times S_\ell$, acting as S_k on $\{1, 2, \ldots, k\}$ and as S_ℓ on $\{n-\ell+1, n-\ell+2, \ldots, n\}$, and fixing the remaining points, for all $n \geq n_0$.*
2. *$X(n)$ is generated by $R(n)$ and the group $S_k \times S_k$ acting as in the previous case (with $k = \ell$) for all $n \geq n_0$.*

Proof. Consider values of n large enough that the conclusions of Lemma 11 hold. Suppose first that $X(n)$ fixes every point from $k+1$ to $n-\ell$. In $X(n+1)$ there are permutations g_i that map the symbol k to each of $1, 2, \ldots, k$ in turn (and fix $k+1$). But then the permutations $g_i - 1 \in X(n)$ fix k and map $k-1$ to each of $1, \ldots, k-1$. Thus $X(n)$ is doubly transitive on its orbit $\{1, \ldots, k\}$. A similar argument with the groups $X(n+2), \ldots, X(n+k-1)$ establishes the k-transitivity of $X(n)$ on $\{1, \ldots, k\}$ so it acts on this orbit as the full symmetric group. In the same way $X(n)$ also acts on the orbit $\{n-\ell+1, \ldots, n\}$ as the full symmetric group. Moreover, as $X(n+\ell) - \{n+1, \ldots, n+\ell\} \subseteq X(n)$ fixes $n-\ell+1, \ldots, n$ and acts as S_k on $\{1, \ldots, k\}$ we have $X(n) = S_k \times S_\ell$.

Now suppose that the second case of Lemma 11 holds. Then $X(n)$ has a subgroup $Y(n)$ of index 2 fixing all of $k+1, \ldots, n-k+1$ and the above arguments prove that $Y(n) = S_k \times S_k$ acting in the natural way on $\{1, \ldots, k, n-k+1, \ldots, n\}$. By Lemma 11 the single permutation of $R(n)$ lies in $X(n) \setminus Y(n)$ and the proof is complete.

4.2 Transitive groups

We denote by Z_n (D_n respectively) the permutation group of degree n generated by the n–cycles $(1, 2, \ldots, n)$ $((1, 2, \ldots, n)$ and $(n, n-1, \ldots, 2, 1))$ respectively). We call these groups the *natural* cyclic and dihedral groups of degree n.

Also, we say that the integers y, z are *consecutive modulo n* if $\{y, z\}$ is a consecutive pair of elements in the list $1, 2, \ldots, n, 1$. Then we observe

Lemma 13. Let G be a transitive group of degree n. Then G is the natural cyclic or dihedral group if and only if for all $g \in G$ and all $1 \leq x < n$, x^g and $(x+1)^g$ are consecutive modulo n.

Lemma 14. A permutation group G of degree n that contains cycles $g = (1, 2, \ldots, n)$ and $h = (1, 2, \ldots, m)$ with $m < n$ contains the alternating group of degree n.

Proof. The group is obviously transitive. Indeed, since it contains all cycles of the form $(i, i+1, \ldots, i+m-1)$ (conjugates of h under powers of g) it is easy to see that the stabilizer of the point n is transitive on the remaining points—that is, G is doubly transitive. The commutator $[g, h]$ is the 3–cycle $(1, 2, m+1)$. But a doubly transitive group containing a 3–cycle necessarily contains A_n.

Lemma 15. Suppose that $g \in X(n)$ and that $1 \leq x \leq n-1$. Let $u = x^g$, $v = (x+1)^g$ and let $h = (g-x)^{-1}(g-(x+1)) \in X(n-1)$. Then h is the cycle $(y, y+1, \ldots, z-1)$ if $y < z$ and the cycle $(z, z+1, \ldots, y-1)$ if $y > z$. In addition, if y and z are not consecutive modulo n then $X(n-2)$ contains a transposition of the form $(t, t+1)$.

Proof. The first statement follows directly from the definitions. For the second part, note that both h and h^{-1} lie in $X(n-1)$ since $X(n-1)$ is a group. Without loss in generality assume that $h = (y, y+1, \ldots, z-1)$. Since y and z are not consecutive modulo n, h is not the identity nor the permutation $(1, 2, \ldots, n-1)$. If $y \neq 1$ then $(y-1)^h = y-1$ and $y^h = y+1$ so, by part 1, $X(n-2)$ contains $(h-(y-1))^{-1}(h-y) = (y-1, y)$. On the other hand, if $z \neq n$ then $z^{h^{-1}} = z$ and $(z-1)^{h^{-1}} = z-2$ so that, again by part 1, $X(n-2)$ contains $(z-2, z-1)$.

Lemma 16. If $X(n)$ is transitive for all n then each $X(n)$ contains the cycle $(1, 2, \ldots, n)$.

Proof. Suppose, for a contradiction, that there is some integer n_1 for which $X(n_1)$ does not contain $(1, 2, \ldots, n_1)$. Let p be any prime larger than n_1 and put $q = p + 2$. Then $X(q)$ also does not contain a cycle $(1, 2, \ldots, q)$ and so it is not the natural cyclic or dihedral group of degree q. Hence, by Lemma 13 $X(q)$ has a permutation g for which there are two points $x, x+1$ with $u = x^g$ and $v = (x+1)^g$ not consecutive modulo q.

However, Lemma 15, now shows that $X(p) = X(q-2)$ contains a transposition of the form $(t, t+1)$. But a transitive group of prime degree containing a transposition is necessarily the full symmetric group; therefore it contains $(1, 2, \ldots, p)$. Therefore, $X(n_1)$ must contain $(1, 2, \ldots, n_1)$, a contradiction.

A permutation group G of degree n with an n–cycle $(1, 2, \ldots, n)$ is said to be *peculiar* if

$$H = \langle g - x \mid g \in G, 1 \leq x \leq n \rangle \neq S_{n-1} \ .$$

The natural cyclic and dihedral groups of degree n are both peculiar (the group H being cyclic or dihedral of degree $n-1$).

Lemma 17. *If G is a peculiar group of degree n and not cyclic or dihedral then n is even, G has a block system {even points}, {odd points} and the stabilizer of a block is the set of even permutations of G. Conversely, any group with a cycle $(1,2,\ldots,n)$ and satisfying these conditions is peculiar. For peculiar groups which are not cyclic or dihedral the group H is the alternating group A_{n-1}.*

Proof. As G contains $c = (1,2,\ldots,n)$ H must contain $c - n = (1,2,\ldots,n-1)$. Now let $g \in G$ and $1 \leq x < n$ and put $u = x^g, v = (x+1)^g$. Let $h(g,x) = (g-x)^{-1}(g-(x+1))$ be a cycle of length $m(g,x) = |u - v|$ (see Lemma 15). Since G is neither cyclic nor dihedral, at least one of these cycle lengths must lie strictly between 1 and $n-1$; therefore, by Lemma 14, H contains A_{n-1}. By hypothesis, however, H is not the full symmetric group and therefore both $n-1$ and all $m(g,x)$ must be odd. The latter condition implies that the elements of $1^g, 2^g, \ldots, n^g$ are alternately even and odd. Hence each element g maps odd points to odd points or maps odd points to even points, giving the required block system.

Since $h(g,x)$ is even, $g - x$ and $g - (x+1)$ have the same parity. This parity is also the parity of $g - n$ and is easily seen to be the parity of g minus the parity of n^g. But this means that g has even parity if and only if it fixes the set of even points.

Theorem 18. *Let X be a group class in which every $X(n)$ is transitive. Then, the sequence of groups $X(n)$ begins with a number of symmetric groups (possibly none or an infinite number), then has at most one alternating group which must then be followed by a peculiar group or has at most one group that properly contains Z_n but is not dihedral, then has a number of natural dihedral groups (possibly none or an infinite number), and finally has a number of natural cyclic groups (none or an infinite number).*

Proof. Since X is closed

$$\langle g - x \mid g \in X(n), 1 \leq x \leq n \rangle \subseteq X(n-1) \text{ for all } n$$

Suppose that not every $X(n)$ is the symmetric group S_n and let $X(m-1)$ be the first that is not. Then $X(m)$ is peculiar and either

1. $X(m)$ is an imprimitive group of the type appearing in Lemma 17 and $X(m-1)$ is the alternating group, or
2. $X(m)$ is the natural cyclic or dihedral group.

In either case, by Lemma 17 again, $X(n)$ is the natural cyclic or natural dihedral group for all $n \geq m+1$. Finally we note that, if $X(n)$ is the natural cyclic group then $X(n+1)$ is also the natural cyclic group.

References

1. M.D. Atkinson, R. Beals: Finiteness conditions on closed classes of permutations, in preparation.

2. M.D. Atkinson, M.J. Livesey, D. Tulley: Permutations generated by token passing in graphs, Theoretical Computer Science 178 (1997), 103-118.
3. M. D. Atkinson: Permutations which are the union of an increasing and a decreasing subsequence, Electronic J. Combinatorics 5 (1998), Paper R6 (13 pp.).
4. D. I. A. Cohen: *Basic Techniques of Combinatorial Theory*, John Wiley & Sons, New York, 1978.
5. A.E. Kézdy, H.S. Snevily, C. Wang: Partitioning permutations into increasing and decreasing subsequences, J. Combinatorial Theory A 74 (1996), 353-359.
6. V.R. Pratt: Computing permutations with double-ended queues, parallel stacks and parallel queues, Proc. ACM Symp. Theory of Computing 5 (1973), 268-277
7. R.E. Tarjan: Sorting using networks of queues and stacks, Journal of the ACM 19 (1972), 341-346.

Process Algebra with Five-Valued Conditions

Jan A. Bergstra[1,2] and Alban Ponse[1]

[1] University of Amsterdam, Programming Research Group, Kruislaan 403,
NL-1098 SJ Amsterdam, The Netherlands. alban@wins.uva.nl
[2] Utrecht University, Department of Philosophy, P.O. Box 80126,
NL-3508 TC Utrecht, The Netherlands.
http://www.phil.uu.nl/eng/home.html

Abstract. We propose a five-valued logic that can be motivated from an algorithmic point of view and from a logical perspective. This logic is combined with process algebra. For process algebra with five-valued logic we present an operational semantics in SOS-style and a completeness result. Finally, we discuss some generalizations.

1 Introduction

Assume P is some simple program or algorithm. Then the initial behaviour of

if ϕ then P else P

depends on evaluation of the condition ϕ: either it yields an immediate error, or it starts performing P, or it diverges in evaluation of ϕ. Note that the second possibility only requires that ϕ is either true or false. The following three non-classical truth values accommodate these intuitions:

Meaningless. Typical examples are errors that are detectable during execution such as a type-clash or division by zero.

Choice or *undetermined.* A typical example is *alternative composition,* i.e. in if ϕ then Q else P either P or Q is executed.

Divergent or *undefined.* Typically, evaluation of a partial predicate can diverge.

We describe a propositional logic that incorporates these three non-classical truth values and discuss its combination with process algebra. Here process algebra is used as a vehicle to specify and analyze concurrent algorithms: a (closed) process term is considered an algebraic notation for an algorithm. We shall use an if_then_else_ construct in which the condition ranges over five-valued propositions. We end the paper with some generalizations and conclusions.

Acknowledgement. We thank Bas Luttik and Piet Rodenburg for discussion, proof reading, and verifying a completeness result (on $\mathbb{K}_4$).

2 Five-valued logic

First we shortly consider the incorporation of each of the previously mentioned non-classical truth values in classical two-valued logic. In [8] it is established that there are only two three-valued logics that satisfy the (nice) algebraic properties defined by the axioms in Table 1, where T stands for "true", F for "false", and $*$ denotes a "third truth value":

Kleene's three-valued logic $\mathbb{K}_3$. This three-valued logic, which we call $\mathbb{K}_3$, is introduced in [18] to model propositional combination of partial predicates. $\mathbb{K}_3$ is defined by the following truth tables:

x	$\neg x$
T	F
F	T
$*$	$*$

$\wedge$	T	F	$*$
T	T	F	$*$
F	F	F	F
$*$	$*$	F	$*$

$\vee$	T	F	$*$
T	T	T	T
F	T	F	$*$
$*$	T	$*$	$*$

and is characterized (cf. [8]) by the axioms in Table 1 and the absorption axiom

$$(\text{Abs}) \quad x \vee (x \wedge y) = x.$$

Strict three-valued logic $\mathbb{S}_3$. This three-valued is due to Bochvar [14]. Citing [8]: "Here, on the theory that one bad apple spoils the barrel, an expression has value $*$ as soon as it has a component with that value". $\mathbb{S}_3$ is defined by

x	$\neg x$
T	F
F	T
$*$	$*$

$\wedge$	T	F	$*$
T	T	F	$*$
F	F	F	$*$
$*$	$*$	$*$	$*$

$\vee$	T	F	$*$
T	T	T	$*$
F	T	F	$*$
$*$	$*$	$*$	$*$

According to [8], $\mathbb{S}_3$ is characterized by the axioms in Table 1 and axioms

$$(\text{S1}) \quad x \vee (\neg x \wedge y) = x \vee y,$$
$$(\text{S2}) \quad * \wedge x = *.$$

The combination of these logics is studied in [8], which also comprises an account of McCarthy's asymmetric connectives.

Table 1. Axioms for three-valued logic.

(1)	$\neg T = F$	(5)	$x \wedge y = y \wedge x$
(2)	$\neg * = *$	(6)	$x \wedge (y \wedge z) = (x \wedge y) \wedge z$
(3)	$\neg \neg x = x$	(7)	$T \wedge x = x$
(4)	$\neg(x \wedge y) = \neg x \vee \neg y$	(8)	$x \wedge (y \vee z) = (x \wedge y) \vee (x \wedge z)$

We observe that two different intuitions for Kleene's non-classical truth value can be distinguished: *choice* or *undetermined*, further written as C, and *divergent* or *undefined*, denoted by D. Incorporation of both C and D leads to a four-valued logic that we call

$$\mathbb{K}_4$$

and that—as far as we know—has not been studied before. It can be argued that the axioms given for $\mathbb{K}_3$ allow at most two distinct elements that satisfy $\neg * = *$, and with C and D in this role imply the identity

$$C \wedge D = F.$$

Adding this identity and replacing (2) in Table 1 by $\neg C = C$ and $\neg D = D$ yields with axiom (Abs) a complete axiomatization for $\mathbb{K}_4$ [20]. Note that S_3 cannot be generalized in a similar fashion because of axiom (S2). Following [8] we set M, called *meaningless*, for the non-classical truth value occurring in S_3.

Combining S_3 and $\mathbb{K}_4$ yields a five-valued logic with constants in $\mathbb{T}_5 = \{M, C, T, F, D\}$. In order to combine this logic with process algebra we shall add McCarthy's asymmetric connectives and conditional composition, and we shall incorporate *fluents* to represent "deterministic conditions".

Asymmetric connectives. With $\wedge\!\!\!\!\wedge$ we denote McCarthy's left to right conjunction (cf. [21]), adopting the asymmetric notation from [8]. First the left argument is evaluated, and if necessary the right argument. From [8] and the intuitions provided for M, C, and D it follows that

$$c \wedge\!\!\!\!\wedge x = c \text{ for } c \in \{M, F, D\} \text{ and } c \wedge\!\!\!\!\wedge x = c \wedge x \text{ for } c \in \{C, T\}.$$

With $\vee\!\!\!\!\vee$ we denote the dual of $\wedge\!\!\!\!\wedge$, called left-sequential disjunction and defined by $x \vee\!\!\!\!\vee y = \neg(\neg x \wedge\!\!\!\!\wedge \neg y)$. So in accordance with the intuition of sequential evaluation, logics with divergence D or meaningless M are asymmetric with respect to these connectives.

We now list the complete truth tables for $\neg, \wedge$, and $\wedge\!\!\!\!\wedge$:

x	$\neg x$
M	M
C	C
T	F
F	T
D	D

$\wedge$	M	C	T	F	D
M	M	M	M	M	M
C	M	C	C	F	F
T	M	C	T	F	D
F	M	F	F	F	F
D	M	F	D	F	D

$\wedge\!\!\!\!\wedge$	M	C	T	F	D
M	M	M	M	M	M
C	M	C	C	F	F
T	M	C	T	F	D
F	F	F	F	F	F
D	D	D	D	D	D

and we define disjunction $\vee$ as usual: $x \vee y = \neg(\neg x \wedge \neg y)$. We denote the resulting five-valued logic by $\Sigma_5(\neg, \wedge, \wedge\!\!\!\!\wedge)$, or shortly Σ_5. Note that the axioms from Table 1 are valid for Σ_5 (with $*$ ranging over $\{M, C, D\}$) and that $\wedge\!\!\!\!\wedge$ and its dual $\vee\!\!\!\!\vee$ are idempotent and associative. The five truth values in $\mathbb{T}_5$ can be arranged in the following partial ordering, reflecting information order and (argumentwise) monotony of $\wedge$ and $\wedge\!\!\!\!\wedge$:

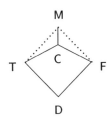

(The outer rhombus represents the original lattice from [8, 13], without C.)

Conditional composition. The expression $x \triangleleft y \triangleright z$, of which the notation stems from [17], denotes if y then x else z. Sequential connectives provide a useful intuition if conditional composition is introduced in the logic:

$$y \underset{\sim}{\wedge} x = x \triangleleft y \triangleright \mathsf{F}.$$

This is plausible because it provides the very underlying intuition of $\underset{\sim}{\wedge}$ (first evaluate y, then, if necessary, x). Similarly, we have $y \overset{\sim}{\vee} x = \mathsf{T} \triangleleft y \triangleright x$. We first define $\_ \triangleleft \_ \triangleright \_$ as a ternary operation:

$$x \triangleleft \mathsf{M} \triangleright y = \mathsf{M}$$
$$x \triangleleft \mathsf{T} \triangleright y = x$$
$$x \triangleleft \mathsf{F} \triangleright y = y$$
$$x \triangleleft \mathsf{D} \triangleright y = \mathsf{D}$$

$\triangleleft \mathsf{C} \triangleright$	M	C	T	F	D
M	M	M	M	M	M
C	M	C	C	C	C
T	M	C	T	C	T
F	M	C	C	F	F
D	M	C	T	F	D

Notice that $x \triangleleft \mathsf{C} \triangleright y$ (as a binary operation) is idempotent, commutative, and associative. This operation can be defined by:

$$x \triangleleft \mathsf{C} \triangleright y = (\mathsf{C} \wedge x) \vee (\mathsf{C} \wedge y) \vee (x \wedge y).$$

Proposition 1. *Conditional composition $x \triangleleft y \triangleright z$ can be defined in Σ_5 by*

$$x \triangleleft y \triangleright z = ((y \vee \mathsf{D}) \underset{\sim}{\wedge} (x \overset{\sim}{\vee} \mathcal{G})) \triangleleft \mathsf{C} \triangleright ((\neg y \vee \mathsf{D}) \underset{\sim}{\wedge} (z \underset{\sim}{\wedge} \mathcal{H})),$$

where $x \triangleleft \mathsf{C} \triangleright y$ is given above, $\mathcal{G} = (y \underset{\sim}{\wedge} x) \vee (\neg y \underset{\sim}{\wedge} z)$, and $\mathcal{H} = (\neg y \overset{\sim}{\vee} x) \wedge (y \overset{\sim}{\vee} z)$.

Fluents. Following McCarthy and Hayes [23], let $f, g, ...$ be names for *fluents*, i.e., objects that in any state (i.e., at each instance of time) may take a deterministic value, thus a value in $\{\mathsf{M}, \mathsf{T}, \mathsf{F}, \mathsf{D}\}$. We write

$$f : \mathsf{DetFluent}$$

to express this, and $f : \mathsf{BoolFluent}$ if fluent f ranges over $\{\mathsf{T}, \mathsf{F}\}$. Fluents are used to model *deterministic conditions*, for example conditions that can occur in an algorithm or a program. Deterministic conditions are further considered in the next section. Let $\mathbb{P}_4$ be a set of fluents of type DetFluent. We write

$$\Sigma_5(\mathbb{P}_4)$$

for the extension of Σ_5 with the fluents in $\mathbb{P}_4$, and we let $\Sigma_5(\mathbb{P}_2)$ denote the extension of Σ_5 with fluents of type BoolFluent in set $\mathbb{P}_2$. In order to equate conditions defined in Σ_5(DetFluent) we use substitution of fluents:

$$[\phi/f]g \stackrel{\triangle}{=} g, \qquad [\phi/f]c \stackrel{\triangle}{=} c \text{ for } c \in \{\mathsf{M},\mathsf{C},\mathsf{T},\mathsf{F},\mathsf{D}\},$$
$$[\phi/f]f \stackrel{\triangle}{=} \phi, \qquad [\phi/f]\neg\psi \stackrel{\triangle}{=} \neg[\phi/f]\psi,$$
$$[\phi/f](\psi_1 \Diamond \psi_2) \stackrel{\triangle}{=} [\phi/f]\psi_1 \Diamond [\phi/f]\psi_2 \text{ for } \Diamond \in \{\wedge, \underline{\wedge}\},$$

and as a proof rule the *excluded fifth rule* (cf. [13]):

$$\frac{\sigma(\phi) = \sigma(\psi) \quad \text{for all } \sigma \in \{[\mathsf{M}/f],[\mathsf{T}/f],[\mathsf{F}/f],[\mathsf{D}/f]\}}{\phi = \psi}.$$

Together with the identities generated by the truth tables this yields a complete evaluation system for equations over $\Sigma_5(\mathbb{P}_4)$. With the associated *excluded third rule* (on substitution of T and F for fluents of type BoolFluent) we find an evaluation system for $\Sigma_5(\mathbb{P}_2)$. We write

$$\Sigma_5(\mathbb{P}_4) \models \phi = \psi$$

if $\phi = \psi$ follows from the system defined above and the truth tables for Σ_5. The identity stated in the following lemma is used later on, and can be easily proved.

Lemma 2. $\Sigma_5(\mathbb{P}_4) \models \phi \vee \mathsf{D} = \phi \underline{\vee} \mathsf{D}$.

3 ACP with five-valued conditions

The axiom system $\mathsf{ACP}(A,\gamma)$ (see e.g., [9, 10, 6]) is parameterized with a set A of constants $a, b, c, ...$ denoting atomic actions (atoms), i.e., processes that are not subject to further division, and that execute in finite time. In $\mathsf{ACP}(A,\gamma)$ there is a constant $\delta \notin A$, denoting the inactive process. We write A_δ for $A \cup \{\delta\}$. The six operations of $\mathsf{ACP}(A,\gamma)$ are

Sequential composition: $X \cdot Y$ denotes the process that performs X, and upon completion of X starts with Y.

Alternative composition: $X + Y$ denotes the process that performs either X or Y.

Merge or *parallel composition:* $X \parallel Y$ denotes the parallel execution of X and Y (including the possibility of synchronization).

Left merge, an auxiliary operator: $X \parallel\!\!\!\_\, Y$ denotes $X \parallel Y$ with the restriction that the first action stems for the left argument X.

Communication merge, an auxiliary operator: $X \mid Y$ denotes $X \parallel Y$ with the restriction that the first action is a synchronization of both X and Y.

Encapsulation: $\partial_H(X)$ (where $H \subseteq A$) renames atoms in H to δ.

We mostly suppress the · in process expressions, and brackets according to the following rules: · binds strongest, and $\|, \mathbin{\|\mkern-6mu\raisebox{0.4ex}{\_}}, |$ all bind stronger than $+$.

In $\mathsf{ACP}(A, \gamma)$ the *communication function* $\gamma : A \times A \to A_\delta$ defines whether actions communicate, and if so, i.e., $\gamma(a, b) \neq \delta$, to what result. In Table 2 we present a slight modification of $\mathsf{ACP}(A, \gamma)$. This modification concerns commutativity of the communication merge | (axiom (CMC), explaining the missing (CM6) and (CM9)). We set $|\restriction_{(A \times A)} = \gamma$.

Table 2. The axiom system $\mathsf{ACP}(A, \gamma)$, where $a, b, c \in A_\delta$, $H \subseteq A$.

(A1)	$X + (Y + Z) = (X + Y) + Z$	(CM1)	$X \parallel Y = (X \mathbin{\|\mkern-6mu\_} Y + Y \mathbin{\|\mkern-6mu\_} X) + X \mid Y$
(A2)	$X + Y = Y + X$	(CM2)	$a \mathbin{\|\mkern-6mu\_} X = aX$
(A3)	$X + X = X$	(CM3)	$aX \mathbin{\|\mkern-6mu\_} Y = a(X \parallel Y)$
(A4)	$(X + Y)Z = XZ + YZ$	(CM4)	$(X + Y) \mathbin{\|\mkern-6mu\_} Z = X \mathbin{\|\mkern-6mu\_} Z + Y \mathbin{\|\mkern-6mu\_} Z$
		(CMC)	$X \mid Y = Y \mid X$
(A5)	$(XY)Z = X(YZ)$	(CM5)	$aX \mid b = (a \mid b)X$
(A6)	$X + \delta = X$	(CM7)	$aX \mid bY = (a \mid b)(X \parallel Y)$
(A7)	$\delta X = \delta$	(CM8)	$(X + Y) \mid Z = X \mid Z + Y \mid Z$
		(D1)	$\partial_H(a) = a$ if $a \notin H$
(C1)	$a \mid b = b \mid a$	(D2)	$\partial_H(a) = \delta$ if $a \in H$
(C2)	$(a \mid b) \mid c = a \mid (b \mid c)$	(D3)	$\partial_H(X + Y) = \partial_H(X) + \partial_H(Y)$
(C3)	$\delta \mid a = \delta$	(D4)	$\partial_H(XY) = \partial_H(X)\partial_H(Y)$

A (very) simple $\mathsf{ACP}(A, \gamma)$ process term is $a \parallel b$, the *interleaving* of two atomic actions a, b, i.e., the setting in which $a \mid b = \delta$. It easily follows that

$$\mathsf{ACP}(A, \gamma) \vdash a \parallel b = ab + ba.$$

A key feature of process algebra is *conditional composition*

$$X \triangleleft \phi \triangleright Y,$$

which represents **if** ϕ **then** X **else** Y where X, Y range over processes and ϕ is a condition. Its introduction in process algebra is described in [3]. In [11–13] we have extended the scope of the condition in conditional composition to various many-valued logics as described in [8], with the intention to model and analyze the occurrence of error-prone conditions in algorithms. Repeated use of conditional composition can lead to cumbersome notation, e.g.,

$$a_1 \cdot X_1 \triangleleft \phi_1 \triangleright (a_2 \cdot X_2 \triangleleft \phi_2 \triangleright (a_3 \cdot X_3 \triangleleft \phi_3 \triangleright a_4 \cdot X_4)),$$

and to laborious inspection of the outer arguments of conditional composition (either processes or again conditions). Therefore we introduce the following alternative notation

$$X +_\phi Y = X \triangleleft \phi \triangleright Y,$$

which has been borrowed from the conventions in probabilistic process algebra [5]. We use association to the right. The above term then reads as

$$a_1 \cdot X_1 +_{\phi_1} a_2 \cdot X_2 +_{\phi_2} a_3 \cdot X_3 +_{\phi_3} a_4 \cdot X_4,$$

which is easier to grasp. A condition in $\Sigma_5(\mathbb{P}_4)$ is called *deterministic* if it does not contain C. There is a fundamental difference between C and the other non-classical constants: the truth values M and D can be established by some external device (e.g., a type checker or a mathematician), whereas C is—on purpose—beyond any means of analysis. We only know it either behaves as T or as F. Of course, a process such as $ab + ba$ can also be described by $ab+_\mathsf{C} ba$ and, more generally, we may consider + as a derived construct if C and conditional composition are available. Stated differently: the alternative composition + of process algebra can be viewed as a notational device which allows one to remove the non-classical truth value C from process expressions involving atoms, sequential composition, and conditional composition (cf. Lemma 6).

Instead of conditional composition we shall often use the *conditional guard construct*

$$\phi :\to X,$$

which (roughly) expresses if ϕ then X. In Table 3 axioms are given for combining $\mathsf{ACP}(A, \gamma)$ with five-valued conditions. Here the constant μ represents the operational contents of M and was introduced in [11, 13]. Furthermore, the ϕ in the conditional guard construct ranges over $\Sigma_5(\mathbb{P}_4)$, so $\phi :\to$ is considered as a *unary* operation and related to conditional composition by axiom (Cond). Later on we show that $\phi :\to X = X \triangleleft \phi \triangleright \delta$. The conditional guard construct binds weaker than $\cdot$ and stronger than $\|$, $\mathbin{\|\mkern-5mu\|}$, and $|$.

Observe that the axioms (GC7) and (GC8) generalize (CM5) and (CM7), respectively. Also observe that $\phi :\to X \mid \psi :\to Y \neq \phi \wedge \psi :\to (X \mid Y)$ (set $\phi :\to X \equiv \mathsf{T} :\to \mu$ and $\psi :\to Y \equiv \mathsf{F} :\to \delta$). We use the acronym

$$\mathsf{ACP}_{\mathsf{C},\mu}(A, \gamma, \mathbb{P}_4)$$

both to refer to the axioms of Tables 2 and 3, and to the signature thus defined.

In order to combine process algebra and five-valued logic, we finally introduce the 'rule of equivalence'

$$\text{(ROE)} \quad \frac{\models \phi = \psi}{\vdash \phi :\to X = \psi :\to X}$$

This rule reflects the 'rule of consequence' in Hoare's Logic (cf. [1]). We write

$$\mathsf{ACP}_{\mathsf{C},\mu}(A, \gamma, \mathbb{P}_4) + \mathrm{ROE}_5 \vdash X = Y,$$

or shortly $\vdash X = Y$, if $X = Y$ follows from the axioms of $\mathsf{ACP}_{\mathsf{C},\mu}(A, \gamma, \mathbb{P}_4)$, the axioms and rules for $\Sigma_5(\mathbb{P}_4)$, and the appropriate rule of equivalence

$$\text{(ROE}_5) \quad \frac{\Sigma_5(\mathbb{P}_4) \models \phi = \psi}{\mathsf{ACP}_{\mathsf{C},\mu}(A, \gamma, \mathbb{P}_4) \vdash \phi :\to X = \psi :\to X}$$

Table 3. Remaining axioms of $\mathsf{ACP}_{\mathsf{C},\mu}(A,\gamma,\mathbb{P}_4)$, $a,b \in A_\delta$, $H \subseteq A$, and $\phi \in \Sigma_5(\mathbb{P}_4)$.

(M1)	$X + \mu = \mu$	(Cond)	$X \triangleleft \phi \triangleright Y = \phi :\to X + \neg\phi :\to Y$
(M2)	$\mu \cdot X = \mu$	(GC1)	$\phi :\to X + \psi :\to X = \phi \vee \psi :\to X$
(M3)	$\mu \mid X = \mu$	(GC2)	$\phi :\to X + \phi :\to Y = \phi :\to (X+Y)$
		(GC3)	$(\phi :\to X)Y = \phi :\to XY$
		(GCL4)	$\phi :\to (\psi :\to X) = \phi \mathbin{\underline{\wedge}} \psi :\to X$
(GM)	$\mathsf{M} :\to X = \mu$	(GC5)	$\phi :\to X \parallel Y = \phi :\to (X \parallel Y)$
(GC)	$\mathsf{C} :\to X = X$	(GC6)	$\phi :\to a \mid \psi :\to b = \phi \wedge \psi :\to a\mid b$
(GT)	$\mathsf{T} :\to X = X$	(GC7)	$\phi :\to aX \mid \psi :\to b = \phi \wedge \psi :\to (a\mid b)X$
(GF)	$\mathsf{F} :\to X = \delta$	(GC8)	$\phi :\to aX \mid \psi :\to bY = \phi \wedge \psi :\to (a\mid b)(X \parallel Y)$
(GD)	$\mathsf{D} :\to X = \delta$	(GC9)	$\partial_H(\phi :\to X) = \phi :\to \partial_H(X)$

We end this section with some useful derivabilities, applied in the remainder of the paper.

Lemma 3. 1. $\mathsf{ACP}_{\mathsf{C},\mu}(A,\gamma,\mathbb{P}_4) + \mathrm{ROE}_5 \vdash \phi :\to X = \phi \mathbin{\underline{\vee}} \mathsf{D} :\to X$,
2. $\mathsf{ACP}_{\mathsf{C},\mu}(A,\gamma,\mathbb{P}_4) + \mathrm{ROE}_5 \vdash \phi \mathbin{\underline{\vee}} \psi :\to X = \phi \vee (\neg\phi \mathbin{\underline{\wedge}} \psi) :\to X$,
3. $\mathsf{ACP}_{\mathsf{C},\mu}(A,\gamma,\mathbb{P}_4) + \mathrm{ROE}_5 \vdash \phi \wedge \psi :\to X = (\phi \mathbin{\underline{\wedge}} \psi) \vee (\psi \mathbin{\underline{\wedge}} \phi) :\to X$.

Proof. As for 1. We apply ROE_5 on the identity proved in Lemma 2:
$$\phi :\to X = \phi :\to X + \delta = \phi :\to X + \mathsf{D} :\to X = \phi \vee \mathsf{D} :\to X \stackrel{2}{=} \phi \mathbin{\underline{\vee}} \mathsf{D} :\to X.$$
As for 2 and 3. By inspection, taking all possible value-pairs for ϕ, ψ, and axioms (GM)–(GD). □

Using the above lemma and (Cond) one easily derives $\phi :\to X = X +_\phi \delta$.

4 Operational semantics and completeness

In this section we provide $\mathsf{ACP}_{\mathsf{C},\mu}(A,\gamma,\mathbb{P}_4)$ with an operational semantics and come up with a completeness result. Of course, interpretations of the conditions occurring at 'top level' in a process expression also determine its semantics. As an example, consider for fluent f and action a the expression $f :\to a$. Depending on the interpretation of f, this process either behaves as μ, as a, or as δ.

Given a (non-empty) set $\mathbb{P}_4$ of fluents, let w range over $\mathcal{W}$, the *valuations* (interpretations) of $\mathbb{P}_4$ in $\{\mathsf{M},\mathsf{T},\mathsf{F},\mathsf{D}\}$. In the usual way we extend w to $\Sigma_5(\mathbb{P}_4)$:

$$w(c) \stackrel{\triangle}{=} c \text{ for } c \in \{\mathsf{M},\mathsf{C},\mathsf{T},\mathsf{F},\mathsf{D}\},$$
$$w(\neg\phi) \stackrel{\triangle}{=} \neg(w(\phi)),$$
$$w(\phi \diamond \psi) \stackrel{\triangle}{=} w(\phi) \diamond w(\psi) \text{ for } \diamond \in \{\wedge, \mathbin{\underline{\wedge}}\}.$$

From the evaluation system defined in Section 2, it follows that

$$\forall w \in \mathcal{W}(\models w(\phi) = w(\psi)) \implies \models \phi = \psi.$$

In Table 4 we define for each $w \in \mathcal{W}$ a unary predicate *meaningless*, notation $\mu(w, \_)$, over process terms in $\mathsf{ACP}_{\mathsf{C},\mu}(A, \gamma, \mathbb{P}_4)$. This predicate defines whether a process expression represents the meaningless process μ under valuation w.

Table 4. Rules for $\mu(w, \_)$ in *panth*-format.

μ	$\mu(w, \mu)$	
$:\to$	$\mu(w, \phi :\to X)$ if $w(\phi) = \mathsf{M}$	$\dfrac{\mu(w, X)}{\mu(w, \phi :\to X)}$ if $w(\phi) \in \{\mathsf{C}, \mathsf{T}\}$
$+, \cdot, \|, \mathbin{\lfloor\!\lfloor}, \mid, \partial_H$	$\dfrac{\mu(w, X)}{\begin{array}{c}\mu(w, X+Y)\\ \mu(w, Y+X)\\ \mu(w, X \cdot Y)\\ \mu(w, \partial_H(X))\end{array}}$	$\dfrac{\mu(w, X)}{\begin{array}{c}\mu(w, X \| Y)\\ \mu(w, Y \| X)\\ \mu(w, X \mathbin{\lfloor\!\lfloor} Y)\\ \mu(w, X \mid Y)\end{array}}$

The axioms and rules for $\mu(w, \_)$ given in Table 4 are extended by axioms and rules given in Table 5, which define transitions

$$\_ \xrightarrow{w,a} \_ \subseteq \mathsf{ACP}_{\mathsf{C},\mu}(A, \gamma, \mathbb{P}_4) \times \mathsf{ACP}_{\mathsf{C},\mu}(A, \gamma, \mathbb{P}_4)$$

and unary "tick-predicates" or "termination transitions"

$$\_ \xrightarrow{w,a} \sqrt{} \subseteq \mathsf{ACP}_{\mathsf{C},\mu}(A, \gamma, \mathbb{P}_4)$$

for all $w \in \mathcal{W}$ and $a \in A$. Transitions characterize under which interpretations a process expression defines the possibility to execute an atomic action, and what remains to be executed (if anything, otherwise $\sqrt{}$ symbolizes successful termination).

The axioms and rules in Tables 4 and 5 yield a structured operational semantics (SOS) with negative premises in the style of Groote [16]. Moreover, they satisfy the so called *panth-format* defined by Verhoef [24] and define the following notion of bisimulation equivalence:

Definition 4. Let $B \subseteq \mathsf{ACP}_{\mathsf{C},\mu}(A, \gamma, \mathbb{P}_4) \times \mathsf{ACP}_{\mathsf{C},\mu}(A, \gamma, \mathbb{P}_4)$. Then B is a *bisimulation* if for all P, Q with PBQ the following conditions hold for all $w \in \mathcal{W}$ and $a \in A$:

- $\mu(w, P) \iff \mu(w, Q)$,
- $\forall P' \, (P \xrightarrow{w,a} P' \implies \exists Q'(Q \xrightarrow{w,a} Q' \wedge P'BQ'))$,
- $\forall Q' \, (Q \xrightarrow{w,a} Q' \implies \exists P'(P \xrightarrow{w,a} P' \wedge P'BQ'))$,
- $P \xrightarrow{w,a} \sqrt{} \iff Q \xrightarrow{w,a} \sqrt{}$.

Two processes P, Q are *bisimilar*, notation $P \leftrightarrow Q$, if there exists a bisimulation containing the pair (P, Q).

Table 5. Transition rules in *panth*-format.

$a \in A$	$a \xrightarrow{w,a} \surd$	
$\cdot, \mathrel{\|\!\_}$	$\dfrac{X \xrightarrow{w,a} \surd}{\begin{array}{c} X \cdot Y \xrightarrow{w,a} Y \\ X \mathrel{\|\!\_} Y \xrightarrow{w,a} Y \end{array}}$	$\dfrac{X \xrightarrow{w,a} X'}{\begin{array}{c} X \cdot Y \xrightarrow{w,a} X'Y \\ X \mathrel{\|\!\_} Y \xrightarrow{w,a} X' \| Y \end{array}}$
$+, \|$	$\dfrac{X \xrightarrow{w,a} \surd \quad \neg\mu(w,Y)}{\begin{array}{c} X + Y \xrightarrow{w,a} \surd \\ Y + X \xrightarrow{w,a} \surd \\ X \| Y \xrightarrow{w,a} Y \\ Y \| X \xrightarrow{w,a} Y \end{array}}$	$\dfrac{X \xrightarrow{w,a} X' \quad \neg\mu(w,Y)}{\begin{array}{c} X + Y \xrightarrow{w,a} X' \\ Y + X \xrightarrow{w,a} X' \\ X \| Y \xrightarrow{w,a} X' \| Y \\ Y \| X \xrightarrow{w,a} Y \| X' \end{array}}$
$a \mid b = c$	$\dfrac{X \xrightarrow{w,a} \surd \quad Y \xrightarrow{w,b} \surd}{\begin{array}{c} X \mid Y \xrightarrow{w,c} \surd \\ X \| Y \xrightarrow{w,c} \surd \end{array}} \; a \mid b = c$	$\dfrac{X \xrightarrow{w,a} \surd \quad Y \xrightarrow{w,b} Y'}{\begin{array}{c} X \mid Y \xrightarrow{w,c} Y' \\ X \| Y \xrightarrow{w,c} Y' \end{array}} \; a \mid b = c$
	$\dfrac{X \xrightarrow{w,a} X' \quad Y \xrightarrow{w,b} \surd}{\begin{array}{c} X \mid Y \xrightarrow{w,c} X' \\ X \| Y \xrightarrow{w,c} X' \end{array}} \; a \mid b = c$	$\dfrac{X \xrightarrow{w,a} X' \quad Y \xrightarrow{w,b} Y'}{\begin{array}{c} X \mid Y \xrightarrow{w,c} X' \| Y' \\ X \| Y \xrightarrow{w,c} X' \| Y' \end{array}} \; a \mid b = c$
∂_H	$\dfrac{X \xrightarrow{w,a} \surd}{\partial_H(X) \xrightarrow{w,a} \surd} \;\text{if}\; a \notin H$	$\dfrac{X \xrightarrow{w,a} X'}{\partial_H(X) \xrightarrow{w,a} \partial_H(X')} \;\text{if}\; a \notin H$
$:\rightarrow$	$\dfrac{X \xrightarrow{w,a} \surd}{\phi :\rightarrow X \xrightarrow{w,a} \surd} \;\text{if}\; w(\phi) \in \{\text{C}, \text{T}\}$	$\dfrac{X \xrightarrow{w,a} X'}{\phi :\rightarrow X \xrightarrow{w,a} X'} \;\text{if}\; w(\phi) \in \{\text{C}, \text{T}\}$

Furthermore, from [16, 24] it easily follows that the transitions and meaningless instances defined by these axioms and rules are uniquely determined. This can be established with help of the following *stratification* S:

$$S(\mu(w, X)) = 0, \ S(X \xrightarrow{w,a} X') = S(X \xrightarrow{w,a} \sqrt{}) = 1.$$

By the main result in [24] it follows that bisimilarity is a *congruence* relation for all operations involved. Notice that conditional guard constructs are considered here as unary operations: for each $\phi \in \Sigma_5(\mathbb{P}_4)$ there is an operation $\phi :\to \_$.

We write $\mathsf{ACP}_{\mathsf{C},\mu}(A, \gamma, \mathbb{P}_4)/_{\underline{\leftrightarrow}} \models P = Q$ whenever $P \underline{\leftrightarrow} Q$ according to the notions just defined, and for $\boldsymbol{X} = X_1, ..., X_n$

$$\mathsf{ACP}_{\mathsf{C},\mu}(A, \gamma, \mathbb{P}_4)/_{\underline{\leftrightarrow}} \models t_1(\boldsymbol{X}) = t_2(\boldsymbol{X})$$

if for all $\boldsymbol{P} = P_1, ..., P_n$ it holds that $t_1(\boldsymbol{P}) = t_2(\boldsymbol{P})$. It is not difficult, but tedious to establish that in the bisimulation model thus obtained all equations of Table 2 are true. Hence we conclude:

Lemma 5. *The system* $\mathsf{ACP}_{\mathsf{C},\mu}(A, \gamma, \mathbb{P}_4) + \mathrm{ROE}_5$ *is sound with respect to bisimulation: if* $\mathsf{ACP}_{\mathsf{C},\mu}(A, \gamma, \mathbb{P}_4) + \mathrm{ROE}_5 \vdash t_1(\boldsymbol{X}) = t_2(\boldsymbol{X})$, *then*

$$\mathsf{ACP}_{\mathsf{C},\mu}(A, \gamma, \mathbb{P}_4)/_{\underline{\leftrightarrow}} \models t_1(\boldsymbol{X}) = t_2(\boldsymbol{X}).$$

Finally, we provide a completeness result for $\mathsf{ACP}_{\mathsf{C},\mu}(A, \gamma, \mathbb{P}_4) + \mathrm{ROE}_5$. Our proof refers to the completeness result in [13], which is based on a representation of closed process terms for which bisimilarity implies derivability in a straightforward way (so called "basic terms"). A crucial observation is that terms over $\mathsf{ACP}_{\mathsf{C},\mu}(A, \gamma, \mathbb{P}_4)$ can be represented without C.

Lemma 6. *In* $\mathsf{ACP}_{\mathsf{C},\mu}(A, \gamma, \mathbb{P}_4)$ *each closed process expression can be proved equal to one in which* C *does not occur.*

Proof. We omit a full proof based on a representation of closed terms not containing ∂_H, $\|$, $\mathbin{\|\mkern-6mu\_}$, $|$, and $\_ \triangleleft \_ \triangleright \_$ (both as a logical connective and as a process constructor, cf. Proposition 1). It can be argued that c need not occur in any guard ϕ in $\phi :\to X$ by induction on the complexity of ϕ. E.g., if $\phi \equiv \phi_1 \wedge \phi_2$ then by Lemma 3.3, $\phi :\to X = (\phi_1 \mathbin{\underline{\wedge}} \phi_2) :\to X + (\phi_2 \mathbin{\underline{\wedge}} \phi_1) :\to X = \phi_1 :\to (\phi_2 :\to X) + \phi_2 :\to (\phi_1 :\to X)$. □

Theorem 7. *The system* $\mathsf{ACP}_{\mathsf{C},\mu}(A, \gamma, \mathbb{P}_4) + \mathrm{ROE}_5$ *is complete with respect to bisimulation: for closed terms* P *and* Q,

$$\mathsf{ACP}_{\mathsf{C},\mu}(A, \gamma, \mathbb{P}_4) + \mathrm{ROE}_5 \vdash P = Q \iff \mathsf{ACP}_{\mathsf{C},\mu}(A, \gamma, \mathbb{P}_4)/_{\underline{\leftrightarrow}} \models P \underline{\leftrightarrow} Q.$$

Proof. By the previous lemma and soundness it is sufficient to prove $\Longleftarrow$ for $\mathsf{ACP}(A, \gamma)$ with four-valued logic over $\{\mathsf{M}, \mathsf{T}, \mathsf{F}, \mathsf{D}\}$ and $\mathbb{P}_4$. A detailed (inductive) proof is spelled out in [13]. □

We end this section with a nice correspondence result.

Proposition 8. *Let $t_1(\boldsymbol{X}, \boldsymbol{x}) = t_2(\boldsymbol{X}, \boldsymbol{x})$ be a process identity with process variables $\boldsymbol{X}$ and condition variables $\boldsymbol{x}$ in which the only constants are in Σ_5 and the only operation is $\_ \triangleleft \_ \triangleright \_$. Then*

$$\mathsf{ACP}_{\mathsf{C},\mu}(A,\gamma,\mathbb{P}_4)/_{\underline{\leftrightarrow}} \models t_1(\boldsymbol{X}, \boldsymbol{x}) = t_2(\boldsymbol{X}, \boldsymbol{x}) \iff \Sigma_5(\mathbb{P}_4) \models t_1'(\boldsymbol{X}, \boldsymbol{x}) = t_2'(\boldsymbol{X}, \boldsymbol{x}),$$

where t_i' is obtained by regarding the process variables of t_i also as condition variables.

5 Generalization of ACP and CpSP

We discuss various systems that generalize $\mathsf{ACP}(A,\gamma)$ [10] to a setting in which alternative composition is a special case of conditional composition, and that provides a parameterized version of the parallel composition operations. Next we provide an algebraic setting for the Cooperating Sequential Processes (CpSP) of Dijkstra [15]. We can do this for all logics that contain C. We define the following operations, where A is the set of atomic actions, Pr is the sort of processes, and $\mathbb{L}$ is the particular logic involved.

— Constants and operations —	— Parametrized operations —
$a \quad : A \subseteq Pr$	$\_+\_\_ : Pr \times \mathbb{L} \times Pr \to Pr$
$\delta \quad : Pr,\ \delta \notin A$	$\_\|\|_\_\_ : Pr \times \mathbb{L} \times \mathbb{L} \times Pr \to Pr$
$\_\|\_ \quad : A_\delta \times A_\delta \to A_\delta$	$\_\|\|\_\_ : Pr \times \mathbb{L} \times \mathbb{L} \times Pr \to Pr$
$\_\cdot\_ \quad : Pr \times Pr \to Pr$	$\_\|\_\_ : Pr \times \mathbb{L} \times \mathbb{L} \times Pr \to Pr$
$\partial_H(\_) : Pr \to Pr \quad (H \subseteq A)$	$\_\lfloor\_\_ : Pr \times \mathbb{L} \times \mathbb{L} \times Pr \to Pr$

We write $G_k(Z)$ for the k-valued generalization of axiomatization Z. We first describe the simplest generalization

$$G_3\bigl(\mathsf{ACP}_\mathsf{C}(A,\gamma,\mathbb{P}_2)\bigr).$$

and write $\Sigma_3^C(\mathbb{P}_2)$ for three-valued logic over $\{\mathsf{C},\mathsf{T},\mathsf{F}\}$ and $\mathbb{P}_2$. The system $G_3\bigl(\mathsf{ACP}_\mathsf{C}(A,\gamma,\mathbb{P}_2)\bigr)$ is defined by the axioms in Table 6, where $\gamma = |\upharpoonright_{(A \times A)}$. Observe that axiom (GA3) is equivalent with

$$X +_\phi X = X,$$

as $\mathsf{T} \triangleleft \phi \triangleright \mathsf{T} = \mathsf{T}$ in $\Sigma_3^C(\mathbb{P}_2)$. However, the formulation used in Table 6 allows straightforward generalizations to systems that contain error-prone conditions (possibly evaluating to M or D). It is easy to see which axioms should be added, e.g., if only D is considered, the axiom

$$(\text{GGD}) \quad X +_\mathsf{D} Y = \delta$$

should be added to Table 6. Involving M gives rise to $\mu \in Pr$ and axioms

$$(\text{GM1}) \quad X +_\mathsf{C} \mu = \mu,$$
$$(\text{GGM}) \quad X +_\mathsf{M} Y = \mu.$$

Observe that $\mu X = \mu$ is derivable from (GGM) and (GA4). Furthermore, $X \,_\phi\!\lfloor_\psi \mu = \mu \,_\phi\!\lfloor_\psi X = \mu$ follows from (GGM) and (GCM8), (GCM9), respectively. The system
$$G_5\big(\mathsf{ACP}_{\mathsf{C},\mu}(A,\gamma,\mathbb{P}_4)\big)$$
is defined as the extension of $G_3\big(\mathsf{ACP}_{\mathsf{C}}(A,\gamma,\mathbb{P}_2)\big)$ with (GM1), (GGM), (GGD), and with conditions ranging over $\Sigma_5(\mathbb{P}_4)$.

Table 6. $G_3\big(\mathsf{ACP}_{\mathsf{C}}(A,\gamma,\mathbb{P}_2)\big)$, $a,b \in A_\delta$, $H \subseteq A$, and $\phi,\psi,\chi \in \Sigma_3^{\mathcal{C}}(\mathbb{P}_2)$.

(GGT)	$X +_{\mathsf{T}} Y = X$
(GA1)	$X +_\phi (Y +_\phi Z) = (X +_\phi Y) +_\phi Z$
(GA2)	$X +_\phi Y = Y +_{\neg\phi} X$
(GA3)	$X +_\phi X = X +_{(\mathsf{T} \triangleleft \phi \triangleright \mathsf{T})} X$
(GA4)	$(X +_\phi Y) Z = X Z +_\phi Y Z$
(GA5)	$(XY)Z = X(YZ)$
(GA6)	$X +_{\mathsf{C}} \delta = X$
(GA7)	$\delta X = \delta$
(C1)	$a \mid b = b \mid a$
(C2)	$(a \mid b) \mid c = a \mid (b \mid c)$
(C3)	$\delta \mid a = \delta$
(GCM1)	$X \,_\phi\!\|_\psi Y = (X \,_\phi\!\mathbin{\lfloor\!\lfloor}_\psi Y +_\psi Y \,_\phi\!\mathbin{\lfloor\!\lfloor}_{\neg\psi} X) +_\phi X \,_\phi\!\lfloor_\psi Y$
(GCM2)	$a \,_\phi\!\mathbin{\lfloor\!\lfloor}_\psi X = aX$
(GCM3)	$aX \,_\phi\!\mathbin{\lfloor\!\lfloor}_\psi Y = a(X \,_\phi\!\|_\psi Y)$
(GCM4)	$(X +_\phi Y) \,_\psi\!\mathbin{\lfloor\!\lfloor}_\chi Z = X \,_\psi\!\mathbin{\lfloor\!\lfloor}_\chi Z +_\phi Y \,_\psi\!\mathbin{\lfloor\!\lfloor}_\chi Z$
(GCMC)	$X \,_\phi\!\lfloor_\psi Y = X \,_\phi\!\lfloor_\psi Y +_\psi Y \,_\phi\!\lfloor_{\neg\psi} X$
(GCM5)	$aX \,_\phi\!\lfloor_\psi Y = a \,_\phi\!\lfloor_\psi (Y \,_\phi\!\mathbin{\lfloor\!\lfloor}_{\neg\psi} X)$
(GCM6)	$a \,_\phi\!\lfloor_\psi b = a \mid b$
(GCM7)	$a \,_\phi\!\lfloor_\psi bX = (a \mid b) X$
(GCM8)	$a \,_\phi\!\lfloor_\psi (X +_\chi Y) = a \,_\phi\!\lfloor_\psi X +_\chi a \,_\phi\!\lfloor_\psi Y$
(GCM9)	$(X +_\phi Y) \,_\psi\!\lfloor_\chi Z = X \,_\psi\!\lfloor_\chi Z +_\phi Y \,_\psi\!\lfloor_\chi Z$
(GD1)	$\partial_H(a) = a$ if $a \notin H$
(GD2)	$\partial_H(a) = \delta$ if $a \in H$
(GD3)	$\partial_H(X +_\phi Y) = \partial_H(X) +_\phi \partial_H(Y)$
(GD4)	$\partial_H(XY) = \partial_H(X) \partial_H(Y)$

Cooperating Sequential Processes, CpSP, in the style of [15] can be abstractly modeled in $G_5\big(\mathsf{PA}_{\delta,\mathsf{C},\mu}(A,\mathbb{P}_4)\big)$ with action history operator and state operator. Here, $\mathsf{PA}_{\delta,\mathsf{C},\mu}(A,\mathbb{P}_4)$ refers to the restriction of parallel composition to interleaving, thus to a setting without communication, and is obtained from

$G_5\big(\mathsf{ACP}_{\mathsf{C},\mu}(A,\gamma,\mathbb{P}_4)\big)$ by restricting $\,_\phi\|_\psi$ to $\,_\mathsf{T}\|_\mathsf{C}$. We further write $\|$ for $\,_\mathsf{T}\|_\mathsf{C}$, and $\mathrel{\|\!\_}$ instead of $\,_\mathsf{T}\mathrel{\|\!\_}_\mathsf{C}$. The axioms of $G_5\big(\mathsf{PA}_{\delta,\mathsf{C},\mu}(A,\mathbb{P}_4)\big)$ are given in Table 7.

Table 7. $G_5\big(\mathsf{PA}_{\delta,\mathsf{C},\mu}(A,\mathbb{P}_4)\big)$, $a \in A_\delta \cup \{\mu\}$, $\sigma \in A^*$, and $\phi \in \Sigma_5(\mathbb{P}_4)$.

(GA1)	$X +_\phi (Y +_\phi Z) = (X +_\phi Y) +_\phi Z$		(GGT)	$X +_\mathsf{T} Y = X$
(GA2)	$X +_\phi Y = Y +_{\neg\phi} X$		(GGD)	$X +_\mathsf{D} Y = \delta$
(GA3)	$X +_\phi X = X +_{(\mathsf{T} \triangleleft \phi \triangleright \mathsf{T})} X$		(GM1)	$X +_\mathsf{C} \mu = \mu$
(GA4)	$(X +_\phi Y) Z = X Z +_\phi Y Z$		(GGM)	$X +_\mathsf{M} Y = \mu$
(GA5)	$(XY)Z = X(YZ)$			
(GA6)	$X +_\mathsf{C} \delta = X$			
(GA7)	$\delta X = \delta$			
(GCM1)	$X \parallel Y = X \mathrel{\|\!\_} Y +_\mathsf{C} Y \mathrel{\|\!\_} X$			
(GCM2)	$a \mathrel{\|\!\_} X = aX$			
(GCM3)	$aX \mathrel{\|\!\_} Y = a(X \parallel Y)$			
(GCM4)	$(X +_\phi Y) \mathrel{\|\!\_} Z = X \mathrel{\|\!\_} Z +_\phi Y \mathrel{\|\!\_} Z$			

Action History Logic, AHL, was introduced in [13] as a natural example of the use of four-valued logic in process algebra. It can be used to express history dependent properties of processes, and comprises the following ingredients:

In, the assertion which is true of the initial state of a process and false thereafter.
$\mathsf{P}_4(\phi)$, the assertion that ϕ is valid in the previous state, i.e., the state before the last action. If there is no such state, $\mathsf{P}_4(\phi) = \mathsf{M}$.
$\mathsf{L}_4(a)$, the condition that expresses that the last action was a. In case the state is initial, $\mathsf{L}_4(a)$ evaluates to M.

Let $\Sigma_5(\mathbb{P}_4)$ be generated from AHL. Writing ϵ for the empty history, the *action history operator* H_ϵ defined in Table 8 memorizes the action history (trace) of a fluent-free process. In order to represent a CpSP-process which involves the interpretation of fluents we consider a data-state space $\mathcal{S} \subseteq \mathcal{T} \times \mathcal{W}$ for some further unspecified set $\mathcal{T}$ and the set $\mathcal{W}$ of interpretations. We use a *state operator* $\lambda_s(\_)$ (see [2]) to model how the execution of actions affects interpretations. Typically, process aX in data-state s is represented as $\lambda_s(aX)$ and satisfies

$$\lambda_s(aX) = a' \cdot \lambda_{s'}(X)$$

where a' is the action (or δ or μ) that occurs as the result of executing a in data-state s, and s' is the data-state which ensues when executing a in s. We assume two given functions describing these effects: $action : A \times \mathcal{S} \to A \cup \{\delta, \mu\}$ and $effect : A \times \mathcal{S} \to \mathcal{S}$. We further set $action(c, s) = c$ for $c \in \{\delta, \mu\}$. Axioms for the state operator are also given in Table 8.

Now $H_\epsilon(\lambda_s(P_1 \parallel \ldots \parallel P_n))$ with P_i not containing history/state operators or $\parallel$, $\mathrel{\|\!\_}$ typically is an algebraic notation for a CpSP-process with (global) initial data-state s.

Table 8. Axioms for history and state operator, $a \in A$, $\sigma \in A^*$, and $\phi, \psi \in \Sigma_5(\mathbb{P}_4)$.

$H_\sigma(X +_\phi Y) = H_\sigma(X) +_{\phi(\sigma)} H_\sigma(Y)$	$\ln(\epsilon) = \mathsf{T}$
$H_\sigma(c) = c$ for $c \in A \cup \{\delta, \mu\}$	$\ln(\sigma a) = \mathsf{F}$
$H_\sigma(a \cdot X) = a \cdot H_{\sigma a}(X)$	$\mathsf{P}_4(\phi)(\epsilon) = \mathsf{M}$
$c(\sigma) = c$ for $c \in \{\mathsf{M}, \mathsf{C}, \mathsf{T}, \mathsf{F}, \mathsf{D}\}$	$\mathsf{P}_4(\phi)(\sigma a) = \phi(\sigma)$
$(\neg \phi)(\sigma) = \neg(\phi(\sigma))$	$\mathsf{L}_4(a)(\epsilon) = \mathsf{M}$
$(\phi \diamond \psi)(\sigma) = \phi(\sigma) \diamond \psi(\sigma)$ for $\diamond \in \{\wedge, \wedge\!\!\!\!\wedge\}$	$\mathsf{L}_4(a)(\sigma b) = a \equiv b \in \{\mathsf{T}, \mathsf{F}\}$
$\lambda_{(t,w)}(X +_\phi Y) = \lambda_{(t,w)}(X) +_{w(\phi)} \lambda_{(t,w)}(Y)$	$\lambda_s(aX) = action(a, s) \cdot \lambda_{s'}(X)$
$\lambda_s(c) = action(c, s)$ for $c \in A \cup \{\delta, \mu\}$	where $s' = \mathit{effect}(a, s)$

6 Conclusions

We observed that Kleene's three-valued logic $\mathbb{K}_3$ allows for two intuitions of the third, non-classical truth value: undetermined and undefined. Indeed, a complete axiomatization of $\mathbb{K}_3$ leaves room for exactly two non-classical constants, notation C and D, and implies C $\wedge$ D = F. The resulting four-valued logic $\mathbb{K}_4$ has a complete, equational axiomatization [20]. The combination of $\mathbb{K}_4$, or one of its sublogics containing T, F, with process algebra yields an equational completeness result (adopting our restriction on the interpretation of fluents, discarding C, and using Lemma 6). This follows from [4, 12]. Adding M (meaningless) to $\mathbb{K}_4$ yields a five-valued logic, which we extended with McCarthy's asymmetric connectives to provide a useful combination with process algebra. We presented a non-equational completeness result (using the 'excluded fifth rule'). Completeness results for all sublogics containing M, T, F follow from [11, 12].

We hope to have indicated that the use of non-classical logics in process theory is interesting in its own right. Expressivity can be enlarged by involving recursive ingredients. For process description we propose the (binary) Kleene star (see [19]), which in process algebra is defined by $X^*Y = X \cdot (X^*Y) + Y$ (see also [7]). In a more general setting, one can define $X^{*\phi}Y = X \cdot (X^{*\phi}Y) +_\phi Y$ and write X^*Y for $X^{*\mathsf{C}}Y$. Examples with recursively defined conditions, such as *schedulers*, are discussed in [13].

References

1. K.R. Apt. Ten years of Hoare's logic, a survey, part I. *ACM Transactions on Programming Languages and Systems*, 3(4):431-483, 1981.
2. J.C.M. Baeten and J.A. Bergstra. Global renaming operators in concrete process algebra. *Information and Computation*, 78(3):205-245, 1988.
3. J.C.M. Baeten and J.A. Bergstra. Process algebra with signals and conditions. In M. Broy, editor, Programming and Mathematical Method, *Proceedings Summer School Marktoberdorf*, 1990 NATO ASI Series F, pages 273-323, Springer-Verlag, 1992.
4. J.C.M. Baeten and J.A. Bergstra. Process algebra with propositional signals. *Theoretical Computer Science*, 177(2):381-406, 1997.

5. J.C.M. Baeten, J.A. Bergstra, and S.A. Smolka. Axiomatizing probabilistic processes: ACP with generative probabilities. *Information and Computation*, 121(2):234-254, 1995.
6. J.C.M. Baeten and W.P. Weijland. *Process Algebra*. Cambridge Tracts in Theoretical Computer Science 18. Cambridge University Press, 1990.
7. J.A. Bergstra, I. Bethke, and A. Ponse. Process algebra with iteration and nesting. *Computer Journal*, 37(4):243-258, 1994.
8. J.A. Bergstra, I. Bethke, and P.H. Rodenburg. A propositional logic with 4 values: true, false, divergent and meaningless. *Journal of Applied Non-Classical Logics*, 5:199-217, 1995.
9. J.A. Bergstra and J.W. Klop. The algebra of recursively defined processes and the algebra of regular processes. In A. Ponse, C. Verhoef, and S.F.M. van Vlijmen, *Algebra of Communicating Processes, Utrecht 1994*, Workshops in Computing, pages 1-25. Springer-Verlag, 1995. An extended abstract appeared in J. Paredaens, editor, *Proceedings 11th ICALP,* Antwerp, volume 172 of *Lecture Notes in Computer Science*, pages 82-95. Springer-Verlag, 1984.
10. J.A. Bergstra and J.W. Klop. Process algebra for synchronous communication. *Information and Control*, 60(1/3):109-137, 1984.
11. J.A. Bergstra and A. Ponse. Bochvar-McCarthy logic and process algebra. Technical Report P9722, Programming Research Group, University of Amsterdam, 1997 (see also http://www.wins.uva.nl/research/prog/reports/reports.html).
12. J.A. Bergstra and A. Ponse. Kleene's three-valued logic and process algebra. *Information Processing Letters*, 67(2):95-103, 1998.
13. J.A. Bergstra and A. Ponse. Process algebra with four-valued logic. Technical Report P9724, Programming Research Group, University of Amsterdam, 1997 (see also http://www.wins.uva.nl/research/prog/reports/reports.html). To appear in *Journal of Applied Non-Classical Logics*.
14. D.A. Bochvar. On a 3-valued logical calculus and its application to the analysis of contradictions (in Russian). *Matématičeskij sbornik*, 4:287-308, 1939.
15. E.W. Dijkstra. Cooperating sequential processes. In F. Genuys, editor, *Programming Languages*, pages 43-112, Academic Press, New York, 1968.
16. J.F. Groote. Transition system specifications with negative premises. *Theoretical Computer Science*, 118(2):263-299, 1993.
17. C.A.R. Hoare, I.J. Hayes, He Jifeng, C.C. Morgan, A.W. Roscoe, J.W. Sanders, I.H. Sorensen, J.M. Spivey, and B.A. Sufrin. Laws of programming. *Communications of the ACM*, 30(8):672-686, August 1987.
18. S.C. Kleene. On a notation for ordinal numbers. *Journal of Symbolic Logic*, 3:150-155, 1938.
19. S.C. Kleene. Representation of events in nerve nets and finite automata. In *Automata Studies*, pages 3-41. Princeton University Press, Princeton NJ, 1956.
20. S.P. Luttik and P.H. Rodenburg. Personal communications, 1998.
21. J. McCarthy. A basis for a mathematical theory of computation. In P. Braffort and D. Hirshberg (eds.), *Computer Programming and Formal Systems*, pages 33-70, North-Holland, Amsterdam, 1963.
22. J. McCarthy. *Formalization of common sense, papers by John McCarthy edited by V. Lifschitz*. Ablex, 1990.
23. J. McCarthy and P.J. Hayes. Some philosophical problems from the standpoint of artificial intelligence. In B. Meltzer and D. Michie, editors, *Machine Intelligence 4*, pages 463-502. Edinburgh University Press, 1969. Reprinted in [22].
24. C. Verhoef. A congruence theorem for structured operational semantics with predicates and negative premises. *Nordic Journal of Computing*, 2(2):274-302, 1995.

A Stability Theorem for Recursive Analysis

Vasco Brattka

Theoretische Informatik I, FernUniversität Hagen, D-58084 Hagen, Germany.
vasco.brattka@fernuni-hagen.de

Abstract. Many structures considered in analysis have the cardinality of the continuum. We formulate a sufficient criterion which guarantees that effectivity over such structures is uniquely determined by the structure itself. This criterion is expressed by virtue of recursiveness over structures. It generalizes Mal'cev's Stability Theorem for finitely generated algebras as well as Pour-El and Richard's Stability Lemma for computable Banach spaces in a certain sense.

1 Introduction

In recursive analysis we are investigating computable operations that occur in analysis. Such operations are typically many-sorted and in general they are neither total nor single-valued. We will give seven instructive examples of such operations:

1. The *addition* $+ : \mathbb{R} \times \mathbb{R} \to \mathbb{R}$ on the real numbers.
2. The partial *limit operation* $\lim \ :\subseteq \mathbb{R}^{\mathbb{N}} \to \mathbb{R}, (x_n)_{n \in \mathbb{N}} \mapsto \lim_{n \to \infty} x_n$ (restricted to rapidly converging Cauchy sequences).
3. The *Mandelbrot function* $f : \mathbb{C} \to \mathbb{C}, z \mapsto z^2 + c$ and its iteration $f^* : \mathbb{C} \times \mathbb{N} \to \mathbb{C}, f(z,n) := f^n(z)$.
4. The *integral operation* $\int : C[0,1] \to \mathbb{R}, f \mapsto \int_0^1 f(x)\,dx$.
5. The *Hausdorff metric* $d_{\mathcal{K}} : \mathcal{K} \times \mathcal{K} \to \mathbb{R}$.
6. The partial *zero operator* $Z :\subseteq C[0,1] \rightrightarrows \mathbb{R}, f \mapsto f^{-1}\{0\}$.
7. The *finite precision test* $< : \mathbb{R} \rightrightarrows \mathbb{N}$.

The zero operator Z is a many-valued function and this many-valuedness is unavoidable in the sense that there are subsets $A \subseteq C[0,1]$ such that there exist only many-valued effective zero operators $Z|_A$ but not even a continuous single-valued selector (cf. also [Wei,Bra95]). The finite precision test $<$ (illustrated in Figure 1) stands for the comparison $x < 0$ which is performed with an area of uncertainty (of length ε), where the result might be "yes" as well as "no" (cf. [Bra96,BH]). In such "hybrid situations" many-valuedness appears quite natural: whenever we want to compute a discrete information n from a continuous value x the computation realizes a many-valued operation $f : X \rightrightarrows \mathbb{N}$ or a trivial function $f : X \to \mathbb{N}$ (by continuity of f). This corresponds to the situation of physical measurements: whenever a system controls a switch depending on

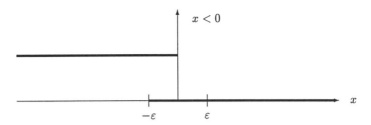

Fig. 1. The finite precision test.

continuous data, then there has to be a certain hysteresis (e.g. in temperature controlled heat systems).

Effectivity of operations in analysis can be either described in *abstract* or *concrete* languages. A concrete language that captures all operations above is "type 2 theory of effectivity" of Weihrauch [Wei87,Wei97,Wei]. The basic idea is to represent sets X by infinite sequences of symbols, i.e. a representation is a surjection $\delta :\subseteq \mathbb{N}^{\mathbb{N}} \to X$. Then a *computable* operation is an operation such that a representing name of the output can be computed from a representing name of the input by a (slightly generalized) Turing machine.

Besides this concrete language we will introduce an abstract language which is based on the idea that effective operations can be effectively constructed from some initial operations. The initial operations will be provided by *structures* and the effective constructions will be made precise by some *recursion operators* (which generalize the classical recursive closure scheme's for μ-recursive functions). Thus, a *recursive operation* will be an operation that can be finitely generated from some initial operations of a structure by virtue of the recursion operators. A different but related version of such an abstract language has been suggested in [Bra96]. For other abstract languages (while programs over algebras) cf. Tucker and Zucker [TZ,TZ98]. Typical structures that arise in analysis are given by the following examples:

1. *Natural numbers* $\mathcal{N} := (\mathbb{N}, 0, n, n+1)$,
2. *Real numbers* $\mathcal{R} := (\mathbb{R}, 0, 1, x+y, -x, x \cdot y, 1/x, \text{Lim}, <)$,
3. *Compact non-empty subsets* $\mathcal{K} := (\mathcal{K}(\mathbb{R}), \{x\}, A, A \cup B, d_{\mathcal{K}}, \text{Lim})$,
4. *Continuous functions* $\mathcal{C} := (\mathcal{C}[0,1], 1, f, f+g, x \cdot f, f \cdot g, ||f||_{\infty}, \text{Lim})$.

We will not define these structures in detail, but we shortly mention that Lim denotes the limit operation of the corresponding metric space for rapidly converging Cauchy sequences. Strictly speaking the above examples only define *substructures*, since some of the operations (like the metric $d_{\mathcal{K}}$) refer to some other set outside the corresponding substructure. According to our terminology the whole collection $\mathcal{S} := (\mathcal{N}, \mathcal{R}, \mathcal{K}, \mathcal{C})$ would be an example of a *(many-sorted) natural structure*.

From a computer science point of view the concrete language of computable operations can be considered as "assembler language" and the abstract language of recursive operations can be considered as "high level programming language over data structures".

The goal of this paper is to formulate a sufficient condition for structures which guarantee that effectivity over such structures is uniquely characterized by the structure itself. Our main result will be the following Stability Theorem.

Theorem 1 (Stability Theorem). *If S is a natural structure which is recursive via a representation δ and which is effective via a natural representation δ', then $\delta \equiv \delta'$.*

The computer science interpretation is roughly speaking the following: whenever we have a data structure S with ordinary natural number objects together with a representation δ of this structure which can be "synthesized" as well as "analyzed" in a high level programming language, and implemented in assembler language, then all such implementations are equivalent. Here "synthesized" means that there is a high level program which given an infinite sequence p as input computes $\delta(p)$ as output and "analyzed" means that there is a high level program which given x as input computes a p such that $\delta(p) = x$.

Structures that admit representations δ, δ' such that the conditions of the Stability Theorem hold will be called *perfect*. The structure $S := (\mathcal{N}, \mathcal{R}, \mathcal{K}, \mathcal{C})$, as mentioned above, is an example of a perfect structure. The effective stability of the structure of the real numbers has directly been proved by Hertling [Her97] who initiated the discussion of stability of structures in analysis.

We end this introduction with a short survey of the organization of this paper. The next section is dedicated to the basic notions that have to be made more precise. We will formally define structures, the recursion operators and recursive operations over structures, our "abstract language". Moreover, we will extend type 2 theory of effectivity to computable (many-valued) operations, our "concrete language". In Section 3 we will formulate our Stability Theorem and we will conclude that the expressiveness of our abstract and concrete language is essentially the same. More precisely, over (strongly) perfect natural structures there is only one type of computability and this computability coincides with recursiveness. Finally, in Section 4 we shortly discuss perfect structures which are induced by recursive metric spaces. Due to lack of space we omit most proofs in this extended abstract. Details can be found in [Bra98].

2 Recursive and computable operations over structures

In this section we will introduce our abstract and concrete languages for effectivity over structures. First, we have to introduce some technical notations. We will denote by $f :\subseteq X \rightrightarrows Y$ partial multi-valued operations which we will call for short *operations* in the following. Here the symbol $\subseteq$ indicates that the operations f is partial, and $\rightrightarrows$ indicates that f is multi-valued. More precisely, an operation $f :\subseteq X \rightrightarrows Y$ is a correspondence $f = (F, X, Y)$, that is $F \subseteq X \times Y$. We will use these objects from an operational point of view, that is X is a space of inputs and Y a space of outputs. Moreover, we will use some notations for operations:

$$\mathrm{dom}(f) := \{x \in X : (\exists y \in Y)\, (x,y) \in F\},$$
$$\mathrm{range}(f) := \{y \in Y : (\exists x \in X)\, (x,y) \in F\}$$

denotes the *domain, range* of f, respectively and

$$f(A) := \{y \in Y : (\exists x \in A)\, (x,y) \in F\},$$
$$f^{-1}(B) := \{x \in X : (\exists y \in B)\, (x,y) \in F\}$$

denotes the *image, preimage* of $A \subseteq X, B \subseteq Y$ under f, respectively. By $f(x) := f\{x\} = \{y \in Y : (x,y) \in F\}$ we denote the *image* of x under f for each $x \in \mathrm{dom}(f)$. The image $f(x)$ is either non-defined in case $x \notin \mathrm{dom}(f)$ (which is denoted by $f(x) = \uparrow$ in the following), or $f(x) \neq \emptyset$ in case $x \in \mathrm{dom}(f)$. In the later case $\mathrm{range}(f) = \bigcup_{x \in \mathrm{dom}(f)} f(x)$. If $f(x)$ is single-valued, i.e. $f(x) = \{y\}$ for some $y \in Y$, then we also write $f(x) = y$, as usual for functions.

To each operation $f = (F, X, Y)$ we associate the *inverse operation* $f^{-1} = (F^{-1}, Y, X)$, which is given by $F^{-1} := \{(y, x) : (x, y) \in F\}$. Thus, $f^{-1}(x) = f^{-1}\{x\}$ if $x \in \mathrm{dom}(f^{-1}) = \mathrm{range}(f)$ and $f^{-1}(x) = \uparrow$ else.

We will write $f \sqsubseteq g$ for operations $f, g :\subseteq X \rightrightarrows Y$, if $\mathrm{dom}(f) \subseteq \mathrm{dom}(g)$ and $f(x) \subseteq g(x)$ for all $x \in \mathrm{dom}(f)$ (or equivalently, if $f = (F, X, Y), g = (G, X, Y)$ and $F \subseteq G$). If X is a set then we denote by $X^{\mathbb{N}}$ the set of all sequences in X. We will write $\alpha = (x_n)_{n \in \mathbb{N}}$ for the sequence in X defined by $\alpha(n) = x_n$ for all $n \in \mathbb{N}$. We will not distinguish the set product $(X \times Y) \times Z$ from $X \times (Y \times Z)$. For each set X we define $X^0 := \{()\}$ and we assume $X^0 \times Y = Y \times X^0 = Y$ for each set Y.

2.1 Recursive operations over structures

We will consider structures $(X, f_1, ..., f_k)$ with a set X and a finite number of operations $f_1, ..., f_k$. If there is a constant $c \in X$ among the operations, then we will consider it in a canonical way as zero-ary constant function $X^0 \to X, () \mapsto c$ with value c. If there is a set $A \subseteq X$ among the operations, then it stands for its *semi-characteristic operation* $c_A : X \rightrightarrows \mathbb{N}$, defined by

$$c_A(x) := \begin{cases} \{0,1\} & \text{if } x \in A \\ \{1\} & \text{else} \end{cases}$$

Since we are interested in topological structures, where X is a topological space and all initial operations are continuous, we cannot use the ordinary characteristic function to describe effectivity of sets in general. At least for connected topological spaces there are no non-trivial continuous characteristic functions. The semi-characteristic operation c_A is a suitable substitute for the characteristic function since it is continuous if A is open (more precisely: lower-semi-continuous as a set-valued function).

The reader should notice that we allow structures $(X, f_1, ..., f_k)$ with arbitrary operations $f_i :\subseteq Y \rightrightarrows Z$, even in the case that Y or Z is not a product of X. We will correct this "defect" by the following definition where we consider many-sorted structures such that each sort used by the operations is available.

Definition 2 (Many-sorted structures). $S = (S_1, ..., S_n)$ is called a *many-sorted structure* with *universe* $X_1 \times ... \times X_n$, if the X_i are sets and there are operations $f_{ij} :\subseteq Y_{ij} \rightrightarrows Z_{ij}$ for each $i = 1, ..., n$, such that $S_i = (X_i, f_{i1}, ..., f_{ik_i})$, $k_i \in \mathbb{N}$, $i = 1, ..., n$, $j = 1, ..., k_i$, and each Y_{ij}, Z_{ij} can be finitely generated from the sets $X_1, ..., X_n$ by finite product and sequence set constructions. The operations f_{ij} are called *initial operations* of S.

More formally, we can inductively define the *class of sets over S* as the smallest class such that: the sets $X_1, ..., X_n$ are sets over S; if Y, Z are sets over S, then so are $Y \times Z$ and $Y^\mathbb{N}$. Moreover, $f :\subseteq Y \rightrightarrows Z$ is an *operation over S*, if Y, Z are sets over S. Using this terminology, all initial operations of S must be operations over S.

In the following we will say for short *structure* instead of many-sorted structure. If $S = (S_1, ..., S_n)$ is a structure, then we will say that $S_1, ..., S_n$ are *substructures*. If some (S_i) is a structure itself, then we will also say, that S_i is a structure. Now we proceed to define recursive operations over structures.

Definition 3 (Recursive operations over structures). The class of *recursive* operations over a structure S is the smallest class of operations which contains all initial operations of S and which is closed under projection, juxtaposition, product, composition, iteration, inversion, evaluation, transposition, and exponentiation.

Now we will precisely define the operators mentioned in this definition. In the following we will assume that U, V, W, X, Y, Z are arbitrary sets. All recursion operators which will be introduced are defined for arbitrary sets, with exception of those places where the set $\mathbb{N}$ of natural numbers is mentioned explicitly.

Definition 4 (Recursion operators). Define the following *recursion operators*:

1. **Projection:** If $f :\subseteq X \rightrightarrows Y \times Z$ is an operation then the *projection* $f_1 :\subseteq X \rightrightarrows Y$ is defined by

$$f_1(x) := \{y : (\exists z)\, (y, z) \in f(x)\}$$

for all $x \in \mathrm{dom}(f_1) := \mathrm{dom}(f)$. Let the second projection $f_2 :\subseteq X \rightrightarrows Z$ be defined correspondingly.

2. **Juxtaposition:** If $f :\subseteq X \rightrightarrows Y$ and $g :\subseteq X \rightrightarrows Z$ are operations then the *juxtaposition* $(f, g) :\subseteq X \rightrightarrows Y \times Z$ is defined by

$$(f, g)(x) := f(x) \times g(x) = \{(y, z) : y \in f(x) \text{ and } z \in g(x)\}$$

for all $x \in \mathrm{dom}(f, g) := \mathrm{dom}(f) \cap \mathrm{dom}(g)$.

3. **Product:** If $f :\subseteq X \rightrightarrows Y$ and $g :\subseteq U \rightrightarrows V$ are operations then the *product* $f \times g :\subseteq X \times U \rightrightarrows Y \times V$ is defined by

$$(f \times g)(x, u) := f(x) \times g(u) = \{(y, v) : y \in f(x) \text{ and } v \in g(u)\}$$

for all $(x, u) \in \mathrm{dom}(f \times g) := \mathrm{dom}(f) \times \mathrm{dom}(g)$.

4. **Composition**: If $f :\subseteq X \rightrightarrows Y$ and $g :\subseteq Y \rightrightarrows Z$ are operations then the *composition* $g \circ f :\subseteq X \rightrightarrows Z$ is defined by

$$(g \circ f)(x) := g(f(x)) := \{z : (\exists y \in f(x))\ z \in g(y)\}$$

for all $x \in \mathrm{dom}(g \circ f) := \{x : f(x) \subseteq \mathrm{dom}(g)\}$.

5. **Iteration**: If $f :\subseteq X \rightrightarrows X$ is an operation then the *iteration* $f^* :\subseteq X \times \mathbb{N} \rightrightarrows X$ is defined by

$$\begin{cases} f^*(x, 0) := \{x\}, \\ f^*(x, n+1) := f \circ f^*(x, n) \end{cases}$$

and abbreviated by $f^n(x) := f^*(x, n)$ for all $x \in X$ and $n \in \mathbb{N}$.

6. **Inversion**: If $f :\subseteq X \times \mathbb{N} \rightrightarrows Y \times \mathbb{N}$ is an operation then the *(twisted) inversion* $f^{\leftrightarrow} :\subseteq X \times \mathbb{N} \rightrightarrows Y \times \mathbb{N}$ is defined by

$$f^{\leftrightarrow}(x, n) := \{(y, k) : (y, n) \in f(x, k)\}$$

for all $(x, n) \in \mathrm{dom}(f^{\leftrightarrow}) := \{(x, n) : (\forall k)\ (x, k) \in \mathrm{dom}(f)$ and $(\exists k)\ n \in f_2(x, k)\}$.

7. **Evaluation**: If $f :\subseteq X \rightrightarrows Y^{\mathbb{N}}$ is an operation then the *evaluation* $f_* :\subseteq X \times \mathbb{N} \rightrightarrows Y$ is defined by

$$f_*(x, n) := \{y : (\exists (y_k)_{k \in \mathbb{N}} \in f(x)) y_n = y\}$$

for all $(x, n) \in \mathrm{dom}(f_*) := \mathrm{dom}(f) \times \mathbb{N}$.

8. **Transposition**: If $f :\subseteq X \times \mathbb{N} \rightrightarrows Y$ is an operation then the *transposition* $[f] :\subseteq X \rightrightarrows Y^{\mathbb{N}}$ is defined by

$$[f](x) := \{(y_n)_{n \in \mathbb{N}} : (\forall n)\ y_n \in f(x, n)\}$$

for all $x \in \mathrm{dom}([f]) := \{x : (\forall n)\ (x, n) \in \mathrm{dom}(f)\}$.

9. **Exponentiation**: If $f :\subseteq X \rightrightarrows Y$ is an operation then the *exponentiation* $f^{\mathbb{N}} :\subseteq X^{\mathbb{N}} \rightrightarrows Y^{\mathbb{N}}$ is defined by

$$f^{\mathbb{N}}((x_n)_{n \in \mathbb{N}}) := \{(y_n)_{n \in \mathbb{N}} : (\forall n)\ y_n \in f(x_n)\}$$

for all $(x_n)_{n \in \mathbb{N}} \in \mathrm{dom}(f^{\mathbb{N}}) := \{(x_n)_{n \in \mathbb{N}} : (\forall n)\ x_n \in \mathrm{dom}(f)\}$.

The only structure which is involved in the definition of the recursion operators is the structure of the natural numbers $\mathcal{N}$. From a certain point of view this is unavoidable since the nature of computation incorporates the natural numbers.

It is easy to see that we can find the classically computable functions and operators (that are "general recursive operators") among the recursive operations over $\mathcal{N}$.

Proposition 5 (Classically computable functions). *Each classically computable function* $f :\subseteq \mathbb{N}^k \to \mathbb{N}$ *or* $f :\subseteq \mathbb{N}^{\mathbb{N}} \to \mathbb{N}^{\mathbb{N}}$ *is recursive over* $\mathcal{N}$.

The classical substitution operator can be constructed by juxtaposition and composition, the primitive recursive operator can be constructed by juxtaposition and iteration and the classical μ-recursion operator can be constructed by primitive recursion and twisted inversion (in the classical setting it is well-known that μ-recursion can be replaced by ordinary inversion in presence of suitable initial functions). The (twisted) inversion operator is the only one which does not preserve functionality. On the other hand, it is not as strange as it looks like on the first sight. If it is, for instance, applied to a function $f : \mathbb{N} \to \mathbb{N}$, then the result is simply the preimage operation $f^{\leftrightarrow} : \mathbb{N} \rightrightarrows \mathbb{N}, n \mapsto f^{-1}\{n\}$.

In the following we will often need structures which include the natural numbers and some essential operations. Such structures will be called natural. With $\mathrm{id}_X : X \to X, x \mapsto x$ we denote the *identity* of X.

Definition 6 (Natural structures). A structure S with universe X is *natural*, if $\mathcal{N}$ is a substructure of S, id_X is recursive over S and there is a recursive constant $c \in X$ over S.

2.2 Computable operations over structures

In this section we will define computability over structures. Therefore we will use the language of type 2 theory of effectivity ([Wei87,Wei97,Wei]). The basic notion is the notion of a representation.

Definition 7 (Representation). A *representation* of a set X is a surjective mapping $\delta :\subseteq \mathbb{N}^{\mathbb{N}} \to X$. In this situation (X, δ) is called a *represented space*.

Now we can use the classical notion of computability of operators $F :\subseteq \mathbb{N}^{\mathbb{N}} \to \mathbb{N}^{\mathbb{N}}$ to lift the notion of computability to represented sets. Figure 2 illustrates the definition.

Definition 8 (Computability). Let (X, δ_X), (Y, δ_Y) be represented spaces. Then $f :\subseteq X \to Y$ is called (δ_X, δ_Y)-computable, if there is a computable function $F :\subseteq \mathbb{N}^{\mathbb{N}} \to \mathbb{N}^{\mathbb{N}}$ such that $f\delta_X(p) = \delta_Y F(p)$ for all $p \in \mathrm{dom}(f\delta_X)$. If additionally $p \not\in \mathrm{dom}(F)$ for all $p \in \mathrm{dom}(\delta_X) \setminus \mathrm{dom}(f\delta_X)$ holds, then f is called *strongly (δ_X, δ_Y)-computable*.

From given representations of spaces we can easily construct representations of finite and countably infinite product spaces according to the following definition. For the definition we will use the pairing functions defined by

$$\langle p, q \rangle(k) := \begin{cases} p(n) & \text{if } k = 2n \\ q(n) & \text{if } k = 2n+1, \end{cases}$$
$$\langle p_0, p_1, ...\rangle\langle n, k\rangle := p_n(k)$$

for all $p, q, p_i \in \mathbb{N}^{\mathbb{N}}, i, k, n \in \mathbb{N}$, where $\langle n, k \rangle := \frac{1}{2}(n+k)(n+k+1) + k$ denotes *Cantor's pairing function* on the natural numbers.

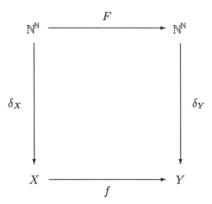

Fig. 2. Computability w.r.t. representations.

Definition 9 (Product and sequence representation). Let $(X, \delta_X), (Y, \delta_Y)$ be represented spaces.

1. The *product representation* $[\delta_X, \delta_Y] :\subseteq \mathbb{N}^{\mathbb{N}} \to X \times Y$ is defined by $[\delta_X, \delta_Y]\langle p, q \rangle := (\delta_X(p), \delta_Y(q))$ for all $p, q \in \mathbb{N}^{\mathbb{N}}$.
2. The *sequence representation* $\delta_X^\infty :\subseteq \mathbb{N}^{\mathbb{N}} \to X^{\mathbb{N}}$ is defined by $\delta_X^\infty \langle p_0, p_1, ... \rangle (n) := \delta_X(p_n)$, for all $p = \langle p_0, p_1, ... \rangle \in \mathbb{N}^{\mathbb{N}}, n \in \mathbb{N}$.

The product of representations can easily be generalized to finite products $[\delta_1, ..., \delta_n]$ by $[\delta] := \delta$ and $[\delta_1, ..., \delta_{n+1}] := [[\delta_1, ..., \delta_n], \delta_{n+1}]$. A suitable tool for the comparison of representation is reducibility.

Definition 10 (Reducibility). Let δ, δ' be representations.

1. δ is *reducible* to δ', or $\delta \leq \delta'$ for short, if there is a computable operator $F :\subseteq \mathbb{N}^{\mathbb{N}} \to \mathbb{N}^{\mathbb{N}}$ such that $\delta(p) = \delta' F(p)$ for all $p \in \text{dom}(\delta)$.
2. δ is *equivalent* to δ', or $\delta \equiv \delta'$ for short, if $\delta \leq \delta'$ and $\delta' \leq \delta$.

It is easy to prove that equivalent representations induce the same (strong) computability on represented sets. Moreover, the product operation on representations is associative up to equivalence, i.e. $[[\delta_1, \delta_2], \delta_3] \equiv [\delta_1, [\delta_2, \delta_3]]$ and equivalent representations have equivalent product and sequence representations.

For the set $\mathbb{N}$ we will use the representation $\delta_{\mathbb{N}} : \mathbb{N}^{\mathbb{N}} \to \mathbb{N}$, which is defined by $\delta_{\mathbb{N}}(p) := p(0)$ for all $p \in \mathbb{N}^{\mathbb{N}}$. Especially, a function $f :\subseteq \mathbb{N} \to \mathbb{N}$ is $(\delta_{\mathbb{N}}, \delta_{\mathbb{N}})$-computable, if and only if it is a restriction of a classically recursive function. Now we will extend the notion of computability to operations.

Definition 11 (Computable operations). Let $(X, \delta_X), (Y, \delta_Y)$ be represented spaces. Then an operation $f :\subseteq X \rightrightarrows Y$ is called (δ_X, δ_Y)-*computable* if there is a computable function $F :\subseteq \mathbb{N}^{\mathbb{N}} \to \mathbb{N}^{\mathbb{N}}$ such that

$$f \delta_X(p) = \{\delta_Y F \langle p, q \rangle : q \in \mathbb{N}^{\mathbb{N}}\}$$

and $\langle p, \mathbb{N}^{\mathbb{N}} \rangle \subseteq \mathrm{dom}(F)$ for all $p \in \mathrm{dom}(f\delta_X)$. Furthermore, f is called *strongly* (δ_X, δ_Y)-*computable* if additionally $\langle p, \mathbb{N}^{\mathbb{N}} \rangle \not\subseteq \mathrm{dom}(F)$ for all $p \in \mathrm{dom}(\delta_X) \setminus \mathrm{dom}(f\delta_X)$. In this situation we will also say that f is (strongly) (δ_X, δ_Y)-computable via F.

Here, $\langle p, \mathbb{N}^{\mathbb{N}} \rangle := \{\langle p, q \rangle : q \in \mathbb{N}^{\mathbb{N}}\}$. It is easy to see that equivalent representations induce the same kind of (strong) computability of operations. Moreover, a function $f :\subseteq X \to Y$ considered as an operation is (δ_X, δ_Y)-computable, if and only if it is (δ_X, δ_Y)-computable as a function (but a corresponding statement for strong computability does not hold). Thus, our notion of computability of operations is a suitable generalization of the notion of computability of functions.

Now we will define representations of structures which will be used in order to define effective structures.

Definition 12 (Representation of structures). Let S be a structure with universe $X = X_1 \times \ldots \times X_n$ and let $\delta_1, \ldots, \delta_n$ be representations of $X_1, \ldots, X_n$, respectively. Then $\delta := [\delta_1, \ldots, \delta_n]$ is called a *representation* of S and (S, δ) is called a *represented structure*.

Now we can also extend the notion of computability to structures.

Definition 13 (Computability in structures). Let S be a structure with universe $X = X_1 \times \ldots \times X_n$ and representation $\delta = [\delta_1, \ldots, \delta_n]$. Let $f :\subseteq Y \to Z$ be an operation over S and let δ_Y, δ_Z be representations of Y, Z, respectively, which are finitely generated from $\delta_1, \ldots, \delta_n$, correspondingly as Y, Z are finitely generated from $X_1, \ldots, X_n$. Then f is called *(strongly) computable* w.r.t. δ, if it is (strongly) (δ_Y, δ_Z)-computable.

If for instance $Y = (X_1 \times X_2)^{\mathbb{N}}$, then the corresponding representation is $\delta_Y = [\delta_1, \delta_2]^{\infty}$.

3 The Stability Theorem for perfect structures

In this section we will define recursive and effective structures and we will compare recursive and computable operations over such structures. Structures which are both, recursive and effective, will be called *perfect*.

3.1 Recursive and effective structures

To define recursive structures we will use the notion of a recursive retraction.

Definition 14 (Recursive retraction). An operation $f :\subseteq X \rightrightarrows Y$ over a structure S is called a *recursive retraction over S*, if it is recursive and it admits a recursive right inverse operation $f^- : Y \rightrightarrows X$ over S, i.e. $f \circ f^- = \mathrm{id}_Y$.

A structure will be called recursive, if it admits a representation which is a recursive retraction. Hence, a recursive structure is a structure with a representation which can be "synthetized" as well as "analyzed" within the structure. Intuitively, a representation of a structure which is a recursive retraction can be "simulated" in the high level language (of recursive operators) over this structure. Formally, these representations correspond to admissible representations in theory of effectivity (cf. [Wei97]).

Definition 15 (Recursive structure). A structure S is called *recursive*, if there is a representation δ of S which is a recursive retraction over S. In this case S is also called *recursive via δ*.

It is easy to prove that the structure of natural numbers is recursive.

Proposition 16. *The structure $\mathbb{N}$ is recursive via $\delta_\mathbb{N}$.*

Since the structure of the natural numbers is recursive it makes sense to apply the notion of recursiveness to natural structures at all. The following theorem shows that strongly computable operations are recursive over recursive structures.

Theorem 17 (Computable operations over recursive structures). *Let S be a natural structure which is recursive via δ. Then each operation over S which is strongly computable w.r.t. δ is also recursive over S.*

The second notion that has to be introduced is the notion of an effective structure.

Definition 18 (Effective structures). A structure S is called *(strongly) effective*, if there is a representation δ of S such that all initial operations of S are (strongly) computable w.r.t. δ. In this situation S is also called *(strongly) effective via δ*.

In the following we will use effective structures with fixed effectivity on natural numbers. We will say that δ is a *natural representation* of a natural structure S if the representation δ' of the natural number component is equivalent to $\delta_\mathbb{N}$. By structural induction one can prove the following theorem.

Theorem 19 (Recursive operations over effective structures). *Let S be a natural structure which is (strongly) effective via a natural representation δ, then each operation which is recursive over S is also (strongly) computable w.r.t. δ.*

3.2 Perfect structures

Over effective structures recursive operations are computable. Over recursive structures computable operations are recursive. Thus, it makes sense to consider structures which are both, recursive and effective. We will see that such structures have very nice properties. The first result states that for such structures there is a unique effectivity which is characterized by the structure itself.

Theorem 20 (Stability theorem). *If S is a natural structure which is recursive via a representation δ and which is effective via a natural representation δ', then $\delta \equiv \delta'$.*

We can deduce that, if a structure S is recursive, then all natural representations δ such that the structure S is effective via δ are equivalent. Thus, the structure S is *effectively categorical*, a notion that has been introduced in a slightly different setting by Hertling ([Her97]). But even more, if a structure is effective via some natural representation, then all representations δ which are recursive retractions over S are also equivalent. Thus, S has a further property which could be called *recursively categorical*.

It should be mentioned that there is no hope for a corresponding result without the restriction to natural representations. As long as we do not demand any evaluation property for the output we can effectivize a structure just by using terms and their evaluations. The situation changes if we do fix effectivity for at least one "output sort". For instance, in the classical setting this can be done by fixing effectivity for the boolean sort, i.e. the "output sort" of predicates. In our setting we have fixed effectivity on the natural numbers.

The proof of the Stability Theorem can be deduced from the following Proposition.

Proposition 21. *Let S be a natural structure which is effective via a natural representation δ' and let δ be another representation of S with right inverse δ^-. Then $\delta'' := (\delta^-)^{-1}$ is a representation of S.*

1. *If δ is recursive over S, then $\delta \leq \delta'$.*
2. *If δ^- is recursive over S, then $\delta' \leq \delta''$.*
3. *Always, $\delta'' \leq \delta$.*

Thus, if δ, δ^- are recursive over S, then $\delta \equiv \delta' \equiv \delta''$.

Proof. Let $X = \mathbb{N} \times X_2 \times ... \times X_n$ be the domain of S and let $\delta' = [\delta'_1, ..., \delta'_n]$. Since δ' is a natural representation $\delta'^\infty_1 \equiv \delta'^\infty_\mathbb{N} \equiv \mathrm{id}_{\mathbb{N}^\mathbb{N}}$.

1. Let $\delta :\subseteq \mathbb{N}^\mathbb{N} \to X$ be recursive over S. By Theorem 19 it follows that δ is computable w.r.t. δ', i.e. δ is $(\delta'^\infty_1, \delta')$-computable. Thus, δ is $(\mathrm{id}_{\mathbb{N}^\mathbb{N}}, \delta')$-computable. But this means $\delta \leq \delta'$.
2. Now let $\delta^- : X \rightrightarrows \mathbb{N}^\mathbb{N}$ be recursive over S. Again we can deduce by Theorem 19 that δ^- is $(\delta', \delta'^\infty_1)$-computable and thus $(\delta', \mathrm{id}_{\mathbb{N}^\mathbb{N}})$-computable, i.e. there is a computable function $F :\subseteq \mathbb{N}^\mathbb{N} \to \mathbb{N}^\mathbb{N}$ such that

$$\delta^-\delta'(p) = \{F\langle p, q\rangle : q \in \mathbb{N}^\mathbb{N}\}$$

for all $p \in \mathrm{dom}(\delta^-\delta') = \mathrm{dom}(\delta')$. Then $G :\subseteq \mathbb{N}^\mathbb{N} \to \mathbb{N}^\mathbb{N}$, defined by $G(p) := F\langle p, p\rangle$ for all $p \in \mathbb{N}^\mathbb{N}$ is computable and

$$\delta'(p) = \delta''G(p)$$

for all $p \in \mathrm{dom}(\delta')$, i.e. $\delta' \leq \delta''$.

3. Since δ^- is a right inverse operation of δ, it follows that δ'' is a restriction of δ and thus $\delta'' \leq \delta$.

The Stability Theorem suggests the following definitions.

Definition 22 (Perfect structures). A natural structure S is called *(strongly) perfect*, if it is recursive and (strongly) effective via a natural representation. In this situation each natural representation δ such that S is recursive via δ is called a *standard representation* of S.

In other words, the Stability Theorem states that all standard representations of a perfect structure and all natural representations which make this structure effective belong to the same equivalence class.

Definition 23 (Computable operations over perfect structures). Let S be a perfect structure with standard representation δ. An operation f over S is called *(strongly) computable over S*, if it is (strongly) computable w.r.t. δ.

Now we can formulate a corollary of Theorem 17 and 19 on effective operations over strongly perfect structures.

Theorem 24 (Effective operations over perfect structures). *An operation over a strongly perfect structure is strongly computable, if and only if it is recursive.*

Furthermore, over perfect structures strongly computable operations are recursive and recursive operations are computable.

Another nice property of strongly perfect structures is the property of "consistent extension". To make this more precise we introduce some definitions. If $S = (S_1, ..., S_k)$, $S' = (S_1, ..., S_n)$ are structures with $n \geq k$, then we will say that S' is an *extension* of S and we will write $S \subseteq S'$.

Now consider a natural structure S with an extension S'. In general the additional substructures of S' can obviously increase the class of recursive operations even over the universe of S. The following theorem states that this cannot happen for perfect structures.

Theorem 25 (Consistent extension). *Let $S \subseteq S'$ be strongly perfect structures and let f be an operation over S. Then f is recursive over S, if and only if f is recursive over S'.*

4 Metric structures

In this section we will discuss some examples of perfect structures. As we have already mentioned in the introduction, it is easy to see that the structure of the natural numbers $\mathcal{N} = (\mathbb{N}, 0, n, n+1)$ is a perfect structure. Here, 0 denotes the constant 0, n denotes the identity function and $n+1$ the successor function. Now we can successively construct other perfect structures. For instance, the structure of the real numbers $\mathcal{R} := (\mathbb{R}, 0, 1, x+y, -x, x\cdot y, 1/x, \mathrm{Lim}, <)$ with the constants $0, 1$, addition, negation, multiplication and inversion on the real numbers, as well as the limit function $\mathrm{Lim} :\subseteq \mathbb{R}^\mathbb{N} \to \mathbb{R}$, $(x_n)_{n\in\mathbb{N}} \mapsto \lim_{n\to\infty} x_n$, which is restricted to rapidly converging Cauchy sequences, more precisely, $\mathrm{dom}(\mathrm{Lim}) := \{(x_n)_{n\in\mathbb{N}} : (\forall n > k)|x_n - x_k| < 2^{-k}\}$, and the semi-characteristic operation $c_<$ of the order relation $\{(x,y) \in \mathbb{R}^2 : x < y\}$, gives rise to a perfect structure $(\mathcal{N}, \mathcal{R})$.

The easiest way to construct further perfect structures is to start with a recursive metric space (cf. [Wei93,Bra99]).

Definition 26 (Recursive metric space). We will call a triple (X, d, α) a *recursive metric space*, if

1. $d : X \times X \to \mathbb{R}$ is a metric on X,
2. $\alpha : \mathbb{N} \to X$ is a sequence which is dense in X,
3. $d \circ (\alpha \times \alpha) : \mathbb{N}^2 \to \mathbb{R}$ is recursive over $(\mathcal{N}, \mathcal{R})$.

It is not very difficult to obtain the following result.

Theorem 27 (Recursive metric spaces). *If (X, d, α) is a recursive metric space, then the structure $\mathcal{M} = (\mathcal{N}, \mathcal{R}, (X, \alpha, \mathrm{id}_X, d, \mathrm{Lim}))$ is a perfect structure.*

Here, $\mathrm{Lim} :\subseteq X^\mathbb{N} \to X$ denotes the limit operation restricted to rapidly converging Cauchy sequences (defined analogously as in case of the real numbers). The perfectness of the structure $\mathcal{S}$, as mentioned in the introduction, can be derived by this theorem (each of the substructures is based on a recursive metric space). A standard representation of the structure $\mathcal{M}$ in the previous theorem can be constructed by the ordinary Cauchy representation of the corresponding metric spaces (cf. [Wei93]). Consequently, we can deduce that the recursive operations over $\mathcal{M}$ are exactly the ordinary computable operations.

Recursive Banach spaces can be defined analogously to recursive metric spaces. Additionally, we demand that the algebraic operations $+, \cdot$ of the Banach space are recursive. If we replace the recursive metric space (X, d, α) in the previous theorem by a recursive Banach space $(X, \|\ \|, e)$, where $e : \mathbb{N} \to X$ is a sequence whose linear span is dense in X, then we can replace $\mathcal{M}$ in the previous theorem by $\mathcal{B} = (\mathcal{N}, \mathcal{R}, (X, e, +, \cdot, \|\ \|, \mathrm{Lim}))$, where $+, \cdot$ denote the algebraic operations of the linear space, and we obtain a perfect structure too. Recursive Banach spaces are very closely related to Banach spaces with computability structures in the sense of Pour-El and Richards [PER89]. The recursive sequences $s : \mathbb{N} \to X$ over a recursive Banach space $\mathcal{B}$ fulfill the axioms of Pour-El and Richards, and vice versa, each Banach space which fulfills these axioms and which, additionally, admits a computable sequence e whose linear span is dense in the space, gives rise to a recursive Banach space.

5 Conclusion

In this paper we have discussed a Stability Theorem for type 2 structures. It provides a sufficient condition for computational stability of such structures, expressed by virtue of an abstract recursive language over such structures. There are some classical results which are related to the Stability Theorem presented here and which express stability conditions without using an abstract recursive language. Our result sheds some new light on these classical theorems. On the one hand there is Mal'cev's stability theorem [Mal71,SHT95].

Theorem 28 (Mal'cev's Stability Theorem). *If A is a finitely generated algebra with canonical term numbering ν such that $\mathrm{dom}(\nu)$ is r.e. and ν' is an effective numbering of A such that $\{(n,k) : \nu'(n) = \nu'(k)\}$ is r.e., then $\nu \equiv \nu'$.*

This theorem can be deduced from our theorem, since under the assumptions of Mal'cev's Stability Theorem the canonical term numbering ν is recursive and surjective since A is finitely generated, and it admits a recursive right inverse over A extended by equality.

A second result which is related to our Stability Theorem is Pour-El and Richard's Stability Lemma for Banach spaces with computable sequence structures [PER89].

Theorem 29 (Pour-El and Richard's Stability Lemma). *If X is a Banach space with computable sequence structures S, S' such that there is a sequence $(e_n)_{n \in \mathbb{N}} \in S \cap S'$ whose linear span is dense in X, then $S = S'$.*

Here, the assumptions of Pour-El and Richard's Stability Lemma guarantee that $(X, \|\ \|, e)$ is a recursive Banach space and hence the standard structure $\mathcal{B}$, as discussed in the previous section, is a perfect structure. Thus, not only the computable sequences $s : \mathbb{N} \to X$ are uniquely determined, as expressed by Pour-El and Richards's result, but all computable operations over this structure are uniquely determined by our Stability Theorem. The same applies to a generalized version of Pour-El and Richard's result for metric spaces, which has been proved by Mori, Tsujii and Yasugi [MTY97].

All these classcial results are quite special. One does cover only finitely generated algebras, the other one separable Banach spaces (separable metric spaces, respectively). Our Stability Theorem generalizes both results in a certain sense and can be applied to other type-2 structures as well. Especially, it covers many structures which are of interest in recursive analysis, including recursive quasi-metric spaces (and thus many interesting T_0-spaces with countable bases). The details are out of scope of this paper and can be found in [Bra98].

References

[BH] Vasco Brattka and Peter Hertling. Feasible real random access machines. *Journal of Complexity*. to appear.

[Bra95] Vasco Brattka. Computable selection in analysis. In Ker-I Ko and Klaus Weihrauch, editors, *Computability and Complexity in Analysis*, volume 190 of *Informatik-Berichte*, pages 125–138. FernUniversität Hagen, September 1995. CCA Workshop, Hagen, August 19-20, 1995.

[Bra96] Vasco Brattka. Recursive characterization of computable real-valued functions and relations. *Theoretical Computer Science*, 162:45–77, 1996.

[Bra98] Vasco Brattka. *Recursive and Computable Operations over Topological Structures*. PhD thesis, FernUniversität Hagen, Fachbereich Informatik, Hagen, Germany, 1998. in preparation.

[Bra99] Vasco Brattka. Computable invariance. *Theoretical Computer Science*, 210:3–20, 1999.

[Her97] Peter Hertling. The real number structure is effectively categorical. CDMTCS Research Report Series 057, University of Auckland, Auckland, September 1997.

[Mal71] Anatolii Ivanovič Mal'cev. Constructive algebras. I. In *The Metamathematics of Algebraic Systems*, pages 148–214, Amsterdam, 1971. North Holland. Collected papers: 1936–1967.

[MTY97] Takakazu Mori, Yoshiki Tsujii, and Mariko Yasugi. Computability structures on metric spaces. In Douglas S. Bridges, Cristian S. Calude, Jeremy Gibbons, Steve Reeves, and Ian H. Witten, editors, *Combinatorics, Complexity, and Logic*, Discrete Mathematics and Theoretical Computer Science, pages 351–362, Singapore, 1997. Springer. Proceedings of DMTCS'96.

[PER89] Marian B. Pour-El and J. Ian Richards. *Computability in Analysis and Physics*. Perspectives in Mathematical Logic. Springer, Berlin, 1989.

[SHT95] V. Stoltenberg-Hansen and J.V. Tucker. Effective algebras. In S. Abramsky, D.M. Gabbay, and T.S.E. Maibaum, editors, *Handbook of Logic in Computer Science, Volume 4*, pages 357–527, Oxford, 1995. Clarendon Press.

[TZ] J.V. Tucker and J.I. Zucker. Computable functions and semicomputable sets on many-sorted algebras. In S. Abramsky, D.M. Gabbay, and T.S.E. Maibaum, editors, *Handbook of Logic in Computer Science, Volume 5*, Oxford. Clarendon Press.

[TZ98] J.V. Tucker and J.I. Zucker. Computation by 'while' programs on topological partial algebras. *Theoretical Computer Science*, 1998. to appear.

[Wei] Klaus Weihrauch. *An Introduction to Computable Analysis*. (Book, to appear 1999).

[Wei87] Klaus Weihrauch. *Computability*, volume 9 of *EATCS Monographs on Theoretical Computer Science*. Springer, Berlin, 1987.

[Wei93] Klaus Weihrauch. Computability on computable metric spaces. *Theoretical Computer Science*, 113:191–210, 1993. Fundamental Study.

[Wei97] Klaus Weihrauch. A foundation for computable analysis. In Douglas S. Bridges, Cristian S. Calude, Jeremy Gibbons, Steve Reeves, and Ian H. Witten, editors, *Combinatorics, Complexity, and Logic*, Discrete Mathematics and Theoretical Computer Science, pages 66–89, Singapore, 1997. Springer. Proceedings of DMTCS'96.

Languages Based on Structural Local Testability

Alessandra Cherubini[1], Stefano Crespi–Reghizzi[2], and Pierluigi San Pietro[2]

[1] Dipartimento di Matematica, Politecnico di Milano,
P.za Leonardo da Vinci 32, I–20133 Milano.
aleche@mate.polimi.it
[2] Dipartimento di Elettronica e Informazione, Politecnico di Milano,
P.za Leonardo da Vinci 32, I–20133 Milano.
{crespi,sanpietr}@elet.polimi.it

Abstract. The generative capacity of context-free grammars is not only insufficient but also misdirected, because it encompasses languages characterised by unnatural mathematical properties. The new Associative Language Description (ALD) model, a combination of locally testable and constituent structure ideas, is proposed, arguing that in practice it equals c.f. grammars in explanatory adequacy, yet it provides a simpler description and it excludes unnatural mathematical sets based on counting properties. The ALD model has been recently proposed as an approach consistent with current views on brain organization. ALD is a "pure", i.e. nonterminal-free definition. Basic formal properties are considered and compared with those of c.f. languages. It is shown that ALD languages enjoy the non-counting property of parenthesized c.f. languages (stencil tree languages). Typical technical languages (Pascal, HTML) can be rather conveniently described by ALD rules.

1 Introduction

In spite of their universal adoption in language reference manuals and compilers, context-free (c.f.) grammars have several shortcomings. A frequently voiced criticism is that they are unable to generate various linguistic constructs, or to handle long distance dependencies. To overcome such limitations, several extended models, known as mildly context-sensitive, have been proposed (e.g. tree adjoining grammars). But the shortcomings of c.f. grammars that we consider are entirely different.

First their generative capacity is not only insufficient, but also misdirected, because it affords languages that are unsuitable for practical use. We have in mind counting languages, which violate the non-counting (n.c.) property, since they characterise the legal strings by some numerical congruence. Clearly nobody has ever proposed a language where grammaticality depends on the number of certain items being odd or even, or more generally congruous to some integer value. Yet c.f. grammars generate all kinds of counting languages. In an attempt to rule out counting, years ago the n.c. c.f. languages have been introduced for parenthesis grammars, and later on reformulated within the theory of tree languages.

A second criticism, originally voiced by Marcus' school of contextual grammars, is that c.f. grammars require an unbounded number of metasymbols, the nonterminals. A "pure" grammar should not use metavariables, but rely instead on structural and distributional properties[1]

The language definition technique to be presented addresses both criticisms, but does not (at present) extend the capacity of c.f. grammars. In essence we have attempted to combine the concepts of local testability and of phrase structure in as simple a way as possible. The objective of this presentation is to formalise the definitions, to highlight the explanatory adequacy by representative examples, and to establish the basic properties.

In Section 2 we introduce the ALD model. In Section 3 we prove its basic properties and compare it with c.f. languages. In Section 4, comparisons are made with n.c. c.f. languages. In the conclusion we discuss related research, including a seemingly analogous model, the semi-contextual (or insertion) grammars of Galiukshov. We terminate mentioning our early experiences on specifying by ALD technical languages such as Pascal or HTML.

2 Basic definitions and introductory examples

Let Σ be a finite alphabet and let $\bot$, the left/right terminator, and Δ, the placeholder, be two new characters. A tree whose internal nodes are labelled by Δ and whose leaves are in Σ will be called a *stencil tree*. Such a tree is composed by juxtaposed subtrees having height one and leaves in $\Sigma \cup \{\Delta\}$, called *constituents*. The *frontier* of a stencil tree T is denoted by $\tau(T)$, while $\tau(K_i)$ denotes the frontier of the constituent K_i.

Definition 1 (left and right contexts).

Let T be a stencil tree. For an internal node i of T, labelled by Δ, let K_i and T_i be respectively the constituent and the maximal subtree of T having root i. Introduce a new symbol $k_i \notin \Sigma \cup \{\bot, \Delta\}$, and consider the tree T_{k_i} obtained replacing in T the subtree T_i with k_i. Consider the frontier of T_{k_i}, $\tau(T_{k_i}) = sk_i t$ with $s, t \in \Sigma^*$. The strings $\bot s$ and $t \bot$ are called, respectively, the left and right contexts of K_i in T and of T_i in T : $\text{left}(K_i, T) = \text{left}(T_i, T) = \bot s$ and $\text{right}(K_i, T) = \text{right}(T_i, T) = t \bot$.

For instance in Fig. 1 the left context of K_3 is $\bot$ acbbbacbcbbcb and the right context of K_1 is $\bot$. Notice that the terminators are automatically prefixed/appended. Next comes the main definition.

---

[1] The idea is in also related with Z. Harrys's linguistic models of word distribution in sentences. Such approaches, also known as Skinner's associative models, were antagonised by Chomsky's generative grammars. Yet associative models on one hand provide an intuitively appealing explanation of many linguistic regularities, on the other they are aligned with current views on information processing in the brain. The first account [CRB98] of the present ALD model was indeed motivated by the want of a brain compatible theory of language [BP92].

Languages based on structural local testability   161

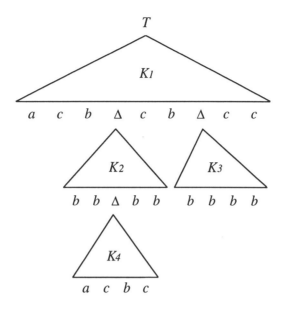

Fig. 1. A stencil tree with four constituents schematized by triangles.

Definition 2 (ALD, pattern, permissible contexts of a rule).

An *Associative Language Description* (ALD) A is a finite collection of triples (x, z, y) or rules [2] where $x \in (\bot \Sigma^* \cup \Sigma^+)$, $y \in (\Sigma^* \bot \cup \Sigma^+)$, and $z \in (\Sigma \cup \{\Delta\})^*$. For a rule, the string z is called the pattern and the strings x and y are called the *permissible left/right contexts*.

As a shorthand, if a left/right permissible context is irrelevant, it can be replaced by the "don't care" symbol '-'. Another shorthand is the new symbol Λ, which may be used to denote the optionality of a Δ, that is to merge two rules $(x, z'\Delta z'', y)$ and $(x, z'z'', y)$ into the rule $(x, z'\Lambda z'', y)$.

An ALD defines a set of constraints or test conditions that a stencil tree must satisfy, in the following sense.

Definition 3 (Valid trees).

A tree T is valid for an ALD A iff for each constituent K_i of T there exists in A a rule $(u, \tau(K_i), v)$ where u is a suffix of the left context of K_i in T and v is a prefix of the right context of K_i in T. In such case, we say that such a rule matches the contexts of K_i in T.

Therefore, an ALD is a device for defining a set of stencil trees and a string language, corresponding to their frontiers.

---

[2] An alternative notation is $x \boxed{z} y$.

Definition 4 (Tree language and string language of an ALD).
The *(stencil) tree language defined by an ALD A*, denoted by $T_L(A)$, is the set of all stencil trees valid for A.
The *(string) language defined by the ALD A*, denoted by $L(A)$ is the set of the strings $x \in \Sigma^*$ such that $x = \tau(T)$ for some tree $T \in T_L(A)$.

The ALDs of some simple languages are shown; more complex ones appear in Section 3.

Examples of ALD

1) The language $\{a^n c b^n \mid n \geq 1\}$ is defined by the ALD rules:

$$(\bot, a\Delta b, \bot), (a, a\Delta b, b), (a, c, b).$$

Actually both contexts can be dropped from the first two rules, and the right (or left but not both) contexts can be omitted from the last rule, obtaining the equivalent description:

$$(-, a\Delta b, -), (a, c, -).$$

where '-' stands for "don't care".
Similarly, the language $L' = \{a^n b^n \mid n \geq 1\}$ is defined by the ALD $(-, a\Lambda b, -)$.

2) The Dyck language (without the empty string ϵ) on the opening parentheses 'b' and the closing parenthesis 'e' is defined by the ALD rules:

$$(-, b\Delta e\Delta, -), (-, b\Delta e, -), (-, be\Delta, -), (-, be, -),$$

where all contexts are "don't care", or shortly by $(-, b\Lambda e\Lambda, -)$.
The phenomenon of *ambiguity* can occur in ALDs much as in c.f. grammars. Without formalizing it, we simply exhibit an ambiguous ALD.

Example of ambiguity

The following rules ambiguously define the Dyck language D_1 on the alphabet $\{b, e\}$

$$(-, \Delta\Delta, -), (-, b\Delta e, -), (-, \epsilon, -)$$

because a sentence like 'bebe' admits distinct tree structures.

Definition 5 (degree, width of an ALD).
We introduce several integer attributes that can be considered for a rule $(x, z, y) \in A$:

- the length of the permissible left/right context, $|x|$ or $|y|$; for a don't care the length is zero.
- the length of the pattern $|z|$, called its width.

For an ALD, the *degree* is the maximum of the lengths of its permissible contexts. The *width* is the maximum of the widths of its patterns.

One-nonterminal context-free grammars are very naturally associated with ALDs.

A *singleton grammar* is a c.f. grammar $G = (\{\Delta\}, \Sigma, P, \Delta)$. For an ALD A, the associated singleton grammar G_{A0} has the production set $P = \{\Delta \to z \mid (x, z, y) \in A\}$.

In a singleton grammar the left part of a production $\Delta \to z$, $z \in (\Sigma \cup \{\Delta\})^*$ can be dropped.

It is obvious that $T_L(G_{A0}) \subseteq T_L(A)$, hence $L(G_{A0}) \subseteq L(A)$. A singleton grammar is essentially the same as an ALD of degree zero, since no permissible contexts are specified.

3 Main properties and comparison with the context-free family

In this section we shall see that the family of ALD languages is strictly included in the c.f. one, yet it owns the hardest context-free language. To prove that certain languages are not ALD we develop an interchange lemma, which can be conveniently combined with the pumping lemma. Then we show that most closure properties no longer hold for ALD and that ALD form a hierarchy w.r.t. pattern width.

The last result relates ALD with non-counting c.f. languages, showing that ALD stencil tree languages enjoy the non-counting property. This confirms the first motivation stated in the introduction.

We start by showing that for any ALD it is possible to construct a structurally equivalent c.f. grammar. For the proof, a few definitions are needed.

Definition 6 (first$_k$, last$_k$, left$_k$, right$_k$).

Let $w \in \Sigma^*$ be a string. The operator first$_k(w)$ denotes the prefix of length k of w if $|w| \geq k$, otherwise it denotes w. Symmetrically for last$_k(w)$: the suffix of w of length k if $|w| \geq k$, otherwise w. The two operators are extended to trees as follows: given a stencil tree T on the terminal alphabet Σ, first$_k(T)$ = first$_k(\tau(T))$, last$_k(T)$ = last$_k(\tau(T))$.

The operators left$_k$ and right$_k$ are defined on pairs of trees: let Z be a stencil tree and T be one of its subtrees whose root is labelled Δ : left$_k(T, Z)$ = last$_k$(left(T, Z)), that is the left context of length k of T in Z (prefixed by $\perp$ if shorter than k). Simmetrically, right$_k(T, Z)$ = first$_k$(right(T, Z)).

Statement 7. For every ALD A there exists a structurally equivalent c.f. grammar.

Proof sketch: Let A be the ALD $\{(x_i, z_i, y_i) \mid 1 \leq i \leq n\}$ on an alphabet Σ and assume for the sake of simplicity that for all i $|x_i| = |y_i| = 1$. The costruction of G in the general case can be done with some technical complications.

We also assume that there is no ϵ-rule in A, that is every z_i is not ϵ. We claim that there exists a c.f. grammar G that is structurally equivalent to A, that is every derivation tree of G is a stencil tree of A and vice versa.

Let $G = (V_N, \Sigma, P, S)$, where P is defined next and the set of nonterminal symbols V_N is $\{< le, fi, la, ri >|\ le, ri \in (\bot \cup \Sigma), fi, la \in \Sigma\}$. P is built so that if there is a subtree T of a stencil tree Z such that:

1. the left and the right contexts of T are, respectively, le and ri: $\text{left}_1(T, Z) = le$, $\text{right}_1(T, Z) = ri$;
2. the first and last symbol of $\tau(T)$ are fi and la respectively; then $< le, fi, la, ri > \Rightarrow^*_G \tau(T)$ and the corresponding derivation tree is structurally equivalent to T. Moreover, there is no other nonterminal symbol from which a structurally equivalent derivation tree may be derived.

Formally, the set P is defined by the following clauses:

1. For every rule $(x_i, z_i, y_i) \in A$, with $z_i \in \Sigma^+$, the production
   $< x_i, first_1(z_i), last_1(z_i), y_i > \to z_i$ is in P.
2. For every rule $(x_i, w_1 \Delta w_2 \Delta \ldots w_m \Delta w_{m+1}, y_i) \in A$, with $m \geq 1, w_j \in \Sigma^*$, then for all $f_i, fi_1, \ldots, fi_m, la, la_1, \ldots, la_m \in \Sigma$ and for all $le, le_1, \ldots, le_m$, $ri, ri_1, \ldots, ri_m \in \Sigma$ the production:
   $< x_i, fi, la, y_i > \to w_1 < le_1, fi_1, la_1, ri_1 > w_2 \ldots w_m < le_m, fi_m, la_m, ri_m > w_{m+1}$
   is in P if all the following conditions hold:
   a. if $w_1 \neq \epsilon$ then $fi = le_1 = last_1(w_1)$, else $f_i = fi_1$ and $le_1 = x_i$;
   b. if $w_{m+1} \neq \epsilon$ then $la = ri_m = first_1(w_{m+1})$, else $la = la_m$ and $ri_m = y_i$;
   c. for $1 < j \leq m$, if $w_j \neq \epsilon$ then $ri_{j-1} = le_j = w_j$, else $ri_{j-1} = fi_j$ and $le_j = la_{j-1}$.
3. For all $fi, la \in \Sigma$, the production $S \to <\bot, fi, la, \bot>$ is in P.
4. No other production is in P.

Clause (1) deals with ALD rules that do not have a Δ. Clause (2) considers the case of rules with $m \geq 1$ occurrences of a Δ, possibly separated by terminal substrings w_j. Condition (2a) deals with the values of fi and le_1: if $w_1 \neq \epsilon$ then the first terminal symbol (i.e., fi) generated by $< x_i, fi, la, y_i >$ is w_1 and the left context (i.e., le_1) of the first nonterminal symbol $< le_1, fi_1, la_1, ri_1 >$ is w_1; otherwise, fi is the first terminal generated by the first nonterminal $< le_1, fi_1, la_1, ri_1 >$ and le_1 is exactly the left context of the ALD rule, x_i. Condition (2b) is the symmetrical, for la and ri_m, of condition (2a). Condition (2c) deals with the intermediate nonterminal symbols $< le_j, fi_j, la_j, ri_j >$. Clause (3) permits to derive from S every nonterminal symbol with initial left and right contexts: these symbols are the real axioms of the grammar.

The proof of structural equivalence procedes then by induction, after noticing that if $< le, fi, la, ri >$ is the root of a subtree α of a derivation tree γ in G then $le = left_k(\alpha, \gamma)$, $fi = first_k(\alpha)$, $la = last_k(\alpha)$, $ri = right_k(\alpha, \gamma)$. □

The next technical lemma permits to interchange two subtrees provided they have similar profiles.

Statement 8 (Swapping lemma).
Let A be an ALD of degree $k \geq 0$ and let Z and Z' be two valid trees of A. Suppose that there exist two subtrees T of Z and T' of Z' such that:
$\text{left}_k(T, Z) = \text{left}_k(T', Z')$, $\text{right}_k(T, Z) = \text{right}_k(T', Z')$, $\text{last}_k(T) = \text{last}_k(T')$, $\text{first}_k(T) = \text{first}_k(T')$.
Then also the stencil tree Z'' obtained from Z by replacing the subtree T with the subtree T' is a valid tree of A.

Proof: Consider the grammar G structurally equivalent to A as defined in the proof of Statement refs3.1. Let T, T', Z and Z' be stencil trees verifying the conditions of the Lemma. The derivation tree of G corresponding to Z has a subtree corresponding to T with a nonterminal symbol X at its root; analogously the subtree T' has a nonterminal symbol Y. By construction of G, it must be $X = <\text{left}_k(T, Z), \text{first}_k(T), \text{last}_k(T), \text{right}_k(T, Z)>$ and $Y = <\text{left}_k(T', Z'), \text{first}_k(T'), \text{last}_k(T'), \text{right}_k(T', Z')>$. Hence, $X = Y$ and it is possible to build a derivation tree of G where the subtree corresponding to T is replaced by the subtree corresponding to T'. □

We can now show that certain c.f. languages are not ALD.

Theorem 9. *The family of ALD languages is strictly included in the family of c.f. languages.*

Proof: We show that the nondeterministic c.f. language $L = L' \cup L'' = \{a^n b^n \mid n \geq 1\} \cup \{a^n b^{2n} \mid n \geq 1\}$ is not in ALD.

By contradiction, let A be an ALD of degree k such that $L = L(A)$.

The traditional Pumping Lemma (see for instance [HU79]) can be applied to $L = L(G)$, where G is the structurally-equivalent c.f. grammar of A: let $m > 1$ be the constant of the Pumping Lemma and let $a^n b^n$ and $a^n b^{2n}$ be two words, for $n > 3km$. Then, $a^n b^n = xuwvy$, $a^n b^{2n} = x'u'w'v'y'$, with $|uwv| < m$ and $|u'w'v'| < m$, uv and $u'v'$ are not ϵ, and both $xu^i wv^i y$ and $x'u'^i w'v'^i y'$ are in L.

We consider the word $z = xu^k wv^k y$, which is in L. Notice that, since $n > 3km$ and $|uwv| < m$, it is $|u^k wv^k| < n$. Similarly, we consider the word $z' = x'u'^k w'v'^k y'$.

Because of the nature of the language L, $u^k wv^k = a^p b^q$, for some p and q, with $k \leq p \leq km$ and $k \leq q \leq km$ (being $|uwv| < m$), and $x = a^{n-p}$ and $y = b^{n-q}$, with $|x| > k$ and $|y| > k$ (being $p \leq km$, $n - p \geq 3km - km > k$).

The Pumping Lemma is based on the fact that $u^k wv^k$ is the yield of a subtree T of the derivation tree Z of z.

Clearly, $\text{left}_k(T, Z) = a^k$, $\text{right}_k(T, Z) = b^k$, $\text{last}_k(T) = b^k$, $\text{first}_k(T) = a^k$.

By applying analogous considerations to z', we can find that $u'^k w' v'^k = a^{p'} b^{q'}$, for some p' and q', $x' = a^{n-p'}$ and $y' = b^{n-q'}$, with $k \leq p' \leq km$ and $k \leq q' \leq km$, and $|x'| > k$, $|y'| > k$; moreover, $u'^k w' v'^k$ is the yield of a subtree T' of the derivation tree Z' of z: $\text{left}_k(T', Z') = a^k$, $\text{right}_k(T', Z') = b^k$, $\text{last}_k(T') = b^k$, $\text{first}_k(T') = a^k$.

We can apply the Swapping Lemma to T and T': the tree Z'' has yield x u'^k w' v'^k y, that is the word $z'' = a^{n-p+p'} cb^{n-q+q'}$. The word z'' cannot be of type $a^v b^{2v}$. If this were the case, then $2(n - p + p') = n - q + q'$, that is $n = 2p' - 2p + q' - q < 2p' + q'$. But being $p' \leq km$ and $q' \leq km$, it is $n < 2p' + q' \leq 3km < n$, a contradiction. The word z'' can neither be of type $a^v b^v$: if this were the case, then $p = p'$ and $q = q'$: however, in this case we could have chosen a different z', namely $x'u'^{k+1} w'v'^{k+1} y'$ and applied the Swapping Lemma to it: again z'' cannot be of type $a^v b^{2v}$ for the same reasons as above (by choosing n suitably large), and it cannot be of type $a^v b^v$ since now $p \neq p'$ or $q \neq q'$. □

Corollary 10. *The ALD family is not closed under union.*

Proof: It follows from the proof of the previous statement, since L' and L'' are easily defined by ALDs. □

Corollary 11. *The ALD family is not closed under (alphabetic, nonerasing) homomorphism.*

Proof: Let L be the language $\{a^n b^n \mid n \geq 1\} \cup \{a^n c^{2n} \mid n \geq 1\}$. L is an ALD, as shown by the following set of rules: $\{(\bot, a \Lambda b, \bot), (a, a \Lambda b, b), (\bot, a \Lambda cc, \bot), (a, a \Lambda cc, c)\}$.

Let h be the homomorphism: $h(a) = a$, $h(b) = b$, $h(c) = b$. Then $h(L)$ is the language $\{a^n b^n \mid n \geq 1\} \cup \{a^n b^{2n} \mid n \geq 1\}$, which is shown in Theorem 9 not to be an ALD. □

Statement 12. 1. *The ALD family is not closed under concatenation*
2. *The ALD family is not closed under Kleene star*

Proof: Let $L = \{a^n b a^n a^{2p} \mid n \geq 1, p \geq 1\}$. L is defined by the ALD: $\{(\bot, a \Delta a \Delta, \bot), (a, a \Delta a, a), (a, b, a), (a, aa \Lambda, \bot)\}$.

To prove part (1), we show that $L \cdot L \notin$ ALD. Suppose $L \cdot L$ is $L(A)$ for an ALD of degree $k \geq 1$. We can apply the Ogden Lemma (in the version given in [HU79]) to the structurally equivalent c.f. grammar G, with m the constant of the Lemma, using a string $z = a^n b a^n a^{2p} a^n b a^n a^{2p} \in L(A)$ such that: $z = \tau(Z)$ where Z is a stencil tree of A, and n and p are greater than $2(k+m) \cdot width(A)$. Let Z be the stencil tree of z. Mark as distinguished every position in the first group $a^n b a^n$ and apply the Ogden Lemma:

$z = xuwvy$ with:

1. $u w v$ containing at most m distinguished positions;
2. $u v$ containing at least one distinguished position;
3. $x u^i w v^i y$ is in $L(A)$ for every $i \geq 0$.

As in the proof of Theorem 9, a subtree T' of Z is identified for a subword $a^r b a^{r'}$ (corresponding to uwv), for some r and r' greater than k (if either r or r' are smaller than k then consider $u^k w v^k$ instead). Analogously, mark as distinguished the positions of the second group $a^n b a^n$: a subtree T'' of Z is identified for a subword $a^t b a^{t'}$, for some $t, t' > k$. Both T' and T'' have $left_k$, $right_k$, $first_k$

Languages based on structural local testability   167

and last_k equal to a^k. Now the substring s between the two substrings frontiers of T' and T'' is of type a^+ with $|s| > 2p + 2(n-m) > 8k \cdot width(A)$. Hence, there is certainly a subtree W of Z such that: W has no nodes in common with T' and T''; $\text{left}_k(W, Z) = \text{right}_k(W, Z) = \text{last}_k(W) = \text{first}_k(W) = a^k$; the frontier of W is a substring of s. Hence, the Interchange Lemma may be applied: the string z''' obtained by replacing the subtree T' with the subtree W is in $L(A)$. But z''' is a string of type $a^+a^nba^na^{2p}$, which is not in $L(A)$: a contradiction. Part (2) follows immediately, too, since $z \in L^*$ but $z''' \notin L^*$. □

Another example which is c.f. but not ALD is the language $L'L'' = \{a^nb^n a^mb^{2m} \mid n, m \geq 1\}$. This can be easily proved with the same technique used in Theorem 9.

A philosophical remark: a common prejudice we wish to question is that any family of languages not closed under union or concatenation is practically useless. Actually, the study of the evolution of natural languages provides evidence to the contrary: the addition of new constructs that are similar to existing ones and could cause confusion, causes their disappearance from the language. But also the older artificial languages, though much simpler, exhibit the same phenomenon. Each revision and extension of a language such as Fortran was severely constrained in the choice of the syntactic constructs to be added to the language by the requirement of "upper-compatibility". In conclusion, any practical language should not be viewed as the union of unrelated and independent sublanguages, but as a harmonious whole.

A further confirmation: the contextual grammars, a powerful, rich family of "pure" devices proposed for describing natural languages, are also rarely closed under basic operations.

The two integer parameters we have introduced, degree and width, classify the ALD family into hierarchies. Here we consider the width hierarchy, to show that it determines a strict hierarchy of languages. This is in contrast with the well-known existence of width-bounded normal forms (Chomsky normal form) for c.f. grammars.

Let $ALD_{W=k}$, $k \geq 0$, be the subfamily of ALD having width k.

Statement 13. For any $k \geq 0$ there exists a language which is in $ALD_{W=k+1}$ but not in $ALD_{W=k}$.

Proof: For $k = 0$ the only language in $ALD_{W=0}$ is $\{\epsilon\}$, giving the strict inclusion $ALD_{W=0} \subset ALD_{W=1}$.

Now consider for $j \geq 1$ the language
$L_j = \{a^{n_1}ba^{n_1}a^{n_2}ba^{n_2}\ldots a^{n_j}ba^{n_j} \mid n_i \geq 1\}$
which is in $ALD_{W=j}$ being defined by the rules:
$(\bot, \Delta^j, \bot)$, $(\bot, a\Delta a, a)$, $(a, a\Delta a, a)$, $(a, a\Delta a, \bot)$, (a, b, a)

Each substring a^nba^n is called a group: groups are ordered from 1 to j. By contradiction suppose that L_j is also defined by A_{j-1} in $ALD_{W=j-1}$. By applying the c.f. pumping lemma to the sentences of the form:
$aba\ldots abaa^nba^naba\ldots aba$

for n large enough, we find that their stencil trees must contain a subtree T' which includes the frontier $a^m b a^m$, with $m \leq n$, unbounded. Since the width is $j-1$, for some string in the language there must be a subtree T'' whose frontier contains two or more groups, say the groups at positions $1 \leq h < k \leq j$, i.e., it contains two or more characters b. From the swapping lemma, no matter how large is the degree of A_{j-1}, the trees T' and T'' have left/right contexts and prefixes/suffixes which cannot be distinguished. But substituting T'' for T' in a sentence the number of groups would exceed j, causing the strings not to be in the language. □

The next result is a piece of evidence that by restricting the c.f. family to the ALD capacity its adequacy is not impaired.

Definition 14 (Hardest Context-Free Language H [H78; §10.5]).

Let D_2 be the semi-Dyck Set[3] on $\{a_1, a_2, a'_1, a'_2\}$, $\Sigma = \{a_1, a_2, a'_1, a'_2, c, \$\}$, $d \notin \Sigma$.

Then $H = \{\epsilon\} \cup \{x_1 c \$ y_1 c z_1 d x_2 c y_2 c z_2 d \ldots x_n c y_n c z_n d \mid n \geq 1, y_1 y_2 \ldots y_n \in D_2, x_i, z_i \in \Sigma^*, y_i \in \{a_1, a_2, a'_1, a'_2\}^*\}$.

Statement 15. The hardest c.f. language H is an ALD language.

Proof: Let A be the following ALD

$$A = \{(\bot, \Lambda c \$ \Lambda c \Lambda d, \bot), (\bot, \epsilon, \bot)\} \cup \{(-, p\Lambda, d) \mid p \in \Sigma\} \cup \{(d, \Lambda p, -) \mid p \in \Sigma\}$$
$$\cup \{(\bot, \Lambda p, -) \mid p \in \Sigma\} \cup_{p,q \in \Sigma} \{\{(p, a_1 \Lambda a'_1 \Lambda, q)\} \cup \{(p, a_2 \Lambda a'_2 \Lambda, q)\}$$
$$\cup \{(p, c \Lambda d \Lambda c \Lambda, q)\} \cup \{(p, \Lambda c \Lambda d \Lambda c, q)\}\}$$

$L(A) = HCFL$. This can be shown by noticing that H is generated by the following grammar G:
$G = (\Sigma \cup \{d\}, \{S, X, Y\}, S, P)$ where P is the set:

$$P = \{S \to X c \$ Y c X d | \epsilon, X \to a_1 X | a'_1 X | a_2 X | a'_2 X | c X | \$ X | \epsilon,$$
$$Y \to \epsilon | a_1 Y a'_1 Y | a_2 Y a'_2 Y | c X d X c Y | Y c X d X c\}.$$

To show that A and G are structurally equivalent, hence $L(A) = L(G)$, we prove by induction that every derivation tree α from S in G has a corresponding ALD tree T_α obtained from α by erasing the nonterminal labels.

If the height of α has length 1, then α is $S \Rightarrow \epsilon$. Clearly, T_α is the empty tree, obtained by means of the ALD rule $(\bot, \epsilon, \bot)$.

If α has height greater than one, then at its root there is an application of the production $S \to X c \$ Y c X d$: the corresponding ALD rule to apply at the root of the stencil tree T_α is $(\bot, \Lambda c \$ \Lambda c \Lambda d, \bot)$. Whenever a production of type $X \to bX$, $b \in \Sigma$ is used in α, the right context of X is d or the left context is in $\{d, \bot\}$: the ALD rules $\{(-, p\Lambda, d) \mid p \in \Sigma\} \cup \{(d, \Lambda p, -) \mid p \in \Sigma\} \cup \{(\bot, \Lambda p, -) \mid p \in \Sigma\}$ can be applied. When a production of the types

---
[3] D_2 is the language generated by the grammar $S \to a_1 S a'_1 S | a_2 S a'_2 S | \epsilon$.

$Y \to a_1 Y a_1' \; Y|a_2 Y a_2' Y|cXdXcY|YcXdXc$ is applied, since both the left context and the right context of every node labeled Y are always in Σ, the ALD rules $\cup_{p,q \in \Sigma}\{\{(p, a_1 \Lambda a_1' \Lambda, q)\} \cup \{(p, a_2 \Lambda a_2' \Lambda, q)\} \cup \{(p, c\Lambda d\Lambda c\Lambda, q)\} \cup \{(p, \Lambda c\Lambda d\Lambda c, q)\}\}$ can be applied.

The converse proof that every ALD tree T_α corresponds to a derivation tree α from S in G is analogous and is omitted. □

Since any c.f. language can be expressed as the inverse homomorphism of the language H [H78; §10.5] and the family of ALD languages is strictly included in the family of c.f. languages, we have the following non-closure result.

Corollary 16. *The family of ALD is not closed under inverse homomorphism.*

Corollary 17. *The problem of parsing ALD languages has the same complexity of parsing c.f. languages.*

Proof: Parsing of any c.f. language is at least as hard as parsing of H [H78]. □

A remark on deterministic parsing. ALD includes both deterministic and non-deterministic c.f. languages. Deterministic ALD subclasses could be easily defined and parsing algorithms for ALD could be obtained by straightforward changes to the classical bottom-up algorithms ($LR(k)$ or precedence).

4 Non-counting property

This section focuses on the non-counting (n.c.) property that is generally viewed as a linguistic universal of human linguistic behaviour,[4] for both natural and artificial languages.

Different notions of non-counting have been proposed in the past for (i) regular string languages, (ii) c.f. parenthesized languages, and (iii) tree languages. Given that ALD are a "pure" definition coping with stencil tree structures, the most natural comparison is with (ii), since parenthesized strings are isomorphic to stencil trees.

We prove the inclusion of the ALD tree language family in the n.c. c.f.. language family of [CRGM78], whose definitions[5] are not reproduced for brevity sake.

Statement 18. *The family of ALD tree language is strictly included in the family of noncounting parenthesized c.f. languages.*

Proof sketch: First we prove inclusion. Let A be an ALD of degree k. Take $n = k + 1$. Let T be a stencil tree with $\tau(T) = xv^n wu^n y$ and let vwu and w be the terminal strings of two subtrees T_1 and T_2 of T. Denote by K_1 and K_2 the constituents generating respectively T_1 and T_2 of T. Suppose that T is a valid

---

[4] Modulo-counting is on the other hand important in other forms of human behaviour such as music and arithmetics.

[5] In [CRGM78] the concept is expressed using parentheses languages, but its reformulation in terms of stencil trees is obvious, see [THO82].

tree for A, then there is a rule $(z, \tau(K_1), t)$ in A matching with the context of K_1 in T, hence z is a suffix of v^{n-1} and t is a prefix of u^{n-1}. This rule matches the context of K_2 of T, hence the tree T' obtained by T replacing T_2 with T_1 is a valid tree with $\tau(T') = xv^{n+1}wu^{n+1}y$. Repeating the same argument we prove that for each positive integer h there is a valid tree $T(h)$ such that $\tau(T(h)) = xv^{n+h}wu^{n+h}y$. Conversely suppose that there is a valid tree T' such that $\tau(T') = xv^{n+1}wu^{n+1}y$. Let vwu and w be the terminal strings of two subtrees T'_1 and T'_2 of T' generated by K'_1 and K'_2; then the rule $(s, \tau(K'_2), r)$ in A matches the context of K'_2 in T', hence s is a suffix of v^n and t is a prefix of u^n. By choice of n, s is a suffix of v^{n-1} and t is a prefix of u^{n-1}, whence this rule matches the context of K'_1 of T', hence the tree T obtained by T' replacing T_1 with T_2 is a valid tree with $\tau(T) = xv^n wu^n y$.

To show that the inclusion is strict, it suffices to reconsider the non-ALD language $L = \{a^n b^n \mid n \geq 1\} \cup \{a^n b^{2n} \mid n \geq 1\}$, used in the proof of Theorem 9, for which a n.c. c.f. parenthesis grammar is: $S \to (X)$, $S \to (Y)$, $X \to (aXb)$, $X \to (ab)$, $Y \to (aXbb)$, $Y \to (abb)$ (of course parentheses do not belong to the terminal alphabet of the ALD). □

Further remarks on the relations between ALD and other n.c. families are in the conclusion.

5 Conclusion

We discuss related work, including non-counting aspects, then some practical work on the specification of Pascal by ALD, to finish with grammar inference issues.

Related work

ALD is a "pure" technique, since it does not use nonterminal symbols. The best known classes of pure generative grammars are the *contextual grammars* of S. Marcus [Ma97] and [EPR97]. Interesting enough, Marcus original motivation was to formalize the findings of distributional or associative linguistics, in contrast to Chomsky's generative approach. The model is much more powerful than ALD, because it allows insertion of two matching words in specified contexts, but a simplified variant looks at first rather similar to ALD. A *semicontextual* or *insertion* grammar, due to B. Galiukshov, studied in [P84], is a collection of triples $< x, z, y >$, of terminal strings. Each triple is interpreted as a rewriting rule $xy \to xzy$ allowing the insertion of the string z in the context $x \ldots y$. Consider now a complex ALD rule (x, Z, y) where Z is the finite set obtained from z optionally inserting the placeholder in any position. At first glance, these ALD rules and the previous insertion rule would appear to have the same effect. But the fundamental difference is that the insertion rule gives rise to a *generative* derivation process, the ALD rule is used for *testing* a given tree. Notice that we have not defined the notion of derivation for ALD models, because, in general, one rule cannot be applied to insert a pattern into a string independently of the other rules. A simple example is the language $\{ab\}$, defined by the ALD: $\{(\bot, \Delta\Delta, \bot), (\bot, a, b), (a, b, \bot)\}$. Clearly the two insertions must be done in parallel, because they interlock. On the other hand the semicontextual grammar $\bot\ b \to \bot\ ab$, $a\ \bot \to ab\ \bot$, with initial set $\{\bot\bot\}$ generates nothing. Moreover it is known that semicontextual grammars generate also non c.f. languages. In conclusion ALD and contextual grammars are quite different entities, apart from both being pure models, and of having very few closure properties. A more accurate comparison remains to be done.

Relations with noncounting and locally testable families

Our model is connected with the mathematical theory of local testability and counter-free languages, that has been investigated starting from the 60's, and later extended to stencil tree and to tree languages.

For the following we will use concepts and terminology given in [MNP71] 2.1, 2.1, 2.3.

For string languages, the simplest form of Local Testability (l.t.) is qualified "in the strict sense". It is immediate to prove that any language that is l.t. in strict sense can be defined by an ALD: the degree of the ALD equals the length of the words occurring in the test set.

The relationships between ALD and the more general l.t. and noncounting (also known as aperiodic) families of regular languages are not completely known.

Moving now to the tree counterparts of l.t. in strict sense (see [GS97], 8, p.19) we obtain incomparable families of tree languages. To make the comparison with ALD meaningful, the tree language must be transformed first into a stencil tree language, by mapping all internal labels onto the placeholder symbol.

There is an essential difference between the "local" test performed by ALD and by l.t. tree languages. The latter extract the subtrees of depth k which occur in the tree, and check that they are included in the acceptance set.

In fact, the tree language T corresponding to the c.f. grammar $G: \{S \to aSb|ab|X, X \to aXbb|abb\}$ is l.t. by means of the set shown in Fig 2. But the string language of T is $\{a^n b^n \mid n \geq 1\} \cup \{a^n b^{2n} \mid n \geq 1\}$ that is shown in the proof of Theorem 9 not to be ALD.

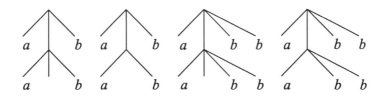

Fig. 2. A set ot trees for local testability of T.

On the other hand the language $c^+ ac^+ b$ is defined by the ALD $A = \{(\bot, \Delta\Delta, \bot), (\bot, c\Delta, -), (c, c\Delta, -), (c, a, c), (-, b, \bot), (a, c\Delta, \bot)\}$, but the tree language defined by A is not l.t. in strict sense.

Less obviously related ideas on the combination of local testability with constituent structures can be found in the concept of *precedence of operators* introduced by Floyd [F63], to make c.f. parsing deterministic. He proposes that pairs of terminal characters (precedence relations) be used to detect the left and right border of a constituent, essentially a form of local test quite similar to the one used in ALD. The non-counting operator precedence grammars have been investigated in [CRGM81].

Empirical convenience of ALD

To what extent existing technical languages can be defined by ALD? Using ALD we [6] have completed the definition of the c.f. syntax of Pascal. For instance the start rules of Pascal are the following:

$$\bot \,\boxed{\text{program id}; \, \Delta \text{ begin } \Delta \text{ end}}$$

$$\bot \,\boxed{\text{program id (id } \Delta); \, \Delta \text{ begin } \Delta \text{ end}}$$

To allow the insertion of more program parameters we use the rule of left degree two:

$$(\text{id} \,\boxed{\Delta, \text{ id}}$$

Another placeholder stands for constant / type / variable / procedure / function declarations, within the permissible context ";...begin". The last placeholder of the start rules stands for a list of statements. We used here the alternative notation for the ALD rules to avoid ambiguity between characters.

The size of the ALD definition is comparable to a c.f. grammar of Pascal using short notations. This line of research is in progress.

Grammar inference

Grammar inference (see [YS97] for a recent survey) is the process of learning a grammar (in our case an ALD) from examples, counterexamples, or other hints provided by an informant.

Inference is straightforward when the given examples are structured, i.e. stencil trees are given, rather than strings. The form of the stencil trees is imposed by the semantics, and the ALD must cope with it: for instance, in arithmetic expressions the order of evaluation from left to right, combined with products taking precedence over additions, determines the stencil trees. In that situation the inference of the ALD is purely mechanical, since it is enough to inspect the given stencil tree and extract the patterns, together with their minimal distinctive contexts.

In fact, a sample of stencil trees functionally determines an ALD grammar, for a fixed degree k. If the inferred ALD overgeneralises to some illegal strings, then the degree must be increased, until necessary. This remark shows that

---

[6] Work done by Matteo Pradella.

language specification by example (as proposed in [CRML73]) should be feasible with ALD. A practical application could be in extensible languages, in order to infer from the examples the new extensions.

On the other hand, inferring an ALD from examples of strings is more complex because it is not trivial to assign a suitable structure to the strings.

The formal study of ALD has just started and several theoretical questions are still open or under investigation, e.g. inclusion of regular set, decidability of ambiguity, minimization w.r.t. degree or width. We hope that the associative language description model could be a simpler competitor of the context-free model both as an explanation of fundamental syntactic phenomena and as a practical technique for language specification. Thanks to its simplicity, the model should be a good basis for extensions and refinements. This is an open line of research that could address the problems of non-context-free constructions of natural languages.

Acknowledgements: Valentino Braitenberg has provided the initial stimulus for formalizing associative language descriptions.

References

[BP92] V. Braitenberg and F. Pulvermüller, *Entwurf einer neurologischen Theorie der Sprache*, Naturwissenschaften **79** (1992), 103-117.

[CRB98] S. Crespi Reghizzi and V. Braitenberg, *Towards a brain compatible theory of language based on local testability*, in V. Braitenberg and F.J. Rdermacher (eds), Interdisciplinary approaches to a new understanding of cognition and consciousness, to appear. Also as Tech. Rept. 98-004, Politecnico di Milano, Dip. Elettronica e Informazione.

[CRGM78] S. Crespi-Reghizzi, G. Guida and D. Mandrioli , *Non-counting context-free languages*, Journ. ACM, **25** (1978), 4, 571-580.

[CRGM81] S. Crespi-Reghizzi, G. Guida and D. Mandrioli [1981], *Operators precedence grammars and the non-counting Property*, SIAM J. Computing, **10** (1981), 174-191.

[CRML73] S.Crespi-Reghizzi, M.A. Melkanoff, and L. Lichten, *The use of grammatical inference for designing programming languages*, Comm.ACM, **16** (1973), 2, 83-90.

[EPR97] A. Ehrenfeucht, G. Pun and G. Rozenberg, *Contextual grammars and formal languages*, Handbook of formal languages (Eds. G.Rozenberg, A.Saloma), Vol.II, Ch.6, Springer (1997), 237-290.

[F63] R. W. Floyd , *Syntactic analysis and operator precedence*, Journ. ACM, **10** (1963), 316-333.

[GS97] F. Gcseg, M. Steinby, *Tree languages*, Handbook of formal languages (Eds. G.Rozenberg, A.Saloma), Vol.III, Ch.1, Springer (1997), 1-61.

[H78] M. Harrison, *Introduction to formal language theory*, Addison Wesley (1978).

[HU79] J. H. Hopcroft, J.D. Ullman, *An Introduction to Automata Theory*, Formal Languages and Computation, Addison Wesley (1979).

[Ma97] S. Marcus, *Contextual grammars and natural languages*, Handbook of formal languages (Eds. G.Rozenberg, A.Saloma), Vol.II, Ch.6, Springer (1997), 215-132.

[MNP71] R. McNaughton, S. Papert, Counter-free Automata, MIT Press (1971).
[P84] G. Pun, *On semicontextual grammars* Bull. Math Soc. Sci. Math. Roumanie, **28** (1984), 76, 63-68.
[THO82] W. Thomas, *On noncounting tree languages*, Grundl. Theor. Informatik, Proc. 1. Int. Workshop, Paderborn 1982, 234-242.
[YS97] Yasubami Sakakibara, *Recent advances of grammatical inference*, TCS 185, 15-45, 1997.

Towards Automatic Bisimilarity Checking in the Spi Calculus

Anders Strandløv Elkjær, Michael Höhle, Hans Hüttel, and Kasper Overgård

BRICS, Department of Computer Science, Aalborg University, Fredrik Bajers Vej 7E,
9220 Aalborg Ø, DENMARK. {ase,hoehle,hans,kaspero}@cs.auc.dk

Abstract. The spi calculus by Abadi and Gordon, an extension of the π-calculus of Milner, Parrow and Walker, is designed to model cryptographic protocols. Classical security properties are easily expressed in spi using the notion of testing equivalence by De Nicola and Hennessy. However, proving processes testing equivalent is a daunting task. Thus framed bisimilarity, a bisimulation method implying testing equivalence, has been proposed by Abadi and Gordon. Unfortunately the definition of framed bisimilarity uses several levels of quantification over infinite domains and is therefore not effective. In this paper we define *fenced bisimilarity*, a concept similar to framed bisimilarity in which one of these quantifiers has been replaced by an effective condition, and show that fenced bisimilarity coincides with framed bisimilarity.

1 Introduction

Cryptographic protocols arise in the context of secure data communication. They only consist of 2-5 messages and are used to identify users and authenticate transactions. Despite their size, quite a number of protocols contain serious design blunders [5] that make it easier for an intruder to attack the protocol rather than to break its message enciphering algorithm [15]. Several approaches to proving correctness of cryptographic protocols exist [6, 1, 2, 12]; some require the behaviour of the attacker to be modelled, others operate on too high levels of abstraction thus making the relation to the actual implementation tenuous.

The spi-calculus [1, 2] is an extension of the π-calculus [11] with cryptographic primitives. It strikes a golden mean in that it does not require any assumptions about the behaviour of an attacker (except that it should be expressible within the calculus) and has a precise operational semantics.

In the spi-calculus, correctness is checked by comparing implementations with their specifications, both expressed within the calculus. Abadi and Gordon suggest using the notion of *testing equivalence*, first introduced by DeNicola and Hennessy, as the correctness relation. Two processes are testing equivalent if they allow the same end observations in all observation contexts. As observers in the spi-calculus setting are potentially malicious, two processes are equivalent if they can resist the same attacks.

It is seldom possible to check directly if two processes are testing equivalent as the definition contains a universal quantification over attackers. Therefore it is necessary to find more tractable notions of equivalence which imply testing equivalence. Variants of the notion of bisimilarity, introduced by Park [14], are popular candidates, as their definition allows one to prove equivalence by means of a coinductive technique.

However, strong bisimulation used in the π-calculus is a too restrictive notion of bisimilarity for cryptographic protocols, as there are numerous examples of processes that are testing equivalent but not strongly bisimilar. A less restrictive notion of bisimulation has been introduced by Abadi and Gordon in [3], the notion of framed bisimilarity. They conjecture that framed bisimilarity can be fully automated. Unfortunately, their definition is not effective as it uses several levels of quantification over infinite domains.

This paper presents a sound and complete characterization of framed bisimilarity, called *fenced bisimilarity*, which replaces one of the infinite existential quantifications of the definition with a terminating algorithm, and therefore is a significant step towards a fully effective characterization of framed bisimilarity.

2 The spi calculus

2.1 Syntax

Terms. The basic building blocks of the spi-calculus are terms, whose syntax is

$$L, M, N ::= 0 \mid x \mid m \mid \operatorname{suc}(M) \mid (M, N) \mid \{M\}_N$$

The presence of the terms 0 and $\operatorname{suc}(M)$ gives us the ability to construct all natural numbers. A major difference from the π-calculus is that we here distinguish between variables (x, y, z) and names (c, m, n, k). A name refers to a key or a channel, whereas variables act as placeholders for arbitrary terms. Tuples of names are written in boldface as e.g. $\boldsymbol{n}$.

Terms M and N can be paired as (M, N). The most important new term constructor, though, is $\{M\}_N$ which describes the *encryption* of a term M with a key N. The resulting ciphertext can only be decrypted if N is known[1].

Processes. Processes are the executable entities of the spi-calculus; their syntax is defined as follows:

$$\begin{aligned}P, Q, R ::= &\ (\nu n)P \mid \overline{M}\langle N\rangle.P \mid M(x).P \mid P \mid Q \\ &\mid\ !P \mid [M \text{ is } N]P \mid \mathbf{0} \mid \text{let } (x, y) = M \text{ in } P \\ &\mid\ \text{case } M \text{ of } 0 : P \ \operatorname{suc}(x) : Q \mid \text{case } L \text{ of } \{x\}_N \text{ in } P\end{aligned}$$

$(\nu n)P$ is the restriction operator. The new name n is bound in and local to P. $\overline{M}\langle N\rangle.P$ denotes output: N is sent out on the channel M. $M(x).P$ is input: a

---

[1] [1] also describes versions of spi with public key encryption and hash functions. In this paper we only consider shared key cryptography, as the notion of framed bisimilarity arose in a shared-key setting.

term is received on the channel M (so the term M must be a channel), and bound to x in P.

let $(x, y) = M$ in P is used to split pairs; the variables x and y are bound in P. case M of $0 : P \operatorname{suc}(x) : Q$ executes P if M is 0 and $Q[N/x]$ if M is the successor of a term N; in this case x is bound in Q.

The most important new construct is *decryption*; case L of $\{x\}_N$ in P is used to decrypt terms; x is bound in P.

Processes are identified up to renaming of bound names and variables. A process without any free variables is said to be *closed* and we let *Proc* denote the set of closed processes. Further, we let $\operatorname{fn}[\![P]\!]$ denote the set of free names in P, and $\operatorname{fv}[\![P]\!]$ the free variables in P.

Agents. In the semantics presented later, a communication request has two stages: a process first *commits* to using a channel and then communicates a term. The process commits to becoming a so-called *agent*. The syntax of agents is

$$A, B ::= P \mid C \mid F$$
$$F, G ::= (x)P$$
$$C, D ::= (\nu m)\langle M \rangle P$$

$(x)P$ is called an *abstraction*. This is the type of agent a process commits to before receiving a term on a channel; the term received is then bound to x. We also use abstractions to denote parameterized processes. For example, if $S \triangleq (x)P$, we let $S(M) = P[M/x]$.

$(\nu m)\langle M \rangle P$ is called a *concretion*. A concretion is immediately able to output the term M. If the concretion does not generate any new names, i.e. $\{m\} = \emptyset$, we write $(\nu)\langle M \rangle P$.

In what follows, Ag will denote the set of closed agents.

2.2 A labelled transition semantics

Our labelled transition semantics is that of [1] and is presented as a commitment semantics in the two-level style of [11]: The rules defining the commitment relation rely on a reduction relation[2].

---

[2] A slight difference from the common presentation of the π-calculus is that the definition of the commitment relation is only based on the reduction relation defined in the following instead of a structural congruence, as is usual.

The reduction relation. The reduction relation describes how processes unfold to prepare for a commitment.

$$!P > P \mid !P$$
$$[M \text{ is } M]P > P$$
$$\text{let}(x,y) = (M,N) \text{ in } P > P[M/x][N/y]$$
$$\text{case } 0 \text{ of } 0 : P \text{ suc}(x) : Q > P$$
$$\text{case suc}(M) \text{ of } 0 : P \text{ suc}(x) : Q > Q[M/x]$$
$$\text{case } \{M\}_N \text{ of } \{x\}_N \text{ in } P > P[M/x]$$

The commitment relation. Commitment transitions are on the form $P \xrightarrow{\alpha} A$ where $P \in \mathit{Proc}$ and $A \in \mathit{Ag}$. The label α is a name, a co-name or the unobservable action τ.

To express the semantics, restriction and parallel composition are extended to agents:

$$(\nu n)(x)P \triangleq (x)(\nu n)P$$
$$Q \mid (x)P \triangleq (x)(Q \mid P) \quad \text{where } x \notin \text{fv}[\![Q]\!]$$
$$(\nu n)(\nu \boldsymbol{m})\langle M\rangle P \triangleq \begin{cases} (\nu n, \boldsymbol{m})\langle M\rangle P & \text{if } n \in \text{fn}[\![M]\!] \\ (\nu \boldsymbol{m})\langle M\rangle(\nu n)P & \text{otherwise} \end{cases}$$
$$Q \mid (\nu \boldsymbol{m})\langle M\rangle P \triangleq (\nu \boldsymbol{m})\langle M\rangle(Q \mid P) \quad \text{where } \{\boldsymbol{m}\} \cap \text{fn}[\![Q]\!] = \emptyset$$

(We assume here that $n \notin \{\boldsymbol{m}\}$.) The dual composition $A \mid Q$ is defined symmetrically.

For an abstraction $F = (x)P$ and concretion $C = (\nu \boldsymbol{n})\langle N\rangle Q$ we define the *interaction* between F and C as

$$C \bullet F \triangleq (\nu \boldsymbol{n})(Q \mid P[N/x]),$$
$$F \bullet C \triangleq (\nu \boldsymbol{n})(P[N/x] \mid Q),$$

when $\{\boldsymbol{n}\} \cap \text{fn}[\![P]\!] = \emptyset$. An interaction can be interpreted as a communication over a given channel between a sender and a receiver.

The commitment relation is now defined by the following rules[3]:

$$m(x).P \xrightarrow{m} (x)P \qquad \overline{m}\langle M\rangle.P \xrightarrow{\overline{m}} (\nu)\langle M\rangle P$$

$$\frac{P \xrightarrow{m} F \quad Q \xrightarrow{\overline{m}} C}{P \mid Q \xrightarrow{\tau} F \bullet C} \qquad \frac{P \xrightarrow{\overline{m}} C \quad Q \xrightarrow{m} F}{P \mid Q \xrightarrow{\tau} C \bullet F}$$

$$\frac{P \xrightarrow{\alpha} A}{P \mid Q \xrightarrow{\alpha} A \mid Q} \qquad \frac{Q \xrightarrow{\alpha} A}{P \mid Q \xrightarrow{\alpha} P \mid A}$$

$$\frac{P \xrightarrow{\alpha} A \quad \alpha \notin \{m, \overline{m}\}}{(\nu m)P \xrightarrow{\alpha} (\nu m)A} \qquad \frac{P > Q \xrightarrow{\alpha} A}{P \xrightarrow{\alpha} A}$$

2.3 Testing equivalence

Two processes are testing equivalent if they can resist the same attacks from any environment. That is, the two processes will be indistinguishable to any attacker—this is essentially the notion of may-testing from [13]. The notion of indistinguishability relies on the concept of a barb.

Definition 1. A *barb* is a name n or a co-name $\overline{n}$.

Definition 2. Two closed processes P and Q are *testing equivalent*, written $P \simeq Q$, iff for all closed processes R and all barbs β:

$$P \mid R \xrightarrow{\tau}^* P' \xrightarrow{\beta} A \text{ for some } P' \text{ and } A$$
$$\text{if and only if}$$
$$Q \mid R \xrightarrow{\tau}^* Q' \xrightarrow{\beta} B \text{ for some } Q' \text{ and } B$$

3 Framed bisimilarity

We now define the notion of *framed bisimulation* introduced by Abadi and Gordon [3].

3.1 Frames, environments and indistinguishability

The basic idea of framed bisimulation is that processes are related with respect to an *environment*.

Definition 3. A *frame* f is a finite set of names. A *theory* t is a finite set of pairs of terms (M, N). An *environment* is a frame-theory pair, $e = (f, t)$.

---

[3] Note that the last rule defines the commitment relation up to reduction on the left side and not up to structural congruence as usual in the presentation of the semantics of the π-calculus.

Intuitively, the elements of the frame are the names of P and Q that the environments knows. If $(M, N) \in t$ this means that the environment cannot distinguish the term M coming from P and the term N coming from Q.

Normally we use the letters e, f and t to denote environments, frames and theories. If the letters have the same marks or indices, e.g. e', f' and t', they are related as $e' = (f', t')$.

Definition 4. Let e be an environment. Two terms M and N are *indistinguishable* under e, written $e \vdash M \leftrightarrow N$, if this can be derived by the following rules:

$$\text{(Ind Zero)} \quad \frac{}{e \vdash 0 \leftrightarrow 0} \qquad \text{(Ind Var)} \quad \frac{}{e \vdash x \leftrightarrow x}$$

$$\text{(Ind Frame)} \quad \frac{n \in f}{e \vdash n \leftrightarrow n} \qquad \text{(Ind Theory)} \quad \frac{(M, N) \in t}{e \vdash M \leftrightarrow N}$$

$$\text{(Ind Pair)} \quad \frac{e \vdash M \leftrightarrow M' \quad e \vdash N \leftrightarrow N'}{e \vdash (M, N) \leftrightarrow (M', N')}$$

$$\text{(Ind Suc)} \quad \frac{e \vdash M \leftrightarrow M'}{e \vdash \text{suc}(M) \leftrightarrow \text{suc}(M')}$$

$$\text{(Ind Enc)} \quad \frac{e \vdash M \leftrightarrow M' \quad e \vdash N \leftrightarrow N'}{e \vdash \{M\}_N \leftrightarrow \{M'\}_{N'}}$$

Environments should be *consistent*. Assume that the theory contains the pair (M, N) because the environment has seen M from process P and N from process Q and assumes that they are identical. Now P sends M, and Q sends L, where $N \neq L$. Now, if we added the pair (M, L) to the theory, the environment would believe M to be the same as two different terms. An environment containing such a pair will clearly not be consistent.

Now consider the case where $(\{M_1\}_{M_2}, \{N_1\}_{N_2}) \in t$ and M_2 is revealed. Again, what the theory contains are elements that the environment thinks are alike. But an attacker would be able to decrypt one of the terms, but not the other. This means that they have different keys and thus are distinguishable. Again, such an environment will be inconsistent.

Finally, only ciphertexts should appear in the theory, as two non-ciphertexts are immediately distinguishable, unless identical.

The above observations motivate to the following definition from [3]:

Definition 5. Environment e is ok, written $e \vdash \text{ok}$, if:

1. $\forall (M, N) \in t$ it must hold that M is closed, $\exists M_1, M_2 : M = \{M_1\}_{M_2}$ and $\nexists N_2 : e \vdash M_2 \leftrightarrow N_2$. The converse must also hold for N.
2. whenever $(M, N) \in t$ and $(M', N') \in t$, $M = M'$ iff $N = N'$.

During the execution of a protocol, the environment never shrinks but may be extended, since all information obtained will continue to be known and new information may appear. We now define what it means for one environment to be an extension of another.

Definition 6. Let e and e' be environments. We say e' *extends* e, written $e \leq e'$ iff $\forall M, N : e \vdash M \leftrightarrow N \Rightarrow e' \vdash M \leftrightarrow N$.

Note that this relation is reflexive and transitive, but neither symmetric nor antisymmetric; there may exist pairs of environments that are not related.

The following property of the extension ordering (from [3]) becomes useful later:

Lemma 7. *Suppose* $e \vdash ok$ *and* $e \vdash M \leftrightarrow N$. *Then*

1. *If* $e \vdash M \leftrightarrow N'$ *then* $N = N'$.
2. *If* $e \vdash M' \leftrightarrow N$ *then* $M = M'$.

3.2 Framed simulations and bisimulations

A *framed process pair* is a quadruple (f, t, P, Q), where $P, Q \in Proc$. If $\mathcal{R}$ is a set of framed process pairs, we write $e \vdash P\mathcal{R}Q$ when $(f, t, P, Q) \in \mathcal{R}$. A *framed relation* is a set $\mathcal{R}$ of framed process pairs, such that $e \vdash ok$ whenever $e \vdash P\mathcal{R}Q$.

Intuitively, for a process Q to be able to simulate a process P, whenever P commits to some action and becomes P', then Q must be able to commit to the same action in such a way that the resulting process Q' is able to simulate P'. In the case of output commitments one needs to make an additional assumption, as the output of a term may provide an observer with additional information about the indistinguishability of terms. Thus, after an output transition the environment in which P and Q must be compared may have to be extended. This leads to the following

Definition 8. A *framed simulation* is a framed relation $\mathcal{S}$ such that, whenever $e \vdash P\mathcal{S}Q$, the following three conditions hold

1. If $P \xrightarrow{\tau} P'$ then there exists a process Q' such that $Q \xrightarrow{\tau} Q'$ and $e \vdash P'\mathcal{S}Q'$.
2. If $P \xrightarrow{c} (x)P'$ and $c \in f$ then there exists an abstraction $(x)Q'$ with $Q \xrightarrow{c} (x)Q'$ and, for all sets $\{n\}$ disjoint from $\text{fn}[\![P]\!] \cup \text{fn}[\![Q]\!] \cup f \cup \text{fn}(t)$ and all closed terms M and N, if $(f \cup \{n\}, t) \vdash M \leftrightarrow N$ then $(f \cup \{n\}, t) \vdash P'[M/x]\mathcal{S}Q'[N/x]$.
3. If $P \xrightarrow{\bar{c}} (\nu m)\langle M\rangle P'$, $c \in f$ and $\{m\} \cap (\text{fn}[\![P]\!] \cup \text{fn}(\pi_1(t)) \cup f) = \emptyset$ then there exists a concretion $(\nu n)\langle N\rangle Q'$ with $Q \xrightarrow{\bar{c}} (\nu n)\langle N\rangle Q'$ and $\{n\} \cap (\text{fn}[\![Q]\!] \cup \text{fn}(\pi_2(t)) \cup f) = \emptyset$. Furthermore $\exists e' : e \leq e'$, $e' \vdash M \leftrightarrow N$, and $e' \vdash P'\mathcal{S}Q'$.

Definition 9. A *framed bisimulation* is a framed simulation $\mathcal{S}$ such that $\mathcal{S}^{-1} = \{e' \vdash Q\mathcal{S}P \mid e \vdash P\mathcal{S}Q \land e' = (f, \{(M, N) \mid (N, M) \in t\})\}$ is also a framed simulation. If processes P and Q are related by some framed bisimulation, we write that $P \sim_f Q$ and say that P and Q are *framed bisimilar*.

Notice that the above definition yields a *late* equivalence in the sense of [11].

Framed bisimilarity implies testing equivalence. This is the major result of [4] and our basis for studying framed bisimilarity.

Theorem 10. *Let P and Q be closed processes and n be a name such that $n \notin fn[\![P]\!] \cup fn[\![Q]\!]$. If $(fn[\![P]\!] \cup fn[\![Q]\!] \cup \{n\}, \emptyset) \vdash P \sim_f Q$ then $P \simeq Q$.*

The converse implication does not hold. Consider a simple example from [3].

$$P \triangleq (\nu c)(\mathbf{0} \mid \mathbf{0})$$
$$Q \triangleq (\nu c)(\bar{c}\langle 0 \rangle.\mathbf{0} \mid c(x).\mathbf{0})$$

Neither P nor Q will exhibit any barbs, since c is restricted. Looking at the definition of test equivalence, we see that if P in parallel with a process exhibits a barb after a series of commitments and communications, so should Q in parallel with the same process. As they do not exhibit any barbs, this holds. So $P \simeq Q$. However, $P \overset{\tau}{\not\rightarrow}$ whereas $Q \overset{\tau}{\rightarrow}$, so clearly $P \not\sim_f Q$..

4 Fenced bisimilarity

As mentioned in the introduction, bisimulations are in general suitable for automatic verification. Unfortunately the definition of framed bisimulation is not effective, as it contains quantifiers that range over infinite domains.

We now eliminate the need for the existential quantifier over environments in condition 3 in Definition 8. We do this by constructing a computable function that, given appropriate input, will return an environment e' that fulfills the given conditions. This allows us to define an alternative characterization of framed bisimilarity, *fenced bisimilarity*.

4.1 Fundamental definitions

Consider an environment e and two processes P and Q for which it holds that $P \overset{\bar{c}}{\rightarrow} (\nu \boldsymbol{m})\langle M \rangle P', c \in f, \{\boldsymbol{m}\} \cap (fn[\![P]\!] \cup fn(\pi_1(t)) \cup f) = \emptyset$, $Q \overset{\bar{c}}{\rightarrow} (\nu \boldsymbol{n})\langle N \rangle Q'$ and $\{\boldsymbol{n}\} \cap (fn[\![Q]\!] \cup fn(\pi_2(t)) \cup f) = \emptyset$. Looking at the definition of framed bisimilarity we see that in order for $e \vdash P \sim_f Q$, it must hold that $\exists e' : e \leq e' \vdash \text{ok}$, $e' \vdash M \leftrightarrow N$ and $e' \vdash P' \sim_f Q'$.

In other words, in order for P and Q to be framed bisimilar under e, it must be possible to expand e such that the new environment e' is ok and cannot distinguish between the output M coming from P and the output N coming from Q. Furthermore under e' the remaining processes P' and Q' need to be framed bisimilar.

We now show that a suitable e' can always be determined effectively by means of a function, called the *expand* function ξ. Furthermore, we define the notion of fenced bisimilarity, which is variant of framed bisimilarity, using ξ. In section 4.2 we prove that the notion of fenced and framed bisimilarity coincide.

Assuming the existence of ξ, we can define fenced bisimilarity as follows.

Definition 11. A *fenced simulation* is a framed relation $\mathcal{T}$ such that, whenever $e \vdash P\mathcal{T}Q$, the following three conditions hold:

1. If $P \xrightarrow{\tau} P'$ then there exists a process Q' such that $Q \xrightarrow{\tau} Q'$ and $e \vdash P'\mathcal{T}Q'$.
2. If $P \xrightarrow{c} (x)P'$ and $c \in f$ then there exists an abstraction $B = (x)Q'$ such that $Q \xrightarrow{c} B$ and, for all sets $\{n\}$ disjoint from $\text{fn}[\![P]\!] \cup \text{fn}[\![Q]\!] \cup f \cup \text{fn}(t)$ and all closed terms M and N, if $(f \cup \{n\}, t) \vdash M \leftrightarrow N$ then $(f \cup \{n\}, t) \vdash P'[M/x]\mathcal{T}Q'[N/x]$.
3. If $P \xrightarrow{\bar{c}} A = (\nu m)\langle M \rangle P'$, $c \in f$ and $\{m\} \cap (\text{fn}[\![Q]\!] \cup \text{fn}(\pi_1(t)) \cup f) = \emptyset$ then there is a concretion $B = (\nu n)\langle N \rangle Q'$ such that $Q \xrightarrow{\bar{c}} B$, the set $\{n\}$ is disjoint from $\text{fn}[\![Q]\!] \cup \text{fn}(\pi_2(t)) \cup f$ and $\xi(e, N, M) \vdash P'\mathcal{T}Q'$ and $\xi(e, N, M) \neq \bot$.

Definition 12. A *fenced bisimulation* is a fenced simulation $\mathcal{T}$ such that $\mathcal{T}^{-1}$ is also a fenced simulation. If processes P and Q are related by some fenced bisimulation, we write that $P \sim_\# Q$ and say that P and Q are *fenced bisimilar*.

We now give an effective definition of the ξ-function in the form of the recursive algorithm found in Table 4.1. The algorithm is presented as an imperative Pascal-style function, where the value returned from a function call is the last one assigned to e_ξ.

The basic underlying idea is to make a minimal extension of the environment by recursive calls. The interesting cases of the algorithm occur when we add something to the environment. That is, in the cases where $M = N = n$ and $M = \{M_1\}_{M_2}$ and $N = \{N_1\}_{N_2}$.

If $M = n = N$ it means that both the processes have sent out n. The only thing we can do is to add n to the frame (Line 5). This is the only way to make the environment unable to distinguish a name. But this might violate the fact that our new environment is ok, so we need to repair the environment after putting n in the frame. This is done by moving all elements of the theory which violate the ok condition to a set of leftover pairs λ (Lines 7-10). Having done this, we are no longer assured that the terms which were indistinguishable in e remain indistinguishable in the new environment, so for each of the pairs in λ we call ξ to make these pairs indistinguishable again (Lines 11-12).

In the case that $M = \{M_1\}_{M_2}$ and $N = \{N_1\}_{N_2}$ there are two possibilities, depending on whether the keys are already indistinguishable or not. If they are, we simply call ξ on M_1 and N_1 (Line 14). That way N and M become indistinguishable. If not, we put the pair in the theory and repair the violations of the ok-condition as before (Line 16-25).

Because a pair is added to the theory, it is now also possible to violate condition 2 of definition 5, thus the lines 16-17 check this condition. By Lemma 7, a ciphertext cannot be indistinguishable from two different ciphertexts in an ok environment.

We define $\bot$ as the environment that is not ok and not related to any other environment. We say that ξ is *undefined* if it returns $\bot$.

1 **function** $\xi(e, M, N)$
2 **if** $(e \vdash M \leftrightarrow N)$ **then** $e_\xi := e$ **else**
3 **case** (M, N) **of**
4    $(suc(L), suc(L'))$: $e_\xi := \xi(e, L, L')$
5   (n, n) : $e_\xi := (f \cup \{n\}, t)$
6      $\lambda := \emptyset$
7   **for all** $(\{M_1\}_{M_2}, \{N_1\}_{N_2}) \in t_\xi$ **do**
8   **if** $\exists L : (e_\xi \vdash M_2 \leftrightarrow L) \vee (e_\xi \vdash L \leftrightarrow N_2)$ **then**
9          $e_\xi := (f_\xi, t_\xi \setminus \{(M, N)\})$
10         $\lambda := \lambda \cup \{(M, N)\}$
11   **for all** $(\{M_1\}_{M_2}, \{N_1\}_{N_2}) \in \lambda$ **do**
12       $e_\xi := \xi(\xi(e_\xi, M_2, N_2), M_1, N_1)$
13   $(\{M_1\}_{M_2}, \{N_1\}_{N_2})$:
14   **if** $(e \vdash M_2 \leftrightarrow N_2)$ **then** $e_\xi := \xi(e, M_1, N_1)$
15   **else**
16   **if** $\exists (O, O') \in t : O = M \not\leftrightarrow O' = N$ **then**
17         $e_\xi := \bot$
18   **else** $e_\xi := (f, t \cup \{(M, N)\})$
19         $\lambda := \emptyset$
20   **for all** $(\{O_1\}_{O_2}, \{O'_1\}_{O'_2}) \in t_\xi$ **do**
21   **if** $\exists L : (e_\xi \vdash O_2 \leftrightarrow L) \vee (e_\xi \vdash L \leftrightarrow O'_2)$ **then**
22             $e_\xi := (f_\xi, t_\xi \setminus \{(O, O')\})$
23             $\lambda := \lambda \cup \{(O, O')\}$
24   **for all** $(\{O_1\}_{O_2}, \{O'_1\}_{O'_2}) \in \lambda$ **do**
25           $e_\xi := \xi(\xi(e_\xi, O_2, O'_2), O_1, O'_1)$
26   $((M_1, M_2), (N_1, N_2))$:
27     $e_\xi := \xi(\xi(e, M_1, N_1), M_2, N_2)$
28   **otherwise** : $e_\xi := \bot$

Table 1. The algorithm defining the expand function.

The if-clause in line 2 catches the two trivial cases $M = N = 0$ and $M = N = x$. Note that the conditions of line 2, 8, 14, 16, and 21 are effective because $\leftrightarrow$ clearly is.

Theorem 13. ξ *is well-behaved, i.e. for all (e, M, N) it is the case that $\xi(e, M, N)$ terminates.*

Proof. Given (e, M, N), let i denote the total number of term constructors used in M and N. The result now follows by a straightforward induction on i, observing that the case $i = 2$ is covered in line 5 of the algorithm for ξ and that i decreases by every recursive call.

After presenting the ξ-function the question arises as to how the notion of fenced bisimilarity relates to that of framed bisimilarity.

4.2 Properties of fenced bisimilarity

We now prove that ξ really gives us the desired extension. First of all, if ξ is well defined on a given input, then it returns a valid environment, as described in Section 4. But ξ is also complete, that is, if there exists a valid environment, then ξ is well-defined. Actually we prove a somewhat stronger result, namely that not only does ξ return a valid extension whenever possible, it actually returns the smallest possible valid extension of the environment.

Soundness. The following lemma is the key to the soundness, as it states that $\xi(e, M, N)$ computes a consistent extension of e identifying M and N.

Lemma 14. *Let $e \vdash ok$. If $\xi(e, M, N) \neq \bot$ then $e \leq \xi(e, M, N) \vdash ok$ and $\xi(e, M, N) \vdash M \leftrightarrow N$.*

Theorem 15 (Soundness). *If $e \vdash P \sim_\# Q$ then $e \vdash P \sim_f Q$*

Proof. We show that $\sim_\#$ is a framed bisimulation. So consider an arbitrary framed process pair (e, P, Q) such that $e \vdash P \sim_\# Q$. It is enough to show that transitions of P can be matched by transitions of Q within $\sim_\#$.

Since the definition of $\sim_\#$ differs from $\sim_f$ only if $P \xrightarrow{\bar{c}} A = (\nu m)\langle M \rangle P', c \in f$ and $\{m\} \cap (\text{fn}\llbracket P \rrbracket \cup \text{fn}(\pi_1(t)) \cup f) = \emptyset$ we shall only consider this case. Since $e \vdash P \sim_\# Q$ we know that $\exists B = (\nu n)\langle N \rangle Q' : Q \xrightarrow{\bar{c}} B, c \in f, \{n\} \cap (\text{fn}\llbracket Q \rrbracket \cup \text{fn}(\pi_2(t)) \cup f) = \emptyset$ and $\xi(e, M, N) \vdash P' \sim_\# Q'$, provided that we can show that $e \leq \xi(e, M, N) \vdash ok$, $\xi(e, M, N) \vdash N \leftrightarrow M$ and $\xi(e, M, N) \vdash P' \sim_f Q'$. But these two conditions follow immediately from Lemma 14.

Completeness. Here, two lemmas are necessary. The first lemma states that ξ yields the smallest valid extension.

Lemma 16. *If $e \vdash ok$ and $\exists e' : e \leq e' \vdash ok$ and $e' \vdash M \leftrightarrow N$ then $\xi(e, M, N) \neq \bot$, $\xi(e, M, N) \vdash ok$ and $\xi(e, M, N) \leq e'$.*

Proof. Let $e \vdash ok$, $e \leq e' \vdash ok$ and $e' \vdash M \leftrightarrow N$. In order to show the lemma it suffices to prove that $\xi(e, M, N) \leq e'$, since this only holds if $\xi(e, M, N) \neq \bot$ which again by Lemma 14 implies $\xi(e, M, N) \vdash ok$. The proof then proceeds by induction on the number of extra calls to ξ used to calculate $\xi(e, M, N)$.

The second lemma states that when two processes are framed bisimilar under an extension of a given environment, then they are also fenced bisimilar under the least extension to this environment.

Lemma 17. *If $e \vdash ok$ and $\exists e' : e \leq e' \vdash ok$, $e' \vdash M \leftrightarrow N$ and $e' \vdash P \sim_f Q$ then $\xi(e, M, N) \vdash P \sim_\# Q$.*

Proof. The framed relation $\mathcal{S} = \{(\xi(e, M, N), P, Q) \mid e \vdash ok, \exists e' : e \leq e', e' \vdash ok, e' \vdash M \leftrightarrow N, e' \vdash P \sim_f Q\}$ is a fenced bisimulation. The proof of this uses Lemma 16.

Theorem 18 (Completeness). *If $e \vdash P \sim_f Q$ then $e \vdash P \sim_\# Q$.*

Proof. We show that $\sim_f$ is a fenced bisimulation. So consider an arbitrary framed process pair (e, P, Q). We must show that transitions of P can be matched by Q within $\sim_f$ according to the definition of fenced bisimulations.

The only interesting case occurs when $P \xrightarrow{\bar{c}} A = (\nu m)\langle M \rangle P'$, $c \in f$ and $\{m\} \cap (\mathrm{fn}[\![P]\!] \cup \mathrm{fn}(\pi_1(t)) \cup f) = \emptyset$. As $e \vdash P \sim_f Q$ we get $e \vdash ok$, $\exists B = (\nu n)\langle N \rangle Q' : Q \xrightarrow{\bar{c}} B$, $\{n\} \cap (\mathrm{fn}[\![Q]\!] \cup \mathrm{fn}(\pi_2(t)) \cup f) = \emptyset$ and $\exists e' : e \leq e' \vdash ok$, $e' \vdash M \leftrightarrow N$ and $e' \vdash P' \sim_f Q'$.

We now only need to show $\xi(e, M, N) \vdash P' \sim_\# Q'$. But this follows immediately from Lemma 17.

5 An example

In order to show how to establish fenced bisimilarity, we present a small example of its use. In what follows, we are able to show *how* a fenced simulation is constructed, providing further evidence that the ξ-function makes fenced bisimilarity more suited for automation.

Our example illustrates the fact that the environment can compare ciphertexts without being able to decrypt them.

Consider the two processes

$$P \triangleq (\nu K)\,\bar{c}\langle(\{M\}_K, \{M\}_K)\rangle$$
$$Q \triangleq (\nu K)\,\bar{c}\langle(\{M'\}_K, \{M''\}_K)\rangle$$

In the above, M, M', M'', K are closed terms. We use the notation $\text{fn}(P, Q, t, f)$ $\triangleq (\text{fn}[\![P]\!] \cup \text{fn}[\![Q]\!] \cup \text{fn}(t) \cup f)$.

Let $e = (\text{fn}[\![P]\!] \cup \text{fn}[\![Q]\!] \cup \{n\}, \emptyset) = (\{c, n\}, \emptyset)$. We now construct a fenced simulation $\mathcal{T}$ by starting out with $e \vdash P\mathcal{T}Q$.

By looking at $e \vdash P\mathcal{T}Q$ we see that both P and Q can make a $\bar{c}$-commitment

$$P \xrightarrow{\bar{c}} (\nu K)\langle(\{M\}_K, \{M\}_K)\rangle.\mathbf{0}$$
$$Q \xrightarrow{\bar{c}} (\nu K)\langle(\{M'\}_K, \{M''\}_K)\rangle.\mathbf{0}$$

and in both cases $K \notin \text{fn}(P, Q, t, f)$, thus we need to consider the framed process pair $e_1 \vdash \mathbf{0}\mathcal{T}\mathbf{0}$ where $e_1 = \xi(e, (\{M\}_K, \{M\}_K), (\{M'\}_K, \{M''\}_K))$. The ξ-call for the first components of the pairs yields $e_2 = (\{c, n\}, \{(\{M\}_K, \{M'\}_K)\})$. The ξ-call of the second components will after line 18 have $e_\xi = (f_2, t_2 \cup \{(\{M\}_K, \{M''\}_K)\})$. If $M' = M''$, then we have $e_1 = e_2$. If $M' \neq M''$ we will never be able to repair the environment, thus $e_1 = \bot$. In this case the two processes are not even testing equivalent. A separating process is

$$R \triangleq c(z).\,\text{let}(x, y) = z \text{ in}[x \text{ is } y].\overline{d}\langle 1 \rangle.$$

R exhibits the differences of P and Q after a single communication. For we have that $R|P \xrightarrow{\tau} \xrightarrow{\bar{d}}$, whereas whenever $R|Q \xrightarrow{\tau} Q'$, then $Q' \xrightarrow{\bar{d}}\!\!\!\!\!/$. This is due to the two terms $\{M'\}_K$ and $\{M''\}_K$ being different. R shows how an attacker can distinguish between ciphertexts even without being able to decrypt them.

6 Conclusions and further work

In this paper we have defined *fenced bisimilarity*, an alternative characterization of the notion of framed bisimilarity and shown it to be sound and complete.

This alternative characterization of framed bisimilarity overcomes the minor of the two obstacles towards automatic proving of framed bisimilarity in the spi calculus. The quantifier in the output condition of framed bisimilarity was replaced by an algorithm. This enables us to construct the extended environment instead of just demanding its existence. A precise complexity analysis of the ξ function is a topic for further work; one immediately notices that in the worst case $\xi(e, M, M)$ may have added pairs for all matching subterms of M and N.

The remaining obstacle on our way to a fully effective characterization is the universal quantifiers in the matching condition for abstractions. Abadi and Gordon believe that framed bisimilarity can be automatically checked based on their conjecture that the set of all possible inputs can be partitioned into finitely many equivalence classes[3].

This partitioning idea is the underlying idea of the symbolic operational semantics for the π-calculus suggested by Boreale and De Nicola [7] and Lin [9] and employed by Dam in [8], so the search for a decision algorithm can probably take advantage of these results.

One should note, though, that one can only hope to find decision algorithms for fragments of the spi calculus because of its universal computing power. A good candidate for such a fragment is the finite-control fragment whose π-calculus counterpart has already been investigated by Lin [9] and Dam [8]. It remains to be seen which encryption protocols are expressible within this subcalculus.

There is yet another issue to be dealt with, if one wants to automate the verification of security protocols.

Generally, at least two properties are desired by the designer of a cryptographic protocol: secrecy and authenticity. A protocol P has the *secrecy* property if $P(M') \simeq P(M')$ for any two messages M, M', i.e. if no environment can tell which message is being sent by P. A protocol P has the *authenticity* property if it only responds to the message being sent, i.e. for a specification P_{spec} which mimics all the communication in the protocol and magically only responds to the intended message, we have that $P(M) \simeq P_{spec}(M)$ for all messages M. As these properties are defined using universal quantification over messages, they cannot be checked directly using an equivalence checking algorithm. We conjecture that an application of the symbolic techniques such as to partition the set of terms into finitely many equivalence classes will prove useful here.

Acknowledgements

Discussions with Anders Rhod Gregersen, Thomas Illum Rasmussen, Thomas Ording and the members of the ν-club at the Department of Computer Science helped refine the ideas in this paper. A copy of the unpublished [4] was provided to us by Martín Abadi.

References

1. M. Abadi and A. D. Gordon. A calculus for cryptographic protocols: The spi calculus. In *Fourth ACM Conference on Computer and Communications Security*, pages 36–47. ACM Press, 1997.
2. Martín Abadi and Andrew D. Gordon. Reasoning about cryptographic protocols in the spi calculus. In A. Mazurkiewicz and J. Winkowski (ed.) *CONCUR '97: Concurrency Theory, 8th International Conference*, Lecture Notes in Computer Science, 1243, pages 59–73, Warsaw, Poland, 1–4 July 1997. Springer-Verlag.
3. M. Abadi and A. D. Gordon. A bisimulation method for cryptographic protocols. In *Programming Languages and Systems: 7th European Symposium on Programming (ESOP '98)*, Lecture Notes in Computer Science, 1381, pages 12–26. Springer-Verlag, 1998.
4. M. Abadi and A. D. Gordon. A bisimulation method for cryptographic protocols. Extended version of [3] containing all proofs. Unpublished manuscript, January 27, 1998.
5. R. Anderson and R. Needham. Programming Satan's computer. In: Leeuwen, J. van (ed.) *Computer Science Today. Recent Trends and Developments*, Lecture Notes in Computer Science, 1000, pages 426–440. Springer-Verlag, 1995.

6. Michael Burrows, Martín Abadi, and Roger Needham. A logic of authentication. *Proceedings of the Royal Society*, Series A, 426, 1871, pages 233-271, December 1989. Also appeared, in a shortened form, in *ACM Transactions on Computer Systems* 8, 1 (February 1990), 18-36.
7. Michele Boreale and Rocco De Nicola. A symbolic semantics for the π-calculus. *Information and Computation*, **126**(1), pages 34–52, 1996.
8. Mads Dam. On the decidability of process equivalences for the π-calculus. *Theoretical Computer Science*, **183**, pages 215–228, 1997.
9. Huimin Lin. Symbolic bisimulation and proof systems for the π-calculus. Technical Report 7/94, School of Cognitive and Computing Sciences, University of Sussex, UK, 1994.
10. Robin Milner. *Communication and Concurrency*. Prentice Hall Europe, 1989.
11. Robin Milner. The polyadic π-calculus: a tutorial. In F. L. Bauer, W. Brauer, and H. Schwichtenberg (ed.), *Logic and Algebra of Specification*. Springer-Verlag, 1993.
12. J. C. Mitchell, M. Mitchell, and U. Stern. Automated analysis of cryptographic protocols using murφ. In *IEEE Symposium on Security and Privacy*, pages 141–151, Oakland, CA, 1997. IEEE Computer Society Press.
13. Rocco De Nicola and Matthew C. B. Hennessy. Testing equivalence for processes. In Josep Díaz (ed.), *Automata, Languages and Programming, 10th Colloquium*, Lecture Notes in Computer Science 154, pages 548–560, Barcelona, Spain, 18–22 July 1983. Springer-Verlag.
14. D.M.R. Park. Concurrency and automata on infinite sequences. In P. Deussen (ed.), *Proceedings of 5th GI Conference*, Lecture Notes in Computer Science 104, pages 167–183. Springer-Verlag, 1981.
15. Bruce Schneier. *Applied Cryptography*. John Wiley & Sons, Inc., 1996.

Three Remarks on SAGBI Bases for Polynomial Invariants of Permutation Groups

Manfred Göbel[*]

International Computer Science Institute, 1947 Center Street (Suite 600), Berkeley, CA 94704-1198, USA. manfredg@icsi.berkeley.edu

Abstract. This note contains three remarks on SAGBI bases for polynomial invariants of permutation groups.
First, let K be a commutative field, let $K[X_1, \ldots, X_n]$ be the polynomial ring in $X_1, \ldots, X_n$ over K, and let Γ be a finite $n \times n$ matrix group, which has a permutation group as a conjugate. We present a transformation technique for computing a small finite basis B of the K-algebra of Γ-invariant polynomials and a polynomial representation of any Γ-invariant polynomial in $K[X_1, \ldots, X_n]$ as a polynomial in the elements of the basis B. Our method is such that B contains only polynomials with a total degree of at most $\max\{n, \frac{n(n-1)}{2}\}$, independent of the order of Γ.
Second, we show that the ring of polynomial invariants $\mathbb{C}[X_1, X_2, X_3]^{A_3}$ of the alternating group A_3 is the smallest invariant ring, which has no finite SAGBI basis w.r.t. any admissible order. "Smallest" refers to the number of variables, which is 3, and to the number of generators of the invariant ring, which is 4.
Third, we prove the existence of an invariant ring $\mathbb{C}[X_1, \ldots, X_n]^G$ generated by elements with a total degree of at most 2, which has no finite SAGBI basis w.r.t. any admissible order. Therefore, 2 is the optimal lower bound for the total degree of generators of invariant rings with such a property.

1 The basic setup

$\mathbb{N}$, $\mathbb{Q}$, $\mathbb{R}$, and $\mathbb{C}$ denote the natural, rational, real, and complex numbers. K is an arbitrary commutative field, $K[X_1, \ldots, X_n]$ is the commutative polynomial ring over K in the indeterminates $X_1, \ldots, X_n$, and T is the set of terms (= power-products of the X_i) in $K[X_1, \ldots, X_n]$. $GL(n, K)$ denotes the general linear group over K. Let $\Gamma \leq GL(n, K)$ be finite, let $\pi = (a_{ij})_{1 \leq i,j \leq n} \in \Gamma$, and let $f \in K[X_1, \ldots, X_n]$. Then $\pi(f)$ is defined as $f(\sum_{i=1}^n a_{1i}X_i, \ldots, \sum_{i=1}^n a_{ni}X_i)$, and f is called Γ-invariant, if $f = \pi(f)$ for all $\pi \in \Gamma$. $K[X_1, \ldots, X_n]^\Gamma$ denotes the K-algebra of Γ-invariant polynomials in $K[X_1, \ldots, X_n]$ and

$$orbit_\Gamma(f) = \sum_{p \in \{\pi(f) | \pi \in \Gamma\}} p$$

---

[*] New Address: Deutsches Fernerkundungsdatenzentrum, Algorithmen & Prozessoren, DLR Oberpfaffenhofen, 82234 Weßling, Germany; Manfred.Goebel@dlr.de

the Γ-invariant orbit of f. $G < GL(n,K)$ denotes any group of permutations, S_n and A_n the symmetric and alternating group, and $\sigma_1 = X_1 + \ldots + X_n, \ldots,$ $\sigma_n = X_1 \ldots X_n$ the elementary symmetric polynomials.

The set of terms T can be ordered in multiple ways. A characterization of all admissible orders $<$, which are such that $t > 1$ for all $1 \neq t \in T$ and $st_1 > st_2$ for all $s, t_1, t_2 \in T$ with $t_1 > t_2$, can be found in [7, 10]. Let $HT(f)$ and $HC(f)$ be the highest term of $f \in K[X_1, \ldots, X_n]$, and the coefficient of $HT(f)$, respectively. A relation $\leq \subseteq T \times T$ is an admissible order on T iff there exists $f_1, \ldots, f_n \in \mathbb{R}[Z]$ with a total degree of at most n, which are linearly independent in $\mathbb{R}[Z]$ as a vector space over $\mathbb{Q}$ and are such that $HC(f_i) >_\mathbb{R} 0$ for $1 \leq i \leq n$. Then $X_1^{e_1} \ldots X_n^{e_n} > X_1^{d_1} \ldots X_n^{d_n}$ iff $HC(\sum_{i=1}^{n}(e_i - d_i)f_i) >_\mathbb{R} 0$.

A SAGBI (Subalgebra Analogue to Gröbner Basis for Ideals [8, 4]) basis B of a subalgebra of $\mathbb{C}[X_1, \ldots, X_n]$ is such that w.r.t. a given admissible order every head term of an element in the subalgebra can be expressed as a product of head terms of the elements in B.

2 Remark I

Our first remark is an appendix to [1], where a rewriting technique for polynomial invariants of permutation groups in $K[X_1, \ldots, X_n]$ was described. We show that any ring of polynomial invariants of a finite matrix group Γ, which has a permutation group as a conjugate, can be treated with the same methods by applying a linear transformation. Moreover, the computation of algebra bases for these invariant rings leads to polynomials with a total degree of at most $\max\{n, \frac{n(n-1)}{2}\}$, independent of the order of Γ.

This result is the foundation for a challenging problem stated in Subsection 2.2, which deals with the finiteness of SAGBI bases for polynomial invariants of conjugates of permutation groups and its algorithmic consequences. We refer to [1–3], which we adopt as a starting point.

2.1 The transformation technique

Emmy Noether [6] proved the following classical theorem:

Theorem 1. *Let $char(K) = 0$, and let $\Gamma < GL(n, K)$ be finite. Then $K[X_1, \ldots, X_n]^\Gamma$ has an algebra basis consisting of at most $\binom{n + |\Gamma|}{n}$ invariants, whose total degree is at most $|\Gamma|$ (Noether's bound). Note that Noether's proof of this theorem is constructive.*

In the case where Γ is a permutation group G, bases for rings of polynomial invariants can be also computed by the algorithm described in [1].

Theorem 2. *Let $G \leq S_n$. Then any $f \in K[X_1, \ldots, X_n]^G$ has a representation as a $K[\sigma_1, \ldots, \sigma_n]$-linear combination of special G-invariant orbits, i.e.*

$$f = \sum_{t \text{ spec.}} p_t(\sigma_1, \ldots, \sigma_n) \cdot orbit_G(t)$$

with $p_t \in K[X_1,\ldots,X_n]$. Note that the number of special terms and special G-invariant orbits in $K[X_1,\ldots,X_n]$ is finite, and any special G-invariant orbit has a total degree of at most $\max\{n, \frac{n(n-1)}{2}\}$.

This bound is in most cases superior to Noether's bound. We show now in the following how Theorem 2 can be extended to invariant rings of finite matrix groups, which have a permutation group as a conjugate.

Lemma 3. Let $\delta \in GL(n,K)$, and let $f \neq g \in K[X_1,\ldots,X_n]$. Then $\delta(f) \neq \delta(g)$ and $\delta^{-1}(f) \neq \delta^{-1}(g)$. Note that the total degrees of f, $\delta(f)$ and $\delta^{-1}(f)$ are equal, because δ is a linear substitution.

Proof. Assume that $\delta(f) = \delta(g)$ or $\delta^{-1}(f) = \delta^{-1}(g)$. It follows that $\delta^{-1}(\delta(f)) = \delta^{-1}(\delta(g))$ or $\delta(\delta^{-1}(f)) = \delta(\delta^{-1}(g))$ and so $f = g$ (contradiction). □

Definition 4. Let $\Gamma \leq GL(n,K)$ be a matrix group, and let $\delta \in GL(n,K)$. Then $\Gamma^\delta = \{\delta^{-1}\pi\delta \mid \pi \in \Gamma\}$ is a conjugate of Γ. Note that $\Gamma^\delta \leq GL(n,K)$ is isomorphic to Γ, and especially $|\Gamma| = |\Gamma^\delta|$.

Lemma 5. Let $\Gamma \leq GL(n,K)$, let $\delta \in GL(n,K)$, let $f \in K[X_1,\ldots,X_n]^\Gamma$, and let $g \in K[X_1,\ldots,X_n]^{\Gamma^\delta}$. Then $\delta(f) \in K[X_1,\ldots,X_n]^{\Gamma^\delta}$ and $\delta^{-1}(g) \in K[X_1,\ldots,X_n]^\Gamma$.

Proof. Let $\pi \in \Gamma^\delta$, and let $\hat\pi \in \Gamma$. Then we have $\delta\pi\delta^{-1} \in \Gamma$ and $f = (\delta\pi\delta^{-1})(f)$. Hence, $\delta(f) = \delta((\delta\pi\delta^{-1})(f)) = \pi(\delta(f))$ and $\delta(f) \in K[X_1,\ldots,X_n]^{\Gamma^\delta}$. Furthermore, we have $\delta^{-1}\hat\pi\delta \in \Gamma^\delta$ and $g = (\delta^{-1}\hat\pi\delta)(g)$. Hence, $\delta^{-1}(g) = \delta^{-1}((\delta^{-1}\hat\pi\delta)(g)) = \hat\pi(\delta^{-1}(g))$ and $\delta^{-1}(g) \in K[X_1,\ldots,X_n]^\Gamma$. □

Lemma 6. Let $\Gamma < GL(n,K)$ be finite, let $\delta \in GL(n,K)$, and let $f \in K[X_1,\ldots,X_n]$. Then $\delta(\text{orbit}_\Gamma(f)) = \text{orbit}_{\Gamma^\delta}(\delta(f))$ and $\delta^{-1}(\text{orbit}_{\Gamma^\delta}(f)) = \text{orbit}_\Gamma(\delta^{-1}(f))$.

Proof. We have

$$\delta(\text{orbit}_\Gamma(f)) = \delta(\sum_{p \in \{\pi(f) \mid \pi \in \Gamma\}} p) = \delta(\sum_{p \in \{(\delta\pi\delta^{-1})(f) \mid \pi \in \Gamma^\delta\}} p)$$

$$= \sum_{p \in \{(\delta\pi)(f) \mid \pi \in \Gamma^\delta\}} p = \sum_{p \in \{\pi(\delta(f)) \mid \pi \in \Gamma^\delta\}} p = \text{orbit}_{\Gamma^\delta}(\delta(f))$$

and

$$\delta^{-1}(\text{orbit}_{\Gamma^\delta}(f)) = \delta^{-1}(\sum_{p \in \{\pi(f) \mid \pi \in \Gamma^\delta\}} p) = \delta^{-1}(\sum_{p \in \{(\delta^{-1}\pi\delta)(f) \mid \pi \in \Gamma\}} p)$$

$$= \sum_{p \in \{(\delta^{-1}\pi)(f) \mid \pi \in \Gamma\}} p = \sum_{p \in \{\pi(\delta^{-1}(f)) \mid \pi \in \Gamma\}} p = \text{orbit}_\Gamma(\delta^{-1}(f)).$$

□

Theorem 7. Let $\Gamma < GL(n,K)$ be finite, and let $\delta \in GL(n,K)$ such that $\Gamma^\delta \leq S_n$. Then there exists an algorithm which represents any $f \in K[X_1,\ldots,X_n]^\Gamma$ as a finite $K[\delta^{-1}(\sigma_1),\ldots,\delta^{-1}(\sigma_n)]$-linear combination of Γ-invariant orbits of linear substitutions of special terms, i.e.

$$f = \sum_{t \text{ spec.}} p_t(\delta^{-1}(\sigma_1),\ldots,\delta^{-1}(\sigma_n)) \cdot \text{orbit}_\Gamma(\delta^{-1}(t)) \tag{1}$$

with $p_t \in K[X_1,\ldots,X_n]$.

Proof. By Lemma 5, we know $\delta(f) \in K[X_1,\ldots,X_n]^{\Gamma^\delta}$, and by Theorem 2, $\delta(f)$ must have a representation as $\delta(f) = \sum_{t \text{ spec.}} p_t(\sigma_1,\ldots,\sigma_n) \cdot \text{orbit}_{\Gamma^\delta}(t)$ with $p_t \in K[X_1,\ldots,X_n]$. The representation can be computed by [1], algorithm 3.12. The theorem follows immediately by applying δ^{-1} to the representation of $\delta(f)$ and by Lemma 6. □

Corollary 8. Let $\Gamma < GL(n,K)$ be finite, and let $\delta \in GL(n,K)$ such that $\Gamma^\delta \leq S_n$. Then $K[X_1,\ldots,X_n]^\Gamma$ is generated by at most $\frac{2^{n-1}|\Gamma|}{|S_n|}$ invariants (cf. [2]) with a total degree of at most $\max\{n, \frac{n(n-1)}{2}\}$. Moreover, the total degrees of the primary invariants $\delta^{-1}(\sigma_1)$, ..., $\delta^{-1}(\sigma_n)$ in Equation 1 are 1, ..., n.

As shown in Figure 1, there are at least two possible methods (cf. [5] for other techniques) to compute a representation for any $f \in K[X_1,\ldots,X_n]^\Gamma$, if the finite $\Gamma < GL(n,K)$ has a conjugate $\Gamma^\delta \leq S_n$: We can use the proof of Noether's theorem. This leads to a representation of f as a polynomial in homogeneous Γ-invariant polynomials with a total degree of at most $|\Gamma|$. And, we can use Theorem 7 to compute a representation of f as a polynomial in Γ-invariant orbits with a total degree of at most $\max\{n, \frac{n(n-1)}{2}\}$. The structure of this representation is as described by Equation 1. The second method leads

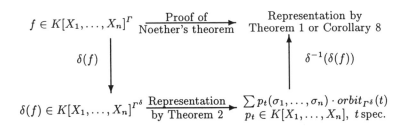

Fig. 1. Representation of Γ-invariant polynomials, if $\Gamma^\delta \leq S_n$.

usually to better bounds for the basis polynomials than the first method, and, in addition, it works also for $char(K) \neq 0$.

Example 9. Let $\alpha = \begin{pmatrix} -\frac{1}{2} & -\frac{\sqrt{3}}{2} & 0 \\ \frac{\sqrt{3}}{2} & -\frac{1}{2} & 0 \\ 0 & 0 & 1 \end{pmatrix}$, let $\delta = \begin{pmatrix} 1 & -\frac{1}{2} & -\frac{1}{2} \\ 0 & -\frac{\sqrt{3}}{2} & \frac{\sqrt{3}}{2} \\ 1 & 1 & 1 \end{pmatrix}$, and let $f = orbit_\Gamma(X_1^4) = \frac{9}{8}X_1^4 + \frac{9}{4}X_1^2X_2^2 + \frac{9}{8}X_2^4 \in \mathbb{C}[X_1,X_2,X_3]^\Gamma$. Then $\Gamma = \{\alpha,\alpha^2,\alpha^3\} < GL(3,\mathbb{C})$ is finite, and the conjugate $\Gamma^\delta < S_3$ is the cyclic permutation group. We proceed as follows to compute a representation of f: First we apply δ to f and obtain $\delta(f) = \frac{9}{8}orbit_{\Gamma^\delta}(X_1^4) - \frac{9}{4}orbit_{\Gamma^\delta}(X_1^3X_2) - \frac{9}{4}orbit_{\Gamma^\delta}(X_1^3X_3) + \frac{27}{8}orbit_{\Gamma^\delta}(X_1^2X_2^2) \in \mathbb{C}[X_1,X_2,X_3]^{\Gamma^\delta}$ by Lemma 5. Then we compute a representation of $\delta(f)$ by Theorem 2 (cf. [1], algorithm 3.12) and obtain $\delta(f) = \frac{9}{8}\sigma_1^4 - \frac{27}{4}\sigma_1^2\sigma_2 + \frac{81}{8}\sigma_2^2$. Finally, we compute $\hat{\sigma}_1 = \delta^{-1}(\sigma_1) = z$ and $\hat{\sigma}_2 = \delta^{-1}(\sigma_2) = -\frac{1}{3}X_1^2 - \frac{1}{3}X_2^2 + \frac{1}{3}X_3^2$, apply δ^{-1} to the representation of $\delta(f)$ and obtain so $f = \frac{9}{8}\hat{\sigma}_1^4 - \frac{27}{4}\hat{\sigma}_1^2\hat{\sigma}_2 + \frac{81}{8}\hat{\sigma}_2^2$ as result.

2.2 A major open problem

Our result so far is — in the most cases — a non-trivial improvement of Noether's bound for polynomial invariants of conjugates of permutation groups, which is certainly worth to be mentioned. The great challenge of our transformation technique is w.r.t. the computation of a *new* kind of finite bases for polynomial invariants of permutation groups G. It is well known that $K[X_1,\ldots,X_n]^G$ has a finite SAGBI basis w.r.t. the lexicographical term order iff G is a direct product of symmetric groups [3]. A transformation $\delta \in GL(n,K)$ in dependence of G might have some positive influence on the finiteness of SAGBI bases (cf. [8], Section 4.13), i.e. we conjecture the following:

Conjecture 10. Let $G \leq S_n$. Then there exists a matrix $\delta \in GL(n,K)$ such that $K[X_1,\ldots,X_n]^{G^\delta}$ has a finite SAGBI basis w.r.t. the lexicographical order.

The trueness of this conjecture would imply that a finite basis for polynomial invariants of permutation groups can be computed with a finite SAGBI basis for polynomial invariants of conjugates of permutations groups and with δ^{-1}. Hence, SAGBI bases would have a much bigger (algorithmic) potential for the treatment of polynomial invariants of permutation group as previously assumed (cf. [1,3]). The construction of a suitable $\delta \in GL(n,K)$ and the characterization of all $\delta \in GL(n,K)$ leading to a finite SAGBI basis for $K[X_1,\ldots,X_n]^{G^\delta}$ are other challenging problems.

3 Remark II

The structure of SAGBI bases for polynomial invariants of permutation groups w.r.t. the lexicographical order $<_{lex}$ with $X_1 >_{lex} \ldots >_{lex} X_n$ was recently investigated in [3]: Roughly speaking only invariant rings of direct products of permutation groups have a finite SAGBI basis, which is then, in addition, multilinear. Our second remark deals with the "smallest" ring of polynomial invariants of a permutation group, which has no finite SAGBI basis w.r.t. any admissible order. "Smallest" refers to the number of variables, which is 3, and to the number of generators of the invariant ring, which is 4. The permutation group in question is the alternating group A_3.

Of course, invariant rings which have no finite SAGBI basis w.r.t. any admissible order can be treated with other techniques to compute a finite basis, see e.g. [1, 5, 6]. Some of these methods lead to total degree bounds for the generators. However, the more technique are available to study invariant rings, which are by no means trivial algebraic objects, the better. We show by example how the concept of finite SAGBI bases can be rescued according to Remark I, even if the invariant ring has no finite SAGBI basis w.r.t. any admissible order: The invariant ring of a conjugate of A_3 has a finite SAGBI basis B w.r.t. $<_{lex}$. B can be used to compute a basis $\hat{B}$ for $\mathbb{C}[X_1, X_2, X_3]^{A_3}$ and to rewrite any A_3-invariant polynomial by a guided SAGBI bases reduction as a polynomial in the polynomials of B.

3.1 All the other cases

We briefly verify that any invariant ring of a permutation group $G \neq A_3$ in three or less variables has a finite SAGBI basis w.r.t. any admissible order.

Lemma 11. *The invariant ring $\mathbb{C}[X_1, \ldots, X_n]^{S_n}$ has the finite SAGBI basis $\{\sigma_1, \ldots, \sigma_n\}$ w.r.t. any admissible order (see [8]).*

The invariant ring $\mathbb{C}[X_1, \ldots, X_n]^{\{id\}}$ has the finite SAGBI basis $\{X_1, \ldots, X_n\}$ w.r.t. any admissible order (obvious).

Lemma 12. *The invariant ring $\mathbb{C}[X_1, X_2, X_3]^{S_2 \times \{id\}}$ has the finite SAGBI basis $\{X_1 + X_2, X_1 X_2, X_3\}$ w.r.t. any admissible order.*

Proof. This is a consequence of Lemma 11. □

Any other possible case $G \neq A_3$ not listed so far follows from Lemma 12 by rearranging some variables. $\mathbb{C}[X_1, X_2, X_3]^{A_3}$ is generated by $\sigma_1, \sigma_2, \sigma_3$ and one additional polynomial, which is, e.g. $X_1^2 X_2 + X_1 X_3^2 + X_2^2 X_3$, or $X_1^2 X_3 + X_1 X_2^2 + X_2 X_3^2$, or $(X_1 - X_2)(X_1 - X_3)(X_2 - X_3)$ [1].

Lemma 13. *Any invariant ring $\mathbb{C}[X_1, \ldots, X_n]^G$, which has no finite SAGBI basis w.r.t. any admissible order, is generated by at least 4 generators.*

Proof. $\mathbb{C}[X_1, \ldots, X_n]^G$ is generated by at most 3 polynomials implies $n \leq 3$. All these cases are covered by Lemma 11 and Lemma 12. □

3.2 The alternating group A_3

Lemma 14. *The invariant ring $\mathbb{C}[X_1, X_2, X_3]^{A_3}$ has no finite SAGBI basis w.r.t. $<_{lex}$.*

Proof. See [1]. For the sake of completeness we redo the proof here:

Assume that $\{\psi_1,\ldots,\psi_k\}$ is a finite SAGBI basis of $\mathbb{C}[X_1,X_2,X_3]^{A_3}$ w.r.t. $<_{lex}$ with $HT(\psi_i) = X_1^{e_{i_1}} X_2^{e_{i_2}} X_3^{e_{i_3}}$ and $d = \max\{e_{i_j} \mid 1 \leq i \leq k, 1 \leq j \leq 3\}$. Note that either $e_{i_1} \geq e_{i_2} \geq e_{i_3}$ or $e_{i_1} > e_{i_3} > e_{i_2}$.

Now let $f = X_1^{d+1} X_3^d + X_1^d X_2^{d+1} + X_2^d X_3^{d+1} \in \mathbb{C}[X_1,X_2,X_3]^{A_3}$. ψ_i is involved in a reduction of f implies that $e_{i_2} = 0$, i.e. either $HT(\psi_i) = X_1^{e_{i_1}}$ with $d \geq e_{i_1} \geq 0$ or $HT(\psi_i) = X_1^{e_{i_1}} X_3^{e_{i_3}}$ with $d \geq e_{i_1} > e_{i_3} > 0$. In any case, we have to multiply at least two terms $X_1^{e_{i_1}} X_3^{e_{i_3}}$ with $d \geq e_{i_1} > e_{i_3} > 0$ for the reduction of f in order to obtain $HT(f) = X_1^{d+1} X_3^d$. Any such product has a difference of at least two in the exponents of X_1 and X_3 which shows that $HT(f)$ cannot be a product of $HT(\psi_i)$ for $1 \leq i \leq k$ (contradiction). □

Corollary 15. *The invariant ring $\mathbb{C}[X_1,X_2,X_3]^{A_3}$ has an infinite SAGBI basis $B = \{\sigma_1,\sigma_2,\sigma_3\} \cup \{X_1^{d+2} X_3^{d+1} + X_1^{d+1} X_2^{d+2} + X_2^{d+1} X_3^{d+2} \mid d \in \mathbb{N}\}$ w.r.t. $<_{lex}$. B is minimal as follows: $B \setminus \hat{B}$ is for any $\emptyset \neq \hat{B} \subset B$ no SAGBI basis of $\mathbb{C}[X_1,X_2,X_3]^{A_3}$ w.r.t. $<_{lex}$.*

The question now is, if there exists a finite SAGBI basis w.r.t. any other admissible order.

Theorem 16. *The invariant ring $\mathbb{C}[X_1,X_2,X_3]^{A_3}$ has no finite SAGBI basis w.r.t. any admissible order $<$.*

Proof. Again, assume that $B = \{\psi_1,\ldots,\psi_k\}$ is a finite SAGBI basis of $\mathbb{C}[X_1,X_2,X_3]^{A_3}$ w.r.t. $<$ with $HT(\psi_i) = X_1^{e_{i_1}} X_2^{e_{i_2}} X_3^{e_{i_3}}$ and $d = \max\{e_{i_j} \mid 1 \leq i \leq k, 1 \leq j \leq 3\}$. We can rearrange the variables w.l.o.g. in such a way that $X_1 > s \in \{X_2, X_3\}$, $X_1 X_2 > s \in \{X_1 X_3, X_2 X_3\}$, because $\sigma_1, \sigma_2, \sigma_3$ is a SAGBI basis of $\mathbb{C}[X_1,X_2,X_3]^{S_n} \subset \mathbb{C}[X_1,X_2,X_3]^{A_3}$ for any admissible order (cf. Lemma 11). Furthermore, we must have $X_2 > X_3$, because otherwise $X_2 < X_3$ implies $X_1 X_2 < X_1 X_3$ (contradiction), i.e. $X_1 > X_2 > X_3$. By a similar reasoning we obtain $X_1 X_2 > X_1 X_3 > X_2 X_3$.

We know now that $\{\sigma_1, \sigma_2, \sigma_3\} \subset B$. Next let $t = X_1^{e_1} X_2^{e_2} X_3^{e_3}$, and consider the following cases:

1. $|\{e_1, e_2, e_3\}| \leq 2$: Then we have $orbit_{A_3}(t) = orbit_{S_3}(t)$. $HT(orbit_{S_3}(t))$ can be reduced by a unique product of σ_1, σ_2 and σ_3.

2. $e_1 > e_2 > e_3 = 0$: Then we have $orbit_{A_3}(t) \neq orbit_{S_3}(t)$, but $HT(orbit_{A_3}(t)) = HT(orbit_{S_3}(t))$, i.e. $orbit_{A_3}(t)$ can be reduced by a unique product of σ_1 and σ_2.

3. $e_1, e_2, e_3 \geq 1$: Then we have $orbit_{A_3}(t) = orbit_{A_3}(X_1^{e_1-1} X_2^{e_2-1} X_3^{e_3-1})\sigma_3$, and $HT(orbit_{A_3}(t)) = HT(orbit_{A_3}(X_1^{e_1-1} X_2^{e_2-1} X_3^{e_3-1}))\sigma_3$.

4. $e_1 > e_3 > e_2 = 0$: Then we have $orbit_{A_3}(t) = X_1^{e_1} X_3^{e_3} + X_1^{e_3} X_2^{e_1} + X_2^{e_3} X_3^{e_1}$. Furthermore, we have $X_2^{e_3} X_3^{e_1} \neq HT(orbit_{A_3}(t))$, because $X_1^{e_1} X_3^{e_3} = X_1^{e_1-e_3}(X_1 X_3)^{e_3} > X_3^{e_1-e_3}(X_2 X_3)^{e_3} = X_2^{e_3} X_3^{e_1}$ and $X_1^{e_3} X_2^{e_1} = X_2^{e_1-e_3}(X_1 X_2)^{e_3} > X_3^{e_1-e_3}(X_2 X_3)^{e_3} = X_2^{e_3} X_3^{e_1}$.

Consequently, $B_{HT} = \{HT(\psi_i) \,|\, 1 \leq i \leq k\}$ has to be a subset of $\{X_1, X_1X_2, X_1X_2X_3\} \cup \{X_1^{e_1}X_3^{e_3}, X_1^{e_3}X_2^{e_1} \,|\, 0 < e_3 < e_1 \leq d\}$. Furthermore, $X_1^{e_1}X_3^{e_3} \in B_{HT}$ implies $X_1^{e_3}X_2^{e_1} \notin B_{HT}$ and $X_1^{e_3}X_2^{e_1} \in B_{HT}$ implies $X_1^{e_1}X_3^{e_2} \notin B_{HT}$. Our goal is now to construct an infinite sequence of head terms $t_0, t_1, t_2, \ldots$ of A_3-invariant orbits — similar as in the proof of Lemma 14 — such that almost all of these terms are not generated by products of terms in B_{HT}.

Let $t_0 = HT(orbit_{A_3}(X_1^2X_3))$, and let $s_0 = X_1X_3$, if $t_0 = X_1^2X_3$, and let $s_0 = X_2$, otherwise. Furthermore, for $i \geq 1$, let $t_i = HT(orbit_{A_3}(t_{i-1}s_{i-1}))$, and let $s_i = s_{i-1}$, if $t_i = t_{i-1}s_{i-1}$, and let $s_i = \begin{cases} X_1^{e_1}X_3^{e_3}, \text{ if } t_{i-1} = X_1^{e_3}X_2^{e_1} \\ X_1^{e_3}X_2^{e_1}, \text{ if } t_{i-1} = X_1^{e_1}X_3^{e_3} \end{cases}$ otherwise. t_i is either $X_1^{e_1}X_3^{e_3}$ or $X_1^{e_3}X_2^{e_1}$ with $e_1 > e_3 > 0$ for all $i \in \mathbb{N}$, the total degree of t_{i_1} is always smaller than the total degree of t_{i_2} for $i_1 < i_2 \in \mathbb{N}$, and s_i is never a head term of an A_3-invariant orbit for all $i \in \mathbb{N}$.

The sequence $t_0, t_1, t_2, \ldots$ has by construction the following properties: First, t_i is never a product of terms in $W_{i-1} = \{X_1, X_1X_2, X_1X_2X_3\} \cup \{t_0, \ldots, t_{i-1}\}$ for all $i \in \mathbb{N}$. Any product of terms in W_{i-1} matching the exponent of X_3 (X_1) is unable to match simultaneously the exponent of X_1 (X_2), if $t_i = X_1^{e_1}X_3^{e_3}$ ($X_1^{e_3}X_2^{e_1}$). Second, all other head terms in $\mathbb{C}[X_1, X_2, X_3]^{A_3}$ have an expression as a product of terms in $W = \{X_1, X_1X_2, X_1X_2X_3\} \cup \{t_0, t_1, t_2, \ldots\}$.

Altogether, this implies that any t_i with a sufficiently large total degree has no expression as a product of terms of the finite set B_{HT}. Hence, there exists no finite SAGBI basis of $\mathbb{C}[X_1, X_2, X_3]^{A_3}$ w.r.t. $<$ (contradiction). □

Corollary 17. *Let the admissible order $<$ and the sequence $t_0, t_1, t_2, \ldots$ be as in the proof of Theorem 16. Then the invariant ring $\mathbb{C}[X_1, X_2, X_3]^{A_3}$ has an infinite SAGBI basis $B = \{\sigma_1, \sigma_2, \sigma_3\} \cup \{orbit_{A_3}(t_i) \,|\, i \in \mathbb{N}\}$ w.r.t. $<$. B is minimal as follows: $B \setminus \hat{B}$ is for any $\emptyset \neq \hat{B} \subset B$ no SAGBI basis of $\mathbb{C}[X_1, X_2, X_3]^{A_3}$ w.r.t. $<$.*

The results in Subsection 3.1 and in this subsection hold not only for the field $\mathbb{C}$ but for any ring R, because our argumentation is based on A_3-invariant orbits.

3.3 A way out (Part I): Conjugates of Permutation Groups

As already mentioned the computation of a finite basis of $\mathbb{C}[X_1, X_2, X_3]^{A_3}$ and the reduction of A_3-invariant polynomials can be done with other techniques. However, there is still a way to do it with SAGBI bases (cf. Subsection 2.2).

Lemma 18. *Let $\delta = \begin{pmatrix} \frac{1}{2} & \frac{1}{2} & 0 \\ -\frac{1}{2} & \frac{1}{2} & 0 \\ 0 & 0 & 1 \end{pmatrix}$. Then $A_3^\delta = \langle \begin{pmatrix} -\frac{1}{2} & \frac{1}{2} & -1 \\ -\frac{1}{2} & -\frac{1}{2} & 1 \\ \frac{1}{2} & \frac{1}{2} & 0 \end{pmatrix} \rangle$ is a conjugate of A_3.*

Proof. Obvious, because $A_3 = \langle \begin{pmatrix} 0 & 1 & 0 \\ 0 & 0 & 1 \\ 1 & 0 & 0 \end{pmatrix} \rangle$. □

Lemma 19. *Let δ be as in Lemma 18. Then the invariant ring $\mathbb{C}[X_1, X_2, X_3]^{A_3^\delta}$ has w.r.t. $<_{lex}$ a finite SAGBI basis $B = \{\psi_1, \psi_2, \psi_3, \psi_4\} = \{X_2+X_3, X_1^2+X_2^2+2X_3^2, X_1^2X_3+\frac{2}{3}X_2^3+X_2^2X_3+2X_2X_3^2+\frac{2}{3}X_3^3, X_1^3-X_1X_2^2+4X_1X_2X_3-4X_1X_3^2\}$. Furthermore, $\hat{B} = \{\delta^{-1}(\psi_1), \delta^{-1}(\psi_2), \delta^{-1}(\psi_3), \delta^{-1}(\psi_4)\} = \{\delta^{-1}(f) \mid f \in B\}$ is a basis of $\mathbb{C}[X_1, X_2, X_3]^{A_3}$. The elements of B and $\hat{B}$ have a total degree of at most 3.*

Proof. First simple calculus shows that $B \subset \mathbb{C}[X_1, X_2, X_3]^{A_3^\delta}$. Furthermore, $\hat{B} = \{X_1+X_2+X_3, 2X_1^2+2X_2^2+2X_3^2, 4X_1^2X_3+4X_1X_2^2+4X_2X_3^2-4X_1^2X_2-4X_2^2X_3-4X_1X_3^2, \frac{2}{3}X_1^3+\frac{2}{3}X_2^3+\frac{2}{3}X_3^3+2X_1^2X_2+2X_1^2X_3+2X_1X_2^2+2X_1X_3^2+2X_2^2X_3+2X_2X_3^2\}$, which is obviously a basis of $\mathbb{C}[X_1, X_2, X_3]^{A_3}$ (cf. [1]). Hence, B is a basis of $\mathbb{C}[X_1, X_2, X_3]^{A_3^\delta}$.

Once we know that B is a basis of $\mathbb{C}[X_1, X_2, X_3]^{A_3^\delta}$, we only have to check, if it's in addition a SAGBI basis (cf. [8]). An accurate analysis of the set of head terms $\{X_2, X_1^2, X_1^2X_3, X_1^3\}$ of the elements of B implies that we only have to verify that $f = \psi_2^3 - \psi_4^2$ can be reduced to zero by means of B to ensure that B is a SAGBI basis. This is the case, because $f = \frac{8}{3}\psi_1^6 + 4\psi_1^4\psi_2 - 16\psi_1^3\psi_3 + 5\psi_1^2\psi_2^2 - 18\psi_1\psi_2\psi_3 + 27\psi_3^2$. It follows that B is indeed a SAGBI basis of $\mathbb{C}[X_1, X_2, X_3]^{A_3^\delta}$. □

Example 20. Let δ, B and $\hat{B}$ be as in Lemma 19, and let $f = X_1^3X_2^2 + X_1^2X_3^3 + X_2^2X_3^3 \in \mathbb{C}[X_1, X_2, X_3]^{A_3}$. Then $\delta(f) = -\frac{1}{32}X_1^5 + \frac{1}{32}X_1^4X_2 + \frac{1}{16}X_1^3X_2^2 + \frac{1}{8}X_1^3X_3^2 + \frac{3}{8}X_1^2X_2X_3^2 - \frac{1}{16}X_1^2X_2^3 + \frac{1}{4}X_1^2X_3^3 - \frac{1}{32}X_1X_2^4 + \frac{3}{8}X_1X_2^2X_3^2 - \frac{1}{2}X_1X_2X_3^3 + \frac{1}{32}X_2^5 + \frac{1}{8}X_2^3X_3^2 + \frac{1}{4}X_2^2X_3^3 \in \mathbb{C}[X_1, X_2, X_3]^{A_3^\delta}$, which has the representation $\delta(f) = -\frac{1}{12}\psi_1^5 - \frac{5}{48}\psi_1^3\psi_2 + \frac{5}{16}\psi_1^2\psi_3 + \frac{1}{16}\psi_1^2\psi_4 + \frac{1}{32}\psi_1\psi_2^2 - \frac{1}{32}\psi_2\psi_4 - \frac{1}{32}\psi_2\psi_3$ in terms of the elements of the SAGBI basis B. Hence, we can read off the representation of f in terms of the elements of $\hat{B}$ by applying δ^{-1} to $\delta(f)$, i.e. $f = -\frac{1}{12}\delta^{-1}(\psi_1)^5 - \frac{5}{48}\delta^{-1}(\psi_1)^3\delta^{-1}(\psi_2) + \frac{5}{16}\delta^{-1}(\psi_1)^2\delta^{-1}(\psi_3) + \frac{1}{16}\delta^{-1}(\psi_1)^2\delta^{-1}(\psi_4) + \frac{1}{32}\delta^{-1}(\psi_1)\delta^{-1}(\psi_2)^2 - \frac{1}{32}\delta^{-1}(\psi_2)\delta^{-1}(\psi_4) - \frac{1}{32}\delta^{-1}(\psi_2)\delta^{-1}(\psi_3)$.

4 Remark III

Our third remark shows the existence of an invariant ring generated only by polynomial invariants with a total degree of at most 2, which has no finite SAGBI basis w.r.t. any admissible order. Hence, 2 is the optimal lower bound, because any invariant ring generated by polynomial invariants with a total degree of at most 1 has for trivial reasons a finite SAGBI basis. In addition, our example has w.r.t. this property the minimal number of variables 4, if we restrict ourself to polynomial invariants of permutation groups, and also the minimal group order 2.

4.1 The optimal lower bound

Lemma 21. *Let $G = \langle(12)(34)\rangle$. Then $\mathbb{C}[X_1, X_2, X_3, X_4]^G$ is generated by $B = \{X_1+X_2, X_1X_2, X_3+X_4, X_3X_4, X_1X_4+X_2X_3\}$.*

Proof. A closer look at the G-invariant orbits of $\mathbb{C}[X_1, X_2, X_3, X_4]^G$ via the reduction technique described in [1] shows that we only have to ensure the representation of $orbit_G(X_1^2 X_3)$, $orbit_G(X_1 X_3^2)$, $orbit_G(X_1^2 X_4)$, and $orbit_G(X_1 X_4^2)$ in terms of the elements of B. We have

$$orbit_G(X_1^2 X_3) = (X_1 + X_2)(X_1 X_3 + X_2 X_4) - (X_1 X_2)(X_3 + X_4),$$
$$orbit_G(X_1 X_3^2) = (X_3 + X_4)(X_1 X_3 + X_2 X_4) - (X_3 X_4)(X_1 + X_2),$$
$$orbit_G(X_1^2 X_4) = (X_1 + X_2)(X_1 X_4 + X_2 X_3) - (X_1 X_2)(X_3 + X_4), \text{ and}$$
$$orbit_G(X_1 X_4^2) = (X_3 + X_4)(X_1 X_3 + X_2 X_4) - (X_3 X_4)(X_1 + X_2)$$

with $X_1 X_3 + X_2 X_4 = (X_1 + X_2)(X_3 + X_4) - (X_1 X_4 - X_2 X_3)$. Furthermore, we have $e_1, e_2 > 0$ or $e_3, e_4 > 0$ for any other non-multilinear special $f = orbit_G(X_1^{e_1} X_2^{e_2} X_3^{e_3} X_4^{e_4})$ not listed so far, i.e. we can rewrite these G-invariant orbits as follows: $f = \begin{cases} (X_1 X_2) orbit_G(X_1^{e_1-1} X_2^{e_2-1} X_3^{e_3} X_4^{e_4}), \text{if } e_1, e_2 > 0 \\ (X_3 X_4) orbit_G(X_1^{e_1} X_2^{e_2} X_3^{e_3-1} X_4^{e_4-1}), \text{if } e_3, e_4 > 0 \end{cases}$ □

Lemma 22. *Let G be as in Lemma 21. Then G has no finite SAGBI basis w.r.t. $<_{lex}$.*

Proof. G is not a direct product of permutation groups. So, following [3], $\mathbb{C}[X_1, X_2, X_3, X_4]^G$ can not have a finite SAGBI basis w.r.t. $<_{lex}$. □

Theorem 23. *Let G be as in Lemma 21. Then $\mathbb{C}[X_1, X_2, X_3, X_4]^G$ has no finite SAGBI basis w.r.t. any admissible order $<$.*

Proof. Assume that $\mathbb{C}[X_1, X_2, X_3, X_4]^G$ has a finite SAGBI basis B w.r.t. $<$, and assume further w.l.o.g. that $X_1 > X_2 > X_3 > X_4$. Then B contains the multilinear G-invariant orbits $\{X_1 + X_2, X_1 X_2, X_1 X_4 + X_2 X_3, X_3 + X_4, X_3 X_4\}$ and a finite number of non-multilinear G-invariant orbits of the form $\psi_{e_1, e_2} = X_1^{e_1} X_4^{e_2} + X_2^{e_1} X_3^{e_2}$ with $e_1 \neq e_2 \geq 1$. Note that the head term of $\psi = X_1 X_4 + X_2 X_3$ is w.r.t. $<$ not determined so far.

Our goal is to construct an infinite sequence of head terms $t_0, t_1, t_2, \ldots$ of G-invariant orbits such that almost all of these terms are not generated by products of head terms of the polynomials in B. Let

$$t_0 = \begin{cases} HT(orbit_G(X_1 X_4^2)), \text{if } HT(\psi) = X_1 X_4 \\ HT(orbit_G(X_2^2 X_3)), \text{otherwise} \end{cases},$$

and let $s_0 = X_4$, if $t_0 = X_1 X_4^2$, $s_0 = X_2 X_3$, if $t_0 = X_2 X_3^2$, $s_0 = X_2$, if $t_0 = X_2^2 X_3$, and $s_0 = X_1 X_4$ otherwise. Furthermore, for $i \geq 1$, define $t_i = HT(orbit_G(t_{i-1} s_{i-1}))$, and let $s_i = s_{i-1}$, if $t_i = t_{i-1} s_{i-1}$, and let $s_i = \begin{cases} X_1^{e_1} X_4^{e_2}, \text{if } t_{i-1} = X_2^{e_1} X_3^{e_2} \\ X_2^{e_1} X_3^{e_2}, \text{if } t_{i-1} = X_1^{e_1} X_4^{e_2} \end{cases}$, otherwise. For all $i \in \mathbb{N}$, we have t_i is $X_1^{e_1} X_4^{e_2}$ or $X_2^{e_1} X_3^{e_2}$ with $1 \leq e_1 < e_2$, if $HT(\psi) = X_1 X_4$, and with $e_1 > e_2 \geq 1$, otherwise. The total degree of t_{i_1} is always smaller than the total degree of t_{i_2} for $i_1 < i_2 \in \mathbb{N}$, and s_i is never a head term of a G-invariant orbit for all $i \in \mathbb{N}$.

The sequence $t_0, t_1, t_2, \ldots$ has by construction the following properties: First, t_i is never a product of terms in $W_{i-1} = \{X_1, X_1X_2, HT(\psi), X_3, X_3X_4\} \cup \{t_0, \ldots, t_{i-1}\}$ for all $i \in \mathbb{N}$. Second, all other head terms in $\mathbb{C}[X_1, X_2, X_3, X_4]^G$ have an expression as a product of terms in $W = \{X_1, X_1X_2, HT(\psi), X_3, X_3X_4\} \cup \{t_0, t_1, t_2, \ldots\}$.

Altogether, this implies that any t_i with a sufficiently large total degree has no expression as a product of head terms of the polynomials in the finite set B. Hence, there exists no finite SAGBI basis of $\mathbb{C}[X_1, X_2, X_3, X_4]^G$ w.r.t. $<$ (contradiction). □

$\mathbb{C}[X_1, \ldots, X_n]^\Gamma$ is generated by polynomials with a total degree of at most 1 implies that Γ is the trivial group, and that the generators are $X_1, \ldots, X_n$. Hence, 2 is the smallest possible and therefore optimal lower bound for the generators of an invariant ring without a finite SAGBI w.r.t. any admissible order. Furthermore, we must have $|\Gamma| \geq 2$ for any $\mathbb{C}[X_1, \ldots, X_n]^\Gamma$ with this property, i.e. our example is minimal w.r.t. the group order, because $|G| = 2$.

Lemma 24. *Let $n < 4$, and let $\mathbb{C}[X_1, \ldots, X_n]^G$ be generated by elements with a total degree of at most 2. Then $\mathbb{C}[X_1, \ldots, X_n]^G$ has a finite SAGBI basis.*

Proof. G is either S_1, $S_1 \times S_1$, S_2, $S_1 \times S_2$, $S_2 \times S_1$, or $S_1 \times S_1 \times S_1$, i.e. $\mathbb{C}[X_1, \ldots, X_n]^G$ has a finite SAGBI basis (cf. [3]). □

Hence, $\mathbb{C}[X_1, X_2, X_3, X_4]^G$ with $G = \langle(12)(34)\rangle$ is, in addition, minimal w.r.t. the number of variables, if we restrict ourself to polynomial invariants of permutation groups.

The results in this subsection hold not only for the field $\mathbb{C}$ but for any ring R, because our argumentation is based on G-invariant orbits.

4.2 A way out (Part II): Conjugates of Permutation Groups

$\mathbb{C}[X_1, X_2, X_3, X_4]^G$ has similar properties as $\mathbb{C}[X_1, X_2, X_3]^{\langle(123)\rangle}$ in Remark II: There exists a conjugate Γ of G such that the corresponding $\mathbb{C}[X_1, X_2, X_3, X_4]^\Gamma$ has a finite SAGBI basis w.r.t. $<_{lex}$ (cf. Subsection 2.2).

Lemma 25. *Let $G = \{id, \pi\}$ be as in Lemma 21, let $\delta = \begin{pmatrix} 1 & 0 & 0 & 0 \\ 1 & 1 & 0 & 0 \\ 0 & 0 & 1 & 0 \\ 0 & 0 & 1 & 1 \end{pmatrix}$, and let $\Gamma = \{id, \delta^{-1}\pi\delta\}$ be a conjugate of G. Then $\mathbb{C}[X_1, X_2, X_3, X_4]^\Gamma$ has a finite SAGBI basis $\hat{B} = \{X_1 + \frac{1}{2}X_2, X_2^2, X_2X_4, X_3 + \frac{1}{2}X_4, X_4^2\}$ w.r.t. $<_{lex}$.*

Proof. First we have to ensure that $\hat{B}$ is a basis of $\mathbb{C}[X_1, X_2, X_3, X_4]^\Gamma$. For this we compute $\{\delta^{-1}(f) \mid f \in \hat{B}\} = \{\psi_1, \ldots, \psi_5\} = \{X_1^2 + X_2^2 - 2X_1X_2, X_1X_3 + X_2X_4 - (X_1X_4 + X_2X_3), \frac{1}{2}X_1 + \frac{1}{2}X_2, X_3^2 + X_4^2 - 2X_3X_4, \frac{1}{2}X_3 + \frac{1}{2}X_4\}$, which is a basis for $\mathbb{C}[X_1, X_2, X_3, X_4]^G$, because (cf. Lemma 21) $X_1 + X_2 = 2\psi_3$, $X_1X_2 = \psi_3^2 - \frac{1}{4}\psi_1$, $X_3 + X_4 = 2\psi_5$, $X_3X_4 = \psi_5^2 - \frac{1}{4}\psi_4$, and $X_1X_4 + X_2X_3 = 2\psi_3\psi_5 - \frac{1}{2}\psi_2$.

Second we have to show that $\hat{B}$ is a SAGBI basis of $\mathbb{C}[X_1, X_2, X_3, X_4]^\Gamma$. Following [8], the analysis of the set of head terms $\{X_1, X_2^2, X_2 X_4, X_3, X_4^2\}$ of $\hat{B}$ reveals that we only have to verify, if $f = (X_2^2 X_4^2) - (X_2 X_4)^2$ can be reduced to zero modulo $\hat{B}$. But this is obviously the case. □

Lemma 25 is the key to analyze $\mathbb{C}[X_1, X_2, X_3, X_4]^G$ and to rewrite its elements by means of SAGBI bases according to Remark I. The finite SAGBI basis $\hat{B}$ of $\mathbb{C}[X_1, X_2, X_3, X_4]^\Gamma$ allows us to reduce any G-invariant polynomial $f \in \mathbb{C}[X_1, \ldots, X_n]^G$ by applying the transformation δ to f, computing the representation of $\delta(f)$ w.r.t. $\hat{B}$, and back-transforming the result by applying δ^{-1}.

5 Conclusion

This paper deals with several aspects of SAGBI bases for polynomial invariants of permutation groups. Most important in the study of these bases is the finiteness question formulated in Conjecture 10. I'm quite sure, that a proof of this challenging statement will give SAGBI bases a similar rôle in the treatment of invariant rings as Gröbner bases have nowadays [9]. On the other hand, if Conjecture 10 is false, the significance of SAGBI bases will be certainly limited to a couple of special cases [3].

References

1. Göbel, M. (1995). Computing Bases for Permutation-Invariant Polynomials. Journal of Symbolic Computation 19, 285–291
2. Göbel, M. (1997). On the Number of Special Permutation-Invariant Orbits and Terms. AAECC 8(6), 505–509
3. Göbel, M. (1998). A Constructive Description of SAGBI Bases for Polynomial Invariants of Permutation Groups. Journal of Symbolic Computation 26, 261–272
4. Kapur, D., Madlener, K. (1989). A Completion Procedure for Computing a Canonical Basis of a k-Subalgebra. In: Kaltofen, E., Watt, S. (eds.), Proceedings of Computers and Mathematics 89. MIT, Cambridge, 1–11
5. Kemper, G. (1996). Calculating Invariant Rings of Finite Groups over Arbitrary Fields. Journal of Symbolic Computation 21, 351–366
6. Noether, E. (1916). Der Endlichkeitssatz der Invarianten endlicher Gruppen. Math. Ann. 77, 89-92
7. Robbiano, L. (1985). Term Orderings on the Polynomial Ring. In: Caviness, B. (ed.), European Conference on Computer Algebra, EUROCAL'85, Proceedings Vol. 2: Research Contributions, volume 204 of *LNCS*, 513–517, Springer
8. Robbiano, L., Sweedler, M. (1990). Subalgebra Bases. In: Bruns, W., Simis, A. (eds.), Commutative Algebra (Lect. Notes Math. 1430). 61–87, Springer
9. Sturmfels, B. (1993). Algorithms in Invariant Theory. Springer
10. Weispfenning, V. (1987). Admissible Orders and Linear Forms. ACM SIGSAM Bulletin 21/2, 16–18

Computing Prüfer Codes Efficiently in Parallel

Raymond Greenlaw[1]* and Rossella Petreschi[2]

[1] Department of Computer Science, Armstrong Atlantic State University,
11935 Abercorn Street, Savannah, Georgia, USA 31419-1997.
`greenlaw@pirates.armstrong.edu`
[2] Department of Computer Science, University of Rome "La Sapienza,"
Via Salaria 113, Rome, 00198, Italy. `petreschi@dsi.uniroma1.it`

Abstract. A *Prüfer code* of a labeled free tree with n nodes is a sequence of length $n - 2$ constructed by the following sequential process: for i ranging from 1 to $n - 2$ insert the label of the neighbor of the smallest remaining leaf into the ith position of the sequence, and then delete the leaf. Prüfer codes provide an alternative to the usual representation of trees. We present an optimal $O(\log n)$ time, $n/\log n$ processor EREW-PRAM algorithm for determining the Prüfer code of an n-node labeled chain and an $O(\log n)$ time, n processor EREW-PRAM algorithm for constructing the Prüfer code of an n-node labeled free tree. This resolves an open question posed by Wang, Chen, and Liu.

1 Introduction

Trees are an important structure in computer science. A wide variety of interesting problems have been defined on trees, for example, involving arrangements, expression evaluation, and graph theoretic computations. Moreover, parallel techniques such as tree contraction [1, 14] and centroid decomposition [5] can be used to solve many of these problems efficiently. We know of no natural problem defined on unweighted trees that is P-complete. The *bandwidth problem*, which does not involve weights, restricted to trees is NP-complete ([6], [16]). If weights are allowed, there are other natural tree problems that are known to be NP-complete (see, for example, [17]). A P-complete problem, whose definition is based solely on the structure of trees, might prove useful in helping to resolve the complexity of a number of open problems, currently not known to be in NC or to be P-complete. A list of such problems is given in [7].

Initial applications of Prüfer codes were to count the number of labeled free trees [11, 13, 15]. Since there is an isomorphism between Prüfer codes and labeled free trees, it is easy to see that there are exactly n^{n-2} labeled free trees of size n. Prüfer codes provide an alternative to the usual representation of trees. It may be easier to compute information about the tree from the Prüfer code.

A labeled free tree is depicted in Figure 1. Its corresponding Prüfer code is $(9, 6, 5, 6, 1, 6, 1, 1)$. Notice that the degree of each node in the tree is one plus

---

* This research was supported in part by a visiting Italian Fellowship at the University of Rome "La Sapienza."

the number of times the node appears in the sequence. This observation is true in general.

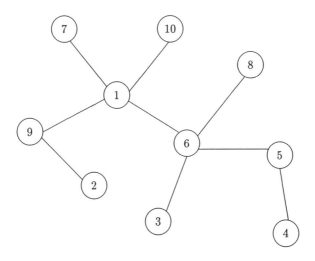

Fig. 1. A labeled free tree whose corresponding Prüfer code is $(9,6,5,6,1,6,1,1)$.

Algorithm Sequential Prüfer, specified below, computes the Prüfer code of a labeled free tree T having n nodes.

Algorithm Sequential Prüfer

Input: A labeled free tree $T = (V, E)$ represented by an edge list, where $V = \{1, 2, \ldots, n\}$ and $n \geq 2$.
Output: An array Pcode that contains the Prüfer code of T.
    begin
        $T' = T$;
        for $i = 1$ **to** $n - 2$ **do**
            $u =$ smallest leaf in T';
            Pcode$[i] =$ value of the neighbor of u in T';
            $T' = T' - \{u\}$;
        for $i = 1$ **to** $n - 2$ **do print** Pcode$[i]$;
    end

The Prüfer code is stored in an array Pcode having $n-2$ entries. The decision about the next leaf to remove seems to depend directly on the previous removal. Despite the apparent highly sequential nature of the problem, we present

1. an optimal $O(\log n)$ time, $n/\log n$ processor EREW-PRAM algorithm for determining the Prüfer code of an n-node labeled chain
2. an $O(\log n)$ time, n processor EREW-PRAM algorithm for constructing the Prüfer code of an n-node labeled free tree.

Both our algorithms make use of a number of well-known parallel techniques and assume the input is coded as an edge list.

2 Preliminaries

Many of the prerequisites for this paper can be found in [9, 10, 19]. We focus on the EREW-PRAM model, which is the weakest of the PRAM models. We make use of several classical theorems involving parallel computation and cite some of the original papers regarding these results. Many of the results can also be found in the previously mentioned books.

Theorem 1 (Brent's Scheduling Principle). *[3, 10]*
If processor allocation is not a problem, then a $w(n)$ work and $t(n)$ time parallel algorithm can be simulated using $w(n)/p(n) + t(n)$ time and $p(n)$ processors.

Theorem 2 (Parallel Prefix Computation). *[12]*
The Parallel Prefix Problem can be solved in $O(\log n)$ time using $n/\log n$ processors on an EREW-PRAM.

Theorem 3 (Euler Tour). *[20]*
An Euler tour of an n-node tree can be computed in $O(\log n)$ time using $n/\log n$ processors on an EREW-PRAM.

Theorem 4 (List Ranking). *[2]*
Given a list with n nodes, the List Ranking Problem can be solved in $O(\log n)$ time using $n/\log n$ processors on an EREW-PRAM.

Theorem 5 (Parallel Sorting). *[4], [9, page 173]*
A list of n elements can be sorted in $O(\log n)$ time using n processors on an EREW-PRAM.

Theorem 6 (Parallel Tree Contraction). *[1, 8, 14, 18]*
If L is a $(t(n), p(n))$-decomposable Bottom-up Algebraic Tree Computation Problem under EREW-PRAM computations, then the associated algebraic expression of any input tree having n leaves can be evaluated in $O(t(n) \log n)$ time using $np(n)/\log n$ processors on an EREW-PRAM.

3 An optimal parallel algorithm for computing the Prüfer Code of a chain

We first present an algorithm for computing Prüfer codes on chains. There are a number of reasons for proceeding in this way. The algorithm on chains

- is easier to understand yet contains many of the key ideas for the general algorithm.
- is optimal, whereas the general algorithm is not. It may be possible to develop an optimal algorithm for trees by exploiting the chain algorithm.
- uses many of the same steps as the algorithm on trees, and so simplifies our exposition.

Computing Prüfer codes efficiently in parallel 205

- can be fully detailed, whereas the general algorithm is too complicated to describe in the same amount of detail.

Algorithm Prüfer Chain given below computes the Prüfer code of an n-node labeled chain.

Algorithm Prüfer Chain

Input: A chain $T = (V, E)$ represented by an edge list, where $V = \{1, 2, \ldots, n\}$ and $n \geq 2$.
Output: An array Pcode that contains the Prüfer code of T.
  begin
/* We assume the chain is oriented with its maximum leaf as the "starting point," and all required array values are initialized (in parallel) to 0. */

/* Compute the position of each node in the chain. */
    1. use parallel list ranking to construct the array Position such that
       Position[i] = v, $1 \leq i \leq n$, where node v has a ranking of i in T

/* Compute the maximum nodes encountered thus far in left-to-right and right-to-left traversals over the chain. */
    2. use parallel prefix to construct the arrays LRMax and RLMax
       with LRMax[0] = 0; LRMax[$n+1$] = $n+1$;
       LRMax[i] = max{Position[j] | $1 \leq j \leq i$}, where $1 \leq i \leq n$ and
       RLMax[0] = 0; RLMax[$n+1$] = $n+1$;
       RLMax[i] = max{Position[j] | $n-i+1 \leq j \leq n$}, where $1 \leq i \leq n$

/* Compute when a node becomes a maximum (if it does) for both left-to-right and right-to-left traversals. */
    3. for $1 \leq i \leq n$ in parallel do
       if LRMax[$i-1$] $\neq$ LRMax[i] then LRStart[LRMax[i]] = i;
       if RLMax[$i-1$] $\neq$ RLMax[i] then RLStart[RLMax[i]] = i;

/* Compute when a node is no longer a maximum (if it was) for left-to-right and right-to-left traversals. */
    4. for $1 \leq i \leq n$ in parallel do
       if LRMax[i] $\neq$ LRMax[$i+1$] then LREnd[LRMax[i]] = i;
       if RLMax[i] $\neq$ RLMax[$i+1$] then RLEnd[RLMax[i]] = i;

/* Compute how many positions a node was maximum for. */
    5. for $1 \leq i \leq n$ in parallel do
       if LRStart[i] $\neq$ 0 then LRSpan[i] = LREnd[i] − LRStart[i] + 1;
       if RLStart[i] $\neq$ 0 then RLSpan[i] = RLEnd[i] − RLStart[i] + 1;

/* Compute how many nodes need to be removed from the left before a given node can be removed. Similarly for the right. */
    6. use parallel prefix to construct the arrays
       LeftClear and RightClear, where for $1 \leq i \leq n$
       LeftClear[i] = LRSpan[0] + $\cdots$ + LRSpan[$i-1$];
       RightClear[i] = RLSpan[0] + $\cdots$ + RLSpan[$i-1$];

/* Removal[i] denotes when the node in Position[i] is removed. */
    7. for $1 \leq i \leq n$ in parallel do

```
                if RLMax[n − i] > LRMax[i − 1]
                   then k = LRMax[i];
                        Removal[i] = i + RightClear[k];
                   else k = RLMax[n − i − 1];
                        Removal[i] = (n − i) + 1 + LeftClear[k];

/* Right neighbor removed later, include in the Prüfer code. */
     8. for 1 ≤ i ≤ n − 1 in parallel do
            if Removal[i] < Removal[i + 1]
               then Pcode[Removal[i]] = Position[i + 1];

/* Left neighbor removed later, include in the Prüfer code. */
     9. for 2 ≤ i ≤ n in parallel do
            if Removal[i] < Removal[i − 1]
               then Pcode[Removal[i]] = Position[i − 1];

/* Print the Prüfer code. */
    10. for 1 ≤ i ≤ n − 2 in parallel do print Pcode[i];
end
```

Example.

Before proving the correctness and complexity of the algorithm, we present an example of how Algorithm Prüfer Chain works. Consider the chain shown in Figure 2.

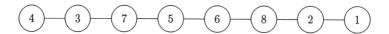

Fig. 2. A labeled chain whose Prüfer code is $(2, 8, 3, 7, 5, 6)$.

In step 1 of the algorithm, the array Position is constructed. Position[i] indicates the node in the chain whose rank is i. (For readability, we list the indices of the array as well.)

$$\begin{array}{|c|c|c|c|c|c|c|c|} \hline 4 & 3 & 7 & 5 & 6 & 8 & 2 & 1 \\ \hline \end{array}$$
$$1\ 2\ 3\ 4\ 5\ 6\ 7\ 8$$

Position[3] = 7 means node 7 has rank 3 (appears third from left-to-right) in the chain.

In step 2 the arrays LRMax and RLMax are constructed. LRMax[i] maintains the maximum node encountered so far in a left-to-right traversal of the chain (with special ending conditions).

$$\begin{array}{|c|c|c|c|c|c|c|c|c|c|} \hline 0 & 4 & 4 & 7 & 7 & 7 & 8 & 8 & 8 & 9 \\ \hline \end{array}$$
$$0\ 1\ 2\ 3\ 4\ 5\ 6\ 7\ 8\ 9$$

RLMax has a similar meaning but for a right-to-left traversal.

0	1	2	8	8	8	8	8	9
0	1	2	3	4	5	6	7	8	9

In step 3 the arrays LRStart and RLStart are constructed. LRStart[i] indicates where in a left-to-right traversal node i becomes a maximum (if it does).

0	0	0	1	0	0	3	6
1	2	3	4	5	6	7	8

RLStart has a similar meaning but for a right-to-left traversal.

1	2	0	0	0	0	0	3
1	2	3	4	5	6	7	8

In step 4 the arrays LREnd and RLEnd are constructed. LREnd[i] indicates where in a left-to-right traversal node i is no longer a maximum (if it was).

0	0	0	2	0	0	5	8
1	2	3	4	5	6	7	8

RLEnd has a similar meaning but for a right-to-left traversal.

1	2	0	0	0	0	0	8
1	2	3	4	5	6	7	8

In step 5 the arrays LRSpan and RLSpan are constructed. LRSpan[i] indicates in a left-to-right traversal how many positions node i was a maximum for (if it was).

0	0	0	0	2	0	0	3	3
0	1	2	3	4	5	6	7	8

RLSpan has a similar meaning but for a right-to-left traversal.

0	1	1	0	0	0	0	6
0	1	2	3	4	5	6	7	8

In step 6 the arrays LeftClear and RightClear are constructed. The element LeftClear[i] specifies how many values i is greater than before encountering a node of size greater than or equal to i in a left-to-right traversal of the chain. For example, LeftClear[6] = 2 indicates that before encountering a node of size greater than or equal to 6, in this case 7, 6 is greater than two nodes, 4 and 3, when traversing from the left end of the chain.

0	0	0	0	2	2	2	5
1	2	3	4	5	6	7	8

RightClear has a similar meaning but in the right-to-left direction.

0	1	2	2	2	2	2	2
1	2	3	4	5	6	7	8

In step 7 the array Removal is constructed. Removal[i] indicates when the node in Position[i] is removed by Algorithm Sequential Prüfer.

3	4	5	6	7	8	2	1
1	2	3	4	5	6	7	8

In steps 8 and 9 the array Pcode containing the Prüfer code is constructed.

2	8	3	7	5	6	8	−
1	2	3	4	5	6	7	8

In step 10 the first $n-2$ values of the array Pcode are printed. In this case the resulting Prüfer code is $(2, 8, 3, 7, 5, 6)$.

Correctness.

Having elaborated on the algorithm via a complete example, we now turn to its analysis. The following lemma is the key to proving correctness.

Lemma 7. *The array Removal constructed in step 7 of Algorithm Prüfer Chain specifies the order in which leaves are eliminated by Algorithm Sequential Prüfer.*

Proof: A node can be removed by Algorithm Sequential Prüfer only if it is a leaf. A leaf in a chain either has no left neighbor or no right neighbor. We consider the case where a node u, in Position[i], gets removed from the "left." A similar argument can be made when a node gets removed from the "right."

By our assumption, before u is removed all of the nodes to its left in the chain will be removed. According to Algorithm Sequential Prüfer, all nodes from the right end of the chain with values smaller than the maximum of u and the nodes to the left of u will be removed prior to u. That is, all nodes with values less than LRMax[i] will be removed from the right end of the chain before u is deleted.

The **then** clause in step 7 of Algorithm Prüfer Chain is executed when a node is removed from the left. The value k contains the maximum value of any node up through position i. The number of nodes less than k from the right end of the chain is given by RightClear[k]. Thus, node u is removed after

$$\underbrace{(i-1)}_{\text{nodes to its left}} + \underbrace{\text{RightClear}[k]}_{\text{nodes removed from the right end}} \quad (1)$$

nodes have been eliminated. So, node u is the $(i + \text{RightClear}[k])$th node removed, as computed by Algorithm Prüfer Chain. □

Theorem 8. *The Prüfer code of an n-node chain can be computed in $O(\log n)$ time using $n/\log n$ processors on an EREW-PRAM.*

Proof: Lemma 7 shows that Algorithm Prüfer Chain correctly computes the order in which leaves are eliminated during the construction of the Prüfer code. Steps 8 and 9 of Algorithm Prüfer Chain store into array Pcode the neighbors of the leaves as they are eliminated. Thus, at the completion of the algorithm, Pcode contains the Prüfer code of the input tree. Pcode is output in step 10.

We now argue the complexity of the algorithm. The initializations, steps 3–5, and steps 8–10 can all be carried out on an EREW-PRAM in $O(\log n)$ time using $n/\log n$ processors. (It is necessary to apply Theorem 1.) Step 7 can also be carried out within these bounds but care must be taken to avoid concurrent reads of the values RightClear[k] and LeftClear[k]. This can be accomplished by fanning out the appropriate number of copies of these values based on the LRSpan and RLSpan arrays. Applying the list ranking algorithm of Theorem 4 to step 1 and the parallel prefix algorithm of Theorem 2 to steps 2 and 6 completes the analysis. □

4 A fast parallel algorithm for computing the Prüfer Code of a tree

Input encoding

The algorithm described in this section takes as input a labeled free tree. When considering both small running times and a limited number of processors on weak parallel models, it is necessary to specify the exact encoding of the input. To be very general, we suppose the input is represented by an (arbitrary) edge list as we have been assuming implicitly up to this point. The input encoding necessary to apply the Euler tour algorithm of Theorem 3, however, is more restricted than this.

In order to apply the Euler tour construction of Theorem 3, the input encoding needs additional pointers [9, pages 108–114]. The algorithm requires adjacency lists that are directed and circular, and have "reverse" edges between adjacency lists. The following algorithm shows how to convert an arbitrary edge list into the form necessary to apply the Euler tour construction.

Algorithm Convert Edge List Encoding

Input: A labeled free tree $T = (V, E)$ represented by an edge list, where $V = \{1, 2, \ldots, n\}$ and $n \geq 2$.
Output: A representation of the tree by directed, circular adjacency lists plus reverse edges between the adjacency lists.
    begin
/* Construct the directed adjacency lists. */
       1. **for each** undirected edge $\{u, v\}$ **do**
           construct the two directed edges (u, v) and (v, u);

       2. sort these edges according to the first component;

       3. construct the circular adjacency list within each "block";

/* Add in the reverse edges. */

  4. **for each** edge (u, v) **do**
    put the index of (u, v) into an array for u in position v;

  5. **for each** edge (u, v) **do**
    read from the array for v in position u the value p and add this pointer
    to the information contained in the adjacency list at position (u, v);
end

Steps 1, 3, 4, and 5 can be implemented in $O(\log n)$ time using $n/\log n$ processors on an EREW-PRAM. By Theorem 5, the sorting in step 2 can be implemented in $O(\log n)$ time using n processors on an EREW-PRAM. In summary, we have

Lemma 9. *The input conversion from an edge list to the special adjacency lists needed for the Euler tour construction can be performed in $O(\log n)$ time and n processors on an EREW-PRAM.*

Algorithm Prüfer Tree

We now specify Algorithm Prüfer Tree for computing the Prüfer code of an n-node labeled free tree. The algorithm essentially generalizes the ideas of Algorithm Prüfer Chain. (In the special case where the input is actually a chain, Algorithm Prüfer Chain could be called.)

Algorithm Prüfer Tree

Input: A labeled free tree $T = (V, E)$ represented by an edge list, where $V = \{1, 2, \ldots, n\}$ and $n \geq 2$.
Output: An array Pcode that contains the Prüfer code of T.
 begin
/* Convert the input to a special form. */
  1. run Algorithm Convert Edge List Encoding on T;

/* Begin processing to root the tree at the maximum labeled node. */
  2. construct an Euler tour starting from the maximum labeled node;

/* Root the tree so all removals of leaves are in an upward direction towards the root. */
  3. using parallel prefix on the Euler tour constructed in step 1
    root the tree and form the array Parent with
    Parent$[i] = i$'s parent in the rooted tree, for $1 \leq i \leq n$;

/* Convert to a binary tree. See text for an explanation. */
  4. transform the rooted tree into a binary tree using Knuth's method,
    dummy nodes get value 0, leaves get "their" value as maximum;

/* Compute the maximum value in each subtree. */
  5. **for** $1 \leq i \leq n$ **in parallel do**
    Maximum$[i]$ = maximum value in the subtree rooted at i
    including the value at i;

/* Associate the maximum value and node name. */

6. **for** $1 \leq i \leq n$ **in parallel do** MaxName[i] = (Maximum[i], i);

/* Arrange based on maximum values so processors can be scheduled appropriately for list ranking in step 8. */

7. sort the elements in MaxName in increasing order based on the first component;

/* Tree is split into chains. */

8. **for** $1 \leq i \leq n$ **in parallel do**
   if Maximum[i] $\neq$ Maximum[Parent[i]]
   then Chain[i] = **nil**;
   else Chain[i] = Parent[i];

/* Set up for sorting. */

9. list rank each "chain";

/* Compute the removal number. */

10. **for** $1 \leq i \leq n$ **in parallel do**
    Removal[i] = the position in sorted order where the sorting is done using the maximum value as the primary key and the list ranking value within each chain as the secondary key;

/* Compute the Prüfer code. */

11. **for** $1 \leq i \leq n - 2$ **in parallel do** Pcode[Removal[i]] = Parent[i];

/* Print the Prüfer code. */

12. **for** $1 \leq i \leq n - 2$ **in parallel do print** Pcode[i];
    end

Example.

Before proving the correctness and complexity of the algorithm, we present an example of how Algorithm Prüfer Tree works. Algorithm Prüfer Tree generalizes Algorithm Prüfer Chain but there are some key differences. The main one being the repeated use of sorting. Consider the tree with root 33 depicted in Figure 3. (All edges are directed downward.) This is the situation after step 3 of the algorithm has been completed.

In step 4 of the algorithm the rooted tree T is converted to a *regular binary tree* T_R using Knuth's method [11, pages 332–345]. Every internal node has exactly two children in a regular binary tree. The idea of Knuth's construction is to replace every node u in T having d children by $d+1$ nodes $u^1, u^2, \ldots, u^{d+1}$. In T_R, u^{i+1} is the right child of u^i. If node v is the ith child of u in T, then v^1 is the left child of u^i in T_R. Once we have a binary tree, the usual parallel tree contraction can be applied in step 5 to compute the maximum node in each rooted subtree. During the contraction phase, we maintain a current maximum and its provenance. During the uncontraction phase, the relative maximum is moved back and utilized at the appropriate location in the tree, so the true maximum can be computed.

In step 6 the array MaxName is formed. MaxName's purpose is to associate the maximum subtree values with nodes names; MaxName is shown below. (For

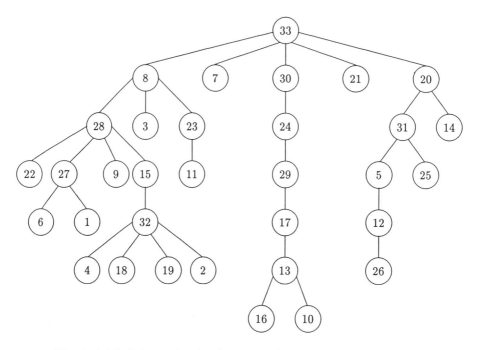

Fig. 3. A labeled tree that has been rooted at its maximum node, 33.

readability, we list the indices of the array as well.)

Max	1	2	3	4	26	6	7	32	9	10	11	26	16	14	32	16	17	18	19	31	21	22	23	29	25	26	27
Name	1	2	3	4	5	6	7	8	9	10	11	12	13	14	15	16	17	18	19	20	21	22	23	24	25	26	27
	1	2	3	4	5	6	7	8	9	10	11	12	13	14	15	16	17	18	19	20	21	22	23	24	25	26	27

Max	32	29	30	31	32	33
Name	28	29	30	31	32	33
	28	29	30	31	32	33

In step 7 the elements in the array MaxName are sorted based on the maximum values. We illustrate one possible sorted order.

Max	1	2	3	4	6	7	9	10	11	14	16	16	17	18	19	21	22	23	25	26	26	26	27	29	29	30	31
Name	1	2	3	4	6	7	9	10	11	14	16	13	17	18	19	21	22	23	25	26	5	12	27	24	29	30	20
	1	2	3	4	5	6	7	8	9	10	11	12	13	14	15	16	17	18	19	20	21	22	23	24	25	26	27

Max	31	32	32	32	32	33
Name	31	8	32	15	28	33
	28	29	30	31	32	33

In step 8 the tree is split into chains. This process is necessary to help determine when a given leaf will be removed. The array Chain, where this data is stored, is shown below. (Figure 3 can be used to infer the `Parent` array.)

nil	nil	nil	nil	nil	nil	nil	nil	nil	nil	nil	5	nil	nil
1	2	3	4	5	6	7	8	9	10	11	12	13	14

28	13	nil	nil	nil	nil	nil	nil	nil	nil	nil	12	nil	8	24	nil	20	15	nil
15	16	17	18	19	20	21	22	23	24	25	26	27	28	29	30	31	32	33

For example, $32 \to 15 \to 28 \to 8 \to \text{nil}$ is one path coded in the array `Chain`.

In step 9 each of these chains is list ranked. The list ranking values allow us to order nodes with the same maximum values relative to one another. For the chain just described, for example, we have

32	→	15	→	28	→	8	→	nil
1		2		3		4		

where the ranking value is shown below each node. This indicates, for example, that of all nodes with maximum value 32, 8 will be the last one removed by Algorithm Sequential Prüfer.

In step 10 the array `Removal` is computed. The array specifies the order in which nodes are removed from the tree. `Removal` has the following value.

1	2	3	4	22	5	6	32	7	8	9	21	12	10	30	11	13	14	15	28	16	17	18	25	19	20	23
1	2	3	4	5	6	7	8	9	10	11	12	13	14	15	16	17	18	19	20	21	22	23	24	25	26	27

31	24	26	27	29	33
28	29	30	31	32	33

In step 11 the array `Pcode` containing the Prüfer code is constructed and in step 12 it is printed. The resulting Prüfer code is

$$(27, 32, 8, 32, 27, 33, 28, 13, 23, 20, 13, 17, 29, 32, 32, 33, 28,$$
$$8, 31, 12, 5, 31, 28, 24, 30, 33, 20, 33, 15, 28, 8)$$

Correctness.

Now that we have gone through an example illustrating how Algorithm Prüfer Tree works, we turn to its analysis. The following lemma is the key to proving correctness.

Lemma 10. *The array* `Removal` *constructed in step 10 of Algorithm Prüfer Tree specifies the order in which leaves are eliminated by Algorithm Sequential Prüfer.*

Proof: By rooting the tree at its maximum node, leaves, including newly created ones, can only be removed in the "upward" direction. Note that the maximum value in a subtree plays a similar role to that played by `LeftMax` and `RightMax` entries.

By definition of the Prüfer code, it is easy to see that before a node u can be removed from the tree, all nodes whose associated maximum value is less than the maximum value of the subtree rooted at u must first be removed. Call these type A nodes with respect to u. Additionally, all nodes having the same maximum value as u but which are descendants of u must also be removed before u (for otherwise, u is not yet a leaf). Call these type B nodes with respect to u. Once all type A and B nodes with respect to u have been removed, u is the lowest numbered remaining leaf and so is the next node to be removed.

Consider the calculation in Algorithm Prüfer Tree of the removal number of node u. The sorting in step 7 based on maximum subtree values places u "after" all type A nodes with respect to u. In step 8 chains of nodes having the same maximum value are formed and in step 9 these chains are list ranked. In step 10 the sorting based on the secondary key places u after all type B nodes with respect to u. Therefore, array Removal contains the order in which leaves are eliminated by Algorithm Sequential Prüfer. □

It is not obvious how all of the steps in Algorithm Prüfer Tree can be carried out efficiently in parallel. In the next theorem we explain their implementions.

Theorem 11. *The Prüfer code of an n-node labeled free tree can be computed in $O(\log n)$ time using n processors on an EREW-PRAM.*

Proof: Lemma 10 shows that Algorithm Prüfer Tree correctly computes the order in which leaves are eliminated during the construction of the Prüfer code. Step 11 records, in the array Pcode, the node adjacent to each leaf as it is removed. In step 12 the array is output. Therefore, the algorithm outputs the Prüfer code of the input tree.

We now analyze the time and processor complexity of the algorithm.

Lemma 9 shows that running Algorithm Convert Edge List Encoding to implement step 1 requires $O(\log n)$ time using n processors on an EREW-PRAM. (Throughout the remainder of the proof, we leave off the EREW-PRAM.)

Applying Theorem 3, step 2 involving Euler tours can be performed in $O(\log n)$ time using $n/\log n$ processors.

Theorem 2 shows that the first half of step 3, requiring a parallel prefix computation, can be performed in $O(\log n)$ time using $n/\log n$ processors. Using Brent's scheduling principle (Theorem 1), the Parent array can also be formed within these bounds.

The conversion to a binary tree in step 4 is straightforward. Using an application of Brent's scheduling principle, step 4 can be implemented in $O(\log n)$ time using $n/\log n$ processors.

By Theorem 6, parallel tree contraction can be used to implement step 5 in $O(\log n)$ time using $n/\log n$ processors.

In step 6 processor i is associated with index i. Applying Brent's scheduling principle, this step can be implemented in $O(\log n)$ time using $n/\log n$ processors.

Using Theorem 5, the parallel sorting done in steps 7 and 10 can be implemented in $O(\log n)$ time using n processors.

In step 8 we must be careful to avoid concurrent reads when looking up the maximum value of a parent. This can be accomplished by doing a sort based on parent value and then a broadcast of the parent's maximum value in the appropriate "block." Because of the sorting, this step requires $O(\log n)$ time and n processors.

Applying Theorem 4, the chains in step 9 can be list ranked in $O(\log n)$ time using $n/\log n$ processors. The sorted version of array MaxName can be used to schedule the processors appropriately.

Using Brent's scheduling principle, both steps 11 and 12 can be implemented in $O(\log n)$ time using $n/\log n$ processors.

From this analysis, the time and processor bounds stated in the theorem follow. □

5 Discussion

We have specified an optimal $O(\log n)$ time, $n/\log n$ processor EREW-PRAM algorithm for determining the Prüfer code of an n-node labeled chain. The generalization of this algorithm to trees that we developed uses both an Euler tour computation and parallel sortings. In order to apply Theorem 3 concerning Euler tours, a special input encoding is necessary. The conversion method we use in Algorithm Convert Edge List Encoding also requires parallel sorting. Since parallel sorting takes $O(\log n)$ time and n processors on an EREW-PRAM, sorting is the bottleneck in our algorithm. All steps not involving sorting can be implemented optimally. It would be interesting to develop an EREW-PRAM algorithm for computing the Prüfer code of a tree that does not require sorting.

In [21] a parallel algorithm is given for converting a Prüfer code into a tree. This is the opposite problem we consider. Wang, Chen, and Liu pose as an open question the problem of determining (quickly in parallel) the Prüfer code from the tree. We resolved this open problem. Interestingly, their algorithm runs in the same bounds as ours and on the same model.

Acknowledgements

Ray thanks the faculty and students at the University of Rome "La Sapienza" for their generous support and kindness. Thanks to Jennifer Zito for preliminary discussions about this problem. Thanks to Tiziana Calamoneri and Joshua Jones for their comments on the paper. We thank the referees for several useful observations and for pointing out the Wang, Chen, and Liu paper [21].

References

1. K. Abrahamson, N. Dadoun, D. G. Kirkpatrick, and T. Przytycka. A simple parallel tree contraction algorithm. *Journal of Algorithms*, 10(2):287–302, 1989.

2. R. J. Anderson and G. L. Miller. Deterministic parallel list ranking. In John H. Reif, editor, *VLSI Algorithms and Architectures, 3rd Aegean Workshop on Computing, AWOC 88*, volume 319 of *Lecture Notes in Computer Science*, pages 81–90, Corfu, Greece, June-July 1988. Springer-Verlag.
3. R. P. Brent. The parallel evaluation of general arithmetic expressions. *Journal of the Association for Computing Machinery*, 21(2):201–206, 1974.
4. R. Cole. Parallel merge sort. *SIAM Journal on Computing*, 17(4):770–785, 1988.
5. R. Cole and U. Vishkin. The accelerated centroid decomposition technique for optimal tree evaluation in logarithmic time. *Algorithmica*, 3(3):329–346, 1988.
6. M. R. Garey, R. L. Graham, D. S. Johnson, and D. E. Knuth. Complexity results for bandwidth minimization. *SIAM Journal on Applied Mathematics*, 34:477–495, 1978.
7. R. Greenlaw, H. J. Hoover, and W. L. Ruzzo. *Limits to Parallel Computation: P-Completeness Theory*. Oxford University Press, 1995.
8. X. He. Efficient parallel algorithms for solving some tree problems. In *Proc. 24th Allerton Conference on Communication, Control and Computing*, pages 777–786, 1986.
9. J. JáJá. *An Introduction to Parallel Algorithms*. Addison-Wesley, 1992.
10. R. M. Karp and V. Ramachandran. Parallel algorithms for shared-memory machines. In J. Van Leeuwan, editor, *Handbook of Theoretical Computer Science*, volume A: Algorithms and Complexity, chapter 17, pages 869–941. MIT Press/Elsevier, 1990.
11. D. E. Knuth. *Fundamental Algorithms*, volume 1 of *The Art of Computer Programming*. Addison-Wesley, second edition, 1973.
12. R. E. Ladner and M. J. Fischer. Parallel prefix computation. *Journal of the ACM*, 27(4):831–838, October 1980.
13. L. Lovasz. *Combinatorial Problems and Exercises*. North Holland, 1993.
14. G. L. Miller and S-H. Teng. Systematic method for tree based parallel algorithm development. In *Second International Conference on Supercomputing*, pages 392–403, 1987.
15. J. W. Moon. *Counting Labelled Trees*. Canadian Mathematical Monographs. William Clowes and Sons, Limited, 1970.
16. C. H. Papadimitriou. The NP-completeness of the bandwidth minimization problem. *Computing*, 16:237–267, 1976.
17. Y. Perl and S. Zaks. On the complexity of edge labelings for trees. *Theoretical Computer Science*, 19:1–16, 1982.
18. T. M. Przytycka. *Parallel Algorithms for Tree Construction and Related Problems*. PhD thesis, The University of British Columbia, November 1990. UBC TR 90-28.
19. J. H. Reif, editor. *Synthesis of Parallel Algorithms*. Morgan Kaufmann, 1993.
20. R. E. Tarjan and U. Vishkin. An efficient parallel biconnectivity algorithm. *SIAM Journal on Computing*, 14(4):862–874, November 1985.
21. Y. Wang, H. Chen, and W. Liu. A parallel algorithm for constructing a labeled tree. *IEEE Transactions of Parallel and Distributed Systems*, 8(12):1236–1240, December 1997.

Completeness Results for a Lazy Conditional Narrowing Calculus

Mohamed Hamada[1], Aart Middeldorp[2]*, and Taro Suzuki[3]

[1] Department of Computer Science and Mathematics, Ain-Shams University,
Cairo, Egypt. hamada@asunet.shams.eun.eg

[2] Institute of Information Sciences and Electronics, University of Tsukuba,
Tsukuba 305-8573, Japan. ami@is.tsukuba.ac.jp

[3] School of Information Science, JAIST, Tatsunokuchi,
Ishikawa 923-12, Japan. t_suzuki@jaist.ac.jp

Abstract. We present two new completeness results for the lazy conditional narrowing calculus LCNC: LCNC is complete for arbitrary confluent conditional rewrite systems without extra variables with respect to normalized solutions and for terminating and level-confluent conditional rewrite systems without any restrictions on the distribution of variables in the rewrite rules.

1 Introduction

Besides being a general method for solving unification problems in equational theories that are presented by confluent term rewriting systems (TRSs for short), narrowing (Hullot [10]) is the underlying computational mechanism of many programming languages that integrate the functional and logic programming paradigms (Hanus [7]). The desirable property of narrowing is completeness: for every solution to a given goal a more general solution can be found by narrowing. Since narrowing is a complicated operation, numerous calculi consisting of a small number of more elementary inference rules that simulate narrowing have been proposed (e.g. [9, 8, 15, 16]).

Completeness issues for the lazy narrowing calculus LNC—which is based on the calculus TRANS of Hölldobler [9]—have been extensively studied in Middeldorp et al. [15] and Middeldorp and Okui [14]. In [15] it is proved that LNC is strongly complete whenever basic narrowing (Hullot [10]) is complete. Strong completeness means that the choice of the equation in goals can be made non-deterministically, resulting in a huge reduction of the search space as well as easing implementations. It is also shown in [15] that LNC is complete for arbitrary confluent TRSs with respect to normalized solutions.

---

* Partially supported by the Grant-in-Aids for Scientific Research for Encouraging Young Researchers A 09780235 and for Basic Research B 08458059 of the Ministry of Education, Science, Sports and Culture of Japan

Hamada and Middeldorp [6] extended one of the main results of [15] to conditional rewriting: LCNC—the conditional variant of LNC—is strongly complete whenever basic conditional narrowing is complete. The latter is known for decreasing and confluent CTRSs without extra variables (Middeldorp and Hamoen [13]), terminating and level-confluent CTRSs with extra variables in the conditions only (Giovannetti and Moiso [5]), and terminating and shallow-confluent CTRSs with extra variables (Werner [18]). In this paper we present two further completeness results for LCNC. The first one—the completeness of LCNC for confluent systems without extra variables with respect to normalized solutions—extends the other main result of [15] to conditional systems. The second one—the completeness of LCNC for arbitrary terminating and level-confluent systems—has no counterpart in [15].

This paper is organized as follows. In the next section we recall some definitions pertaining to rewriting and narrowing. In Section 3 we show the completeness of LCNC for the class of terminating and level-confluent CTRSs. The other main result, the completeness of LCNC for the class of confluent CTRSs without extra variables with respect to normalized solutions, is presented in Section 4.

2 Preliminaries

We assume familiarity with the basics of (conditional) term rewriting and narrowing. Surveys can be found in [1, 2, 12, 13].

A conditional term rewriting system (CTRS for short) over a signature $\mathcal{F}$ is a set $\mathcal{R}$ of (conditional) rewrite rules of the form $l \to r \Leftarrow c$ where the conditional part c is a (possibly empty) sequence $s_1 \approx t_1, \ldots, s_n \approx t_n$ of equations. All terms $l, r, s_1, \ldots, s_n, t_1, \ldots, t_n$ must belong to $\mathcal{T}(\mathcal{F}, \mathcal{V})$ and we require that l is not a variable. Following [13], CTRSs are classified according to the distribution of variables in rewrite rules. A 1-CTRS contains no extra variables (i.e., $\mathcal{V}\mathrm{ar}(r, c) \subseteq \mathcal{V}\mathrm{ar}(l)$ for all rewrite rules $l \to r \Leftarrow c$), a 2-CTRS may contain extra variables in the conditions only ($\mathcal{V}\mathrm{ar}(r) \subseteq \mathcal{V}\mathrm{ar}(l)$ for all rewrite rules $l \to r \Leftarrow c$), and a 3-CTRS may also have extra variables in the right-hand sides provided these occur in the corresponding conditions ($\mathcal{V}\mathrm{ar}(r) \subseteq \mathcal{V}\mathrm{ar}(l, c)$ for all rewrite rules $l \to r \Leftarrow c$). Extra variables enable a more natural and efficient style of writing specifications of programs. For instance, without using extra variables it is not so easy to write the following efficient specification of the computation of Fibonacci numbers:

$$\begin{cases} 0 + y \to y \\ s(x) + y \to s(x + y) \\ \mathit{fib}(0) \to \langle 0, s(0) \rangle \\ \mathit{fib}(s(x)) \to \langle z, y + z \rangle \Leftarrow \mathit{fib}(x) \approx \langle y, z \rangle \end{cases}$$

We assume that every CTRS contains the rewrite rule $x \approx x \to \mathbf{true}$. Here $\approx$ and $\mathbf{true}$ are function symbols that do not occur in the other rewrite rules. These symbols may only occur at the root position of terms. Let $\mathcal{R}$ be a CTRS.

We inductively define unconditional TRSs $\mathcal{R}_n$ for $n \geqslant 0$ as follows:

$$\begin{aligned}\mathcal{R}_0 &= \{\, x \approx x \to \text{true}\,\}\;, \\ \mathcal{R}_{n+1} &= \{\, l\theta \to r\theta \mid l \to r \Leftarrow c \in \mathcal{R} \text{ and } c\theta \to^*_{\mathcal{R}_n} \top\,\}\;.\end{aligned}$$

Here $\top$ stands for any sequence of trues. We define $s \to_{\mathcal{R}} t$ if and only if there exists an $n \geqslant 0$ such that $s \to_{\mathcal{R}_n} t$. The minimum such n is called the depth of $s \to_{\mathcal{R}} t$. In the following we abbreviate $\to_{\mathcal{R}_n}$ to $\to_n$. Our CTRS are known as *join* CTRSs in the term rewriting literature.

A CTRS $\mathcal{R}$ is level-confluent (Giovannetti and Moiso [5]) if every $\mathcal{R}_n$ is confluent, i.e., $^*_n\!\leftarrow \cdot \to^*_n \;\subseteq\; \to^*_n \cdot\; ^*_n\!\leftarrow$ for all $n \geqslant 0$, and shallow-confluent if $^*_m\!\leftarrow \cdot \to^*_n \;\subseteq\; \to^*_n \cdot\; ^*_m\!\leftarrow$ for all $m,n \geqslant 0$. Shallow-confluent CTRSs are level-confluent but the reverse is not true. A CTRS $\mathcal{R}$ over a signature $\mathcal{F}$ is decreasing (Dershowitz et al. [4]) if there exists a well-founded order $\succ$ on $\mathcal{T}(\mathcal{F},\mathcal{V})$ with the following three properties: $\succ$ contains $\to_{\mathcal{R}}$, $\succ$ has the subterm property (i.e., $\rhd \subseteq \succ$ where $s \rhd t$ if and only if t is proper subterm of s), and $l\theta \succ s\theta, t\theta$ for every rewrite rule $l \to r \Leftarrow c$ of $\mathcal{R}$, every equation $s \approx t$ in c, and every substitution θ. Note that according to this definition 2-CTRSs and 3-CTRSs are never decreasing. Decreasing CTRSs are terminating and, when there are finitely many rewrite rules, have a decidable rewrite relation. Sufficient syntactic conditions for level-confluence of 3-CTRSs are presented in Suzuki et al. [17].

An equation is a term of the form $s \approx t$. The constant true is also viewed as an equation. A goal is a sequence of equations. A substitution θ is a ($\mathcal{R}$-)solution of a goal G if $G\theta \to^*_{\mathcal{R}} \top$. For confluent $\mathcal{R}$ this is equivalent to validity of the equations in $G\theta$ in all models of the underlying conditional equational system of $\mathcal{R}$ (Kaplan [11]). The level level(e) of an equation $e \in G\theta$ is the smallest n such that $e \to^*_n \text{true}$. We say that the rewrite sequence $G\theta \to^*_{\mathcal{R}} \top$ is *level-minimal* if the depths of its rewrite steps do not exceed the level of the originating equations. Clearly, every rewrite sequence $G\theta \to^*_{\mathcal{R}} \top$ can be transformed into a level-minimal one (which may be longer than the given sequence).

The conditional narrowing calculus CNC consists of the following inference rule:

$$\dfrac{G', e, G''}{(G', e[r]_p, c, G'')\theta} \quad \begin{array}{l}\text{if there exist a fresh variant } l \to r \Leftarrow c \text{ of a rewrite rule} \\ \text{in } \mathcal{R}, \text{ a non-variable position } p \text{ in } e, \text{ and a most general} \\ \text{unifier } \theta \text{ of } e_{|p} \text{ and } l.\end{array}$$

In the above situation we write $G', e, G'' \leadsto_\theta (G', e[r]_p, c, G'')\theta$. This is called a CNC-step. A sequence $G_1 \leadsto_{\theta_1} \cdots \leadsto_{\theta_{n-1}} G_n$ of CNC-steps is called a CNC-derivation and abbreviated to $G_1 \leadsto^*_\theta G_n$ where $\theta = \theta_1 \cdots \theta_{n-1}$. We use the symbol Π (and its derivatives) to denote CNC-derivations. For a CNC-derivation $\Pi\colon G \leadsto^*_\theta G'$, $\Pi\theta$ denotes the corresponding rewrite sequence $G\theta \to^* G'$. A CNC-derivation which ends in $\top$ is called a CNC-refutation. If $G \leadsto^*_\theta \top$ then θ is a solution of G. This is known as the soundness of conditional narrowing. A survey of completeness results for CNC can be found in Middeldorp and Hamoen [13].

For a substitution θ and a set of variables V, we denote $(V \setminus \mathcal{D}(\theta)) \cup \mathcal{I}(\theta\!\upharpoonright_V)$ by $\mathcal{V}ar_V(\theta)$. Here $\mathcal{D}(\theta) = \{x \in \mathcal{V} \mid \theta(x) \neq x\}$ denotes the domain of θ, which is always assumed to be finite, and $\mathcal{I}(\theta\!\upharpoonright_V) = \bigcup_{x \in \mathcal{D}(\theta) \cap V} \mathcal{V}ar(x\theta)$ the set of variables introduced by the restriction of θ to V. In the sequel we frequently make use of the equivalence of $\theta_1 = \theta_2 \; [\mathcal{V}ar_V(\theta)]$ and $\theta\theta_1 = \theta\theta_2 \; [V]$ for all substitutions θ, θ_1, θ_2 and sets of variables V.

The lazy conditional narrowing calculus LCNC (Hamada and Middeldorp [6]) consists of the following five inference rules:

[o] *outermost narrowing*

$$\frac{G', f(s_1, \ldots, s_n) \simeq t, G''}{G', s_1 \approx l_1, \ldots, s_n \approx l_n, r \approx t, c, G''}$$

if there exists a fresh variant $f(l_1, \ldots, l_n) \to r \Leftarrow c$ of a rewrite rule in $\mathcal{R}$,

[i] *imitation*

$$\frac{G', f(s_1, \ldots, s_n) \simeq x, G''}{(G', s_1 \approx x_1, \ldots, s_n \approx x_n, G'')\theta}$$

if $\theta = \{x \mapsto f(x_1, \ldots, x_n)\}$ with $x_1, \ldots, x_n$ fresh variables,

[d] *decomposition*

$$\frac{G', f(s_1, \ldots, s_n) \approx f(t_1, \ldots, t_n), G''}{G', s_1 \approx t_1, \ldots, s_n \approx t_n, G''},$$

[v] *variable elimination*

$$\frac{G', s \simeq x, G''}{(G', G'')\theta}$$

if $x \notin \mathcal{V}ar(s)$ and $\theta = \{x \mapsto s\}$,

[t] *removal of trivial equations*

$$\frac{G', x \approx x, G''}{G', G''}.$$

In the rules [o], [i], and [v], $s \simeq t$ stands for $s \approx t$ or $t \approx s$. The only difference between LCNC and the calculus LNC of [15] is in the outermost narrowing rule: in LCNC the conditional part c of the applied rewrite rule to the new goal. If G and G' are the upper and lower goal in the inference rule $[\alpha]$ ($\alpha \in \{\text{o}, \text{i}, \text{d}, \text{v}, \text{t}\}$), we write $G \Rightarrow_{[\alpha]} G'$. This is called an LCNC-step. The applied rewrite rule or substitution may be supplied as subscript, that is, we write things like $G \Rightarrow_{[\text{o}], l \to r \Leftarrow c} G'$ and $G \Rightarrow_{[\text{i}], \theta} G'$. A finite LCNC-derivation $G_1 \Rightarrow_{\theta_1} \cdots \Rightarrow_{\theta_{n-1}} G_n$ may be abbreviated to $G_1 \Rightarrow^*_\theta G_n$ where $\theta = \theta_1 \cdots \theta_{n-1}$. An LCNC-refutation is an LCNC-derivation ending in the empty goal $\square$.

Example 1. Consider the 3-CTRS

$$\mathcal{R} = \begin{cases} 0 + y \to y \\ s(x) + y \to s(x + y) \\ \mathit{fib}(0) \to \langle 0, s(0) \rangle \\ \mathit{fib}(s(x)) \to \langle z, y + z \rangle \Leftarrow \mathit{fib}(x) \approx \langle y, z \rangle \end{cases}$$

The following LCNC-refutation computes the solution $\{x \mapsto s(0)\}$ for the goal $\mathit{fib}(x) \approx \langle x, x \rangle$ (the selected equations are indicated by underlining):

$$\underline{\mathit{fib}(x) \approx \langle x, x \rangle}$$
$$\Downarrow_{[o],\ \mathit{fib}(s(x_1)) \to \langle z_1, y_1+z_1 \rangle \Leftarrow \mathit{fib}(x_1) \approx \langle y_1, z_1 \rangle}$$
$$x \approx s(x_1),\ \underline{\langle z_1, y_1 + z_1 \rangle \approx \langle x, x \rangle},\ \mathit{fib}(x_1) \approx \langle y_1, z_1 \rangle$$
$$\Downarrow_{[d]}$$
$$x \approx s(x_1),\ \underline{z_1 \approx x},\ y_1 + z_1 \approx x,\ \mathit{fib}(x_1) \approx \langle y_1, z_1 \rangle$$
$$\Downarrow_{[v],\ \{z_1 \mapsto x\}}$$
$$x \approx s(x_1),\ y_1 + x \approx x,\ \underline{\mathit{fib}(x_1) \approx \langle y_1, x \rangle}$$
$$\Downarrow_{[o],\ \mathit{fib}(0) \to \langle 0, s(0) \rangle}$$
$$x \approx s(x_1),\ y_1 + x \approx x,\ \underline{x_1 \approx 0},\ \langle 0, s(0) \rangle \approx \langle y_1, x \rangle$$
$$\Downarrow_{[v],\ \{x_1 \mapsto 0\}}$$
$$x \approx s(0),\ y_1 + x \approx x,\ \underline{\langle 0, s(0) \rangle \approx \langle y_1, x \rangle}$$
$$\Downarrow_{[d]}$$
$$x \approx s(0),\ y_1 + x \approx x,\ \underline{0 \approx y_1},\ s(0) \approx x$$
$$\Downarrow_{[v],\ \{y_1 \mapsto 0\}}$$
$$\underline{x \approx s(0)},\ 0 + x \approx x,\ s(0) \approx x$$
$$\Downarrow_{[v],\ \{x \mapsto s(0)\}}$$
$$0 + s(0) \approx s(0),\ \underline{s(0) \approx s(0)}$$
$$\Downarrow_{[d]}$$
$$0 + s(0) \approx s(0),\ \underline{0 \approx 0}$$
$$\Downarrow_{[d]}$$
$$\underline{0 + s(0) \approx s(0)}$$
$$\Downarrow_{[o],\ 0 + y_2 \to y_2}$$
$$0 \approx 0,\ \underline{s(0) \approx y_2},\ y_2 \approx s(0)$$
$$\Downarrow_{[v],\ \{y_2 \mapsto s(0)\}}$$
$$\underline{0 \approx 0},\ \underline{s(0) \approx s(0)}$$
$$\Downarrow_{[d]\ *}$$
$$\square$$

3 Terminating and level-confluent CTRSs

In this section we prove the completeness of LCNC for the class of terminating and level-confluent (3-)CTRSs. We use induction with respect to the well-founded order on rewrite sequences defined below.

Definition 2. Let $\mathcal{R}$ be a terminating CTRS. Let G be a goal and θ a substitution such that $\mathcal{R} \vdash G\theta$. With every equation $e = s \approx t \in G\theta$ we associate the pair $|e| = (\text{level}(e), \{s, t\})$ whose second component is the multiset which contains both sides of e. We equip the set of these pairs with the order $\sqsupset = \mathrm{lex}(>, \succ_{\mathsf{mul}})$ where $>$ denotes the standard order on natural numbers and $\succ_{\mathsf{mul}}$ denotes the multiset extension of $\succ \ = \ (\to_{\mathcal{R}} \cup \triangleright)^+$, i.e., $M \succ_{\mathsf{mul}} N$ for finite multisets M, N if and only if there exist multisets X and Y such that $\varnothing \neq X \subseteq M$, $N = (M - X) \uplus Y$, and for every $y \in Y$ there exists an $x \in X$ with $x \succ y$. Here $-$ and $\uplus$ denote multiset difference and sum. The multiset consisting of all $|e\theta|$ for $e \in G$ is denoted by $(G; \theta)$. Let $\varPi \colon G \to^* H$ and $\varPi' \colon G' \to^* H'$ be rewrite sequences. We write $\varPi \gg \varPi'$ if $(G; \epsilon) \sqsupset_{\mathsf{mul}} (G'; \epsilon)$.

Since $\to_{\mathcal{R}}$ is closed under contexts, the relation $\succ$ is a well-founded order for every terminating CTRS $\mathcal{R}$. From [3] we know that $\succ_{\mathsf{mul}}$ inherits well-foundedness from $\succ$. Consequently, the lexicographic product of $>$ and $\succ_{\mathsf{mul}}$ is a well-founded order and hence $\gg$ is a well-founded order on rewrite sequences. Note that this order depends only on the initial goals of rewrite sequences. In the proof below we make use of the well-known equivalence of $M \succ_{\mathsf{mul}} N$ and $M - N \succ_{\mathsf{mul}} N - M$.

Theorem 3. *Let $\mathcal{R}$ be a terminating and level-confluent CTRS, For every solution θ of a goal G there exists an LCNC-refutation $G \Rightarrow^*_\sigma \square$ such that $\sigma \leqslant_{\mathcal{R}} \theta$ [$\mathrm{Var}(G)$].*

Proof. We use well-founded induction with respect the order $\gg$ on rewrite sequences. Let $\varPi \colon G\theta \to^* \top$ be a rewrite sequence from $G\theta$ to $\top$. In order to make the induction work we prove $\sigma \leqslant_{\mathcal{R}} \theta$ [$\mathcal{V}$] for a finite set of variables $\mathcal{V}$ that includes $\mathrm{Var}(G)$. The base case is trivial since G must be the empty goal (and thus we can take $\sigma = \epsilon$). For the induction step we proceed as follows. First we transform $\varPi$ into a level-minimal rewrite sequence. (This does not affect the complexity $(G; \theta)$.) By rearranging the order of the equations in $\varPi$, we can assume that the level of the leftmost equation in $G\theta$ is minimal. Write $G = e, H$ and $e = s \approx t$. Since it does not effect level-minimality, we may swap rewrite steps that take place in different sides of $e\theta$ at will. This will simplify the notation in some of the cases below. For the same reason we may assume that in $\varPi$ always the leftmost equation different from $\mathtt{true}$ is rewritten. Let $\ell = \mathsf{level}(e\theta)$. So $\mathsf{level}(e'\theta) \geqslant \ell$ for all $e' \in H$. We will show the existence of an LCNC-step $\varPsi_1 \colon G \Rightarrow_{\sigma_1} G_1$ and a rewrite sequence $\varPi_1 \colon G_1 \theta_1 \to^* \top$ such that $\sigma_1 \theta_1 \leqslant_{\mathcal{R}} \theta$ [$\mathcal{V}$] and $\varPi \gg \varPi_1$. We distinguish the following cases.

1. Suppose $s, t \notin \mathcal{V}$. We distinguish two further cases, depending on what happens to $e\theta$ in $\varPi$.
   (a) Suppose no reduct of $e\theta$ is rewritten at position 1 or 2. We may write $s = f(s_1, \ldots, s_n)$ and $t = f(t_1, \ldots, t_n)$. For $1 \leqslant i \leqslant n$ let e_i be the equation $s_i \approx t_i$. Let $G_1 = e_1, \ldots, e_n, H$. We have $\varPsi_1 \colon G \Rightarrow_{[d]} G_1$. Let $\sigma_1 = \epsilon$ and $\theta_1 = \theta$. The rewrite sequence $\varPi$, which must be of the form

   $$G\theta \to^*_\ell f(u_1, \ldots, u_n) \approx f(u_1, \ldots, u_n), H\theta \to \mathtt{true}, H\theta \to^* \top \ ,$$

   can be transformed into $\varPi_1$:

   $$G_1 \theta_1 \to^* u_1 \approx u_1, \ldots, u_n \approx u_n, H\theta \to^* \top, H\theta \to^* \top \ .$$

Clearly $\sigma_1\theta_1 = \theta$. It remains to show that $\Pi \gg \Pi_1$. We have $(G;\theta) - (G_1;\theta_1) = \{|e\theta|\}$ and $(G_1;\theta_1) - (G;\theta) = \{|e_1\theta|,\ldots,|e_n\theta|\}$. For all $1 \leq i \leq n$ we have $s\theta \triangleright s_i\theta$ and $t\theta \triangleright t_i\theta$ and thus also $s\theta \succ s_i\theta$ and $t\theta \succ t_i\theta$. Hence $\{s\theta, t\theta\} \succ_{\mathsf{mul}} \{s_i\theta, t_i\theta\}$ and therefore it suffices to show that level$(e\theta) \geq$ level$(e_i\theta)$. We have $s_i\theta \to_\ell^* u_i$ and $t_i\theta \to_\ell^* u_i$. Therefore $\mathcal{R}_\ell \vdash e_i\theta$ and hence the level of $e_i\theta$ does not exceed the level of $e\theta$. We conclude that $\Pi \gg \Pi_1$.

(b) Suppose a reduct of $e\theta$ is rewritten at position 1 or 2. Without loss of generality we assume that the first such reduct is rewritten at position 1, using the fresh variant $l \to r \Leftarrow c$ of a rewrite rule in $\mathcal{R}$ with $l = f(l_1,\ldots,l_n)$. We have $s = f(s_1,\ldots,s_n)$. The rewrite sequence Π is of the form

$$G\theta \to_\ell^* f(s'_1,\ldots,s'_n) \approx t\theta, H\theta \to_\ell r\tau \approx t\theta, H\theta \to^* \mathsf{T}$$

with $f(s'_1,\ldots,s'_n) = l\tau$ and $c\tau \to_{\ell-1}^* \mathsf{T}$. For $1 \leq i \leq n$ let $e_i = s_i \approx l_i$ and define $G_1 = e_1,\ldots,e_n, r \approx t, c, H$. We have $\Psi_1: G \Rightarrow_{[o]} G_1$. Let $\sigma_1 = \epsilon$ and define θ_1 as the disjoint union of θ and τ. Because

$$G_1\theta_1 = s_1\theta \approx l_1\tau,\ldots, s_n\theta \approx l_n\tau, r\tau \approx t\theta, c\tau, H\theta$$

we obtain the rewrite sequence Π_1:

$$G_1\theta_1 \to_\ell^* s'_1 \approx l_1\tau,\ldots, s'_n \approx l_n\tau, r\tau \approx t\theta, c\tau, H\theta$$
$$\to^* \mathsf{T}, r\tau \approx t\theta, c\tau, H \to^* \mathsf{T} \ .$$

Since $V \cap \mathcal{D}(\tau) = \varnothing$ we have $\sigma_1\theta_1 = \theta\ [V]$. In order to conclude that $\Pi \gg \Pi_1$ we have to show that $(G;\theta) - (G_1;\theta_1) = \{|e\theta|\} \sqsupset_{\mathsf{mul}} \{|e_1\theta_1|,\ldots,|e_n\theta_1|,|r\tau \approx t\theta|\} \uplus (c;\tau) = (G_1;\theta_1) - (G;\theta)$. For all $1 \leq i \leq n$ we have $s\theta \triangleright s_i\theta \to^* s'_i = l_i\tau$ and consequently $\{s\theta, t\theta\} \succ_{\mathsf{mul}} \{s_i\theta, l_i\tau\}$. Furthermore, the level of $e_i\theta_1$ does not exceed ℓ. Hence $|e\theta| \sqsupset |e_1\theta_1|,\ldots,|e_n\theta_1|$. Next consider the equation $r\tau \approx t\theta$. Since $r\tau \approx t\theta \to_\ell^*$ true, the level of $r\tau \approx t\theta$ is at most ℓ. Also, $s\theta \to^* l\tau \to r\tau$ and thus $\{s\theta, t\theta\} \succ_{\mathsf{mul}} \{r\tau, t\theta\}$. Hence $|e\theta| \sqsupset |r\tau \approx t\theta|$. Finally, from $c\tau \to_{\ell-1}^* \mathsf{T}$ we infer that the level of every equation in $c\tau$ is less than ℓ. Therefore also $\{|e\theta|\} \sqsupset_{\mathsf{mul}} (c;\tau)$.

2. Suppose $s \notin \mathcal{V}$ and $t \in \mathcal{V}$. We distinguish two further cases.

(a) Suppose no reduct of $e\theta$ is rewritten at position 1. Write $s = f(s_1,\ldots,s_n)$. Let $\sigma_1 = \{t \mapsto f(x_1,\ldots,x_n)\}$ and $e_i = x_i \approx s_i$ for $1 \leq i \leq n$. Here $x_1,\ldots,x_n$ are fresh variables. Define $G_1 = (e_1,\ldots,e_n, H)\sigma_1$. We have $\Psi_1: G \Rightarrow_{[i],\sigma_1} G_1$. The rewrite sequence Π is of the form

$$G\theta \to_\ell^* f(s'_1,\ldots,s'_n) \approx f(s'_1,\ldots,s'_n), H\theta \to \mathsf{true}, H\theta \to^* \mathsf{T} \ .$$

Define θ_1 as the disjoint union of θ and $\{x_1 \mapsto s'_1,\ldots,x_n \mapsto s'_n\}$. We have

$$G_1\theta_1 = s'_1 \approx s_1\sigma_1\theta_1,\ldots, s'_n \approx s_n\sigma_1\theta_1, H\sigma_1\theta_1 \ .$$

Because $t\theta \to_\ell^* f(s_1', \ldots, s_n') = f(x_1, \ldots, x_n)\theta_1 = t\sigma_1\theta_1$ and $x\theta = x\sigma_1\theta_1$ for variables x different from $x_1, \ldots, x_n, t$, we have $x\theta \to_\ell^* x\sigma_1\theta_1$ for all variables $x \in \mathcal{V}$. Hence $\sigma_1\theta_1 =_\mathcal{R} \theta \ [\mathcal{V}]$, $H\theta \to_\ell^* H\sigma_1\theta_1$, and $s_i\sigma_1\theta_1 \stackrel{*}{_\ell\leftarrow} s_i\theta \to_\ell^* s_i'$ for all $1 \leqslant i \leqslant n$. Since the level of every equation in $H\theta$ is at least ℓ, level-confluence yields $H\sigma_1\theta_1 \to^* \top$ such that for every equation $e' \in H$ the level of $e'\sigma_1\theta_1$ is less than or equal to the level of $e'\theta$. In addition, we obtain terms $u_1, \ldots, u_n$ such that $s_i\sigma_1\theta_1 \to_\ell^* u_i \stackrel{*}{_\ell\leftarrow} s_i'$ for all $1 \leqslant i \leqslant n$. Hence we obtain the rewrite sequence Π_1:

$$G_1\theta_1 \to_\ell^* u_1 \approx u_1, \ldots, u_n \approx u_n, H\sigma_1\theta_1 \to^* \top \ .$$

We show that $\Pi \gg \Pi_1$. We have $(G; \theta) = \{|e\theta|\} \uplus (H; \theta)$ and $(G_1; \theta_1) = \{|e_1\theta_1|, \ldots, |e_n\theta_1|\} \uplus (H\sigma_1; \theta_1)$. First we show that $|e\theta| \sqsupset |e_1\theta_1|, \ldots, |e_n\theta_1|$. For all $1 \leqslant i \leqslant n$ we have $s\theta \rhd s_i\theta \to^* s_i'$ and $s\theta \rhd s_i\theta \to^* s_i\sigma_1\theta_1$ and thus also $s\theta \succ s_i'$ and $s\theta \succ s_i\sigma_1\theta_1$. Hence $\{s\theta, t\theta\} \succ_{\mathrm{mul}} \{s_i', s_i\sigma_1\theta_1\}$. ¿From $s_i\sigma_1\theta_1 \to_\ell^* u_i \stackrel{*}{_\ell\leftarrow} s_i'$ we infer that the level of $e_i\theta_1$ does not exceed ℓ and therefore $|e\theta| \sqsupset |e_1\theta_1|, \ldots, |e_n\theta_1|$. To conclude that $\Pi \gg \Pi_1$ it now suffices to show that $(H; \theta) \sqsupset_{\mathrm{mul}}^= (H\sigma_1; \theta_1)$. Here $\sqsupset_{\mathrm{mul}}^=$ denotes the reflexive closure of $\sqsupset_{\mathrm{mul}}$. Let $e' = u \approx v \in H$. We already observed that $\mathrm{level}(e'\theta) \geqslant \mathrm{level}(e'\sigma_1\theta_1)$. Furthermore, $e'\theta \to^* e'\sigma_1\theta_1$ and thus $\{u\theta, v\theta\} \succeq_{\mathrm{mul}}^= \{u\sigma_1\theta_1, v\sigma_1\theta_1\}$. Hence $|e'\theta| \sqsupset^= |e'\sigma_1\theta_1|$, implying the desired $(H; \theta) \sqsupset_{\mathrm{mul}}^= (H\sigma_1; \theta_1)$.

(b) Suppose a reduct of $e\theta$ is rewritten at position 1. In this case we proceed as in case 1(b).
3. Suppose $s \in \mathcal{V}$ and $t \notin \mathcal{V}$. This case is similar to case 2.
4. Suppose $s, t \in \mathcal{V}$. We distinguish two further cases.
    (a) Suppose $s = t$. Let $G_1 = H$. We have $\Psi_1 : G \Rightarrow_{[t]} G_1$. Let $\sigma_1 = \epsilon$ and $\theta_1 = \theta$. From Π we extract the rewrite sequence $\Pi_1 : G_1\theta_1 \to^* \top$. We clearly have $\sigma_1\theta_1 = \theta$ and $\Pi \gg \Pi_1$ as $(G_1; \theta_1) \subset (G; \theta)$.
    (b) Suppose $s \neq t$. Let $\sigma_1 = \{s \mapsto t\}$ and $G_1 = H\sigma_1$. We have $\Psi_1 : G \Rightarrow_{[v], \sigma_1} G_1$. Let θ_1 be the $\mathcal{R}_\ell$-normal form of θ, i.e., $\theta_1(x) = x\theta{\downarrow}_\ell$ for all variables $x \in \mathcal{V}$. Here $x\theta{\downarrow}_\ell$ denotes the unique normal form of $x\theta$ with respect to the confluent and terminating TRS $\mathcal{R}_\ell$. Since $e\theta \to_\ell^* \mathsf{true}$, we have $s\theta{\downarrow}_\ell = t\theta{\downarrow}_\ell$. Hence $s\theta \to_\ell^* t\theta{\downarrow}_\ell = s\sigma_1\theta_1$. Because $x\theta \to_\ell^* x\theta{\downarrow}_\ell = x\sigma_1\theta_1$ for variables x different from s, we obtain $x\theta \to_\ell^* x\sigma_1\theta_1$ for all variables $x \in \mathcal{V}$. Therefore $\sigma_1\theta_1 =_\mathcal{R} \theta \ [\mathcal{V}]$. From Π we extract the rewrite sequence $H\theta \to^* \top$. Since $H\theta \to_\ell^* H\sigma_1\theta_1 = G_1\theta_1$, level-confluence yields a rewrite sequence $\Pi_1 : G_1\theta_1 \to^* \top$ such that for every equation $e' \in H$ the level of $e'\sigma_1\theta_1$ is less than or equal to the level of $e'\theta$. Hence we obtain $(H; \theta) \sqsupset_{\mathrm{mul}}^= (G_1; \theta_1)$ as in case 2(a) and thus $(G; \theta) = \{|e\theta|\} \uplus (H; \theta) \sqsupset_{\mathrm{mul}} (G_1; \theta_1)$. We conclude that $\Pi \gg \Pi_1$.

Let $\mathcal{V}_1 = \mathcal{V}ar_\mathcal{V}(\sigma_1) \cup \mathcal{V}ar(G_1)$. Clearly $\mathcal{V}ar(G_1) \subseteq \mathcal{V}_1$. Hence we can apply the induction hypothesis to $\Pi_1 : G_1\theta_1 \to^* \top$. This yields an LCNC-refutation $\Psi' : G_1 \Rightarrow_{\sigma'}^* \square$ such that $\sigma' \leqslant_\mathcal{R} \theta_1 \ [\mathcal{V}_1]$. Define $\sigma = \sigma_1\sigma'$. From $\sigma_1\theta_1 \leqslant_\mathcal{R} \theta \ [\mathcal{V}]$, $\sigma' \leqslant_\mathcal{R} \theta_1 \ [\mathcal{V}_1]$, and $\mathcal{V}ar_\mathcal{V}(\sigma_1) \subseteq \mathcal{V}_1$ we infer that $\sigma \leqslant_\mathcal{R} \theta \ [\mathcal{V}]$. Concatenating the LCNC-step Ψ_1 and the LCNC-refutation Ψ' yields the desired LCNC-refutation Ψ. □

Let us illustrate the above proof on a small example. Consider the CTRS

$$\mathcal{R} = \begin{cases} a \to b & \Leftarrow f(x) \approx g(b) \\ f(b) \to g(x) & \Leftarrow f(x) \approx g(b) \\ f(a) \to g(b) & \end{cases}$$

of Werner [18, Beispiel 9.1]. This 3-CTRS is level-confluent and terminating. Consider the goal $G = f(b) \approx g(b)$. The empty substitution ϵ is a solution of G because we have the rewrite sequence $\Pi: G\epsilon = f(b) \approx g(b) \to_2 g(a) \approx g(b) \to_2 g(b) \approx g(b) \to_0$ true. In both the first and the second rewrite step the instantiated condition of the applied rewrite rule is $f(a) \approx g(b)$ which rewrites to true by applying the third rewrite rule. Figure 1 shows a possible LCNC-refutation Ψ constructed in the above proof. (Note that in general there are several possibilities for Ψ as the equation in $G\theta$ of minimal level is not uniquely determined.) The selected equations are underlined. The numbers on the right refer to the various cases in the proof.

It is interesting to note that basic conditional narrowing fails to solve the goal $f(b) \approx g(b)$ (Werner [18]). Hence the completeness of LCNC for the above CTRS does not follow from the results of Hamada and Middeldorp [6] (because of the incompleteness of basic conditional narrowing). An interesting question for future research is to establish or refute the strong completeness of LCNC for the class of terminating and level-confluent CTRSs and, if strong completeness fails to hold, to identify complete selection functions.

4 Confluent CTRSs without extra variables

In this section we show the completeness of LCNC for arbitrary confluent 1-CTRSs with respect to normalized solutions. So we dispense with the termination requirement of the preceding section and weaken level-confluence to confluence, at the expense of having to exclude extra variables in the rewrite rules. Because of lack of space, some of the proof details in this section have been omitted.

Middeldorp et al. [15] have shown the completeness of LNC with *leftmost selection* for arbitrary confluent TRSs with respect to normalized solutions. Unfortunately, we have not been able to extend the (complicated) proof in [15] to 1-CTRSs. Hence we have to find another way to select equations. Recall that the completeness proof of the preceding section selects an equation such that the level of its instantiation is minimal. For non-terminating CTRSs we need a different selection mechanism.

Definition 4. Let $\Pi: G \leadsto^* \top$ be a CNC-refutation. An equation $s \approx t$ in G is called a *candidate* if one of the following conditions is satisfied:

- $s, t \notin \mathcal{V}$,
- $s \in \mathcal{V}, t \notin \mathcal{V}$, and in Π no descendant of $s \approx t$ is narrowed at position 1,
- $s \notin \mathcal{V}, t \in \mathcal{V}$, and in Π no descendant of $s \approx t$ is narrowed at position 2, or
- $s, t \in \mathcal{V}$ and in Π no descendant of $s \approx t$ is narrowed at a non-root position.

$$f(b) \approx g(b)$$
$$\Downarrow_{[\text{o}], f(b) \to g(x_1) \Leftarrow f(x_1) \approx g(b)} \quad 1(\text{b})$$
$$\underline{b \approx b, g(x_1) \approx g(b)}, f(x_1) \approx g(b)$$
$$\Downarrow_{[\text{d}]} \quad 1(\text{a})$$
$$g(x_1) \approx g(b), \underline{f(x_1) \approx g(b)}$$
$$\Downarrow_{[\text{o}], f(a) \to g(b)} \quad 1(\text{b})$$
$$g(x_1) \approx g(b), \underline{x_1 \approx a}, g(b) \approx g(b)$$
$$\Downarrow_{[\text{i}], \{x_1 \mapsto a\}} \quad 3(\text{a})$$
$$\underline{g(a) \approx g(b)}, g(b) \approx g(b)$$
$$\Downarrow_{[\text{d}]} \quad 1(\text{a})$$
$$g(a) \approx g(b), \underline{b \approx b}$$
$$\Downarrow_{[\text{d}]} \quad 1(\text{a})$$
$$g(a) \approx g(b)$$
$$\Downarrow_{[\text{d}]} \quad 1(\text{a})$$
$$\underline{a \approx b}$$
$$\Downarrow_{[\text{o}], a \to b \Leftarrow f(x_2) \approx g(b)} \quad 1(\text{b})$$
$$\underline{b \approx b}, f(x_2) \approx g(b)$$
$$\Downarrow_{[\text{d}]} \quad 1(\text{a})$$
$$\underline{f(x_2) \approx g(b)}$$
$$\Downarrow_{[\text{o}], f(a) \to g(b)} \quad 1(\text{b})$$
$$\underline{x_2 \approx a}, g(b) \approx g(b)$$
$$\Downarrow_{[\text{i}], \{x_2 \mapsto a\}} \quad 3(\text{a})$$
$$\underline{g(b) \approx g(b)}$$
$$\Downarrow_{[\text{d}]} \quad 1(\text{a})$$
$$\underline{b \approx b}$$
$$\Downarrow_{[\text{d}]} \quad 1(\text{a})$$
$$\Box$$

Fig. 1. An example illustrating the proof of Theorem 3.

Example 5. Consider the NC-refutation

$$a \approx x, \; x \approx b \; \leadsto \; \text{true}, \; a \approx b \; \leadsto \; \text{true}, \; b \approx b \; \leadsto \; \text{true}, \; \text{true} \; .$$

with respect to the TRS $\{a \to b\}$. In the second step the descendant $a \approx b$ of $x \approx b$ is narrowed at position 1. Hence $x \approx b$ is not a candidate. The equation $a \approx x$ is a candidate as no descendant of it is narrowed at position 2.

The following five lemmata provide transformations on narrowing derivations that are used in the completeness proof. Here Π denotes a CNC-refutation $G \leadsto_\theta^+ \top$ with $G = G', s \approx t, G''$ and V denotes a finite set of variables that includes all variables in the initial goal G of Π. The statements and (some of) the proofs are different from the ones in [15, 6] because we don't require that $s \approx t$ is selected in the first step of Π.

Lemma 6. *Suppose narrowing is applied to a descendant of $s \approx t$ in Π at position 1. If $l \to r \Leftarrow c$ is the employed rewrite rule in the first such step then there exists a CNC-refutation $\phi_{[\mathrm{o}]}(\Pi): G', s \approx l, r \approx t, c, G'' \rightsquigarrow^*_{\theta_1} \top$ such that $\theta_1 = \theta\,[V]$.* □

Lemma 7. *Let $s = f(s_1, \ldots, s_n)$, $t = f(t_1, \ldots, t_n)$, and suppose that narrowing is never applied to a descendant of $s \approx t$ in Π at position 1 or 2. There exists a CNC-refutation $\phi_{[\mathrm{d}]}(\Pi): G', s_1 \approx t_1, \ldots, s_n \approx t_n, G'' \rightsquigarrow^*_{\theta_1} \top$ such that $\theta_1 \leqslant \theta\,[V]$.* □

Lemma 8. *Let $s = f(s_1, \ldots, s_n)$ and $t \in \mathcal{V}$. If $\mathrm{root}(t\theta) = f$ then there exists a CNC-refutation $\phi_{[\mathrm{i}]}(\Pi): G\sigma_1 \rightsquigarrow^*_{\theta_1} \top$ such that Π subsumes $\phi_{[\mathrm{i}]}(\Pi)$, $\Pi\theta = \phi_{[\mathrm{i}]}(\Pi)\theta_1$, and $\sigma_1\theta_1 = \theta\,[V]$. Here $\sigma_1 = \{t \mapsto f(x_1, \ldots, x_n)\}$ with $x_1, \ldots, x_n \notin V$.* □

Lemma 9. *Let $t \in \mathcal{V}$, $s \neq t$, and suppose that narrowing is never applied to a descendant of $s \approx t$ in Π at a non-root position. There exists a CNC-refutation $\phi_{[\mathrm{v}]}(\Pi): (G', G'')\sigma_1 \rightsquigarrow^*_{\theta_1} \top$ with $\sigma_1 = \{t \mapsto s\}$ such that $\sigma_1\theta_1 \leqslant \theta\,[V]$.* □

Lemma 10. *Let $t \in \mathcal{V}$, $s = t$, and suppose that narrowing is never applied to a descendant of $s \approx t$ in Π at a non-root position. There exists a CNC-refutation $\phi_{[\mathrm{t}]}(\Pi): G', G'' \rightsquigarrow^*_{\theta_1} \top$ such that $\theta_1 \leqslant \theta\,[V]$.* □

In order to mimic the completeness proof of LNC ([15, Lemma 36, Theorem 39]), we define a class of CNC-refutations with the following three properties:

- it contains all CNC-refutations that produce normalized solutions,
- it is closed under the five transformations, and
- every CNC-refutation that it contains admits a candidate.

Our class is defined as the conjunction of two predicates.

Definition 11. *A CNC-refutation $\Pi: G \rightsquigarrow^* \top$ is called semi-normal if for all steps $\Pi': G', s \approx t, G'' \rightsquigarrow_{\theta_1} H$ in Π such that narrowing is applied to $s \approx t$ the following condition is satisfied: if narrowing takes place at a subterm of s (t) in Π' then $(\theta_1\theta_2)\!\upharpoonright_{\mathrm{Var}(s)}$ $((\theta_1\theta_2)\!\upharpoonright_{\mathrm{Var}(t)})$ is normalized. Here θ_2 is the substitution produced in the subrefutation $H \rightsquigarrow^* \top$ of Π. A CNC-refutation $\Pi: G \rightsquigarrow^* \top$ is said to be well-behaved if the introduced conditions after every application of a conditional rewrite rule are solved before any narrowing steps on other equations take place. Formally, Π is well-behaved if for every equation e occurring in it, Π has the form $G \rightsquigarrow^* G', e, G'' \rightsquigarrow_{\sigma_1, l \to r \Leftarrow c} G'\sigma_1, e', c\sigma_1, G''\sigma_1 \rightsquigarrow^*_{\sigma_2} G'\sigma_1\sigma_2, e'\sigma_2, \top, G''\sigma_1\sigma_2 \rightsquigarrow^* \top$. A CNC-refutation which is both semi-normal and well-behaved will be called well-normal.*

The next lemma states the closure of the class of well-normal CNC-refutations under the five transformations.

Lemma 12. *Let Π be a well-normal CNC-refutation. For every $\alpha \in \{\mathrm{o}, \mathrm{i}, \mathrm{d}, \mathrm{v}, \mathrm{t}\}$ the CNC-refutation $\phi_{[\alpha]}(\Pi)$ is well-normal whenever it is defined.* □

Lemma 13. *Let $\Pi: G \leadsto_{\theta_1}^* H \leadsto_{\theta_2}^* \top$ be a semi-normal CNC-refutation such that in the subderivation from G to H no narrowing step takes place at the root of an equation. Then $\theta_1\lceil_{\mathcal{V}ar(G)}$ is normalized.* □

Lemma 14. *For every non-empty well-normal CNC-refutation $\Pi: G \leadsto^+ \top$ there exists a candidate in G.*

Proof. Suppose to the contrary that G contains no candidates. Write Π as

$$G \leadsto_{\theta_1}^* G', e, G'' \leadsto_{\theta_2} G'\theta_2, \text{true}, G''\theta_2 \leadsto_{\theta_3}^* \top$$

such that narrowing is never applied to a descendant of an equation in G at the root position in the subderivation, denoted by Π_1, producing θ_1. Let e' be the equation in G of which e is a descendant. Since e' is not a candidate, Π_1 must contain a CNC-step in which narrowing is applied to a descendant of e' at a position below or equal to a variable position in e'. Since Π is well-behaved all conditional equations that were introduced during narrowing steps on descendants of e' are solved before the narrowing step that produces θ_2. This implies that a variable in e' is instantiated by θ_1 to a reducible term. This contradicts the previous lemma. □

Next we show that every CNC-refutation that produces a normalized solution is semi-normal. For this result to hold it is essential that we restrict ourselves to 1-CTRSs.

Lemma 15. *Let $\Pi: G \leadsto_\theta^+ \top$ be a CNC-refutation with respect to a 1-CTRS $\mathcal{R}$ such that $\theta\lceil_{\mathcal{V}ar(G)}$ is normalized. Every subrefutation of Π produces a normalized solution.* □

Corollary 16. *Every CNC-refutation that produces a normalized solution with respect to a 1-CTRS is semi-normal.* □

Using a switching lemma on rewrite sequences and a lifting lemma [13, Lemma 6.11], semi-normality in the above statement can be strengthened to well-normality.

Lemma 17. *Every CNC-refutation that produces a normalized solution with respect to a 1-CTRS is well-normal.* □

In the proof of the main theorem below, we use induction on well-normal CNC-refutations with respect to the order defined below. This order is the same as the one used in the completeness proofs of [15].

Definition 18. *The complexity $|\Pi|$ of a CNC-refutation $\Pi: G \leadsto_\theta^* \top$ is defined as the triple consisting of (1) the number of applications of narrowing in Π at non-root positions, (2) the multiset $|\mathcal{MV}ar(G)\theta|$, and (3) the number of occurrences of symbols different from $\approx$ and true in G. Here $\mathcal{MV}ar(G)$ denotes the multiset of variable occurrences in G, and for any multiset $M = \{t_1, \ldots, t_n\}$ of terms,*

$M\theta$ and $|M|$ denote the multisets $\{t_1\theta,\ldots,t_n\theta\}$ and $\{|t_1|,\ldots,|t_n|\}$, respectively. The well-founded order $\ggg$ on CNC-refutations is defined as follows: $\Pi_1 \ggg \Pi_2$ if $|\Pi_1|$ lex$(>,>_{\mathsf{mul}},>)$ $|\Pi_2|$. Here lex$(>,>_{\mathsf{mul}},>)$ denotes the lexicographic product of $>$, $>_{\mathsf{mul}}$, and $>$.

Lemma 19. *Let Π be a CNC-refutation and $\alpha \in \{\mathsf{o},\mathsf{i},\mathsf{d},\mathsf{v},\mathsf{t}\}$. We have $\Pi \ggg \phi_{[\alpha]}(\Pi)$ whenever the latter is defined.* □

Before we are ready to present the main result of this section, we need the easy transformation stated in the next lemma.

Lemma 20. *For every well-normal CNC-refutation $\Pi: G', s \approx t, G'' \leadsto_\theta^* \top$ there exists a well-normal CNC-refutation $\phi_{\mathsf{swap}}(\Pi): G', t \approx s, G'' \leadsto_\theta^* \top$ with the same complexity.*

Proof. Simply swap the two sides in every descendant of $s \approx t$. This does affect neither semi-normality, well-behavedness, nor the complexity. □

Theorem 21. *Let $\mathcal{R}$ be a confluent 1-CTRS. For every normalized solution θ of a goal G there exists an LCNC-refutation $G \Rightarrow_\sigma^* \square$ such that $\sigma \leqslant \theta$ $[\mathcal{V}\mathrm{ar}(G)]$.*

Proof. By the well-known completeness of CNC (e.g. [13, Theorem 6.14]) there exists a CNC-refutation $\Pi: G \leadsto_{\theta'}^* \top$ such that $\theta' \leqslant \theta$ $[\mathcal{V}\mathrm{ar}(G)]$. According to Lemma 17, Π is well-normal. We use well-founded induction on the complexity of Π. In order to make the induction work we prove $\sigma \leqslant \theta'$ $[V]$ for a finite set of variables V that includes $\mathcal{V}\mathrm{ar}(G)$. The base case is trivial since G must be the empty goal (and thus we can take $\sigma = \epsilon$). For the induction step we proceed as follows. We prove the existence of an LCNC-step $\Psi_1: G \Rightarrow_{\sigma_1} G_1$ and a well-normal CNC-refutation $\Pi_1: G_1 \leadsto_{\theta_1}^* \top$ such that $\sigma_1\theta_1 \leqslant \theta'$ $[V]$. According to Lemma 14 G contains a candidate e. Write $G = G', e, G''$. We distinguish the following cases, depending on what happens to $e = s \approx t$ in Π.

1. Suppose narrowing is never applied to a descendant of e at position 1 or 2. We distinguish four further cases.
   (a) Suppose $s, t \notin \mathcal{V}$. We may write $s = f(s_1,\ldots,s_n)$ and $t = f(t_1,\ldots,t_n)$. Let $G_1 = G', s_1 \approx t_1,\ldots,s_n \approx t_n, G''$. We have $\Psi_1: G \Rightarrow_{[\mathsf{d}]} G_1$. Lemma 7 yields a CNC-refutation $\phi_{[\mathsf{d}]}(\Pi): G_1 \leadsto_{\theta_1}^* \top$ such that $\theta_1 \leqslant \theta$ $[V]$. Take $\sigma_1 = \epsilon$.
   (b) Suppose $t \in \mathcal{V}$ and $s = t$. Let $G_1 = G', G''$. We have $\Psi_1: G \Rightarrow_{[\mathsf{t}]} G_1$. Because e is a candidate, the first application of narrowing to a descendant of e must take place at the root position. Hence Lemma 10 is applicable, which yields a CNC-refutation $\phi_{[\mathsf{t}]}(\Pi): G_1 \leadsto_{\theta_1}^* \top$ such that $\theta_1 \leqslant \theta$ $[V]$. Take $\sigma_1 = \epsilon$.
   (c) Suppose $t \in \mathcal{V}$ and $s \neq t$. We distinguish two further cases, depending on what happens to e in Π.
      i. Suppose the first application of narrowing to a descendant of e takes place at the root position. Let $\sigma_1 = \{t \mapsto s\}$ and $G_1 = (G', G'')\sigma_1$. We have $\Psi_1: G \Rightarrow_{[\mathsf{v}], \sigma_1} G_1$. Lemma 9 yields a CNC-refutation $\phi_{[\mathsf{v}]}(\Pi): G_1 \leadsto_{\theta_1}^* \top$ such that $\sigma_1\theta_1 \leqslant \theta$ $[V]$.

ii. Suppose the first application of narrowing to a descendant of e does not take place at the root position. This implies that $s \notin \mathcal{V}$ because e is a candidate. Hence we may write $s = f(s_1, \ldots, s_n)$. Let $\sigma_1 = \{t \mapsto f(x_1, \ldots, x_n)\}$, $G_1 = (G', s_1 \approx x_1, \ldots, s_n \approx x_n, G'')\sigma_1$, and $G_2 = G\sigma_1$. Here $x_1, \ldots, x_n$ are fresh variables. We have $\Psi_1 \colon G \Rightarrow_{[\mathsf{i}], \sigma_1} G_1$. Because narrowing is never applied to a descendant of e at position 2 (by definition of candidate), the root symbol of $t\theta$ equals f. ¿From Lemma 8 we obtain a CNC-refutation $\Pi_2 = \phi_{[\mathsf{i}]}(\Pi) \colon G_2 \leadsto^*_{\theta_2} \top$ such that $\sigma_1 \theta_2 = \theta \ [V]$. Let $V_2 = V \cup \{x_1, \ldots, x_n\}$. Clearly $\mathcal{V}\mathrm{ar}(G_2) \subseteq V_2$. An application of Lemma 7 to Π_2 results in a CNC-refutation $\Pi_1 = \phi_{[\mathsf{d}]}(\Pi_2) \colon G_1 \leadsto^*_{\theta_1} \top$ such that $\theta_1 \leqslant \theta_2 \ [V_2]$. Using the inclusion $\mathcal{V}\mathrm{ar}_V(\sigma_1) \subseteq V_2$ we obtain $\sigma_1 \theta_1 \leqslant \sigma_1 \theta_2 = \theta \ [V]$.

(d) In the remaining case we have $t \notin \mathcal{V}$ and $s \in \mathcal{V}$. This case reduces to case 1(c) by an appeal to Lemma 20.

2. Suppose narrowing is applied to a descendant of e at position 1. Let $l = f(l_1, \ldots, l_n) \to r \Leftarrow c$ be the employed rewrite rule the first time this happens. Because e is a candidate, s cannot be a variable. Hence we may write $s = f(s_1, \ldots, s_n)$. Let $G_1 = G', s_1 \approx l_1, \ldots, s_n \approx l_n, r \approx t, c, G''$ and $G_2 = G', s \approx l, r \approx t, c, G''$. We have $\Psi_1 \colon G \Rightarrow_{[\mathsf{o}]} G_1$. ¿From Lemma 6 we obtain a CNC-refutation $\Pi_2 = \phi_{[\mathsf{o}]}(\Pi) \colon G_2 \leadsto^*_{\theta_2} \top$ such that $\theta_2 = \theta \ [V]$. Let $V_2 = V \cup \mathcal{V}\mathrm{ar}(l \to r \Leftarrow c)$. Clearly $\mathcal{V}\mathrm{ar}(G_2) \subseteq V_2$. An application of Lemma 7 to Π_2 results in a CNC-refutation $\Pi_1 = \phi_{[\mathsf{d}]}(\Pi_2) \colon G_1 \leadsto^*_{\theta_1} \top$ such that $\theta_1 \leqslant \theta_2 \ [V_2]$. Using $V \subseteq V_2$ we obtain $\theta_1 \leqslant \theta \ [V]$. Take $\sigma_1 = \epsilon$.

3. Suppose narrowing is applied to a descendant of e at position 2. This case reduces to the previous one by an appeal to Lemma 20.

In all cases we obtain Π_1 from Π by applying one or two transformation steps $\phi_{[\mathsf{o}]}$, $\phi_{[\mathsf{i}]}$, $\phi_{[\mathsf{d}]}$, $\phi_{[\mathsf{v}]}$, $\phi_{[\mathsf{t}]}$ together with an additional application of ϕ_{swap} in case 1(d) and 3. According to Lemmata 12 and 20 Π_1 is well-normal. According to Lemmata 19 and 20 Π_1 has smaller complexity than Π. Let $V_1 = \mathcal{V}\mathrm{ar}_V(\sigma_1) \cup \mathcal{V}\mathrm{ar}(G_1)$. Clearly $\mathcal{V}\mathrm{ar}(G_1) \subseteq V_1$. Hence we can apply the induction hypothesis to $\Pi_1 \colon G_1 \theta_1 \to^* \top$. This yields an LCNC-refutation $\Psi' \colon G_1 \Rightarrow^*_{\sigma'} \square$ such that $\sigma' \leqslant \theta_1 \ [V_1]$. Define $\sigma = \sigma_1 \sigma'$. From $\sigma_1 \theta_1 \leqslant \theta' \ [V]$, $\theta' \leqslant \theta \ [V]$, $\sigma' \leqslant \theta_1 \ [V_1]$, and $\mathcal{V}\mathrm{ar}_V(\sigma_1) \subseteq V_1$ we infer that $\sigma \leqslant \theta \ [V]$. Concatenating the LCNC-step Ψ_1 and the LCNC-refutation Ψ' yields the desired LCNC-refutation Ψ. □

Note that because we transform narrowing derivations rather than rewrite sequences, the proof of the above result is more complicated than the one of Theorem 3. Actually, this is a bit of surprise since for CNC the situation is exactly the opposite (cf. Middeldorp and Hamoen [13]).

As mentioned before, the above proof does not yield the completeness of LCNC with respect to leftmost selection. In future research we plan to investigate whether such a result can be obtained by putting the conditional part c of the applied rewrite rule $l \to r \Leftarrow c$ in the outermost narrowing rule [o] before the body equation $r \approx t$. In light of the restriction to well-behaved CNC-refutation above, this seems promising.

References

1. F. Baader and Nipkow. T. *Term Rewriting and All That*. Cambridge University Press, 1998.
2. N. Dershowitz and J.-P. Jouannaud. Rewrite systems. In J. van Leeuwen, editor, *Handbook of Theoretical Computer Science*, volume B, pages 243–320. North-Holland, 1990.
3. N. Dershowitz and Z. Manna. Proving termination with multiset orderings. *Communications of the ACM*, 22:465–476, 1979.
4. N. Dershowitz, M. Okada, and G. Sivakumar. Canonical conditional rewrite systems. In *Proceedings of the 9th International Conference on Automated Deduction*, volume 310 of *LNCS*, pages 538–549, 1988.
5. E. Giovannetti and C. Moiso. A completeness result for e-unification algorithms based on conditional narrowing. In *Proceedings of the Workshop on Foundations of Logic and Functional Programming*, volume 306 of *LNCS*, pages 157–167, 1986.
6. M. Hamada and A. Middeldorp. Strong completeness of a lazy conditional narrowing calculus. In *Proceedings of the 2nd Fuji International Workshop on Functional and Logic Programming*, pages 14–32. World Scientific, Singapore, 1997.
7. M. Hanus. The integration of functions into logic programming: From theory to practice. *Journal of Logic Programming*, 19 & 20:583–628, 1994.
8. M. Hanus. Lazy unification with simplification. In *Proceedings of the 5th European Symposium on Programming*, volume 788 of *LNCS*, pages 272–286, 1994.
9. S. Hölldobler. *Foundations of Equational Logic Programming*, volume 353 of *LNAI*. 1989.
10. J.-M. Hullot. Canonical forms and unification. In *Proceedings of the 5th Conference on Automated Deduction*, volume 87 of *LNCS*, pages 318–334, 1980.
11. S. Kaplan. Conditional rewrite rules. *Theoretical Computer Science*, 33:175–193, 1984.
12. J.W. Klop. Term rewriting systems. In S. Abramsky, D. Gabbay, and T. Maibaum, editors, *Handbook of Logic in Computer Science*, volume 2, pages 1–116. Oxford University Press, 1992.
13. A. Middeldorp and E. Hamoen. Completeness results for basic narrowing. *Applicable Algebra in Engineering, Communication and Computing*, 5:213–253, 1994.
14. A. Middeldorp and S. Okui. A deterministic lazy narrowing calculus. *Journal of Symbolic Computation*, 25(6):733–757, 1998.
15. A. Middeldorp, S. Okui, and T. Ida. Lazy narrowing: Strong completeness and eager variable elimination. *Theoretical Computer Science*, 167(1,2):95–130, 1996.
16. C. Prehofer. *Solving Higher-Order Equations: From Logic to Programming*. Progress in Theoretical Computer Science. Birkäuser, 1998.
17. T. Suzuki, A. Middeldorp, and T. Ida. Level-confluence of conditional rewrite systems with extra variables in right-hand sides. In *Proceedings of the 6th International Conference on Rewriting Techniques and Applications*, volume 914 of *LNCS*, pages 179–193, 1995.
18. A. Werner. *Untersuchung von Strategien für das logisch-funktionale Programmieren*. Shaker-Verlag Aachen, March 1998.

The Pagenumber of de Bruijn and Kautz Digraphs

Toru Hasunuma*

Department of Applied Mathematics and Physics,
Graduate School of Informatics, Kyoto University, Japan.
hasunuma@kuamp.kyoto-u.ac.jp

Abstract. The pagenumber of a class S of digraphs is the minimum number of pages in which any digraph in S can be embedded. It is known that the pagenumber of the binary de Bruijn digraphs (the binary Kautz digraphs) is 3. We show that for any $d \geq 3$, the pagenumber of the d-ary de Bruijn digraphs (the d-ary Kautz digraphs) is $(d+1)$.

1 Introduction

Let G be a digraph. Then the vertex set and the arc set of G are denoted by $V(G)$ and $A(G)$, respectively. In this paper, we investigate embeddings of digraphs in structures called books. A *book* consists of a line called the *spine* and half planes called *pages*, sharing the spine as a common boundary. An embedding of G in a book (called a *bookembedding* of G) is defined by an assignment of the vertices to points on the spine (i.e., a linear ordering of $V(G)$) and an assignment of the arcs to pages such that any two arcs on the same page do not cross. The *pagenumber* of G is the minimum number of pages in which G can be embedded. Let S be a class of digraphs. Then the pagenumber of S is the minimum number of pages in which any digraph in S can be embedded. An embedding of G in d pages is called a *d-page bookembedding* of G. A bookembedding of a graph is similarly defined.

The bookembedding problem has been motivated by several areas of computer science (see [6]) such as VLSI-layout and fault-tolerant computing. So far, bookembeddings of various classes of graphs or digraphs have been studied; complete graphs [5], complete bipartite graphs [18], butterflies [10], meshes, X-trees, hypercubes [6] [15], de Bruijn digraphs, Kautz digraphs, shuffle-exchange graphs [13], planar graphs [21], and graphs with a given genus [17]. Among these classes of graphs or digraphs, their pagenumbers have been determined except for the d-ary de Bruijn digraph, the d-ary Kautz digraph ($d \geq 3$), and the complete bipartite graph. (For genus g graphs, the pagenumber is given with O-notation.)

In this paper, we treat the de Bruijn digraphs and the Kautz digraphs. The *de Bruijn digraph* denoted by $B(d, D)$ is a digraph whose vertices are the words of length D on an alphabet of d letters (for example, $\{0, 1, \ldots, d-1\}$). There

---
* Research Fellow of the Japan Society for the Promotion of Science.

is an arc from a vertex $(v_0, v_1, \ldots, v_{D-1})$ to a vertex $(v_1, \ldots, v_{D-1}, \alpha)$, where $\alpha \in \{0, 1, \ldots, d-1\}$. The *Kautz digraph* denoted by $K(d, D)$ is a digraph whose vertices are the words of length D on an alphabet of $(d+1)$ letters with no two consecutive identical letters. There is an arc from a vertex $(v_0, v_1, \ldots, v_{D-1})$ to a vertex $(v_1, \ldots, v_{D-1}, \alpha)$, where $\alpha \in \{0, 1, \ldots, d\}$ and $\alpha \neq v_{D-1}$. In this paper, the d-ary de Bruijn digraphs (resp., the d-ary Kautz digraphs) mean the class $\{B(d, D) \mid D \geq 1\}$ (resp., $\{K(d, D) \mid D \geq 1\}$).

It is known that de Bruijn and Kautz digraphs can be also defined as iterated line digraphs ([9]). The *line digraph* $L(G)$ of G is defined as the digraph whose vertex set is $A(G)$ and in which there is an arc from (u, v) to (w, x) iff $v = w$. We say that $L(G)$ is obtained from G by applying the *line digraph operation*. The *m-iterated line digraph* $L^m(G)$ of G is the digraph obtained from G by applying the line digraph operation m times. Then the de Bruijn digraph is defined by $L^{D-1}(K_d^\circ)$, where K_d° is the complete digraph with d vertices. The Kautz digraph is similarly defined by $L^{D-1}(K_{d+1}^*)$, where K_{d+1}^* is the complete symmetric digraph with $(d+1)$ vertices.

The de Bruijn and Kautz digraphs have been noticed as interconnection networks of massively parallel computers (see [4],[16]) because of their nice properties such as bounded degree, small diameter, and easy routing. So far, their various properties such as connectivity [3], [8], symmetry [20], chromatic index [2], VLSI-decomposition [7], certain spanning trees [1], [11],[14] have been studied. On the bookembedding of de Bruijn and Kautz digraphs, it has been shown in [13] that $B(d, D)$ and $K(d, D)$ can be embedded in $(d + 1)$ pages. When $d = 2$, the $(d+1)$-page bookembeddings are optimal with respect to the number of pages. That is, the pagenumber of the binary de Bruijn digraphs (the binary Kautz digraphs) is 3. However, except for the case $d = 2$, it has been unknown whether $(d + 1)$ pages is necessary for bookembeddings of these digraphs. In this paper, we show that for any $d \geq 3$, the pagenumber of the d-ary de Bruijn digraphs (the d-ary Kautz digraphs) is $(d + 1)$.

The outline of our proof is as follows. We first give an upper bound on the number of arcs of a d-page-embeddable digraph for $d \geq 3$. Using the result and the properties of line digraphs, we derive a necessary condition for a d-page embeddable d-regular line digraph. Then we prove that for any d-regular digraph G, there exists a constant c_G such that if $m > c_G$, then $L^m(G)$ cannot be embedded in d pages, i.e., the pagenumber of d-regular iterated line digraphs is greater than d. From this result, the pagenumber of the d-ary de Bruijn digraphs (the d-ary Kautz digraphs) is determined to be $(d+1)$ for any $d \geq 3$.

This paper is organized as follows. In Section 2, definitions, notations and several basic results used in the paper are given. In Sections 3 and 4, we give upper bounds on the numbers of two types of arcs in a bookembedding of a digraph. In Section 5, we present a necessary condition for a d-page embeddable d-regular line digraph. In Section 6, we show that the pagenumber of d-regular iterated line digraphs is greater than d. Also, we present a constant c such that if $D > c$ then $B(d, D)$ and $K(d, D)$ cannot be embedded in d pages.

2 Preliminaries

2.1 Definitions and notations

In this paper, a digraph may have loops but not multiarcs. Let G be a digraph. Then, the *underlying graph* $U(G)$ of is the graph obtained from G by replacing each arc with the corresponding edge, deleting loops and replacing multiedges with a single edge. The numbers of vertices and arcs of G are denoted by n_G and m_G, respectively. Similarly, for a graph H, n_H and m_H denote the numbers of vertices and edges of H, respectively. We use the notation m_G^* instead of $m_{U(G)}$ to denote the number of edges of $U(G)$. Let $v \in V(G)$. Then $\Gamma_G^+(v)$ and $\Gamma_G^-(v)$ denote the sets $\{w \in V(G) \mid (v,w) \in A(G)\}$ and $\{w \in V(G) \mid (w,v) \in A(G)\}$, respectively. The outdegree of v is $|\Gamma_G^+(v)|$ and the indegree of v is $|\Gamma_G^-(v)|$. If for any vertex v of G, both the outdegree and the indegree of v are equal to d, then G is called *d-regular*.

A *walk* W of length t in G is an alternate sequence of vertices and arcs $(v_0, e_1, v_1, e_2, \ldots, e_t, v_t)$, where $e_i = (v_{i-1}, v_i) \in A(G)$ for $1 \leq i \leq t$. (Since we treat a digraph without multiarcs, it is sufficient to write only the vertices.) When $v_0 = v_t$, W is called *closed*. Let $W = (v_0, v_1, \ldots, v_{t-1}, v_0)$ be a closed walk. If there exists an integer $0 < l < t$ such that $v_i = v_{i+l(\text{mod } t)}$ for $0 \leq i < t$, then we call W a *periodic closed walk*. Otherwise, W is called a *nonperiodic closed walk*. Let $W = (v_0, v_1, \ldots, v_t)$ be a walk. If $v_i \neq v_j$ for $i \neq j$, then we call W a *path*. Also, if $v_i \neq v_j$ for $0 \leq i < j < t$ and $v_0 = v_t$, then W is called a *cycle*. A path and a cycle of length t are called a *t-path* and a *t-cycle*, respectively. In a graph, these terminologies are similarly defined. It is easily checked that a vertex of $L^m(G)$ corresponds to a walk of length m in G.

An undirected walk in G means a walk in $U(G)$. A walk W in G is denoted by $(v_0, v_1, \ldots, v_t)$. On the other hand, an undirected walk in G is denoted by $\langle v_0, v_1, \ldots, v_t \rangle$. The number of t-cycles in G is denoted by $C_G(t)$. Also, $C_G^*(t)$ denotes the number of undirected t-cycles in G. Let $N_G(t)$ be the number of nonperiodic closed walks of length k in G. (Note that the number of closed walks of a certain length means the number of equivalent classes of closed walks of such length with respect to rotation, although we define closed walks as sequences.) Then, $C_G(t) \leq N_G(t)$ since a t-cycle is a nonperiodic closed walk of length t. The number of nonperiodic closed walks of length t is invariant under the line digraph operation (see [12]). That is, $N_{L(G)}(t) = N_G(t)$. Thus, $N_{L^m(G)}(t) = N_G(t)$. Therefore, $C_{L^m(G)}(t) \leq N_G(t)$.

2.2 Bookembeddings

In a bookembedding of a digraph G, loops are useless and double arcs (cycles of length 2) can be treated as one arc (i.e., double arcs are assigned to the same page). Thus, a bookembedding of G is corresponding to a bookembedding of $U(G)$. Since we need to use several properties of line digraphs, we treat bookembeddings of digraphs as it is. However, note that in the propositions for upper bounds on the numbers of several kinds of arcs, we assume that loops are not counted and double arcs are counted as a single arc (see $N_S(\mathcal{B})$ and $N_I(\mathcal{B})$ defined below.)

Consider a bookembedding $\mathcal{B}$ of G. Let $u, v \in V(G)$. If there is no vertex between u and v on the spine, then we say that u and v are *adjoin* on the spine. (The spine can be treated as a circle. In the definition of an adjoin pair of vertices, we consider the spine as a circle.) Let $S_{\mathcal{B}}(G)$ denote the set of adjoin pairs of vertices. An adjoin pair of $S_{\mathcal{B}}(G)$ is denoted as $[u,v]$. Note that $[u,v]$ is an unordered pair. Let $(u,v) \in A(G)$. If $[u,v] \in S_{\mathcal{B}}(G)$, then (u,v) is called a *spinal arc*. Otherwise, (u,v) is called an *internal arc*. We can consider that a spinal arc is assigned to an interval between vertices on the spine rather than a particular page. For $\mathcal{B}$, $N_S(\mathcal{B})$ denotes the number of spinal arcs on the condition that double arcs are counted as one arc. Also, $N_I(\mathcal{B})$ denotes the number of internal arcs on the condition that loops are not counted and double arcs are counted as one arc. Therefore,

$$m_G = N_S(\mathcal{B}) + N_I(\mathcal{B}) + C_G(1) + C_G(2).$$

A graph which can be embedded in one page is called an *outerplanar graph*. It is known that the number of edges of an outerplanar graph H is at most $(2n_H - 3)$. Clearly, for a bookembedding of a graph H, the number of spinal edges is at most n_H. Thus, the number of internal edges of one page is at most $(n_H - 3)$. Therefore, for any d-page bookembedding $\mathcal{B}$ of G,

$$N_S(\mathcal{B}) + N_I(\mathcal{B}) \leq (d+1)n_G - 3d.$$

From this upper bound, it is shown that the pagenumber of the d-ary de Bruijn digraphs (the d-ary Kautz digraphs) is greater than $(d-2)$.

3 An upper bound on the number of spinal arcs

In this section, we present an upper bound on the number of spinal arcs in a d-page bookembedding of a digraph.

Lemma 1. *Let H be a planar graph such that $n_H \geq 4$. Then,*

$$m_H \leq 2(n_H - 2) + \frac{1}{2}C_H(3).$$

Proof. Suppose that H is connected. Consider an embedding of H on the plane. Let $r_H(i)$ be the number of regions with boundary length i. Also, let $r_H = \sum_{i \geq 3} r_H(i)$. Then, it holds that $\sum_{i \geq 3} i \cdot r_H(i) = 2m_H$ since each edge is counted two times in the left summation. If $C_H(3) \leq r_H$, then the following inequality holds.

$$3C_H(3) + 4(r_H - C_H(3)) \leq 2m_H.$$

If $C_H(3) > r_H$, then $3C_H(3) + 4(r_H - C_H(3)) < 3r_H$. Thus, the above inequality always holds. By Euler's formula, $n_H - m_H + r_H = 2$. Thus,

$$4(m_H - n_H + 2) - C_H(3) \leq 2m_H.$$

Therefore,
$$m_H \leq 2(n_H - 2) + \frac{1}{2}C_H(3).$$

For the case that H is disconnected, we can add new edges to make H connected so that the number of 3-cycles does not increase. Let H' be the resultant connected plane graph. Then,
$$m_H < m_{H'} \leq 2(n_{H'} - 2) + \frac{1}{2}C_{H'}(3) = 2(n_H - 2) + \frac{1}{2}C_H(3).$$

□

Lemma 2. *Let $\mathcal{B}$ be a d-page bookembedding of a digraph G such that $n_G \geq 4$ and $d \geq 3$. Then,*
$$N_S(\mathcal{B}) \leq \frac{2}{(d-2)}(dn_G - m_G^* + \frac{1}{2}C_G^*(3) - 2d).$$

Proof. Let H_i be the digraph obtained from $\mathcal{B}$ by considering only two pages P_i and $P_{(i+1)(mod\ d)}$, i.e., the digraph arc-induced by the spinal arcs and the internal arcs assigned to the pages P_i and P_{i+1}. Let $S_i^*(3)$ be the set of undirected 3-cycles on the page P_i. Also, let $S_{i,i+1}^*(3)$ be the set of undirected 3-cycles on both P_i and P_{i+1}. Let $C_i^*(3) = |S_i^*(3)|$ and $C_{i,i+1}^*(3) = |S_{i,i+1}^*(3)|$. Then, from Lemma 1, the following inequality holds.
$$m_{H_i}^* \leq 2(n_{H_i} - 2) + \frac{1}{2}(C_i^*(3) + C_{i+1}^*(3) + C_{i,i+1}^*(3)).$$

Thus,
$$\sum_{0 \leq i < d} m_{H_i}^* \leq 2\sum_{0 \leq i < d}(n_{H_i} - 2) + \frac{1}{2}\sum_{0 \leq i < d}(C_i^*(3) + C_{i+1}^*(3) + C_{i,i+1}^*(3))$$
$$\leq 2d(n_G - 2) + \sum_{0 \leq i < d} C_i^*(3) + \frac{1}{2}\sum_{0 \leq i < d} C_{i,i+1}^*(3).$$

Since $n_G \geq 4$, there is no undirected 3-cycle constructed by only spinal arcs. Thus, $S_i^*(3) \cap S_j^*(3) = \emptyset$ for $i \neq j$. Also, $S_i^*(3) \cap S_{j,j+1}^*(3) = \emptyset$ for any i and j, and $S_{i,i+1}^*(3) \cap S_{j,j+1}^*(3) = \emptyset$ for $i \neq j$. Hence $\sum_{0 \leq i < d} C_i^*(3) + \frac{1}{2}\sum_{0 \leq i < d} C_{i,i+1}^*(3) \leq C_G^*(3)$. Thus,
$$\sum_{0 \leq i < d} m_{H_i}^* \leq 2d(n_G - 2) + C_G^*(3).$$

Let $m^*(P_i)$ be the number of internal arcs on the page P_i, where loops are not counted and double arcs are counted as one arc. Then $\sum_{0 \leq i < d} m_{H_i}^* = 2\sum_{0 \leq i < d} m^*(P_i) + dN_S(\mathcal{B})$. Thus, the following inequality holds.
$$\sum_{0 \leq i < d} m^*(P_i) + \frac{d}{2}N_S(\mathcal{B}) \leq d(n_G - 2) + \frac{1}{2}C_G^*(3).$$

Here, $m_G^* = \sum_{0 \leq i < d} m^*(P_i) + N_S(\mathcal{B})$. Therefore,
$$m_G^* + (\frac{d}{2} - 1)N_S(\mathcal{B}) \leq d(n_G - 2) + \frac{1}{2}C_G^*(3).$$

Hence,
$$N_S(\mathcal{B}) \leq \frac{2}{(d-2)}(dn_G - m_G^* + \frac{1}{2}C_G^*(3) - 2d). \quad \square$$

If G is d-regular, then $m_G = dn_G$. Thus, $dn_G - m_G^* = C_G(1) + C_G(2)$. Therefore, the following corollary is obtained.

Corollary 3. *Let $\mathcal{B}$ be a d-page bookembedding of a d-regular digraph G such that $n_G \geq 4$ and $d \geq 3$. Then,*
$$N_S(\mathcal{B}) \leq \frac{2}{(d-2)}(C_G(1) + C_G(2) + \frac{1}{2}C_G^*(3) - 2d).$$

4 An upper bound on the number of internal arcs

In this section, we give an upper bound on the number of internal arcs in a d-page bookembedding of a digraph.

Definition 4. Let $\mathcal{B}$ be a bookembedding of a digraph G. Let $[u,v] \in S_\mathcal{B}(G)$. A region of $[u,v]$ is a region surrounded by internal arcs on a page and intervals on the spine which contains the interval between u and v on the spine. The number of nontriangle regions of $[u,v]$ is denoted by $\beta([u,v])$.

Note that a region of $[u,v]$ is not always a "real" region. When we assume that there is an arc between any adjoin pair, a region of $[u,v]$ is a "real" region. For a d-page bookembedding, there are d regions of $[u,v]$ for each adjoin pair $[u,v]$. Several adjoin pairs may have a common region on a page.

Let $\mathcal{B}$ be a d-page bookembedding of G. Then, a trivial upper bound on $N_I(\mathcal{B})$ is $d(n_G - 3)$. Using the values $\beta([u,v])$, the upper bound can be improved as follows.

Lemma 5. *Let $\mathcal{B}$ be a d-page bookembedding of G such that $n_G \geq 5$. Then,*
$$N_I(\mathcal{B}) \leq d(n_G - 3) - \frac{1}{3} \sum_{[u,v] \in S_\mathcal{B}(G)} \beta([u,v]).$$

Proof. Let $\mathcal{R}$ be the set of nontriangle regions of adjoin pairs, i.e., $\mathcal{R} = \{R \mid$ there exists $[u,v] \in S_\mathcal{B}(G)$ such that R is a nontriangle region of $[u,v]\ \}$. Let $|R|$ denote the boundary length of R. For each nontriangle region R, we can add $(|R| - 3)$ new internal arcs without crossing. This means that
$$N_I(\mathcal{B}) \leq d(n_G - 3) - \sum_{R \in \mathcal{R}} (|R| - 3).$$

Suppose that any page has at least one internal arc in $\mathcal{B}$. Then, for any $R \in \mathcal{R}$, R has at least one internal arc as its boundary arc. That is, R has at most $(|R|-1)$ spinal intervals as its boundary intervals. Therefore,
$$\sum_{R \in \mathcal{R}} (|R| - 1) \geq \sum_{[u,v] \in S_\mathcal{B}(G)} \beta([u,v]).$$

Since $|R| \geq 4$ for any $R \in \mathcal{R}$,

$$\sum_{R \in \mathcal{R}} (|R| - 3) \geq \frac{1}{3} \sum_{R \in \mathcal{R}} (|R| - 1).$$

Hence, the proposition holds.

Suppose that in $\mathcal{B}$, there are t empty pages (pages in which there is no internal arc). Consider the $(d-t)$-(nonempty)page bookembedding $\mathcal{B}'$ induced from $\mathcal{B}$. Then,

$$N_I(\mathcal{B}') \leq (d-t)(n_G - 3) - \frac{1}{3} \sum_{[u,v] \in S_{\mathcal{B}'}(G)} \beta([u,v]).$$

Since $n_G \geq 5$, $(n_G - 3) \geq \frac{1}{3} n_G$. Thus, in this case, the proposition also holds. □

Definition 6. Let $\mathcal{B}$ be a bookembedding of G. Let $[u,v] \in S_{\mathcal{B}}(G)$. If for any page, there exists an undirected 2-path $\langle u, w, v \rangle$ on the page, then $[u,v]$ is called a *triangulated pair*, and the corresponding arcs in $\langle u, w, v \rangle$ are called *triangulated arcs* of $[u,v]$. An adjoin pair $[u,v]$ which is not a triangulated pair is called a *non-triangulated pair*.

Clearly, if $[u,v]$ is a triangulated pair, then $\beta([u,v]) = 0$. Note that the converse does not always hold.

Definition 7. Let $[u,v]$ be a non-triangulated pair in a bookembedding of G. If $\beta([u,v]) \geq 1$, then $[u,v]$ is called a *proper non-triangulated pair*. If $\beta([u,v]) = 0$, then $[u,v]$ is called a *quasi-triangulated pair*.

Let $Q_{\mathcal{B}}(G)$ be the set of quasi-triangulated pairs in a bookembedding $\mathcal{B}$ of G. Let $[u,v] \in Q_{\mathcal{B}}(G)$. Then there exists $[v,w] \in S_{\mathcal{B}}(G)$ such that there is an arc between u and w but not between v and w. (If $\beta([u,v]) = 0$ and there is no such pair $[v,w]$, then $[u,v]$ must be a triangulated pair.) For such $[u,v]$ and $[v,w]$, we write $[u,v] \vdash [v,w]$. Let

$Q'_{\mathcal{B}}(G) = \{[u,v] \in Q_{\mathcal{B}}(G) \mid $ there exists $[v,w] \in S_{\mathcal{B}}(G)$ such that $[u,v] \vdash [v,w]$ and $\beta([v,w]) = 0\}$

Lemma 8. *For a d-page bookembedding $\mathcal{B}$ of G $(d \geq 3)$,*

$$|Q'_{\mathcal{B}}(G)| \leq \frac{2}{(d-2)^2} (C^*_G(3) + C^*_G(5)).$$

Proof. Let $[u,v] \in Q'_{\mathcal{B}}(G)$. Then, there exists $[v,w] \in S_{\mathcal{B}}(G)$ such that $[u,v] \vdash [v,w]$ and $\beta([v,w]) = 0$. Since $\beta([v,w]) = 0$, there are at least $(d-2)$ undirected 2-path $\langle v, x_i, w \rangle$. Similarly, there are at least $(d-2)$ undirected 2-path $\langle u, y_j, v \rangle$. This means that u and w are contained in an undirected 5-cycle (if $x_i \neq y_j$) or an undirected 3-cycle (if $x_i = y_j$). By this correspondence between $[u,v]$ and an undirected 5 or 3-cycle, we can consider the bipartite graph with partite sets $Q'_{\mathcal{B}}(G)$ and the set of undirected 5 or 3-cycles in which the adjacency is given by the correspondence.

Suppose that $[u,v]$ is corresponding to an undirected 5-cycle $\langle u, y_j, v, x_i, w, u\rangle$. Then, there are two configurations on $\langle u, y_j, v, x_i, w, u\rangle$. One is the case that the undirected 2-paths $\langle u, y_j, v\rangle$ and $\langle v, x_i, w\rangle$ do not cross. In this case, no element in $Q'_\mathcal{B}(G)$ corresponds to $\langle u, y_j, v, x_i, w, u\rangle$ except for $[u,v]$ and $[v,w]$. Consider the case that $\langle u, y_j, v\rangle$ and $\langle v, x_i, w\rangle$ cross. In this case, clearly, arc between u and w, $\langle u, y_j, v\rangle$, and $\langle v, x_i, w\rangle$ are on the different pages. Now, assume that $[x_i, u] \in Q'_\mathcal{B}(G)$ such that $[x_i, u]$ corresponds to $\langle u, y_j, v, x_i, w, u\rangle$. Then, the undirected 2-path $\langle x_i, w, u\rangle$ must be on a page. However, this contradicts the fact that the arc between u and w, and the undirected 2-path $\langle v, x_i, w\rangle$ are on the different pages. Similarly, it is shown that any element in $Q'_\mathcal{B}(G)$ different from $[u, v]$ and $[v, w]$ cannot correspond to $\langle u, y_j, v, x_i, w, u\rangle$. Also, for the case that $[u,v] \in Q'_\mathcal{B}(G)$ is corresponding to an undirected 3-cycle $\langle u, x_i, w, u\rangle$, we can similarly prove that any element in $Q'_\mathcal{B}(G)$ different from $[u, v]$ and $[v, w]$ cannot correspond to $\langle u, x_i, w, u\rangle$. Therefore, for an undirected 5 or 3-cycle, at most two elements in $Q'_\mathcal{B}(G)$ can correspond to it. Hence, the number of edges in the bipartite graph is at most $2(C^*_G(3) + C^*_G(5))$. For any $[u,v] \in Q'_\mathcal{B}(G)$, the degree of $[u, v]$ in the bipartite graph is at least $(d-2)^2$. Therefore, the cardinality of $Q'_\mathcal{B}(G)$ is at most $\frac{2}{(d-2)^2}(C^*_G(3) + C^*_G(5))$. □

Lemma 9. *Let $\mathcal{B}$ be a d-page bookembedding of G ($d \geq 3$). Let N_{non} be the number of non-triangulated pairs in $\mathcal{B}$. Then,*

$$N_I(\mathcal{B}) \leq d(n_G - 3) - \frac{1}{9}N_{non} + \frac{2}{9(d-2)^2}(C^*_G(3) + C^*_G(5)).$$

Proof. Let S_{non} be the set of non-triangulated pairs in $\mathcal{B}$. Also, let $S'_{non} = S_{non} \setminus Q'_\mathcal{B}(G)$.

Let $[u, v] \in S'_{non}$ such that $\beta([u, v]) = 0$. Since $[u, v]$ is a quasi-triangulated pair, there is an adjoin pair $[v, w]$ such that $[u, v] \vdash [v, w]$. If $\beta([v, w]) = 0$, then $[u, v] \in Q'_\mathcal{B}(G)$. Therefore, $\beta([v, w]) \geq 1$. For such an adjoin pair $[v, w]$, there are at most two quasi-triangulated pairs $[x, y]$ such that $[x, y] \vdash [v, w]$. Hence,

$$\sum_{[u,v]\in S_\mathcal{B}(G)} \beta([u,v]) = \sum_{[u,v]\in S'_{non}} \beta([u,v]) \geq \frac{1}{3}|S'_{non}| = \frac{1}{3}(N_{non} - |Q'_\mathcal{B}(G)|).$$

Thus, the inequality in the proposition follows from Lemmas 5 and 8. □

5 A d-page bookembedding of a d-regular line digraph

In this section, we present an upper bound on the number of triangulated pairs in a d-page bookembedding of a d-regular line digraph. As a corollary, we can obtain a lower bound on the number of non-triangulated pairs. From this lower bound and the results in the previous sections, an upper bound on the number of vertices of a d-page embeddable d-regular line digraph is induced. Also, we show that the number of small undirected cycles in a line digraph can be bounded by the number of small cycles.

5.1 An upper bound on the number of triangulated pairs

Definition 10. Let $u, v \in V(L(G))$. Also let $u = (u_1, u_2) \in A(G)$ and $v = (v_1, v_2) \in A(G)$. If $u_2 = v_2$ (resp., $u_1 = v_1$), then we say that u and v are *head-compatible* (resp., *tail-compatible*). Also we say that u and v are *compatible* if u and v are either head-compatible or tail-compatible.

It is easily checked that u and v are head-compatible (resp., tail-compatible) in $L(G)$ iff $\Gamma^+(u) = \Gamma^+(v)$ (resp., $\Gamma^-(u) = \Gamma^-(v)$). Also, for any two vertices u and v of $L(G)$, if $\Gamma^+(u) \cap \Gamma^+(v) \neq \emptyset$, then $\Gamma^+(u) = \Gamma^+(v)$. That is, either $\Gamma^+(u) = \Gamma^+(v)$ or $\Gamma^+(u) \cap \Gamma^+(v) = \emptyset$. The similar property holds for $\Gamma^-(u)$ and $\Gamma^-(v)$.

Lemma 11. *In a d-page bookembedding of a d-regular line digraph $L(G)$ ($d \geq 3$), if $[u, v]$ is a triangulated pair, then u and v are compatible.*

Proof. Let $[u, v]$ be a triangulated pair in a d-page bookembedding of a d-regular line digraph. Then, there are d vertices $w_0, w_1, \ldots, w_{d-1}$ such that there exists an undirected 2-path $\langle u, w_i, v \rangle$ on the page P_i. Suppose that u and v are not compatible. Then, $\Gamma^+(u) \cap \Gamma^+(v) = \emptyset$ and $\Gamma^-(u) \cap \Gamma^-(v) = \emptyset$. Thus, either $(u, w_i), (w_i, v) \in A(L(G))$ or $(v, w_i), (w_i, u) \in A(L(G))$ for $0 \leq i < d$. If $(u, w_i), (w_i, v) \in A(L(G))$, then it is impossible that $(u, w_j), (w_j, v) \in A(L(G))$ for $j \neq i$, because the existence of two 2-paths from u to v implies the existence of multiarcs in G. Similarly, if $(v, w_i), (w_i, u) \in A(L(G))$, then it is impossible that $(v, w_j), (w_j, u) \in A(L(G))$ for $j \neq i$. Thus, there are at most two undirected 2-paths between u and v, which contradicts our assumptions that $[u, v]$ is a triangulated pair and $d \geq 3$. □

Let $[u, v]$ be a triangulated pair in a d-page bookembedding of a d-regular line digraph. Suppose u and v are head-compatible. If $(u, w), (w, v) \in A(L(G))$, then $(v, w) \in A(L(G))$ since $\Gamma^+(u) = \Gamma^+(v)$. Any double arcs can be assigned to the same page. Thus, in the rest of this section, for a triangulated pair $[u, v]$ and an undirected 2-path $\langle u, w_i, v \rangle$ on a page P, we consider $(u, w_i), (v, w_i)$ as triangulated arcs of $[u, v]$. Similarly, when u and v are tail-compatible, we consider $(w_i, u), (w_i, v)$ as triangulated arcs of $[u, v]$.

Suppose that $[u, v]$ is a triangulated pair, and u and v are head-compatible (tail-compatible). Then, there is a one-to-one correspondence φ between $\Gamma^+(u)$ ($\Gamma^-(u)$) and the set of pages $\{P_0, \ldots, P_{d-1}\}$ such that if $\varphi(x) = i$, then the undirected 2-path $\langle u, x, v \rangle$ is on the page P_i. Thus, if a vertex w has a loop, then any adjoin pair containing w cannot be a triangulated pair.

Lemma 12. *For any two triangulated pairs in a d-page bookembedding of a d-regular line digraph $L(G)$ ($d \geq 3$), their triangulated arcs are disjoint.*

Proof. Let $[u, v]$ be a triangulated pair in a d-page bookembedding of a d-regular line digraph. From Lemma 11, u and v are compatible. Suppose that u and v are head-compatible. (The case u and v are tail-compatible can be similarly proved.)

Let $[x, y]$ be a triangulated pair different from $[u, v]$. Assume that $[u, v]$ and $[x, y]$ have a common triangulated arc (v, w) on a page P. Since $[x, y]$ has (v, w) as its triangulated arc, $|\{x, y\} \cap \{v, w\}| = 1$. (If $|\{x, y\} \cap \{v, w\}| = 2$, then either x or y must have a loop. In such a case, $[x, y]$ cannot be a triangulated pair.) Without loss of generality, we can assume that $|\{x\} \cap \{v, w\}| = 1$.

Case 1: (v, w) is an internal arc.

Case 1.1: $v = x$.

In this case, any two vertices of $\{u, v, y\}$ are head-compatible. Thus, for any page P_i, there is a vertex $w_i \in \Gamma^+(u)$ such that $(u, w_i), (v, w_i), (y, w_i)$ are on the page. In such a situation, we cannot assign any arc (z, v) incident to v without crossing on the page P_i unless $z = w_i$. Thus, $\Gamma^-(v) = \{w_0, w_1, \ldots, w_{d-1}\}$. (If $u \in \Gamma^-(v)$ or $y \in \Gamma^-(v)$, then v must have a loop.) Hence, there are d double arcs containing v. However, this implies that there are multiarcs in G, which contradicts our assumption.

Case 1.2: $w = x$ and $u \neq y$.

In this case, $(v, y) \in A(L(G))$ such that (v, y) is assigned to the page P since $[x, y]$ is a triangulated pair. Thus, we cannot use $(u, y), (v, y) \in A(L(G))$ as an undirected 2-path for $[u, v]$ on the other page unless (v, y) is a spinal arc. Therefore, (v, y) is a spinal arc. Similarly, it is induced that (u, x) is a spinal arc. Hence, $n_{L(G)} = 4$. However, $L(G)$ is a d-regular line digraph such that $d \geq 3$. Thus, this case produces a contradiction.

Case 1.3: $w = x$ and $u = y$.

In this case, $(v, u) \in A(L(G))$. This implies that there is a loop at u. Thus, $[u, v]$ cannot be a triangulated pair.

Case 2: (v, w) is a spinal arc.

Now, $w = x$. Since (v, w) is a triangulated arc of $[x, y]$, $(v, y) \in A(L(G))$. Thus, $(u, y) \in A(L(G))$. Here, (u, y) must be assigned to the page P because (u, w) is assigned to the page P. This means that $[u, v]$ and $[x, y]$ have (u, w) as a common triangulated arc. Therefore, this case is reduced to Case 1.2. □

Lemma 13. *In a d-page bookembedding $\mathcal{B}$ of a d-regular line digraph $L(G)$ ($d \geq 3$), the number of triangulated pairs is at most*

$$\frac{1}{2}(n_{L(G)} - 3) + \frac{1}{2d}(N_S(\mathcal{B}) + C_{L(G)}(2)).$$

Proof. Let T be the number of triangulated pairs. Since triangulated arcs of triangulated pairs are disjoint and each triangulated pair has two triangulated arcs on a page, there are at least $2dT$ arcs. The number of arcs except for loops in $\mathcal{B}$ is at most $d(n_{L(G)} - 3) + N_S(\mathcal{B}) + C_{L(G)}(2)$. Thus, the number of triangulated pairs is at most $\frac{1}{2}(n_{L(G)} - 3) + \frac{1}{2d}(N_S(\mathcal{B}) + C_{L(G)}(2))$. □

Corollary 14. *In a d-page bookembedding $\mathcal{B}$ of a d-regular line digraph $L(G)$ ($d \geq 3$), the number of non-triangulated pairs is at least*

$$\frac{1}{2}(n_{L(G)} + 3) - \frac{1}{2d}(N_S(\mathcal{B}) + C_{L(G)}(2)).$$

Using the results in Sections 3 and 4, and the above corollary, we can obtain a necessary condition for a d-page embeddable d-regular line digraph.

Corollary 15. *Let $L(G)$ be a d-regular line digraph $(d \geq 3)$. Suppose that $L(G)$ can be embedded in d pages. Then, $n_{L(G)} \leq X(d, L(G))$, where*

$$X(d, L(G)) = \left(\frac{18d^2+2}{d(d-2)}\right) C_{L(G)}(1) + \left(\frac{18d+1}{d-2}\right) C_{L(G)}(2)$$
$$+ \left(\frac{18d^2-31d-2}{d(d-2)^2}\right) C^*_{L(G)}(3) + \left(\frac{4}{(d-2)^2}\right) C^*_{L(G)}(5)$$
$$- \left(54d + 3 + \frac{72d+4}{d-2}\right).$$

Proof. Let $\mathcal{B}$ be a d-page bookembedding of $L(G)$. By Corollary 3, Lemma 9 and Corollary 14,

$$\begin{cases} N_S(\mathcal{B}) \leq \frac{2}{(d-2)}(C_{L(G)}(1) + C_{L(G)}(2) + \frac{1}{2}C^*_{L(G)}(3) - 2d), \\ N_I(\mathcal{B}) \leq d(n_{L(G)} - 3) - \frac{1}{9}N_{non} + \frac{2}{9(d-2)^2}(C^*_{L(G)}(3) + C^*_{L(G)}(5)), \\ N_{non} \geq \frac{1}{2}(n_{L(G)} + 3) - \frac{1}{2d}(N_S(\mathcal{B}) + C_{L(G)}(2)). \end{cases}$$

From these inequalities,

$$m_{L(G)} = N_S(\mathcal{B}) + N_I(\mathcal{B}) + C_{L(G)}(1) + C_{L(G)}(2)$$
$$\leq dn_{L(G)} - \frac{1}{18}n_{L(G)} + X(d, L(G)).$$

Since $m_{L(G)} = dn_{L(G)}$, the inequality in the proposition is obtained. $\square$

5.2 The number of small undirected cycles

Next, we show that the number of small undirected cycles can be bounded by the number of small cycles.

Lemma 16.

$$\begin{cases} C^*_{L(G)}(3) \leq (d-1)^2 C_{L(G)}(1) + C_{L(G)}(3), \\ C^*_{L(G)}(5) \leq (d-1)^2(2d^2 - 4d + 4)C_{L(G)}(1) + 3(d-1)^2 C_{L(G)}(3) + C_{L(G)}(5). \end{cases}$$

Proof. Consider an undirected 3-cycle $\langle v_0, v_1, v_2, v_0 \rangle$ in $L(G)$. There are two patterns (with respect to isomorphism) on the orientation of edges in $\langle v_0, v_1, v_2, v_0 \rangle$. One is the case that $\langle v_0, v_1, v_2, v_0 \rangle$ is corresponding to a cycle in $L(G)$. Clearly, the number of undirected 3-cycles in such a case is $C_{L(G)}(3)$.

Suppose that $\langle v_0, v_1, v_2, v_0 \rangle$ does not correspond to a cycle in $L(G)$. Then, from the properties of line digraphs, it is easily checked that one of $\{v_0, v_1, v_2\}$ has a loop. Let u be a vertex of $L(G)$ such that u has a loop. Then, for any $w^+ \in \Gamma^+(u)$ and any $w^- \in \Gamma^-(u)$, $(w^-, w^+) \in A(L(G))$. Thus, the number of undirected 3-cycles induced by u is $(d-1)^2$. Therefore, the number of undirected 3-cycles in this case is $(d-1)^2 C_{L(G)}(1)$. (Note that an undirected 3-cycle in $L(G)$ may have both two patterns since an edge in the undirected cycle may correspond to double arcs.)

Next, consider an undirected 5-cycle $\langle v_0, v_1, v_2, v_3, v_4, v_0 \rangle$ in $L(G)$. There are four patterns (with respect to isomorphism) on the orientation of edges in the undirected 5-cycle

1. $(v_0,v_1),(v_1,v_2),(v_2,v_3),(v_3,v_4),(v_4,v_0)$.
2. $(v_0,v_1),(v_1,v_2),(v_2,v_3),(v_3,v_4),(v_0,v_4)$.
3. $(v_0,v_1),(v_1,v_2),(v_2,v_3),(v_4,v_3),(v_0,v_4)$.
4. $(v_0,v_1),(v_1,v_2),(v_3,v_2),(v_3,v_4),(v_0,v_4)$.

An undirected 5-cycle isomorphic to the above undirected 5-cycle numbered by i is call an undirected 5-cycle of type i. Clearly, the number of undirected 5-cycles of type 1 is $C_{L(G)}(5)$.

An undirected 5-cycle of type 2 induces a 3-cycle. (For example, in the above undirected cycle numbered by 2, (v_1,v_2,v_3,v_1) is induced.) Consider a 3-cycle (x,y,z,x). Then, for any $w^- \in \Gamma^-(x)$ and any $w^+ \in \Gamma^+(z)$, $(w^-,w^+) \in A(L(G))$. Thus, there are $3(d-1)^2 C_{L(G)}(3)$ undirected 5-cycles of type 2.

An undirected 5-cycle induces an undirected 3-cycle. (In the above undirected cycle numbered by 3, there exists a vertex x with a loop such that $(v_1,x),(x,v_2) \in A(L(G))$.) Let x be a vertex with a loop. Let $y \in \Gamma^+(x)$ and $z \in \Gamma^-(x)$. Then, $\langle x,y,z,x \rangle$ forms an undirected 3-cycle. The number of such undirected 3-cycles is $(d-1)^2 C_{L(G)}(1)$ as shown before. For the undirected 3-cycle $\langle x,y,z,x \rangle$, there exists a vertex v such that for any $w^- \in \Gamma^-(z)$ and any $w^+ \in \Gamma^+(y)$, $(w^-,v),(v,w^+) \in A(L(G))$. Thus, there are d^2 undirected 5-cycles of type 3 which induces $\langle x,y,z,x \rangle$. Therefore, the number of undirected 5-cycles of type 3 is $d^2(d-1)^2 C_{L(G)}(1)$.

One of the vertices of an undirected 5-cycle of type 4 has a loop. (In the above undirected cycle numbered by 4, v_1 has a loop.) Let x be a vertex with a loop in $L(G)$. For any two vertices $y_1,y_2 \in \Gamma^+(x)$ and any two vertices $z_1,z_2 \in \Gamma^-(x)$, there are 4 undirected 5-cycles of type 4 whose vertices are x,y_1,y_2,z_1,z_2. Thus, there are $4\binom{d-1}{2}\binom{d-1}{2}$ undirected 5-cycles of type 4 containing the vertex x. Therefore, the number of undirected 5-cycles of type 4 is $(d-1)^2(d-2)^2 C_{L(G)}(1)$. □

From Corollary 15 and Lemma 16, we can obtain the following upper bound on the number of vertices of a d-page embeddable d-regular line digraph.

Corollary 17. *Let $L(G)$ be a d-regular line digraph ($d \geq 3$). Suppose that $L(G)$ can be embedded in d pages. Then, $n_{L(G)} \leq Y(d,L(G))$, where*

$$Y(d,L(G)) = \left(\frac{8d^5-14d^4+7d^3-6d^2-9d-6}{d(d-2)^2}\right) C_{L(G)}(1)$$
$$+ \left(\frac{18d+1}{d-2}\right) C_{L(G)}(2) + \left(\frac{12d^3-6d^2-19d-2}{d(d-2)^2}\right) C_{L(G)}(3)$$
$$+ \left(\frac{4}{(d-2)^2}\right) C_{L(G)}(5) - \left(54d+3+\frac{72d+4}{d-2}\right)$$

6 The pagenumber of de Bruijn and Kautz digraphs

In this section, we show that the pagenumber of d-regular iterated line digraphs ($d \geq 3$) is greater than d. From this result, the pagenumber of the d-ary de Bruijn digraphs (the d-ary Kautz digraphs) is determined. Also, we present a constant c such that if $D > c$, then $B(d,D)$ and $K(d,D)$ cannot be embedded in d pages.

Theorem 18. *Let G be a d-regular digraph ($d \geq 3$). Then there exists a constant c_G such that if $m > c_G$ then $L^m(G)$ cannot be embedded in d pages.*

Proof. As stated in Section 2, $C_{L^m(G)}(t) \leq N_G(t)$. (In particular, there exists a value p which depends on G and t such that if $m > p$, then $C_{L^m(G)}(t) = N_G(t)$ ([12]).) Let $Z(d,G)$ be the value obtained from $Y(d, L^m(G))$ by replacing $C_{L^m(G)}(t)$ with $N_G(t)$ for $t = 1, 2, 3, 5$. Then, $Y(d, L^m(G)) \leq Z(d, G)$. Note that $Z(d,G)$ is independent of m. From Corollary 17, if $L^m(G)$ is d-page embeddable, then $n_{L^m(G)} = d^m n_G \leq Y(d, L^m(G)) \leq Z(d, G)$. Thus, if $m > \log_d \frac{Z(d,G)}{n_G}$, then $L^m(G)$ cannot be embedded in d pages. □

It has been shown in [13] that $B(d, D)$ and $K(d, D)$ can be embedded in $(d+1)$ pages. From Theorem 18, there exists a constant c such that if $D > c$, then $B(d, D)$ ($K(d, D)$) cannot be embedded in d pages. Therefore, the following theorem is obtained.

Theorem 19. *For any $d \geq 3$, the pagenumber of the d-ary de Bruijn digraphs (the d-ary Kautz digraphs) is $(d+1)$.*

To see a constant c such that if $D > c$ then $B(d, D)$ (resp., $K(d, D)$) cannot be embedded in d pages, we need to know the number of nonperiodic closed walks in the complete digraph K_d° (resp., the complete symmetric digraph K_{d+1}^*).

If we know the cardinality of the set of closed walks of length k, then we can obtain the number of nonperiodic closed walks using the Möbius inversion formula. Let X_k be the cardinality of the set of closed walks of length k. (Note that each closed walk is treated as a sequence.) Let Y_k be the cardinality of the set of nonperiodic closed walks of length k. Any closed walk of length k consists of iterations of a nonperiodic closed walk of length l, where l is a divisor of k. Thus, $X_k = \sum_{l | k} Y_l$, where $l \mid k$ means that l divides k. Let N_l denote the number of nonperiodic closed walks of length l. Then $Y_l = l N_l$. (Note that the number of nonperiodic closed walks is the equivalence classes with respect to rotation.) Therefore,

$$X_k = \sum_{l \mid k} l N_l.$$

Using the Möbius inversion formula,

$$N_k = \frac{1}{k} \sum_{l \mid k} \mu\left(\frac{k}{l}\right) X_l,$$

where the Möbius function μ is defined as follows.

$$\mu(n) = \begin{cases} 1 & \text{if } n \text{ is a product of an even number of distinct primes,} \\ -1 & \text{if } n \text{ is a product of an odd number of distinct primes,} \\ 0 & \text{otherwise.} \end{cases}$$

Lemma 20.
$$\begin{cases} N_{K_d^\circ}(k) = \frac{1}{k}\sum_{l|k} \mu\left(\frac{k}{l}\right) d^l, \\ N_{K_{d+1}^*}(k) = \frac{1}{k}\sum_{l|k} \mu\left(\frac{k}{l}\right)(d^l + (-1)^l d). \end{cases}$$

Proof. The cardinality of the set of closed walks of length k in K_d° is d^k, since any sequence of vertices of length k corresponds to a closed walk in K_d°.

Let $X(k)$ denote the cardinality of closed walks of length k in K_{d+1}^*. A sequence in which there are consecutive identical vertices does not correspond to a walk in K_{d+1}^*. Thus, there are $(d+1)d^{k-1}$ sequences of vertices of length k corresponding to walks. In such a sequence, if the last vertex is different from the first vertex, then it corresponds to a closed walk of length k. Otherwise, it corresponds to a closed walk of length $(k-1)$. Conversely, any closed walk of length k or $(k-1)$ is counted in the $(d+1)d^{k-1}$ sequences. Therefore, $X(k) + X(k-1) = d^k + d^{k-1}$. It is clear that $X(1) = 0$. From this recurrence, it is induced that $X(k) = d^k + (-1)^k d$. □

Corollary 21.
$$\begin{cases} N_{K_d^\circ}(1) = d, \ N_{K_{d+1}^*}(1) = 0, \\ N_{K_d^\circ}(2) = \frac{1}{2}(d^2 - d), \ N_{K_{d+1}^*}(2) = \frac{1}{2}(d^2 + d), \\ N_{K_d^\circ}(3) = N_{K_{d+1}^*}(3) = \frac{1}{3}(d^3 - d), \\ N_{K_d^\circ}(5) = N_{K_{d+1}^*}(3) = \frac{1}{5}(d^5 - d). \end{cases}$$

Using the values in Corollary 21 as upper bounds on the number of cycles, we can compute a constant c such that if $D > c$ then $B(d, D)$ $(K(d, D))$ cannot be embedded in d pages.

Theorem 22. *For any $d \geq 3$, $B(d, D)$ and $K(d, D)$ cannot be embedded in d pages if $D > 6$.*

Acknowledgment

This research was partially supported by the Scientific Grant-in-Aid from Ministry of Education, Science, Sports and Culture of Japan.

References

1. J.-C. Bermond and P. Fraignaud, Broadcasting and gossiping in de Bruijn networks, *SIAM J. Comput.* 23 (1994) 212–225.
2. J.-C. Bermond and P. Hell, On even factorizations and the chromatic index of the Kautz and de Bruijn digraphs, *J. Graph Theory* 17 (1993) 647–655.
3. J.-C. Bermond, N. Homobono and C. Peyrat, Connectivity of Kautz networks, *Discrete Math.* 114 (1993) 51–62.

4. J.-C. Bermond and C. Peyrat, de Bruijn and Kautz networks: a competition for the hypercube? in: F. André, J.P. Verjus (Eds.), Hypercube and Distributed Computers, North-Holland, Amsterdam, 1989, pp. 279–293.
5. F. Bernhart and P.C. Kainen, The book thickness of a graph, *J. Combin. Theory Ser. B* 27 (1979) 320–331.
6. F.R.K. Chung, F.T. Leighton, and A.L. Rosenberg, Embedding graphs in books: a layout problem with application to VLSI design, *SIAM J. Algebraic Discrete Methods* 8 (1987) 33–58.
7. O. Collins, S. Dolinar, R. McEliece and F. Pollara, A VLSI decomposition of the deBruijn graph, *J. Assoc. Comput. Mach.* 39 (1992) 931–948.
8. A.H. Esfahanian and S.L. Hakimi, Fault-tolerant routing in de Bruijn communication networks, *IEEE Trans. Comput.* 34 (1985) 777–788.
9. M.A. Fiol, J.L.A. Yebra and I. Alegre, Line digraph iteration and the (d, k) digraph problem, *IEEE Trans. Comput.* 33 (1984) 400–403.
10. R.A. Games, Optimal book embeddings of the FFT, Benes, and barrel shifter networks, *Algorithmica* 1 (1986) 233–250.
11. Z. Ge and S.L. Hakimi, Disjoint rooted spanning trees with small depths in de Bruijn and Kautz graphs, *SIAM J. Comput.* 26 (1997) 79–92.
12. T. Hasunuma and Y. Shibata, Counting small cycles in generalized de Bruijn digraphs, *Networks*, 29 (1997) 39–47.
13. T. Hasunuma and Y. Shibata, Embedding de Bruijn, Kautz and shuffle-exchange networks in books, *Discrete Appl. Math.* 78 (1997) 103–116.
14. T. Hasunuma and H. Nagamochi, Independent spanning trees with small depths in iterated line digraphs, Australian Computer Science Communications, Vol. 20, No. 3, Springer-Verlag, pp. 21–35, (Proceedings of 4th Australasian Theory Symposium, CATS'98).
15. M. Konoe, K. Hagiwara and N. Tokura, On the pagenumber of hypercubes and cube-connected cycles, *IEICE Trans.* J71-D (3) (1988) 490–500 (in Japanese).
16. F.T. Leighton, *Introduction to Parallel Algorithms and Architectures: Array, Trees, Hypercubes*, Morgan Kaufmann Publishers, San Mateo, CA 1992.
17. S.M. Malitz, Genus g graphs have pagenumber $O(\sqrt{g})$, *J. Algorithms* 17 (1994) 85–109.
18. D.J. Muder, M.L. Weaver, and D.B. West, Pagenumber of complete bipartite graphs, *J. Graph Theory* 12 (1988) 469–489.
19. J.H. van Lint and R.M. Wilson, *A Course in Combinatorics*, Cambridge University Press, Cambridge 1992.
20. J.L. Villar, The underlying graph of a line digraph, *Discrete Appl. Math.* 37/38 (1992) 525–538.
21. M. Yannakakis, Embedding planar graphs in four pages, *J. Comput. System Sci.* 38 (1989) 36–67.

Formal Synthesis for Pipeline Design

Holger Hinrichsen, Hans Eveking, and Gerd Ritter

Department of Electrical and Computer Engineering,
Darmstadt University of Technology, D-64283 Darmstadt, Germany.
{hinrichsen,eveking,ritter}@rs.tu-darmstadt.de

Abstract. A method of formally correct synthesis is presented and applied to the automatic construction of pipelined processors. The approach is based on a small set of correctness-preserving transformations that are efficiently cross-checked by an independent formal verification tool. Basic pipeline strategies as well as automatic post-synthesis verification are provided.

1 Introduction

Pipelining is used in modern processor architectures to achieve a better throughput. The time of execution of one instruction is not reduced, but issuing the instructions in parallel reduces the time of computation of the whole program. The processing ought not to be speeded up by increasing the number of functional units but by improving the utilization rate of the given resources. The instructions are divided in several stages in order to overlap different stages of as many of instructions as possible. The instructions are issued in regular intervals, called latency. Fig. 1 illustrates this technique.

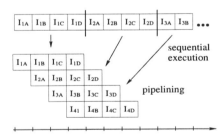

Fig. 1. Pipelining instructions with a latency of one.

In this paper we present a method of formally correct synthesis of pipelined architectures. From a sequential specification a pipeline system is derived that describes the behavior of the pipeline including its filling and its flushing. Since the whole process is automated, a quick design space exploration is possible. Formal synthesis (a survey is given by [KBES96]) uses correctness-preserving transformations to design circuits. In our approach, the transformational derivation is based on a general-purpose calculus. The synthesis tool possesses only a small set of axioms and rules to ensure the correctness of the transformations, because the more the core grows the more the security is reduced. Since faults creep in the implementation of the tools an external equivalence checker verifies the derived descriptions after each transformation as illustrated by Fig. 2. Although a post-synthesis verification is much more difficult than verification accompanying the transformation process, the equivalence checker also proves finally the correctness of the implementation with respect to the specification after synthesis.

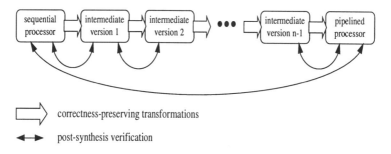

Fig. 2. Formally correct synthesis.

If a pipeline system is created, it has to be decided for a given set of resources at which time the next instruction can be issued. Since even quite simple scheduling problems are NP-complete, many heuristics exist to resolve such problems. Generally they are based on DFG's or CFG's.

Techniques as *As Soon As Possible (ASAP)* or as *As Late As Possible (ALAP)* [MLD92] order the operations according to their data dependencies represented by a DFG. If there are different mappings for an operation, the task is executed as soon as possible or as late as possible. In addition *List-Scheduling* [MLD92] uses the topological order given by a priority criterion or by the mobility of a task. The mobility of an operation is the difference between the earliest and the latest possible schedule. Jobs possessing a higher priority or a lower mobility are preferred. The pipeline synthesis tool *SEHWA* introduced by [PP88] for example uses a modified List-Scheduling to schedule an operation whenever all its predecessors have been scheduled and a functional unit of the proper type is available. The priority of the operations is considered by the order the operations are scheduled. Since the above mentioned methods do not deal with loops or branches they are used to schedule operations inside such constructions.

Path-Based Scheduling (PBS) also called *As Fast As Possible (AFAP)* [Cam90] processes loops and branches and maps the operations to the control states in such a manner that the best schedule is found independently for every possible paths of the corresponding CFG. The computed schedules are overlapped by solving the clique-covering problem. The application of *AFAP* is limited because the number of paths can explode for complex examples.

This paper presents *Case-Based Scheduling (CBS)*, a technique to formally construct a pipeline system. *CBS* uses scheduling techniques in order to determine which instructions or which instruction stages have to be executed in the pipeline in parallel under constraints given by the hardware. Therefore *CBS* is rather similar to *AFAP*. In contrast to *AFAP*, *CBS* is used to schedule instruction stages, e.g. instruction fetch or instruction decode. While *AFAP* computes all paths of the CFG, *CBS* computes the paths on which different hazards between different stages appear in the pipeline. *CBS* separates these paths by introducing new branches. According to *AFAP* each path is then scheduled with respect to the time.

Although the number of paths that must be considered is reduced, techniques such as forwarding or introduction of pipeline registers are applied to resolve dependencies between stages. Operations are only mapped to different control states either if they belong to different stages of the same instruction or if the dependencies cannot be resolved. Therefore *CBS* does not work with DFG's or CFG's because these graphs are not adequate to decide whether a conflict can be resolved or not. The presented technique works directly with the internal data structure of the description.

Before we present our technique to derive pipelines in Section 3 we define the core transformations of the synthesis tool. While no constraints given by the hardware are considered in the simple example of Section 3, the technique will be refined in Section 4 to be applied to the DLX-architecture of [HP96] in Section 5. The verification results of the examples are given in Section 6. A conclusion completes the paper.

2 Correctness preserving transformations

A language of synchronous transition systems called *Language of labelled Segments (LLS)* (for more details see [EHR98]) is used as an input language for formal synthesis and verification tools. The segments identify parts with acyclic flow of control and the labels are used to direct control between segments. Data is manipulated by conditional synchronously parallel assignments.

Many concepts are developed to describe the relation of a specification and its implementation. Two ideas of equivalence are defined as follows:

- Two descriptions $\mathcal{DES}_1$ and $\mathcal{DES}_2$ are *trace-equivalent* $\mathcal{DES}_1 \stackrel{\sim}{=}|_V \mathcal{DES}_2$, if all infinite runs coincide restricted to a subset V of common variables at each step of time.
- Two single LLS segments S_1 and S_2 are *computational equivalent* $S_1 \simeq|_V S_2$, if both produce the same final values on the same initial values of a subset V of common variables.

Correctness-preserving transformations are used to derive a pipeline system from an unpipelined specification. These transformations can be divided into two classes:

- transformations based on a theory of microprogram schemata [Glu65] which maintain *trace-equivalence* $\widetilde{\cong}|_V$ and
- content-dependent transformations which preserve the correctness according to *computational equivalence* $\simeq|_V$.

2.1 Formal transformation of microprograms

The transformations of this class guarantee the *trace-equivalence* between the original description $\mathcal{DES}$ and the modified one. Note that using these transformations does not change the timing properties of the description $\mathcal{DES}$.

$$\text{A0}: \quad \mathcal{DES}\,(\text{a;}) \;\cong\; \mathcal{DES}\begin{pmatrix} \text{IF p} \\ \text{THEN a;} \\ \text{ELSE a;} \end{pmatrix}$$

$$\text{A1}: \quad \mathcal{DES}\begin{pmatrix} \text{IF p} \\ \text{THEN a;} \\ \text{ELSE b;} \end{pmatrix} \;\cong\; \mathcal{DES}\begin{pmatrix} \text{IF } \neg\text{p} \\ \text{THEN b;} \\ \text{ELSE a;} \end{pmatrix}$$

$$\text{A2}: \quad \mathcal{DES}\begin{pmatrix} \text{IF p} \\ \quad \text{THEN IF q} \\ \quad\quad\quad\quad \text{THEN a;} \\ \quad\quad\quad\quad \text{ELSE b;} \\ \text{ELSE c;} \end{pmatrix} \;\cong\; \mathcal{DES}\begin{pmatrix} \text{IF (p}\wedge\text{q)} \\ \text{THEN a;} \\ \text{ELSE IF (p}\wedge\neg\text{q)} \\ \quad\quad\quad \text{THEN b;} \\ \quad\quad\quad \text{ELSE c;} \end{pmatrix}$$

$$\text{A3}: \quad \mathcal{DES}\begin{pmatrix} \text{IF p} \\ \text{THEN a;} \\ \text{ELSE IF q} \\ \quad\quad\quad \text{THEN b;} \\ \quad\quad\quad \text{ELSE c;} \end{pmatrix} \;\cong\; \mathcal{DES}\begin{pmatrix} \text{IF p}\vee\text{q} \\ \text{THEN IF p} \\ \quad\quad\quad \text{THEN a;} \\ \quad\quad\quad \text{ELSE b;} \\ \text{ELSE c;} \end{pmatrix}$$

$$\text{A4}: \quad \mathcal{DES}\begin{pmatrix} \text{IF p}\wedge\text{q} \\ \text{THEN a;} \\ \text{ELSE b;} \end{pmatrix} \;\cong\; \mathcal{DES}\begin{pmatrix} \text{IF p} \\ \text{THEN IF q} \\ \quad\quad\quad \text{THEN a;} \\ \quad\quad\quad \text{ELSE b;} \\ \text{ELSE b;} \end{pmatrix}$$

$$\text{A5}: \quad \mathcal{DES}\begin{pmatrix} \text{IF p}\vee\text{q} \\ \text{THEN a;} \\ \text{ELSE b;} \end{pmatrix} \;\cong\; \mathcal{DES}\begin{pmatrix} \text{IF p} \\ \text{THEN a;} \\ \text{ELSE IF q} \\ \quad\quad\quad \text{THEN a;} \\ \quad\quad\quad \text{ELSE b;} \end{pmatrix}$$

$$\text{A6}: \quad \mathcal{DES}\begin{pmatrix} \text{IF p} \\ \text{THEN a;} \\ \text{ELSE b;} \\ \text{c;} \end{pmatrix} \;\cong\; \mathcal{DES}\begin{pmatrix} \text{IF p} \\ \text{THEN a;c;} \\ \text{ELSE b;c;} \end{pmatrix}$$

$$\text{A7}: \quad \mathcal{DES}\begin{pmatrix} \text{a;} \\ \text{IF p} \\ \text{THEN b;} \\ \text{ELSE c;} \end{pmatrix} \;\cong\; \mathcal{DES}\begin{pmatrix} \text{IF }\{a\}\text{p} \\ \text{THEN a;b;} \\ \text{ELSE a;c;} \end{pmatrix}$$

$$\text{A8}: \quad \mathcal{DES}\begin{pmatrix} \text{IF p} \\ \text{THEN a;} \\ \text{ELSE b;} \end{pmatrix} \;\cong\; \mathcal{DES}\begin{pmatrix} \text{IF p} \\ \text{THEN IF p} \\ \quad\quad\quad \text{THEN a;} \\ \quad\quad\quad \text{ELSE c;} \\ \text{ELSE b;} \end{pmatrix}$$

The expression $\{a\}p$ of axiom [**A7**] is called a virtual predicate and expresses that the condition p has to be checked on condition that a is virtually performed. If the sequence a does not change any values of variables of p, the virtual predicate $\{a\}$ can be dropped. Otherwise the common variables have to be substituted in p by the assigned expressions found in a.

Axioms [**A0**] to [**A8**] make it possible to move and modify sequences inside a segment. On the other hand the rule R1 states that the Label L of a segment S is equivalent to the segment itself. Therefore they can substitute themselves.

R1 : $\dfrac{L:S}{\mathcal{DES}(L) \cong \mathcal{DES}(S)}$

The correct application can be cross-checked by the transformation of the original description and the transformed one into fully labelled form and the proof of a bisimulation relation [EHR98].

2.2 Content-dependent transformations

The axioms of Section 2.1 do not change the timing properties of the transformed descriptions. The transformations of the second class permit to speed up descriptions by manipulating the assignments. Content-dependent transformations are used to parallelize or to serialize sequences of assignments. They are correct according to the above defined *computational equivalence*.

The following sets, functions and styles are used to define the transformations:

sets:
- $\mathcal{Z}$ as set of assignments
- $\mathcal{V}$ as set of variables
- $\mathcal{EX}$ as set of all correct expressions using variables of $\mathcal{V}$
- $\mathcal{D} \subseteq \mathcal{V}$ as set of all variables used as destination of an assignment
- $\mathcal{S} \subseteq \mathcal{V}$ as set of all variables used as source of an assignment

functions:
- $A : \mathcal{Z} \times \mathcal{D} \to \mathcal{Z}$ to compute all assignments of set $\mathcal{Z}$ with destinations found in $\mathcal{D}$
- $D : \mathcal{Z} \to \mathcal{V}$ to compute all variables used as destination of assignments of set $\mathcal{Z}$
- $S : \mathcal{Z} \to \mathcal{V}$ to compute all variables used as source of assignments of set $\mathcal{Z}$
- $Q : \mathcal{Z} \times \mathcal{D} \to \mathcal{EX}$ to compute all expressions assigned by assignments of $\mathcal{Z}$ with destinations of $\mathcal{D}$

styles:
- $\mathcal{M}//b \mapsto a \hat{=}$ expression b substitutes a in set $\mathcal{M}$
- (a,b) denotes the parallel and $a;b$ the sequential composition

If a set of assignments $\mathcal{Z}_2$ should be executed in parallel to $\mathcal{Z}_1$ in order to change the sequence $\mathcal{Z}_1;(\mathcal{Z}_2,\mathcal{Z}_3);$ to $(\mathcal{Z}_1,\mathcal{Z}_2);\mathcal{Z}_3;$ the Bernstein rules [Ber66] have to be observed:

- $D(\mathcal{Z}_1) \cap D(\mathcal{Z}_2) = \emptyset$
- $D(\mathcal{Z}_1) \cap S(\mathcal{Z}_2) = \emptyset$
- $D(\mathcal{Z}_2) \cap S(\mathcal{Z}_3) = \emptyset$

If one of these conditions does not hold the conflict can be resolved by one or more of the following techniques:

C1 Elimination of redundant assignments :
$\forall v \in D(\mathcal{Z}_1) \cap D(\mathcal{Z}_2) : \mathcal{Z}_1; (\mathcal{Z}_2, \mathcal{Z}_3); \simeq (\mathcal{Z}_1, \mathcal{Z}_2'); \mathcal{Z}_3;$
with $\mathcal{Z}_2' = \mathcal{Z}_2 \backslash A(\mathcal{Z}_2, \{v\})$

C2 Forwarding :
$\forall v \in D(\mathcal{Z}_1) \cap S(\mathcal{Z}_2) : \mathcal{Z}_1; (\mathcal{Z}_2, \mathcal{Z}_3); \simeq (\mathcal{Z}_1, \mathcal{Z}_2'); \mathcal{Z}_3;$
with $S(\mathcal{Z}_2') = S(\mathcal{Z}_2) // Q(\mathcal{Z}_1, \{v\}) \mapsto v$

C3 Introduction of pipeline registers :
$\forall v \in D(\mathcal{Z}_2) \cap S(\mathcal{Z}_3) : \mathcal{Z}_1; (\mathcal{Z}_2, \mathcal{Z}_3); \simeq (\mathcal{Z}_1, \mathcal{Z}_2'); \mathcal{Z}_3';$
with $\mathcal{Z}_2' = \mathcal{Z}_2 \cup \{p_v \leftarrow v\}$ and $S(\mathcal{Z}_3') = S(\mathcal{Z}_3) // p_v \mapsto v$

An example of the transformations is given by Fig. 3 where the goal is to push y←x*a to the preceding assignments.

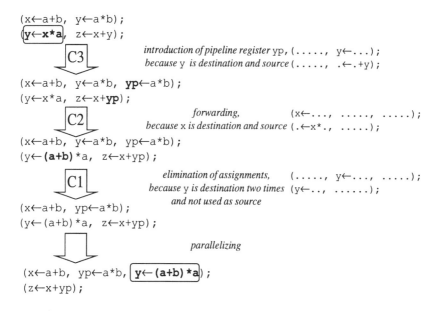

Fig. 3. Parallelizing using the transformations [C1-C3].

According to the techniques explained above more transformations (for more details see [Hin98]) are defined to make it possible that a set of assignments $\mathcal{Z}_2$ can be parallelized with $\mathcal{Z}_3$ to transform $(\mathcal{Z}_1, \mathcal{Z}_2); \mathcal{Z}_3;$ to $\mathcal{Z}_1; (\mathcal{Z}_2, \mathcal{Z}_3);$.

3 Basic pipeline techniques

3.1 Creating a pipeline system

The description of the specification from which the pipeline is derived must be given in a particular format illustrated by Fig. 4. The description contains a single segment L_0 consisting of one IF-sequence I_0. The instruction set is given by the sequences S_0^E to S_n^E of the ELSE-part I_0^E, whereby the sequence S_i^E represents the i-th instruction stage. The operations performed by different instructions in the same stage are combined to one instruction stage. The label L_0 at the end of the ELSE-part I_0^E directs the control to the beginning of the segment. The termination of the execution of the processor is performed by the THEN-part I_0^T comprising the sequences S_0^T to S_m^T.

$$
\begin{array}{ll}
L0: & \text{IF FLUSH THEN} \\
& \quad S_0^T; ...; S_m^T; \\
& \quad L_{end}; \\
& \text{ELSE} \quad S_0^E; ...; S_n^E; \\
& \quad L0;
\end{array}
$$

Fig. 4. The initial segment.

Algorithm *CBS* creates a pipeline out of the specification, Fig. 6 gives an example.

Algorithm 1. *CASE-BASED SCHEDULING*

INPUT *the initial segment L_0*

1. **LET** $x = 0$;
2. **WHILE** $|I_x^E| > 1$ **DO**
3.   *Derive a new segment L_{x+1} that receives the sequences S_{x+1}^E to S_n^E of I_x^E so that $S_{0\,to\,x}^E$ remains in I_x^E* **[R1]**;
4.   *Replace the label L_x at the end of the segment L_{x+1} by the corresponding sequences in order to get a segment containing the sequences S_{x+1}^E to S_n^E and an IF-sequence I_{x+1} that satisfies $I_{x+1} \equiv I_x$* **[R1]**;
5.   *Move the sequences S_{x+1}^E to S_n^E into the IF-sequence I_{x+1}* **[A7]**;
6.   *Move the sequence $S_{0\,to\,x}^E$, situated at the end of the ELSE-part I_{x+1}^E, in front of the sequences S_{x+2}^E to S_n^E and parallelize it with S_{x+1}^E, situated at the beginning of the ELSE-part I_{x+1}^E of I_{x+1}* **[C1-C3]**;
7.   *If it is not the first iteration parallelize the sequences $S_{i=0}^T$ to $S_{i=n-x-1}^T$ and S_{x+1}^E to S_n^E in pairs* **[C1-C3]**;
8.   **LET** $x = x + 1$;
9. **ENDWHILE**;

RETURN *segments L_0 to L_n*;

In line 3 a new segment L_{x+1} is generated that gets the sequences S_{x+1}^E to S_n^E from the ELSE-part I_x^E of the IF-sequence I_x in L_x. The sequence $S_{0\,to\,x}^E$ remaining in I_x^E performs $x+1$ stages of instructions in parallel and describes the filling of the pipeline. It represents the sequences S_0^E to S_x^E that are executed in parallel. The exit label L_x in the segment L_{x+1} is replaced by the corresponding sequences (line 4) in order to move the sequences of L_{x+1} into the IF-sequence I_{x+1} (line 5). The sequence is derived by parallelizing in line 6 that executes

$x+2$ stages of instructions in parallel. The loop (lines 2-9) terminates because each iteration reduces the number of sequences in I_{x+1}^E by one. The iteration of this loop describes the filling of the pipeline. If only one sequence remains in segment L_n the pipeline state is reached. Fig. 5 illustrates the gradual filling.

$$I_0^E : S_0^E ; S_1^E ; ... ; S_{n-1}^E ; S_n^E ; \overset{1.}{\Longrightarrow} I_1^E : (S_0^E, S_1^E); ... ; S_{n-1}^E ; S_n^E ; \overset{2.}{\Longrightarrow} ... \overset{i-th}{\Longrightarrow}$$

$$I_i^E : (S_0^E, S_1^E, ..., S_i^E); ... ; S_{n-1}^E ; S_n^E ; \overset{i+1-th}{\Longrightarrow} ... \overset{n-th}{\Longrightarrow} I_n^E : (S_0^E, S_1^E, ..., S_i^E, ..., S_{n-1}^E, S_n^E);$$

Fig. 5. Iteration over the sequences of the ELSE-part I_i^E.

The THEN-part I_{x+1}^T describes the flushing of the pipeline. In line 5 the sequences S_{x+1}^E to S_n^E are moved into the IF-sequence I_{x+1}. These sequences are required to terminate another instruction. One sequence less is provided in each iteration of the loop (lines 2-9) because the pipeline is filled about one sequence more. In line 7 the added sequences are pipelined by parallelizing with $S_{i=0}^T$ to

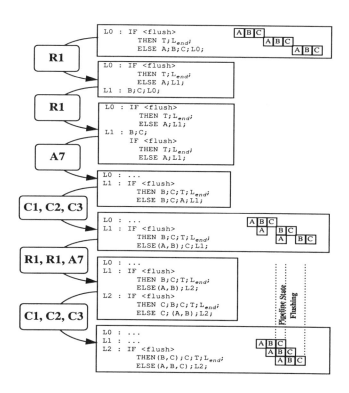

Fig. 6. Creating a pipeline system.

$S^T_{i=n-x-1}$, the first $n-x$ sequences of I^T_{x+1}.

The Fig. 6 illustrates the derivation of a pipeline consisting of three states. Note that in the first iteration in line 7 no transformations are executed because in the case of flushing only one instruction has to be terminated. In the next iteration transformations are applied also in step 7 because the segment L_2 describes the flushing of two instructions.

3.2 A simple example of a pipeline with three stages

In [Cyr96] a simple example of a pipeline with three stages is given and its correctness concerning the specification is shown:

```
L0 :
    (IR<-MEM[BITINT(PC);],
    PC<-INC(PC));
    IF ~IR[0]=0B1 THEN
        (WBREG<-RF[BITINT(IR[6:10]);]);
        (RF[BITINT(IR[1:5]);]<-WBREG);
    ENDIF;
    L0;
```

After the first iteration of the algorithm *CBS* the first and the second stage of the instruction are executed in parallel. A pipeline register P0 is also introduced to store the value of the instruction register IR. In the next iteration the first two stages are overlapped with the third one. L2 describes only the full pipeline state. For a shorter presentation the parts describing the filling and flushing are omitted.

```
L2 : (IR<-MEM[BITINT(PC);],
    PC<-INC(PC),
    IF ~(IR[0]=0B1)&~(~(P0[0]=0B1)&(P0[1:5]=IR[6:10])) THEN
        WBREG<-RF[BITINT(IR[6:10]);];
    ENDIF,
    P0<-IR,
    IF ~(P0[0]=0B1) THEN
        RF[BITINT(P0[1:5]);]<-WBREG;
    ENDIF);
    L2;
```

The synthesis tool produces the result in 18 steps. It takes about 2 seconds to compute the result (300 MHz, SUN Ultra2). As a consequence of the transformation process the description encloses the filling as well as the flushing of the pipeline. Therefore abstraction mappings used by [Cyr96] to verify the pipeline are not required to get a correctness proof by an equivalence prover.

4 Advanced strategies for pipelining

4.1 Possible constraints

The transformations are applied to hardware descriptions. Therefore the available resources restrict the application of them. If, for example, only one adder is provided then the assignments (pc←add(pc,1)); (c←add(a,b)); cannot be parallelized.

The following constraints can be defined by the user of the synthesis tool:

- the number of functional units and memories,
- the number of parallel read and write accesses to memories,
- the connections between registers and memories,
- the latency, and
- certain address modes of the memories can be forbidden.

4.2 Pipelining under constraints

Since dependencies between instructions are resolved by techniques like forwarding, it is possible that some of the given constraints, introduced in Section 4.1, cannot be satisfied after parallelizing. Therefore the steps 6 and 7 of the algorithm CBS must be refined in order to permit that the stages are parallelized only if the dependencies cause no violations. These transformations cannot be

Fig. 7. Introduction of an IF-sequence $I_{conflict}$ with undefined condition.

executed, if a violation of the constraints is detected while transforming the sequences S_0^{part} to S_k^{part} given by I_x^{part} of an IF-sequence I_x . A new IF-sequence $I_{conflict}$ is introduced and replaces S_0^{part} to S_k^{part} in order to copy these sequences in the THEN- as well as in the ELSE-part of the new IF-sequence $I_{conflict}$ according to axiom [**A0**].

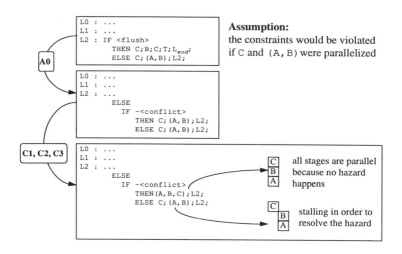

Fig. 8. Resolving violations of constraints.

Then the condition $C_{no\,conflict}$ of the new IF-sequence $I_{conflict}$ must be determined in a way that it is possible to parallelize the sequences in the THEN-part $I^T_{conflict}$ observing the specified constraints. $C_{no\,conflict}$ is composed of the conditions that cause hazards in the pipeline. Section 4.3 explains how $C_{no\,conflict}$ is constructed.

The THEN- and ELSE-parts of $I_{conflict}$ contain also branches with conditions becoming increasingly complex applying [**C1-C3**]. The conditions can be simplified or even decided using $C_{no\,conflict}$ or $\sim C_{no\,conflict}$ respectively. CBS uses binary decision diagrams [Bry86] for this.

The sequences of the THEN-part $I^T_{conflict}$ can be transformed after the simplification without violating the constraints because this part describes the behavior of the pipeline without hazards. On the other hand, the sequences in the ELSE-part are transformed so that the constraints are not violated at the cost of stalling the pipeline. Fig. 8 illustrates how Fig. 6 has to be changed if a hazard appears.

4.3 Deriving conditions for conflict resolution

The condition $C_{no\,conflict}$ is chosen by transforming S_0^{part} to S_k^{part} and fetching every condition that implies a violation. These conditions are negated and combined by conjunction so that the conflicts cannot appear anymore if the resulting condition is valid. Fig. 9 gives an example.

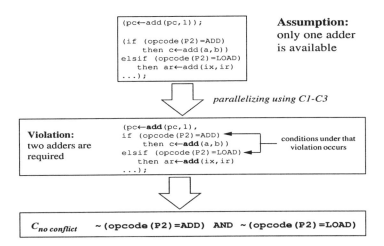

Fig. 9. Constructing the condition.

Note that forwarding must be applied to a condition which cause a conflict while shifting the condition along all preceding sequences. The conflict is caused by variables that are used as destinations of assignments and that can be found in the condition. Therefore, the common variables have to be substituted in the condition by the source expressions of the assignments (cf. [**C2**]). In Fig. 10, for example, the condition opcode(P2)=ADD is transformed to opcode(IR)=ADD.

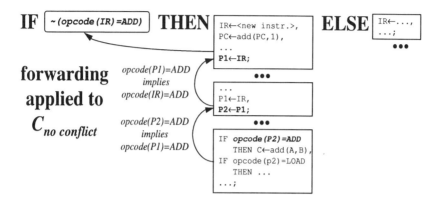

Fig. 10. Forwarding applied to the detected condition.

5 The DLX-processor

The DLX architecture of [HP96] is used as an example of the refined algorithm. The architecture is much more complex than the simple example of Section 3.2. We choose this processor in order to illustrate the main features of RISC architectures like MIPS 2000, MIPS R3000, Intel i860 and Motorola M88000. Several pipelined implementations of the DLX architecture were verified, for instance in [BD94,Cyr96,BM97], and a proof of a superscalar DLX version is given in [Bur96].

5.1 Structural, data and control hazards

Since *CBS* does not add any functional units, the pipeline will stall to resolve structural hazards. Therefore the number of structural hazards is minimized by applying simple scheduling techniques while computing the pipeline stages.

Overlapping dependent instructions causes data and control hazards in the pipeline. The conflicts are resolved as follows:

- introduction of pipeline registers
- forwarding or
- stalling the pipeline.

While the pipeline system is derived by the synthesis tool, pipeline registers are introduced automatically in order to prevent violations of the Bernstein rules [Ber66]. Analogous to the introduction of pipeline registers, *CBS* applies forwarding if parallelizing would be rejected.

Data or control hazards cannot be resolved by forwarding if the required results are not available or, in other words, forwarding neglects the given constraints of the resources. *CBS* refuses to forward any results and stalls the pipeline until the data are available for the waiting instructions.

5.2 Performance of the DLX pipeline

The measure of *Clock cycles per Instructions (CPI)* indicates how many cycles the execution of an instruction takes on average. The difficulty of the measure arises by calculating the probability of an instruction because their probabilities depend on the considered programs.

The ideal CPI of a pipeline is one because no stalls are performed and in each stage a new instruction is issued. In real systems the average number of stalls per instruction must be added. But different programs will cause different numbers of stalls. If the code is optimized by techniques as loop unrolling, software pipelining or trace scheduling [HP96] the number of stalls can be reduced considerably. Therefore only an estimation for the DLX based on the figures given by [HP96] is presented.

For the unpipelined DLX a CPI of 3.88 is computed. The pipeline derived by *CBS* satisfies a CPI of

$$1.08 \leq \text{CPI}_{\text{PIPE}} \leq 1.28$$

and a speedup of

$$3.03 = \frac{3.88}{1.28} \leq \text{SPEEDUP} \leq \frac{3.88}{1.08} = 3.59$$

depending on the considered program.

As a summary, the transformation system synthesized automatically a pipelined version of the DLX, whose stall-behavior is equivalent to the pipeline of [HP96]. The cpu-time for the transformation process was 13 minutes (300 MHz, SUN Ultra2). Most of the time was spent for detecting and resolving resource-conflicts and for simplification of the description.

6 Independent post-synthesis verification

Although our synthesis method is based on sound mathematical concepts, the derived pipelines are verified by an independent equivalence checker for ground-equational logic with uninterpreted functions in order to preclude any faults caused by the *implementation* of the synthesis tool. The verification tool detected some bugs of our synthesis tool (for more details see [Hin98]).

The proof is performed according to the technique introduced by [BD94]. Since the description of the automatically synthesized pipeline contains the filling and the flushing of the pipeline no additional functions have to be computed as in [BD94] and no abstraction function has to be provided as in [Cyr96]. The proof is executed fully machine assisted.

While deriving a pipeline from a specification all possible combinations of parallel instruction stages are computed. The correctness has to be proved for all possible paths through these instruction stages to verify the result. The equivalence checker verifies all possible combinations of instructions and hazards in the pipeline. The number of paths depends on the order of conditions in the description. The proof of correctness has been provided successfully for the three stage pipeline of Section 3.2 in 0.94 seconds. The cpu-time for the fully automatic formal verification of all segments of the DLX pipeline was 46 minutes (300 MHz, SUN Ultra2) (for more details see [EHR98]).

7 Conclusions

Although most high-level synthesis tools are based on DFG's or CFG's in order to avoid over-specifications, it was demonstrated that the approach of combining a theory of program schemata with techniques of forwarding and pipelining, which works directly on algorithmic descriptions, allows for the automated synthesis of very complex pipelined processor schedules. The method is combined with current powerful verification tools like decision-diagram packages and provers for ground-equational logic with uninterpreted functions. This allows for formally correct synthesis, which is a mathematically precise construction technique, and permits also cross-checks by independent verifiers to protect against implementation errors of the synthesis tool.

Acknowledgement

The authors would like to thank the anonymous reviewers for helpful comments.

References

[Ber66] A.J. Bernstein, "Analysis of programs for parallel processing", IEEE Trans. on Electronic Computers, VOL. EC-15, No. 5, pages 757-763, 1966
[BM97] E. Börger, S. Mazzanti, "A Practical Method for Rigorously Controllable Hardware Design", in J.P. Bowen, M.G. Hinchey, D. Till, eds., ZUM'97: The Z Formal Specification Notation, Springer LNCS 1212, pages 151-187, 1997
[Bry86] R.E. Bryant, "Graph-Based Algorithms for Boolean Function Manipulation", IEEE Trans. on Computers, Vol. C-35, No. 8, pages 6777-691,1986
[Bur96] J.R. Burch, "Techniques for verifying superscalar microprocessors", Proc. DAC'96, pages 552-557, 1996
[BD94] J.R. Burch, D.L. Dill, "Automatic verification of pipelined microprocessor control", Proc. Computer Aided Verification '94, Springer LNCS 818, 1994
[Cam90] R. Camposano, "Path-Based Scheduling for Synthesis", IEEE Trans. on CAD, Vol. 10, No. 1, pages 85-93, 1991
[Cyr96] D. Cyrluk, "Inverting the abstraction mapping: a methodology for hardware verification", Proc. FMCAD'96, Springer LNCS 1166, 1996
[EHR98] H. Eveking, H. Hinrichsen, G. Ritter, "Formally Correct Construction of Pipelined Processors", Technical Report, Darmstadt University of Technology, Dept. of Electrical and Computer Engineering, 1998

[Glu65] V.M. Glushko, "*Automata theory and formal microprogram transformations*", Kibernetika, Vol. 1, No. 5, pages 1-9, 1965
[HP96] J.L. Hennessy, D.A. Patterson, "*Computer Architecture: a Quantitative Approach*", Morgan Kaufman Publisher, second edition, 1996
[Hin98] H. Hinrichsen, "*Formally correct construction of a pipelined DLX architecture*", Technical Report, Darmstadt University of Technology, Dept. of Electrical and Computer Engineering, 1998
[KBES96] R. Kumar, C. Blumenröhr, D. Eisenbiegler, D. Schmid, "*Formal synthesis in circuit design - a classification and survey*", in Proc. FMCAD'96, Springer LNCS 1166, 1996
[MLD92] P. Michel, U. Lauther, P. Duzy, "*The synthesis approach to digital system design*", Kluwer Academic Publishers, 1992
[PP88] N. Park, A.C. Parker, "*Sehwa: a software package for synthesis of pipelines from behavioral specifications*", IEEE Trans. on CAD, Vol. 7, No. 3, pages 356-370, 1988

An $O(n \log n)$ Algorithm for Computing all Maximal Quasiperiodicities in Strings

Costas S. Iliopoulos[*,1,2] and Laurent Mouchard[**,3]

[1] Dept. of Computer Science, King's College London, Strand, London, U.K.
[2] School of Computing, Curtin University, Perth, WA, Australia.
csi@dcs.kcl.ac.uk
[3] LIFAR - ABISS, Université de Rouen, France. lm@dir.univ-rouen.fr

Abstract. A typical regularity, the period u of a given string x, grasps the repetitiveness of x since x is a prefix of a string constructed by concatenations of u. A factor w of x is called a *cover* of x and x is called *quasiperiodic* if x can be constructed by concatenations and superpositions of w. The notion "cover" is a generalization of periods in the sense that superpositions as well as concatenations are considered to define it, whereas only concatenations are considered for periods. It is shown here that all maximal quasiperiodic factors of a string x of length n can be detected in time $O(n \log n)$, improving the worst-case time upper bound achieved by the Apostolico and Ehrenfeucht algorithm for the same problem by a factor of $O(\log n)$.

1 Introduction

In recent study of repetitive structures of strings, generalized notions of periods have been introduced. A typical regularity, the period u of a given string x, grasps the repetitiveness of x since x is a prefix of a string constructed by concatenations of u. A factor w of x is called a *cover* of x if x can be constructed by concatenations and superpositions of w. A factor w of x is called a *seed* of x if there exists a superstring of x which is constructed by concatenations and superpositions of w. For example, *abc* is a period of *abcabcabca*, *abca* is a cover of *abcabcaabca*, and *abca* is a seed of *abcabcaabc*. The notions "cover" and "seed" are generalizations of periods in the sense that superpositions as well as concatenations are considered to define them, whereas only concatenations are considered for periods. A variant of the covering problem (see [DS96]) studied here, was shown to have applications to DNA sequencing by hybridization using oligonucleotide probes.

In computation of covers, two problems have been considered in the literature. The *shortest-cover* problem (also known as the superprimitivity test) is

---
[*] Partially supported by the SERC grant GR/J17944, the NATO grant CRG 900293, the 7141 ESPRIT BRA grant for ALCOM II, and the MRC grant G 9115730.
[**] Partially supported by the Conseil Régional de Haute-Normandie and Programme Génomes (CNRS).

that of computing the shortest cover of a given string of length n, and the *all-covers* problem is that of computing all covers of a given string. Apostolico et al in [AFI91] introduced the notion of covers and gave a linear-time algorithm for the shortest-cover problem. Breslauer [Bre92] presented a linear-time on-line algorithm for the same problem. Moore and Smyth [MS94] presented a linear-time algorithm for the all-covers problem. In parallel computation, Breslauer [Bre92] gave two algorithms for the shortest-cover problem. The first one is an optimal $O(\alpha(n) \log \log n)$-time algorithm, where $\alpha(n)$ is the inverse Ackermann function, and the second one is a non-optimal algorithm that requires $O(\log \log n)$ time and $O(n \log n)$ work. Breslauer [Bre92] also obtained an $\Omega(\log \log n)$ lower bound on the time complexity of the shortest-cover problem from the lower bound of string matching. Iliopoulos and Park [IP94] gave an optimal $O(\log \log n)$-time (thus work-time optimal) algorithm for the shortest-cover problem. Li and Smyth [LS98] gave a $O(n)$ algorithm for finding all the covers of all prefixes of a string x.

Iliopoulos, Moore and Park [IMP93] introduced the notion of seeds and gave an $O(n \log n)$-time algorithm for computing all seeds of a given string of length n. For the same problem Ben-Amram, Berkman, Iliopoulos and Park [BBIP94] presented a parallel algorithm that requires $O(\log n)$ time and $O(n \log n)$ work.

Apostolico and Ehrenfeucht [AE93] considered another variant of the covering problem; in [AE93] they presented an $O(n \log^2 n)$ algorithm for finding all maximal quasiperiodic factors (local covers) of a given string, i.e. find all longest coverable factors of a word. Informally, quasiperiodic factor z is maximal, if no extension z could be covered by either the same word w covering z or by an extension wa of w. The algorithm in [AE93] shadows the Apostolico and Preparata [AP83] algorithm for detection of all squares in a string. It is not difficult to see the association between the two problems: the starting position of every quasiperiodic factor is also the starting position of a square. For example, uvuvu is a quasiperiodic factor of uuuvuvuw and the starting position of uvuvu is also the starting position of the square uvuv.

Our approach, in a similar fashion to the Apostolico and Ehrenfeucht algorithm, also shadows another square detection algorithm. Crochemore in [Cro81] presented an $O(n \log n)$ algorithm for detecting all repetitions of a given string. Crochemore's method was based on set partitioning together with "the smaller half" trick that speeded up the method from $O(n^2)$ to $O(n \log n)$ time. Our method is using the partitioning "blocks" of Crochemore's algorithm together with several data structures for gap monitoring (a gap between two occurrences of a string w in x is the factor between two consecutive occurrences of w). Here we present an algorithm that detects all maximal quasiperiodic factors of a string in time $O(n \log n)$. Hence this improves the Apostolico and Ehrenfeucht time upper bound for the same problem by a factor of $O(\log n)$. The computation of maximal quasiperiodicities has direct application in music analysis (see [CIR98]) and molecular biology (see [FLSS92,MJ93,KMGL88]).

In Section 2 we present basic definitions and facts as well as previously known results which we build our algorithm upon. In Section 3 we describe Crochemore's partitioning, in Section 4 we present the main data structures used and in Section 5 we present the fast-local-covers algorithm. In Section 6 we conclude.

2 Preliminaries

A *string* is a sequence of zero or more symbols from an alphabet Σ. The set of all strings over the alphabet Σ is denoted by Σ^*. A string x of length n is represented by $x_1 \cdots x_n$, where $x_i \in \Sigma$ for $1 \leq i \leq n$. A string w is a *factor* of x if $x = uwv$ for $u, v \in \Sigma^*$; we equivalently say that the string w occurs at position $|u| + 1$ of the string x. A string w is a *prefix* of x if $x = wu$ for $u \in \Sigma^*$. Similarly, w is a *suffix* of x if $x = uw$ for $u \in \Sigma^*$.

The string xy is a *concatenation* of two strings x and y. The concatenations of k copies of x is denoted by x^k. For two strings $x = x_1 \cdots x_n$ and $y = y_1 \cdots y_m$ such that $x_{n-i+1} \cdots x_n = y_1 \cdots y_i$ for some $i \geq 1$, the string $x_1 \cdots x_n y_{i+1} \cdots y_m = x_1 \cdots x_{n-i} y_1 \cdots y_m$ is a *superposition* of x and y (the superposition of x and y with i overlaps).

Let x be a string of length n. A prefix $x_1 \cdots x_p$, $1 \leq p < n$, of x is a *period* of x if $x_i = x_{i+p}$ for all $1 \leq i \leq n - p$. The period of a string x is the shortest period of x. A string b is a *border* of x if b is a prefix and a suffix of x. The empty string and x itself are *trivial* borders of x. A string x is said to be *quasiperiodic* if there is another string $w \neq x$ such that every position of x falls within some occurrence of w in x. For example, if $x = abaababaabab$ then x is quasiperiodic, since every position x is within and occurrence of $w = aba$.

Let x be a string of n symbols. Let $z = x[i..m]$ be a quasiperiodic factor of x. A quasiperiodicity z of x is identified by the triplet $(i, w, |z|)$, where i is its starting position, w is the quasiperiod of z and $|z|$ is its *span*. We say that the quasiperiodicity $(i, w, |z|)$ is maximal if and only if

1. Let $a = x[i + |z|]$. The string wa does not cover za.
2. There is no quasiperiodicity $(j, w, |z'|)$ of x, with $j \leq i$, such that $|z'| > i - j + |z|$.

Consider the string x of Fig. 1. For $i = 9$, we have the quasiperiodicities $q_1 = (9, aba, 9)$, $q_2 = (9, abaa, 10)$.
The quasiperiodicity q_1 is not maximal because $w_1.a = abaa$ covers $x[9..17].a$, the quasiperiodicity q_2 is not maximal because is embedded in another larger quasiperiodicity: $q_3 = (2, abaa, 17)$. The quasiperiodicity q_3 is maximal. For properties and facts about quasiperiodicities see [AE93]. Note that the span of $q_4 = (1, aabaabaa, 18)$ is greater than the span of q_3, but q_3 and q_4 are both maximal according to the definition of maximal quasiperiodicity.

In the sequel we make use of the following "Incremental split-find" result.

Theorem 1 (Imai and Asano—1987 [IA87]). *Let $a[1], \ldots, a[n]$ be a doubly linked list. There exists an algorithm that preprocess the list a in such way that after a number of deletions in the list a, one can find the nearest $a[j]$ to the left of $a[i]$ with $a[j] \leq a[i]$ in constant time.*

3 Crochemore's partitioning

For each factor w of x, the *start-set* of w is the set of start positions of all occurrences of w in x. For each start-set, we maintain its elements in ascending order.

$$\begin{array}{c}\overset{1}{}\overset{5}{}\overset{10}{}\overset{15}{}\overset{19}{}\\ x=a\ a\ b\ a\ a\ b\ a\ a\ a\ b\ a\ a\ b\ a\ a\ b\ a\ a\ a\end{array}$$

factors w of x	start-set of w
$\{b, ba, baa\}$	$\{3, 6, 10, 13, 16\}$
$\{ab, aba, abaa\}$	$\{2, 5, 9, 12, 15\}$
$\{aab, aaba, aabaa\}$	$\{1, 4, 8, 11, 14\}$
$\{baab, baaba, baabaa\}$	$\{3, 10, 13\}$
$\{abaab, abaaba, abaabaa\}$	$\{2, 9, 12\}$
$\{aabaab, aabaaba, aabaabaa\}$	$\{1, 8, 11\}$

Fig. 1. Some factors of x and their associated start-sets.

To find the start-sets, we define equivalence relations E_l for $1 \leq l < n$. E_l is defined on the positions $\{1, 2, \ldots, n - l + 1\}$ of x: $iE_l j$ if $x_i \cdots x_{i+l-1} = x_j \cdots x_{j+l-1}$. Now we maintain equivalence classes for each E_l. Note that the start-set of a factor of length l is an equivalence class of E_l. If a start-set A is an equivalence class of $E_l, \ldots, E_{l'}$, the equi-set associated with A is the set of strings of length l to l' whose start positions are the elements of A.

We now describe how to find all start-sets. It is easy to see that E_{l+1} is a refinement of E_l, excluding the position $n - l + 1$. The equivalence classes of E_1 can be computed by scanning x in $O(n \log n)$ time when the alphabet is general. Then we compute $E_2, E_3, \ldots$ successively until all classes are singleton sets or $l = p$ (p is the period of x).

At stage l of the refinements, we compute E_{l+1} from E_l. The refinement is based on:

$$iE_{l+1}j \quad \text{if and only if} \quad iE_l j \text{ and } (i+1)E_l(j+1).$$

That is, i and j in an equivalence class of E_l belong to the same equivalence class of E_{l+1} if and only if $(i+1)E_l(j+1)$.

An easy solution is:

1. take each class C of E_l;
2. partition C so that $i, j \in C$ go to the same class of E_{l+1} if and only if $(i+1)E_l(j+1)$.

This method leads to $O(n^2)$ time, since each refinement requires $O(n)$ time and there can be $O(n)$ stages of refinements.

We do the refinement more efficiently as follows.

1. Take a class C of E_l.

2. Instead of partitioning C, we partition with respect to C those classes D of E_l which has at least one i such that $i + 1 \in C$, and one j such that $j + 1 \notin C$. That is, each D is partitioned into classes $\{i \in D \mid i + 1 \in C\}$ and $\{i \in D \mid i + 1 \notin C\}$.

Note that at the end of stage l, for each class D of E_{l+1} there exists one class A of E_l such that for all $i \in D$, $i + 1 \in A$. This fact can be more easily observed in terms of strings: if aw for $a \in \Sigma$ and $w \in \Sigma^+$ is the string whose start-set is D, then w is the string whose start-set is A. We call D a *preimage class* of A.

1. If a class A of E_l is not split at stage l, we need not partition the preimage classes of A with respect to A at stage $l + 1$.
2. If A is split into $C_1, \ldots, C_r$ at stage l, we need to partition the preimage classes of A at stage $l + 1$. For a preimage class D, let $D_s = \{i \in D \mid i+1 \in C_s\}$ for $1 \leq s \leq r$. We can partition the preimage class D with respect to any $r - 1$ classes of $C_1, \ldots, C_r$, and the result will be the same because $D_s = D - (D_1 + \cdots + D_{s-1} + D_{s+1} + \cdots + D_r)$. Since we can choose any $r - 1$ classes, we partition with respect to $r - 1$ small ones except the largest at stage $l + 1$.

procedure PARTITION
compute E_1;
SMALL $\leftarrow$ all classes of E_1;
$l \leftarrow 1$;
while $l < p$ and there is a non-singleton class of E_l **do**
    copy classes in SMALL into QUEUE;
    empty SMALL;
    $l \leftarrow l + 1$;
    while QUEUE not empty **do**
        extract a class C from QUEUE;
        partition with respect to C;
        for each split class D, maintain its new subclasses;
    enddo
    for each split class D (into r subclasses) **do**
        put $r - 1$ small subclasses of D into SMALL;
    enddo
enddo
end

Fig. 2. Procedure PARTITION.

Procedure PARTITION in Fig. 2 shows the partitioning algorithm. Since there are $O(n)$ classes, we represent each class by a number k for $1 \leq k \leq cn$. Each class is implemented by a doubly linked list of its elements in ascending order. When we partition with respect to C in Fig. 2, the classes D which are partitioned with respect to C can be easily identified because D is a class that contains i such that $i+1$ is in C. See [Cro81] for more details of implementation, where the time complexity of PARTITION is shown to be proportional to the sum of the sizes of classes C with respect to which the partitioning is made. Initially ($l = 1$), all classes of E_1 are in SMALL; i.e., all positions in x belong to SMALL. Consider a position i in a class D of E_l. Suppose that D is split at stage l and the subclass D' containing i is put in SMALL. Then $|D'| \leq |D|/2$. Therefore, one position cannot belong to SMALL more than $\log n$ times. Since there are n positions, procedure PARTITION takes $O(n \log n)$ time.

This partitioning is similar to the single function partitioning due to Hopcroft [Hop71]. The former can be viewed as a special case of the latter in which $f(i) = i + 1$ with the following two exceptions.

1. One position is excluded at each stage.
2. Each stage must be separated from another, because each stage deals with equivalence relation E_l.

Crochemore [Cro81] used this partitioning for computing all repetitions in a string. Although the approach of Apostolico and Preparata [AP83] computes the same equivalence classes with their elements sorted, the partitioning method in Fig. 2 is somewhat simpler ([AP83] uses complex data structures).

4 Data structures for gap monitoring

Let $C = \{i_1, i_2, ..., i_k\}$ be a start-set of a string x, with $i_j \in \{1..|x|\}, j \in \{1..k\}$. We define the lists

$$\textbf{left\_gap}(i_m) = i_m - i_{m-1} \quad \text{for all} \ \ i_m \in \{i_2, ..., i_k\}$$

$$\textbf{right\_gap}(i_m) = i_{m+1} - i_m \quad \text{for all} \ \ i_m \in \{i_1, ..., i_{k-1}\}$$

$$\textbf{left\_gap}(i_1) = 0 \ \ \text{and} \ \ \textbf{right\_gap}(i_k) = \infty$$

Now consider a start-set C of positions of a string x that is a member of the equivalence class E_l.
A position $i_m \in C$ is said to be a **head** and we set the flag **head**(i_m) =ON, if and only if

$$\textbf{left\_gap}(i_m) > l \geq \textbf{right\_gap}(i_m) \quad \text{for all} \ \ i_m \in \{i_2, ..., i_{k-1}\}$$

$$l \geq \textbf{right\_gap}(i_m) \quad \text{for} \ \ i_m = i_1 \tag{4.1}$$

$$\textbf{left\_gap}(i_m) > l \quad \text{for} \ \ i_m = i_k$$

A position $i_m \in C$ is said to be a **tail** and we set the flag **tail**(i_m) =ON if and only if

$$\textbf{left\_gap}(i_m) \leq l < \textbf{right\_gap}(i_m) \quad \text{for all} \ \ i_m \in \{i_1, ..., i_{k-1}\} \tag{4.2}$$

$i_m = i_k$ (We always have $\text{tail}(i_k) = \text{ON}$)

For each $j \in C$, we define $\text{span}(i_m)$ in E_l to be $i_r + l - i_m$ if and only if

$\text{head}(i_m) = \text{ON}$, $\text{tail}(i_p) = \text{OFF}$ for all $m < p < r$ and $\text{tail}(i_r) = \text{ON}$ (4.3)

Otherwise $\text{span}(i_m)$ is set to 0.

i_m		2	5	9	12	15
$\text{left\_gap}(i_m)$		0	3	4	3	3
$\text{right\_gap}(i_m)$		3	4	3	3	∞
	$l=2$	OFF	OFF	OFF	OFF	ON
$\text{head}(i_m)$	$l=3$	ON	OFF	ON	OFF	OFF
	$l=4$	ON	OFF	OFF	OFF	OFF
	$l=2$	ON	OFF	OFF	OFF	ON
$\text{tail}(i_m)$	$l=3$	OFF	ON	OFF	OFF	ON
	$l=4$	OFF	OFF	OFF	OFF	ON
	$l=2$	0	0	0	0	0
$\text{span}(i_m)$	$l=3$	6	0	9	0	0
	$l=4$	17	0	0	0	0

Fig. 3. Head, tail and span.

In Fig. 3 we give the values of **left_gap**, **right_gap**, the flags **head** and **tail** and the values of **span** for the start-set $C = \{2, 5, 9, 12, 15\}$ obtained from the string x of Fig. 1.

Theorem 2. *Let C be a start-set of a string x that belongs to an equivalence class E_l. Let $i \in C$ and $\text{span}(i) = \hat{i}_l$ for E_l for $i \in \{1..n\}$. Let $w = x[i..i+l-1]$. The triplet $(i, w, \hat{i}_l)$ is a maximal quasiperiodicity of x at i if*

1. *The $\text{span}(i)$ is reduced in E_l, i.e,*

$$\hat{i}_{l-1} > \hat{i}_l \quad (4.4)$$

2. *The position i is not overtaken by another larger quasiperiodicity in E_l, i.e*

$$\text{head}(i) = \text{OFF} \text{ and } \text{tail}(i) = \text{OFF} \quad (4.5)$$

Proof. If $\text{head}(i) = \text{ON}$ in E_l, then w covers $x[i..\hat{i}_l + i - 1]$. From (4.4) follows that wa does not cover $x[i..\hat{i}_{l-1} + i - 1]$, where $a = x_{i+l}$. Thus the position i satisfies the first condition of the definition of maximal quasiperiodicity. If there was another quasiperiodicity $(j, w, |z'|)$, with $j \leq i$ and $|z'| > i - j + |z|$, then we would have had $\text{head}(i) = \text{OFF}$ and $\text{tail}(i) = \text{OFF}$ but (4.5) forbids it. Thus the position i satisfies the second condition of the definition of maximal quasiperiodicity. □

5 The fast local covers algorithm

The algorithm presented below creates and maintains the lists **left_gap**, **right_gap**, **head**, **tail** and **span** throughout the computation of the equivalent classes computation of Crochemore's partitioning of Section 3. Every time that **span**(i) of a position i of x satisfies (4.4) and condition (4.5) holds, then we report a maximal quasiperiodicity occurring at i of the length defined in Theorem 2.

Crochemore's partitioning creates two types of blocks at each round of equivalences classes E_l: *old* and *new*. A block is said to be *old*, if it has been created by deleting (zero or more) elements from its parent block. Furthermore a block is said to be *new*, if it is a collection of deleted elements of its parent block. Note that the creation of an old block requires time proportional to the number of elements deleted and not time proportional to its size. The creation of new block requires time proportional to its size. Below we show how to update the data structures of Section 4 in time proportional to the sizes of the new blocks and in time proportional to deletions for the old blocks. For example, let $B = \{1, 2, 3, 4, 5, 6, 7, 8, 9\}$ be a block, if we delete $3, 4$ from B, then we get a "new" set $\{3, 4\}$ and the "old" set $\{1, 2, 5, 6, 7, 8, 9\}$. The "new" blocks are the ones created by partitioning with respect to classes in SMALL (see Fig. 2). The "old" blocks are either blocks that remain the same in consecutive classes E_l and E_{l+1} or the remainder of the parent block after the removal of the "new" siblings (see Fig. 4).

First we need to introduce a queueing system for turning the **heads** and **tails** on at the appropriate E_l class.

We add i in the list **turn_head_on**[l] if and only if

$$\textbf{left\_gap}(i) > l \geq \textbf{right\_gap}(i). \tag{5.1}$$

We add i in the list **turn_tail_on**[l] if and only if

$$\textbf{left\_gap}(i) \leq l < \textbf{right\_gap}(i). \tag{5.2}$$

NEW BLOCKS

In the case of new set of size m, Crochemore's partitioning requires $O(m)$ time. Here the update of **left_gap** and **right_gap** can easily be done in $O(m)$ time (conditions (4.1) and (4.2)). In the same time one can update the flags **head** and **tail**. The list **span** can be computed from these two flags (condition (4.3)) in the same time. We report a maximal quasiperiodicity, if the **span**(i) of a position i is reduced, and the condition (4.5) of Theorem 2 is met.

We also need to update the queue for turning the **head** and **tail** on in future equivalence classes. One can easily test the conditions (5.1) and (5.2) and update the queues **turn_head_on** and **turn_tail_on** in $O(m)$ time.

From above one can see that the computation of **span** only increases the time requirements of Crochemore's partitioning algorithm by constant factor.

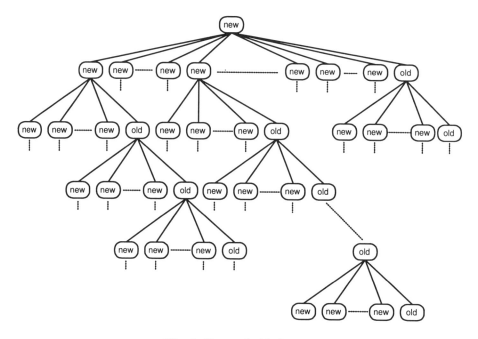

Fig. 4. New and old classes.

OLD BLOCKS

An old block is created by the deletion of, say d, elements of its parent block. Whenever we delete an element i from a block, then we can update the **right_gap** and **left_gap** of its neighbouring elements, say j, k in constant time, since the elements of each block are kept in descending order. If i was in either **turn_head_on** or **turn_tail_on** queues, then we delete i from those queues also in constant time using associate addressing. A new gap has been created between j and k, we check whether **tail** or **head** needs to be turned on (conditions (4.1) and (4.2)), a test that also takes constant time. Otherwise we add i and/or j in the queues **turn_head_on**(l) and **turn_tail_on**(l), with l determined by (5.1) or (5.2) also in constant time.

Finally, we check the queues **turn_head_on** and **turn_tail_on**, for the current l, and we turn the required **head** and **tail** on. One can see that this operation also requires constant time. We then update **span**(i) and we report maximal quasiperiodicities subject to the conditions to Theorem 2. But updating **span** "on-line" with Crochemore's partitioning, in time proportional to d requires some more work.

From Fig. 4 one can see that each "new" block always has one "old" block as a child. Consider one of those pairs, and let B to be the new block of size m and C be its child old block of size k in E_l. We create a doubly linked list

$H[i_1], H[i_2], \ldots, H[i_k]$ for each position $i_j \in C, j \in \{1..k\}$ such that

$$H[i_j] = \begin{cases} l, & \text{if } \mathbf{head}(i_j) = \text{ON in } E_l; \\ r, & i_j \in \mathbf{turn\_the\_head}(r). \end{cases}$$

We create a similar list T for the **tail**. We preprocess the lists H and T according to Theorem 1.

Now we can update the $\mathbf{span}(i)$ of a position i in an "old" block as follows: if i is in the $\mathbf{turn\_the\_head}(r)$ queue, then we will look in the tail list T of the leading "old" block to find the entry $T[j]$ to the right of i which is less than l. If the position i is in the $\mathbf{turn\_the\_tail}$ list, then we find the entry $H(j)$ to the left of i in the list H, which is less than l. We update $\mathbf{span}(j)$ according to (4.3). These operations take time at most k, the size of the first old block in the chain.

Similarly whenever an element is deleted from an "old block", we update $\mathbf{span}(i)$ using the lists H and T as above in a similar manner. This also requires $O(d)$, where d is the total number of deletion throughout the chain of old blocks.

Overall, for each chain of "old" blocks we need $O(m + k + d)$ time, which bounded by $O(m)$, where m is the size of the parent "new" node. Thus we have the following theorem:

Theorem 3. *The above algorithm computes all maximal quasiperiodicities of a string x of length n in $O(n \log n)$ time.*

6 Conclusion

Here we presented an $O(n \log n)$ algorithm for computing the maximal quasiperiodicities. In view of the fact that the number of squares of a Fibonacci string is $O(n \log n)$ but they can be reported in linear time (see [IMS97]), an interesting open problem is to investigate whether a linear representation of all maximal quasiperiodicities exists.

Variants of the maximal quasiperiodicity problem are of interest in music analysis. For example, find all repeated factors of a music score that are at most k positions apart, or compute all repeated non-overlapping factors of lengths greater than k, for some given integer k.

References

[AE93]   A. Apostolico and A. Ehrenfeucht. Efficient detection of quasiperiodicities in strings. *Theor. Comput. Sci.*, 119(2):247–265, 1993.

[AFI91] A. Apostolico, M. Farach, and C. S. Iliopoulos. Optimal superprimitivity testing for strings. *Inf. Process. Lett.*, 39(1):17–20, 1991.

[AP83]   A. Apostolico and F. P. Preparata. Optimal off-line detection of repetitions in a string. *Theor. Comput. Sci.*, 22(3):297–315, 1983.

[BBIP94] A. Ben-Amram, O. Berkman, C. S. Illiopoulos, and K. Park. The subtree max gap problem with application to parallel string covering. In *Proceedings of the 5th ACM-SIAM Annual Symposium on Discrete Algorithms*, pages 501–510, Arlington, VA, 1994.

[Bre92] D. Breslauer. An on-line string superprimitivity test. *Inf. Process. Lett.*, 44(6):345–347, 1992.

[CIR98] T. Crawford, C.S. Iliopoulos, and R. Raman. String matching techniques for musical similarity and melodic recognition. To appear in Computing in Musicology, 1998.

[Cro81] M. Crochemore. An optimal algorithm for computing the repetitions in a word. *Inf. Process. Lett.*, 12(5):244–250, 1981.

[DS96] A.M. Duval and W. M. Smyth. Covering a circular string with substrings of fixed length. *Int. J. Found. Comp. Sci.*, 7(1):87–93, 1996.

[FLSS92] V. A. Fischetti, G. M. Landau, J. P. Schmidt, and P. H. Sellers. Identifying periodic occurrences of a template with applications to protein struture. In A. Apostolico, M. Crochemore, Z. Galil, and U. Manber, editors, *Proceedings of the 3rd Annual Symposium on Combinatorial Pattern Matching*, number 664 in Lecture Notes in Computer Science, pages 111–120, Tucson, AZ, 1992. Springer-Verlag, Berlin.

[Hop71] J. E. Hopcroft. An $n \log n$ algorithm for minimizing states in a finite automaton, pages 189–196. in Kohavi and Paz, ed., New-York, 1971.

[IA87] H. Imai and T. Asano. Dynamic orthogonal segment intersection search. *J. Algorithms*, 8:1–18, 1987.

[IMP93] C. S. Iliopoulos, D. W. G. Moore, and K. Park. Covering a string. In A. Apostolico, M. Crochemore, Z. Galil, and U. Manber, editors, *Proceedings of the 4th Annual Symposium on Combinatorial Pattern Matching*, number 684 in Lecture Notes in Computer Science, pages 54–62, Padova, Italy, 1993. Springer-Verlag, Berlin.

[IMS97] C. S. Iliopoulos, D. W. G. Moore, and W. F. Smyth. A characterization of the squares of a fibonacci string. *Theor. Comput. Sci.*, 172:281–291, 1997.

[IP94] C.S. Iliopoulos and K. Park. An optimal $o(\log \log n)$-time algorithm for parallel superprimitivity testing. *J. Korea Information Science Society*, 21(8):1400–1404, 1994.

[KMGL88] S. Karlin, M. Morris, G. Ghandour, and M.-Y. Leung. Efficient algorithms for molecular sequence analysis. *Proc. Natl. Acad. Sci. U.S.A.*, 85:841–845, 1988.

[LS98] Yin Li and W. F. Smyth. An optimal on-line algorithm to compute all the covers of a string. Submitted, 1998.

[MJ93] A. Milosavljevic and J. Jurka. Discovering simple dna sequences by the algorithmic significance method. *Comput. Appl. Biosci.*, 9:407–411, 1993.

[MS94] D. Moore and W. F. Smyth. Computing the covers of a string in linear time. In *Proceedings of the 5th ACM-SIAM Annual Symposium on Discrete Algorithms*, pages 511–515, Arlington, VA, 1994.

The Graphs of Finite Monadic Semi–Thue Systems Have a Decidable Monadic Second–Order Theory

Teodor Knapik[1] and Hugues Calbrix[2]

[1] IREMIA, Université de la Réunion, BP 7151,
97715 SAINT DENIS Messageries Cedex 9, Réunion, France.
knapik@univ-reunion.fr
[2] College Jean Lecanuet, BP 1024, 76171 ROUEN Cedex, France

Abstract. The monadic second order theory of graphs of finite monadic semi–Thue systems is shown to be decidable. Limitations of monadic semi–Thue systems as graph description technique are discussed.

1 Introduction

Early in the 20th century Axel Thue formulated the word problem that he was tackling using finite semigroup presentations [19]. The latter have been named after as Thue systems. Semi–Thue systems provide a more general approach when unidirectional rules are used instead of equations. Together with term–rewriting systems [8], semi–Thue systems became a basic tool for symbolic computation, especially after the discovery of the Knuth–Bendix procedure [10]. Symbolic computation and more precisely automated deduction have in turn important applications in formal verifications, for instance in the area of algebraic specifications [20]. Besides the latter, finite transition systems [1] are another elegant approach to formal verifications but their finiteness strongly limits their expressive power. This is one of the reasons why infinite transition systems or more generally infinite graphs are the subject of intensive investigations (see e.g. [5, 11, 16] for recent results) especially in connection with monadic second–order logic that is an elementary but powerful language for expressing graph properties.

In the authors' opinion, semi–Thue systems firstly provide a theoretical framework that brings together the above–mentioned two approaches to formal verifications, namely equational logic and (possibly infinite) transition systems. Secondly, since semi–Thue systems have strong links with formal languages (see [14] for an overview), the study of graphs of semi–Thue systems can help establishing connections between monadic second–order logic and formal languages. In the authors' view these two points are motivating enough for studying the interplay between semi–Thue systems and graphs with regard to monadic second–order logic. This is the direction followed in the present paper.

Infinite graphs may be associated to semi–Thue systems in several ways [3, 4, 15]. In [3], the authors consider strongly (resp. unitary) reduction–bounded semi–Thue systems. A semi–Thue system $\mathcal{S}$ is strongly reduction–bounded if there exists $n \in \mathbb{N}$ such that for each irreducible word w and each letter a, the length of any reduction of wa is less than n. Integer n is then called a bound of $\mathcal{S}$. When 1 is a bound of $\mathcal{S}$, semi–Thue system $\mathcal{S}$ is said unitary reduction–bounded. It has been shown that the graphs of strongly reduction bounded semi–Thue systems have decidable monadic second–order theory [3].

In the present paper this approach is carried out further. It is shown that the requirement of the strong reduction boundedness may be dropped out for monadic semi–Thue systems. In other words, it is established that the monadic second–order theory of the graphs of monadic semi–Thue systems is decidable.

2 Preliminaries

Assuming a smattering of formal languages, monadic second–order logic and string–rewriting, several basic definitions and facts are reminded in this section. An introductory material on the latter topics may be found in e.g. [13], [18] and [2].

Words

Given a finite set A called *alphabet*, the elements of which are called *letters*, A^* (resp. A^+) stands for the *free monoid* (resp. *free semigroup*) over A. The elements of A^* (resp. A^+) are all *words* over A, including the *empty word*, written ε (resp. all nonempty words over A). The subsets of A^* are *languages* over A. Each word u is mapped to its *length*, written $|u|$, via the unique monoid homomorphism from A^* onto $(\mathbb{N}, 0, +)$ that maps all letters of A to 1. A word w is a *suffix* (resp. *proper suffix*) of a word u if there exists a word (resp. nonempty word) v such that $u = vw$. The set of suffixes (resp. proper suffixes) of u is written $\mathrm{suff}(u)$ (resp. $\mathrm{suff}^+(u)$). This notation is extended to sets in the usual way:

$$\mathrm{suff}(L) = \bigcup_{u \in L} \mathrm{suff}(u) \qquad (\text{resp. } \mathrm{suff}^+(L) = \bigcup_{u \in L} \mathrm{suff}^+(u))$$

for any language L.

Semi–Thue systems

A semi–Thue system $\mathcal{S}$ (an *sts* for short) over A is a subset of $A^* \times A^*$. A pair (l, r) of $\mathcal{S}$ is called *(rewrite) rule*, the word l (resp. r) is its lefthand (resp. righthand) side. As any binary relation, $\mathcal{S}$ has its domain (resp. range) written $\mathrm{Dom}(\mathcal{S})$ (resp. $\mathrm{Ran}(\mathcal{S})$). Throughout this paper, only finite semi–Thue systems are considered.

The *single-step reduction relation* induced by $\mathcal{S}$ on A^*, is the binary relation $\underset{\mathcal{S}}{\rightarrow} = \{(xly, xry) \mid x, y \in A^*, (l, r) \in \mathcal{S}\}$. The reflexive-transitive closure of $\underset{\mathcal{S}}{\rightarrow}$ is the *reduction relation* induced by $\mathcal{S}$ on A^*, written $\underset{\mathcal{S}}{\overset{*}{\rightarrow}}$. The inverse of the reduction relation is the *derivation relation*, written $\xleftarrow{*}_{S}$, and the reflexive–symmetric–transitive closure of $\xrightarrow{}_{S}$ is the *Thue congruence*, written $\xleftrightarrow{*}_{S}$.

A word v is *irreducible* with respect to (w.r.t. for short) S when v does not belong to $\mathcal{D}om(\xrightarrow{}_{S})$. Otherwise v is *reducible* w.r.t. S. It is easy to see that the set of all irreducible words w.r.t. S, written $\mathfrak{Irr}(S)$, is rational whenever $\mathcal{D}om(S)$ is, since $\mathcal{D}om(\xrightarrow{}_{S}) = A^*(\mathcal{D}om(S))A^*$. A word v is a *normal form* of a word u, when v is irreducible and $u \xrightarrow{*}_{S} v$.

An sts S is *confluent* when for all words $u, v, v' \in A^*$ such that $u \xrightarrow{*}_{S} v$ and $u \xrightarrow{*}_{S} v'$ there exists a word $w \in A^*$ such that $v \xrightarrow{*}_{S} w$ and $v' \xrightarrow{*}_{S} w$. If in addition S is *terminating*, i.e. there is no infinite chain $u_0 \xrightarrow{}_{S} u_1 \xrightarrow{}_{S} \cdots$, then S is said *convergent*. In a convergent sts, each word possesses one unique normal form.

Monadic semi–Thue systems

A semi–Thue system S is *monadic* if $|l| > |r|$ and $|r| = 1$ or $|r| = 0$ for each rule (l, r) of S. Monadic semi–Thue systems enjoy a lot of interesting properties some of which are mentioned below. For any finite monadic sts S

- the extended word problem is solvable in polynomial time,
- the closure under reduction $\xrightarrow{*}_{S}$ of any rational language is rational,
- the closure under derivation $\xleftarrow{*}_{S}$ of any context–free language is context–free.
- If in addition S is confluent then the closure under the Thue congruence $\xleftrightarrow{*}_{S}$ of any rational language is deterministic context–free.

(See [2, 9] for details and further references.)

Graphs

Given an alphabet A, a *simple directed edge-labeled graph* G over A is a subset of $D \times A \times D$ where D is an arbitrary set. Given $d, d' \in D$, an edge from d to d' labeled by $a \in A$ is written $d \xrightarrow{a} d'$. Thus $\xrightarrow{a}$ is a binary relation on D for each $a \in A$. A (finite) *path* in G from some $d \in D$ to some $d' \in D$ is a sequence of edges of the following form: $d_0 \xrightarrow{a_1} d_1, \ldots, d_{n-1} \xrightarrow{a_n} d_n$, such that $d_0 = d$ and $d_n = d'$. The word $a_1 \ldots a_n$ is then the *label* of the path. Given a language L over (resp. a letter a of) of A, a path of G is an *L-path* (resp. *a-path*) if the label of the path belongs to L (resp. a^*). When graphs over $A \cup \{\varepsilon\}$ are considered confusion should be avoided between an empty path and an ε–path. The latter one is a (possibly nonempty) path each edge of which is labeled by ε.

For the purpose of this paper, isolated vertices need not to be considered. The set V_G of the *vertices* of a graph $G \subseteq D \times A \times D$ is therefore defined as $\bigcup_{a \in A}(\mathcal{D}om(\xrightarrow{a}) \cup \mathcal{R}an(\xrightarrow{a}))$. Moreover, the interest lies only in graphs, the vertices of which are all accessible from some distinguished vertex. Thus, a graph

$G \subseteq D \times A \times D$ is said to be *rooted on a vertex* $e \in D$ if there exists a path from e to each vertex of G. The following assumption is made for the sequel. Whenever, in a definition of a graph, a vertex e is distinguished as root, then the maximal subgraph rooted on e is understood.

A graph $G \subseteq D \times A \times D$ is *deterministic* when for each $a \in A$ the relation $\stackrel{a}{\to}$ of G is a function.

Monadic Second–order theory of graphs

Any graph $G \subseteq D \times A \times D$ rooted on some vertex $e \in D$ may be seen as the model theoretic structure $\langle D, (\stackrel{a}{\to})_{a \in A}, e \rangle$ and used for interpreting *monadic second-order formulae*. Such formulae are constructed using *vertex variables* written x, y, x', x_1, etc., *vertex-set variables* written X, Y, X', X_1, etc., the binary predicate symbols $\mathbf{s}_a$ for each letter a of A, "$\in$" and "$=$", the unary predicate symbol $\mathbf{r}$, as well as the classical connectives and quantifiers. An *assignment* maps each vertex variable to a vertex of G and each vertex–set variable to a set of vertices of G. The predicate symbols $\mathbf{s}_a$, $\mathbf{r}$, "$\in$" and "$=$" are interpreted on G resp. as $\stackrel{a}{\to}$, the singleton $\{e\}$, the membership of a vertex in a set of vertices and the equality of vertices. Using classical connectives, from atomic formulae that are of the form $\mathbf{s}_a(x, y)$, $x \in X$, or $x = y$, monadic second order formulae are constructed in the usual way. Quantification is allowed over both vertex variables and vertex–set variables.

The set of valid sentences, i.e. valid formulae with no free variable, of a graph G is called *the monadic second-order theory of G* and is written $\mathrm{MSOTh}(G)$.

Pushdown machines and context–free graphs

An important class of graphs with decidable monadic second–order theory is characterized in [12]. The graphs of this class are called *context-free* and may be defined by means of pushdown machines.

A *pushdown machine* over A (a *pdm* for short) is a triple $\mathcal{P} = (Q, Z, T)$ where Q is the set of *states*, Z is the *stack alphabet* and T is a finite subset of $A \cup \{\varepsilon\} \times Q \times Z \times Z^* \times Q$, called the set of *transition rules*. A is the *input alphabet*.

An *internal configuration* of a pdm $\mathcal{P}$ is a pair $(q, h) \in Q \times Z^*$. To any pdm $\mathcal{P}$ together with an internal configuration ι, one may associate a simple directed edge-labeled graph $\mathcal{G}(\mathcal{P}, \iota)$ defined as follows. The vertices of the graph are all internal configurations accessible from the configuration ι. The latter is the root of the graph. There is an edge labeled by $a \in A \cup \{\varepsilon\}$ from (q_1, h_1) to (q_2, h_2) whenever there exists a letter $z \in Z$ and two words $g_1, g_2 \in Z^*$ such that $h_1 = g_1 z$, $h_2 = g_1 g_2$ and $(q_1, a, z, g_2, q_2) \in T$.

The following theorem (see [12] for the proof) states that the graphs of pushdown machines have decidable monadic second–order theory.

Theorem 1 (Muller and Schupp). *For any pushdown machine $\mathcal{P}$ and its internal configuration ι, the monadic second-order theory of the graph $\mathcal{G}(\mathcal{P}, \iota)$ is decidable.*

3 Graphs of semi–Thue systems

In the same way as Cayley graph is associated to a group presentation, a simple directed edge–labeled graph may be associated to each semi–Thue system.

Given an sts $\mathcal{S}$ over A, the *graph generated by* $\mathcal{S}$, written $\mathcal{G}(\mathcal{S})$, has $\mathrm{Irr}(\mathcal{S})$ as the set of vertices and, for each $a \in A$, the edge relation is defined as

$$\xrightarrow{a} = \{(u,v) \in \mathrm{Irr}(\mathcal{S}) \times \mathrm{Irr}(\mathcal{S}) \mid v \text{ is a normal form of } ua\} \ .$$

Given in addition a word u of $\mathrm{Irr}(\mathcal{S})$, the *graph generated by* $\mathcal{S}$ *from* u, written $\mathcal{G}(\mathcal{S}, u)$, is the graph $\mathcal{G}(\mathcal{S})$ rooted on u, viz the maximal subgraph of $\mathcal{G}(\mathcal{S})$, all vertices of which are accessible from its root u.

Example 2. Over the alphabet $A_1 = \{a, b\}$, consider a single–rule sts $\mathcal{S}_1 = \{(ba, ab)\}$. Obviously $\mathcal{S}_1$ is convergent and the set of irreducible words is a^*b^*. The graph $\mathcal{G}(\mathcal{S}_1, \varepsilon)$ (see Fig. 1) is isomorphic to the grid $\omega \times \omega$. Consequently its monadic second–order theory is not decidable [17]. Observe in addition that $\mathcal{S}_1$ is not strongly reduction–bounded. Indeed for an irreducible word b^n, the length of the unique reduction of $b^n a$ into its normal form ab^n equals to n.

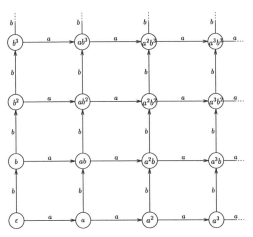

Fig. 1. Graph $\mathcal{G}(\mathcal{S}_1, \varepsilon)$.

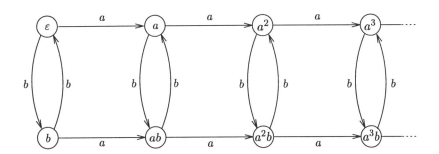

Fig. 2. Graph $\mathcal{G}(\mathcal{S}_2, \varepsilon)$.

Example 3. Consider sts $\mathcal{S}_2 = \mathcal{S}_1 \cup \{(b^2, \varepsilon)\}$. Obviously $\mathcal{S}_2$ is terminating and, since all critical pairs resolve, $\mathcal{S}_2$ is confluent [10]. The set of irreducible words w.r.t. $\mathcal{S}_2$ is the set $a^*(b + \varepsilon)$. Sts $\mathcal{S}_2$ is unitary reduction–bounded. Indeed for any $n \in \mathbb{N}$, $a^n ba$ (resp. $a^n b^2$) reduces in one step into $a^{n+1}b$ (resp. a^n) that is irreducible. Therefore, according to [3], the monadic second–order theory of $\mathcal{G}(\mathcal{S}_2, \varepsilon)$ (depicted on Fig. 2) is decidable.

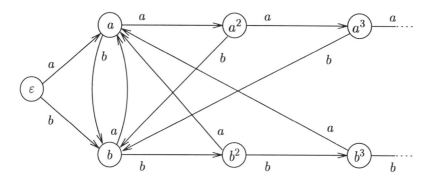

Fig. 3. Graph $\mathcal{G}(\mathcal{S}_3, \varepsilon)$.

Example 4. Consider monadic sts $\mathcal{S}_3 = \{(ab, b), (ba, a)\}$ over A_1. Observe that $\mathcal{S}_3$ is not confluent, since e.g. a and aa are two normal forms of aba. Moreover this system is not strongly reduction–bounded. Indeed, the length of the reduction of $b^n a$ into its normal form a equals to n. However, as established in the sequel, since $\mathcal{S}_3$ is monadic, the monadic second–order theory of the graph $\mathcal{G}(\mathcal{S}_3, \varepsilon)$ (see Fig. 3) is decidable. It may be observed in addition that the in–degree of the graph is infinite.

It may be useful to notice that the graphs of semi–Thue systems are complete. Thus, for each $v \in A^*$, there is a path in $\mathcal{G}(\mathcal{S}, u)$ labeled by v from u to some vertex that is a normal form of uv. From the existence and uniqueness of normal forms for convergent semi–Thue systems, it follows that any graph of a convergent sts is deterministic. However the converse does not hold. The reader may note that graph $\mathcal{G}(\mathcal{S}_3, \varepsilon)$ is deterministic whereas $\mathcal{S}_3$ is not confluent.

4 Main result

In this section the decidability of monadic second–order theory of the graphs of monadic semi–Thue systems is established. The proof is based on a construction of a pushdown machine $\mathcal{P}_\mathcal{S}$ and its internal configuration ι_u for arbitrary finite monadic semi–Thue system $\mathcal{S}$ over A and a word u of $\mathrm{Irr}(\mathcal{S})$. It is then shown that the graphs $\mathcal{G}(\mathcal{S}, u)$ and $\mathcal{G}(\mathcal{P}_\mathcal{S}, \iota_u)$ fulfill the following property.

Property 5. For each monadic second–order sentence φ one may effectively construct a monadic second–order sentence φ' such that that φ belongs to MSOTh($\mathcal{G}(\mathcal{S}, u)$) if and only if φ' is in MSOTh($\mathcal{G}(\mathcal{P}_\mathcal{S}, \iota_u)$).

The construction of the pushdown machine $\mathcal{P}_\mathcal{S}$ and its internal configuration ι_u is now developed. Let $\mathcal{S}$ be a finite monadic semi–Thue system over A and u an irreducible word w.r.t. $\mathcal{S}$. Let $m = \max_{l \in \mathrm{Dom}(\mathcal{S})} |l|$ be the maximum length of the lefthand sides of $\mathcal{S}$. A is the input alphabet of $\mathcal{P}_\mathcal{S}$ and the stack alphabet Z

is the disjoint union of A and $\{\$\}$ where $\$$ is the bottom symbol. The (disjoint) union $Q \uplus P$ of the following sets forms the set of states of $\mathcal{P}_S$:

$$Q = \{q_w \mid w \in \mathrm{Irr}(\mathcal{S}), |w| < m\},$$
$$P = \{p_w \mid w \in \mathrm{Irr}(\mathcal{S}), |w| < m\} \cup$$
$$\{p_{v\#w} \mid w \in \mathrm{Dom}(\mathcal{S}), vw \in \mathrm{suff}(\mathrm{Dom}(\mathcal{S})),$$
$$\forall x, y \in A^+ \ (v = xy \Rightarrow yw \notin \mathrm{Dom}(\mathcal{S}))\} \ .$$

Here $\#$ is an additional symbol that does not belong to A. The set T of transition rules of $\mathcal{P}_S$ is given in Fig. 4. On the whole $\mathcal{P}_S = (Q \uplus P, Z, T)$.

Internal configuration ι_u corresponding to u is defined as follows. Either $|u| < m - 1$ and then $\iota_u = (q_u, \$)$ or $u = vw$ for some $v, w \in A^*$ such that $|w| = m - 1$. In the latter case, $\iota_u = (q_w, \$v)$. Configuration ι_u is referred to as *starting configuration*.

$$T = T_1 \cup T_2 \cup \ldots \cup T_{11}$$

$T_1 = \{(a, q_w, z, z, q_{wa}) \mid a \in A, \ w, wa \in \mathrm{Irr}(\mathcal{S}), \ |w| < m - 1, \ z \in Z\},$
$T_2 = \{(a, q_{bw}, z, zb, q_{wa}) \mid a, b \in A, \ aw, wb \in \mathrm{Irr}(\mathcal{S}), \ |aw| = m - 1, \ z \in Z\},$
$T_3 = \{(a, q_{vw}, z, zv, p_{\#wa}) \mid a \in A, \ vw \in \mathrm{Irr}(\mathcal{S}), \ |vw| < m, \ z \in Z,$
$\qquad\qquad wa \in \mathrm{Dom}(\mathcal{S}), \ \mathrm{suff}^+(wa) \cap \mathrm{Dom}(\mathcal{S}) = \emptyset\},$
$T_4 = \{(\varepsilon, p_{v\#w}, a, \varepsilon, p_{av\#w}) \mid vw \notin \mathrm{Dom}(\mathcal{S}), \ w \in \mathrm{Dom}(\mathcal{S}), \ a \in A,$
$\qquad\qquad avw \in \mathrm{suff}(\mathrm{Dom}(\mathcal{S}))\},$
$T_5 = \{(\varepsilon, p_{v\#w}, z, z, p_{\#vw}) \mid vw, w \in \mathrm{Dom}(\mathcal{S}), \ z \in Z\},$
$T_6 = \{(\varepsilon, p_{\#w}, z, z, p_{\#a}) \mid (w, a) \in \mathcal{S}, \ a \in \mathrm{Dom}(\mathcal{S}), \ z \in Z\},$
$T_7 = \{(\varepsilon, p_{\#w}, z, z, p_x) \mid (w, x) \in \mathcal{S}, \ x \notin \mathrm{Dom}(\mathcal{S}), \ z \in Z\},$
$T_8 = \{(\varepsilon, p_w, a, \varepsilon, p_{\#aw}) \mid w \in \mathrm{suff}(\mathrm{Dom}(\mathcal{S})) \setminus \mathrm{Dom}(\mathcal{S}), \ a \in A, \ aw \in \mathrm{Dom}(\mathcal{S})\},$
$T_9 = \{(\varepsilon, p_w, a, \varepsilon, p_{aw}) \mid a \in A, \ aw \in \mathrm{Irr}(\mathcal{S}), \ |aw| < m - 1\},$
$T_{10} = \{(\varepsilon, p_w, a, \varepsilon, q_{aw}) \mid a \in A, \ aw \in \mathrm{Irr}(\mathcal{S}), \ |aw| = m - 1\},$
$T_{11} = \{(\varepsilon, p_w, \$, \$, q_w) \mid w \in \mathrm{Irr}(\mathcal{S}), \ |w| < m - 1\}.$

Fig. 4. Transition rules of pdm $\mathcal{P}_S$.

The running of $\mathcal{P}_S$ requires some comments. The input word is scanned trough a sliding window the size of which equals to $m - 1$. The indexes of the states of Q represent all possible contents of the window. The sliding of the window one letter forward consists in removing the leftmost letter from the window, piling it on the stack of $\mathcal{P}_S$ and appending on right, to the contents of the window, the input letter read. If a is the input letter read in an internal configuration $(q_w, \$h)$ and the word wa has a suffix that is a lefthand side of a rewrite rule then ε-transitions corresponding to all possible reductions of the word hwa are fired non deterministically. For each resulting normal form of hwa, its maximal suffix — the length of which is less than $m - 1$ — is put into the sliding window. The usual scanning of the input word is then resumed.

Having a look on the transition rules of $\mathcal{P}_\mathcal{S}$ (see Fig. 4), the running of $\mathcal{P}_\mathcal{S}$ may be detailed as follows. If $\mathcal{P}_\mathcal{S}$ starts from a configuration where the sliding window is not completely filled, viz $(q_u, \$)$ and $|u| < m - 1$, the window is filled first (cf. T_1) except when a lefthand side of a rewrite rule is detected (cf. T_3). The sliding of the window is implemented by the rules of T_2. When a lefthand side wa (minimal w.r.t. the suffix ordering) of a rewrite rule is detected in the sliding window (cf. T_3) one may either apply corresponding rewrite rules (cf. T_6, T_7) or look for another rewrite rule such that wa is a suffix of its lefthand side (cf. T_4, T_5).

At this point, the reader may note that a state $p_{v \# w}$ codes the following situation: w is the maximal proper suffix of vw such that w is a lefthand side of a rewrite rule and vw is a suffix of a lefthand side of a rewrite rule. If vw belongs to $\text{Dom}(\mathcal{S})$, then $\mathcal{P}_\mathcal{S}$ moves from state $p_{v \# w}$ to state $p_{\# vw}$ (cf. T_5). The search of another lefthand side is done by checking the topmost symbol of the stack (cf. T_4).

T_6 and T_7 concern the application of a rewrite rule having a word w as lefthand side. Two cases have to be distinguished. In the first case, the righthand side of a rewrite rule is reducible (cf. T_6). Since $\mathcal{S}$ is monadic, it may happen only when the righthand side is a letter say a and there exists rewrite rule (a, ε). Thus, $\mathcal{P}_\mathcal{S}$ moves from $p_{\# w}$ to $p_{\# a}$. As in the case of T_3, there is a nondeterministic choice between reducing by rule (a, ε) or looking for a rewrite rule the lefthand side of which has a as suffix.

The second case of the application of a rewrite rule happens when the righthand side of the rule is irreducible (cf. T_7). Once the rewrite rule is applied, starting from its righthand side x, the search of a lefthand side of a rewrite rule is performed (cf. T_9). Then, either the lefthand side of a rewrite rule is found (cf. T_8) or the search fails due to one of the following reasons: the size of the sliding window is reached (cf. T_{10}) or the bottom stack symbol is reached (cf. T_{11}). In both cases $\mathcal{P}_\mathcal{S}$ moves to the corresponding state of Q.

The reader may observe that a state p_w with $w \in A^*$ codes the situation when w is irreducible but there may be a possibility that by prefixing w with symbols popped out of the stack, a lefthand side of a rule may be formed. Such a possibility disappears when $|w| = m - 1$ or there is no more letters of A in the stack.

Next lemma follows from the construction of $\mathcal{P}_\mathcal{S}$.

Lemma 6. *Let $(q_{w_1}, \$h_1)$ and $(q_{w_2}, \$h_2)$ be two vertices of $\mathcal{G}(\mathcal{P}_\mathcal{S}, \iota_u)$ such that there exists a vertex d, an edge $(q_{w_1}, \$h_1) \xrightarrow{a} d$ of $\mathcal{G}(\mathcal{P}_\mathcal{S}, \iota_u)$ and an ε-path from d to $(q_{w_2}, \$h_2)$. Then $h_2 w_2$ is a normal form of $h_1 w_1 a$.*

Consider now the following map of the vertices of $\mathcal{G}(\mathcal{S}, u)$ into the vertices of $\mathcal{G}(\mathcal{P}_\mathcal{S}, \iota_u)$:

$$f(w) = \begin{cases} (q_w, \$) & \text{if } |w| < m - 1, \\ (q_y, \$x) & \text{if } w = xy \text{ for some } x, y \in A^* \text{ such that } |y| = m - 1. \end{cases}$$

Note that f is injective. Now, taking into account the definition of f, the next lemma follows from the construction of $\mathcal{P}_\mathcal{S}$.

Lemma 7. *Let $w \xrightarrow{a} v$ be an edge of $\mathcal{G}(\mathcal{S}, u)$. Then there exists a vertex d of $\mathcal{G}(\mathcal{P}_\mathcal{S}, \iota_u)$ such that $f(w) \xrightarrow{a} d$ is an edge of $\mathcal{G}(\mathcal{P}_\mathcal{S}, \iota_u)$ and there is an ε-path from d to $f(v)$.*

The main result is stated as follows.

Theorem 8. *For each finite monadic semi-Thue system $\mathcal{S}$ over A and each irreducible word u w.r.t.$\mathcal{S}$, the monadic second order theory of $\mathcal{G}(\mathcal{S}, u)$ is decidable.*

The proof is based upon the existence of a functional definable graph transduction Δ such that $\Delta(\mathcal{G}(\mathcal{P}_\mathcal{S}, \iota_u))$ is isomorphic to $\mathcal{G}(\mathcal{S}, u)$. For this purpose the general definition of definable hypergraph transduction of [7] is adapted to the case of noncopying graph transduction, $(1, 1)$-definable without parameters. This case is referred to in the sequel as *simply definable graph transduction*.

Definition 9. *Let Γ be the following set of predicate symbols: $\{(\mathbf{s}_a)_{a \in A}, \mathbf{r}\}$. A simple Γ-definition scheme is a tuple Δ of monadic second-order formulae $\Delta = \langle \psi(x), (\theta_{\mathbf{s}_a}(x, y))_{a \in A}, \theta_\mathbf{r}(x) \rangle$ such that x is the unique free variable of both $\psi(x)$ and $\theta_\mathbf{r}(x)$ and, the only free variables of $\theta_{\mathbf{s}_a}(x, y)$ are x and y. Given a graph $G = \langle D_G, (\xrightarrow[G]{a})_{a \in A}, e_G \rangle$, a graph $H = \langle D_H, (\xrightarrow[H]{a})_{a \in A}, e_H \rangle$ is defined by Δ whenever*

$$D_H = \{d \in D_G \mid G \models \psi(d)\},$$
$$\xrightarrow[H]{a} = \{(d, d') \in D_G \times D_G \mid G \models \theta_{\mathbf{s}_a}(d, d')\} \quad \text{for each } a \in A,$$
$$\{e_H\} = \{d \in D_G \mid G \models \theta_\mathbf{r}(d)\}.$$

The graph H is then written $\Delta(G)$. A function ξ of simple directed edge-labeled rooted graphs is *simply Γ-definable (graph transduction)* if there exists a simple Γ-definition scheme Δ such that $\xi = \lambda G.\Delta(G)$.

The following proposition is an adaptation of the Proposition 3.1 of [7].

Proposition 10. *Let ξ be a simply Γ-definable graph transduction. Then for each monadic second-order sentence φ one may effectively construct a monadic second-order sentence φ_ξ such that for each simple directed edge-labeled rooted graph G the following holds: $G \models \varphi_\xi$ if and only if $\xi(G) \models \varphi$.*

With the help of the above result, the proof of Theorem 8 may readily be developed.

Proof (Theorem 8). Let $\mathcal{S}$ be a finite monadic semi-Thue system $\mathcal{S}$ over A and $u \in \mathfrak{Irr}(\mathcal{S})$. It is claimed that Property 5 holds.
Define a predicate $\mathbf{s}_\varepsilon^*(x, y)$ the meaning of which is "there is an ε-path from x to y":

$$\forall X \, ((x \in X \wedge (\forall x' \, \forall y' \, ((x' \in X \wedge \mathbf{s}_\varepsilon(x', y')) \Rightarrow y' \in X))) \Rightarrow y \in X) \ .$$

Let $\Delta = \langle \psi(x), (\theta_{\mathbf{s}_a}(x,y))_{a \in A}, \theta_{\mathbf{r}}(x) \rangle$ be the following simple Γ-definition scheme

$$\psi(x) = \exists x' \bigvee_{a \in A} \mathbf{s}_a(x, x'),$$

$$\theta_{\mathbf{s}_a}(x,y) = \exists x' \, (\mathbf{s}_a(x,x') \wedge \mathbf{s}_\varepsilon^*(x',y)) \quad \text{for each } a \in A,$$

$$\theta_{\mathbf{r}}(x) = \mathbf{r}(x).$$

Observe that the set of vertices of $\Delta(\mathcal{G}(\mathcal{P}_S, \iota_u))$ is the set of all vertices of $\mathcal{G}(\mathcal{P}_S, \iota_u)$ that have an outgoing edge labeled by a letter of A. This is precisely the set of all vertices of the form $(q_w, \$v)$ or, in other words, the set $f(\mathrm{Irr}(\mathcal{S}))$, where f is the map defined between Lemmas 6 and 7. Observe in addition that there is an edge $d_1 \stackrel{a}{\to} d_2$ in $\Delta(\mathcal{G}(\mathcal{P}_S, \iota_u))$ if and only if the following holds: $d_1, d_2 \in f(\mathrm{Irr}(\mathcal{S}))$ and there is an edge $d_1 \stackrel{a}{\to} d$ in $\mathcal{G}(\mathcal{P}_S, \iota_u)$ for some vertex d that is the origin of an ε-path ended by d_2. Now, using Lemmas 6 and 7 one concludes that $\mathcal{G}(\mathcal{S}, u)$ and $\Delta(\mathcal{G}(\mathcal{P}_S, \iota_u))$ are isomorphic. Hence Property 5 holds due to Proposition 10. Theorem 1 allows finally to conclude that the monadic second-order theory of $\mathcal{G}(\mathcal{S}, u)$ is decidable. □

5 Limitations of monadic semi–Thue systems

As shown so far, finite monadic semi–Thue systems may be used as descriptions of infinite graphs. However the expressive power of finite monadic semi–Thue systems in this context is rather limited. First of all, only complete graphs are handled. But among complete graphs of pushdown machines not all may be described by finite monadic semi–Thue systems. In fact, the situation is even worse. There exist finite graphs, hence finite–state machines that are isomorphic to no graph of monadic sts. In order to persuade yourself about it, consider graph G depicted on Fig. 5. The following proposition may be stated.

Proposition 11. *There is no monadic sts $\mathcal{S}$ and no word $u_0 \in \mathrm{Irr}(\mathcal{S})$ such that the graph $\mathcal{G}(\mathcal{S}, u_0)$ is isomorphic to G.*

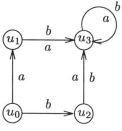

Fig. 5. Graph G.

Proof. Assume by contradiction that there exists a monadic sts $\mathcal{S}$ and $u_0, u_1, u_2, u_3 \in \mathrm{Irr}(\mathcal{S})$ that form $\mathcal{G}(\mathcal{S}, u_0)$ isomorphic to G. Obviously the alphabet of $\mathcal{S}$ is $\{a, b\}$.

Note that $(a, \varepsilon) \notin \mathcal{S}$ and $(b, \varepsilon) \notin \mathcal{S}$. Otherwise one should have loops on u_0, u_1 and u_2. Thus the length of the lefthand side of each rule of $\mathcal{S}$ is at least 2. Consequently, due to the loop on u_3, there exists $v_3 \in \{a, b\}^*$ and $e_3 \in \{a, b\}$ such that $u_3 = v_3 e_3$ and

$$(e_3 a, e_3), (e_3 b, e_3) \in \mathcal{S} \ . \tag{1}$$

Observe that both $u_1 a$ and $u_1 b$ (resp. $u_2 a$ and $u_2 b$) are reducible. Indeed, the assumption that both $u_1 a$ and $u_1 b$ (resp. $u_2 a$ and $u_2 b$) are irreducible leads to the contradiction $u_3 = u_1 a \neq u_1 b = u_3$ (resp. $u_3 = u_2 a \neq u_2 b = u_3$) and, the assumption that one of these two words is reducible and the other one is not leads to the contradiction $|u_3| < |u_3|$. Consequently, taking into account that the length of the lefthand side of each rule of $\mathcal{S}$ is at least 2, there exist $v_1, v_2 \in \{a,b\}^*$ and $e_1, e_2 \in \{a,b\}$ such that

$$u_1 = v_1 e_1 \quad \text{and} \quad u_2 = v_2 e_2 . \tag{2}$$

Observe that if $e_1 = e_3$ (resp. $e_2 = e_3$) the rules of (1) would generate loops over u_1 (resp. u_2). Therefore $e_1 \neq e_3$ and $e_2 \neq e_3$. Hence $e_1 = e_2$. Depending on the value of e_1 and e_2 two cases are distinguished.

- Case $e_1 = e_2 = a$.
  Then $e_3 = b$. Consequently (1) becomes $(ba, b), (bb, b) \in \mathcal{S}$. Since $u_1 = v_1 a$ and $u_2 = v_2 a$ (cf. (2)) are irreducible, $v_1 = a^{i_1}$ and $v_2 = a^{i_2}$ for some $i_1, i_2 \in \mathbb{N}$. But as established above, $u_1 a = a^{i_1+2}$ is reducible. Consequently, there is a rule $(l_1, r_1) \in \mathcal{S}$ such that l_1 is a factor of a^{i_1+2}. Now, l_1 cannot be a factor of a^{i_1+1} because $a^{i_1+1} = u_1$ is irreducible. Hence $l_1 = a^{i_1+2}$. Similarly, there is a rule (l_2, r_2) such that $l_2 = a^{i_2+2}$. But $i_1 \neq i_2$ because u_1 and u_2 are two different vertices of $\mathcal{G}(\mathcal{S}, u_0)$. If $i_1 < i_2$ then $i_1 + 2 \leq i_2 + 1$, hence l_1 is a factor of u_2. This contradicts the irreducibility of u_2. Similarly, $i_2 < i_1$ contradicts irreducibility of u_1.
- Case $e_1 = e_2 = b$.
  Then $e_3 = a$ and by symmetry, this case also yields a contradiction. □

6 Conclusion

As mentioned in Sect. 2, monadic semi–Thue systems enjoy a number of interesting properties. This paper adds a new one, namely the decidability of the monadic second–order theory of the graphs of finite monadic semi–Thue systems.

The expressive power of the monadic semi–Thue systems as description of finite or infinite graphs is rather limited in comparison to e.g. regular systems of equations [6]. Indeed, the graphs of semi–Thue systems are always complete. But this drawback may be overcome to some extent thanks to the feature of rational restrictions on paths considered in [5]. The latter may also be expressed as a particular case of noncopying monadic second–order graph transductions.

In sect. 4 it is shown that the class of the graphs of finite monadic semi–Thue systems is included (up to isomorphism) in the class of complete graphs obtained from the graphs of pushdown machines by glueing together the states connected by ε–paths. Sect. 5 shows that this inclusion is proper. This points out an additional limitation of semi–Thue systems as graph descriptions technique.

Acknowledgments

The authors thank Bruno Courcelle for his explanations about monadic second–order definable graph transductions, Serge Burckel and Étienne Payet for careful proof reading and Aurelia Segatti that helped us in improving the English of this paper.

References

1. A. Arnold. *Finite Transition Systems*. Prentice–Hall International, 1994.
2. R. V. Book and F. Otto. *String–Rewriting Systems*. Texts and Monographs in Computer Science. Springer–Verlag, 1993.
3. H. Calbrix and T. Knapik. A string–rewriting characterization of context–free graphs. In V. Arvind and R. Ramanujam, editors, *18th International Conference on Foundations of Software Technology and Theoretical Computer Science*, Chennai, Dec. 1998.
4. D. Caucal. On the regular structure of prefix rewriting. *Theoretical Comput. Sci.*, 106:61–86, 1992.
5. D. Caucal. On infinite transition graphs having a decidable monadic second–order theory. In F. M. auf der Heide and B. Monien, editors, *23th International Colloquium on Automata Languages and Programming*, LNCS 1099, pages 194–205, 1996.
6. B. Courcelle. Graph rewriting: An algebraic and logic approach. In J. van Leeuwen, editor, *Formal Models and Semantics*, volume B of *Handbook of Theoretical Computer Science*, pages 193–242. Elsevier, 1990.
7. B. Courcelle. Monadic second–order definiable graph transductions: a survey. *Theoretical Comput. Sci.*, 126:53–75, 1994.
8. N. Dershowitz and J.-P. Jouannaud. Rewrite systems. In J. van Leeuwen, editor, *Formal Models and Semantics*, volume B of *Handbook of Theoretical Computer Science*, pages 243–320. Elsevier, 1990.
9. M. Jantzen. *Confluent String Rewriting*, volume 14 of *EATCS Monographs on Theoretical Computer Science*. Springer–Verlag, 1988.
10. D. Knuth and P. Bendix. Simple word problems in universal algebra. In J. Leech, editor, *Computational Problems in Abstract Algebra*, pages 263–297. Pergamon Press, 1970.
11. F. Moller, editor. Infinity'97: 2nd international workshop on verification of infinite state systems. Technical Report 148, Uppsala Computing Science Department, June 1997.
12. D. E. Muller and P. E. Schupp. The theory of ends, pushdown automata and second–order logic. *Theoretical Comput. Sci.*, 37:51–75, 1985.
13. G. Rozenberg and A. Salomaa, editors. *Handbook of Formal Languages*, volume 1. Springer–Verlag, 1997.
14. G. Sénizergues. Formal languages and word rewriting. In *Term Rewriting: French Spring School of Theoretical Computer Science*, LNCS 909, pages 75–94, Font-Romeu, May 1993.
15. C. C. Squier, F. Otto, and Y. Kobayashi. A finiteness condition for rewriting systems. *Theoretical Comput. Sci.*, 131(2):271–294, 1994.
16. B. Steffen and T. Margaria, editors. INFINITY international workshop on verification of infinite state systems. Technical Report MIP–9614, Fakultät für Mathematik und Informatik, Universität Passau, Aug. 1996.
17. W. Thomas. Automata on infinite objects. In J. van Leeuwen, editor, *Formal Models and Semantics*, volume B of *Handbook of Theoretical Computer Science*, pages 133–191. Elsevier, 1990.
18. W. Thomas. Languages, automata and logic. In G. Rozenberg and A. Salomaa, editors, *Beyond Words*, volume 3 of *Handbook of Formal Languages*, pages 389–455. Springer–Verlag, 1997.

19. A. Thue. Probleme über veränderungen von zeichenreihen nach gegebenen regeln. *Skr. Vid. Kristiania, I Mat. Natuv. Klasse*, 10:34 pp, 1914.
20. M. Wirsing. Algebraic specification. In J. van Leeuwen, editor, *Formal Models and Semantics*, volume B of *Handbook of Theoretical Computer Science*, pages 675–788. Elsevier, 1990.

Low$_n$, High$_n$, and Intermediate Subrecursive Degrees

Lars Kristiansen

University of Oslo, Department of Mathematics,
Boks 1053 Blindern, 0316 Oslo, Norway. larskri@math.uio.no

1 Introduction

In this paper we study a degree structure which in many respects is similar to the structure of Turing degrees in computability-theory. Our degrees are called honest subrecursive degrees or, in short, just honest degrees. The notion of honesty bears relation to the notions of tape, space, and time constructibility in complexity theory (see Machthey [11, 12]), and the structure of honest degrees is obviously related to the subrecursive hierarchies of Wainer et al. Moreover, the degree theory developed in this paper has applications in a proof-theoretic study of Peano Arithmetic. On this point see Beklemishev [2, 3] and Kristiansen [7].

Honest subrecursive degrees were studied in the early and mid-seventies by, among others, Meyer and Ritchie [13], Basu [4], and Machtey [10–12]. Machtey proved that the structure of honest degrees is a lattice, and a lot of density, embedding, minimal pair, and splitting results were also proved. Kristiansen [5] introduces and investigates a jump operator on honest degrees. This jump operator is analogous to the jump operator on the Turing degrees, and yields canonical degrees $\mathbf{0}, \mathbf{0}', \mathbf{0}'', \ldots$. Thus we can define low$_n$ and high$_n$ honest degrees analogous to the low$_n$ and high$_n$ Turing degrees. In this paper we prove that nontrivial low$_n$ and high$_n$ honest degrees exist for any n. We also prove that intermediate honest degrees exist. (This solves a problem stated as open in Meyer and Ritchie [13].) A degree $\mathbf{b}$ caps to a degree $\mathbf{a}$ if there exists a degree $\mathbf{c} > \mathbf{a}$ such that $\mathbf{a}$ is the g.l.b. of $\mathbf{b}$ and $\mathbf{c}$. We conclude this paper by some results on the "caps to"-relation in the structure of honest degrees.

Our reducibility relation is "being (Kalmar) elementary in", but we believe that our results and proof methods can be generalized to other subrecursive reducibilities, e.g. "being primitive recursive in" and "being k-recursive in", but not "being polynomial time computable in the *length of input*". "Being polynomial time computable in the *input*" might work, at least in some respects.

2 General preliminaries and definitions

We assume the reader is familiar with the most basic concepts of classical computability-theory (recursion-theory). An introduction and survey can be found several places, e.g. [14] or [17]. We also assume some acquaintance with subrecursion and, in particular, with the elementary functions. An introduction to this subject can be found in [15] or [16]. Here we just state some important basic facts and definitions. See [15] and [16] for proofs.

The *initial elementary functions* are the successor ($\mathcal{S}$), projections ($\mathcal{I}_i^n$), zero (0), addition (+), and modified subtraction ($\dot{-}$) functions. The *elementary schemes* are *composition*, i.e. $f(x) = h(g_1(x), \ldots, g_m(x))$ and bounded *sum* and *product*, i.e. $f(x,y) = \sum_{i<y} g(x,i)$ and $f(x,y) = \prod_{i<y} g(x,i)$. The class of *(Kalmar) elementary functions* is the least class which contains the initial elementary functions and is closed under the elementary schemes. A *relation* or a *predicate* $R(x)$ *is elementary* when there exists an elementary function f with range $\{0,1\}$ such that $f(x) = 0$ iff $R(x)$ holds. That a function f has an *elementary graph* means that the relation $f(x) = y$ is elementary. If we can define a function g from the function f plus the initial elementary functions by the elementary schemes, we say that g *is elementary in* f.

The definition scheme $(\mu z < x)[\ldots]$ is called the *bounded μ-operator*, and $(\mu z < y)[R(x,z)]$ denotes the least $z < y$ such that the relation $R(x,z)$ holds. Let $(\mu z < y)[R(x,z)] = 0$ if no such z exists. The elementary functions are closed under the bounded μ-operator. If f is defined by primitive recursion over g and h and $f(x,y) \leq j(x,y)$, we say that f is a *limited recursion* over g, h and j. (It is convenient to think about limited recursion as a scheme with g, h and j as parameters, although the j is actually not used to generate f.) The elementary functions are closed under limited recursion, but not under primitive recursion. (So if f is a limited recursion over g, h and j where g, h and j are elementary, then f is elementary.) Moreover, the elementary relations are closed under the operations of the propositional calculus and under bounded quantification, i.e. $(\forall x < y)[R(x)]$ and $(\exists x < y)[R(x)]$.

Rose [16] proves that the class of the elementary functions equals the third Grzegorczyk class $\mathcal{E}^3$, and $\mathcal{E}^3$ is the class $\{0, \mathcal{S}, \mathcal{I}_i^n, 2^x, \max\}$ closed under composition and limited recursion.

All the closure properties of the elementary functions can be proved by using Gödel numbering and standard coding techniques. Uniform systems for coding the finite sequences of natural numbers are available inside the class of elementary functions. Let $F_f(x)$ be the code for the sequence $\langle f(0), f(1), \ldots f(x) \rangle$. Then F_f belongs to the elementary functions if f does. We indicate the use of coding functions with the notation $\langle \ldots \rangle$. Our coding system is monotone, i.e. $\langle x_0, \ldots, x_n \rangle < \langle x_0, \ldots, x_n, y \rangle$ for every value of y, and $\langle x_0, \ldots, x_i, \ldots, x_n \rangle < \langle x_0, \ldots, x_i + 1, \ldots, x_n \rangle$.

The function f^k is the k'th iterate of the unary function f, i.e. $f^0(x) = x$ and $f^{k+1}(x) = ff^k(x)$.

3 The honest elementary degrees and the jump operator

Definition 1. A function f is *honest* when (i) f is unary, (ii) $f(x) \geq 2^x$, (iii) f is monotone, i.e. $f(x) \leq f(x+1)$, and (iv) f has an elementary graph.

It is clause (iv) in the definition which is essential. We require that an honest function shall have elementary graph because we want no "hidden complexity" in the function. We want the growth of the function to mirror the computational complexity of the function. The structure of honest degrees would be the same if an honest function were not required to satisfy (i), (ii), and (iii), but those requirements are needed for other purposes. Note that when f is honest, we have $f^{y+1}(x) > f^y(x)$, but we do not necessarily have $f(x+y) > f(x)$.

Definition 2. Let $f \leq_E g$ denote that f is elementary in g, let $f <_E g$ denote that $f \leq_E g$ and $g \not\leq_E f$, and let $f \equiv_E g$ denote that $f \leq_E g$ and $g \leq_E f$. The equivalence classes induced by $\equiv_E$ are *the elementary degrees*. We let $\deg(f) \stackrel{\text{def}}{=} \{g \mid g \equiv_E f\}$ and refer to $\deg(f)$ as *the degree of f*. We use $<, \leq$ for the ordering induced on the degrees by $<_E, \leq_E$. An elementary degree **a** is *honest* when $\mathbf{a} = \deg(f)$ for some honest f. We will use small bold-faced letters early in the Latin alphabet, i.e. **a, b, c,** ..., to denote honest elementary degrees. If $\mathbf{a} \leq \mathbf{b} \leq \mathbf{c}$ then **a** is a degree *below* **b**, and **b** is a degree *above* **a**, and **b** is a degree *between* **a** and **c**. Most degrees mentioned in this paper are *honest elementary degrees*. If we just say *degree* or *honest degree*, we do really mean honest elementary degree. From now on we reserve the letters f, g and h, with or without subscripts, to denote honest functions only.

Theorem 3 (The Growth Theorem). *Let f and g be honest functions. Then*

$$g \leq_E f \iff g(x) \leq f^k(x) \text{ for some fixed } k.$$

Proof. Suppose $g \leq_E f$. Then g can be generated from f and some initial functions by composition, bounded sum, and bounded product. Use induction on such a generation of g to prove that there exists a fixed k such that $g(x) \leq f^k(x)$. (Recall that $f(x) \geq 2^x$.) Now suppose that $g(x) \leq f^k(x)$. Since g is honest, the relation $g(x) = y$ is elementary. We have $g(x) = (\mu y \leq f^k(x))[g(x) = y]$. Hence $g \leq_E f$ since the elementary functions are closed under composition and the bounded μ-operator. □

The previous theorem eliminates the need for the traditional computability-theoretic constructions in the sequel. To prove that a degree, for instance a Turing degree, is incomparable or lies above another degree usually requires diagonalization. We manage this without explicit diagonalization when we deal with the honest elementary degrees. To prove that $g \leq_E f$ we just provide a fixed k such that $g(x) \leq f^k(x)$; to prove that $g \not\leq_E f$ we prove that such a k does not exist.

Definition 4. Let f be an honest function. We define the function f' by $f'(x) \stackrel{\text{def}}{=} f^{x+1}(x)$. We call $\cdot'$ the *jump operator* (on honest functions).

The function f' is honest whenever f is honest. Further, the jump operator is monotonic on honest function, i.e. $g \leq_E f \Rightarrow g' \leq_E f'$ for honest functions f and g. (See [5] for proofs.) It follows that the operator $\cdot'$ induces a jump operator on the honest elementary degrees, i.e. the monotonicity of the jump operator implies that $f' \equiv_E g'$ whenever f and g are honest functions such that $f \equiv_E g$. Hence the next definition does not depend on the choice of honest f in the degree a.

Definition 5. Let a be an honest elementary degree. Then $\mathbf{a}' \stackrel{\text{def}}{=} \deg(f')$ where f is some honest function in a. We call $\cdot'$ the *jump operator* (on honest degrees). Let $\mathbf{0} \stackrel{\text{def}}{=} \deg(2^x)$, i.e. $\mathbf{0}$ is the class of elementary functions.

Assume that we have some standard enumeration of the unary functions elementary in f, and let $[e]^f$ denote the e'th function elementary in f under this enumeration. Further let $\mathcal{J}(f)(\langle y, z \rangle) \stackrel{\text{def}}{=} [y]^f(z)$. Then $\mathcal{J}(\cdot)$ is a operator (on total computable functions) which is similar to the Turing jump in classical computability-theory. We can prove that $\mathcal{J}(f) \equiv_E f'$ for any honest f. The details can be found in [5]. Thus our jump operator $\cdot'$ is indeed equivalent to a subrecursive version of the Turing jump.

The definitions and comments above show that the theory of elementary honest degrees bears a strong resemblance to traditional Turing degree theory. Moreover, the theory of the honest elementary degrees is a theory about computational complexity. Roughly speaking, if $\Phi_f(x)$ is a function giving the number of steps required to compute the honest function f on the argument x, then Φ_f is honest and has the same elementary degree as f. Thus the theory of the honest degrees can be viewed as a theory of computational complexity. If a computable function f is not reducible to an honest function g, it is because a computation of f requires to much resources, i.e. more resources than g affords. In contrast, a theory about subrecursive reducibility between computable sets can be viewed as a theory of information. If a set A is not reducible to a computable set B, it is because B does not contain enough information to decide membership in A within a certain limited amount of time or space. (Subrecursive degrees of computable sets are investigated in a number of places, e.g. Ladner [9] and Ambos-Spies [1]. The relationship between such set-degrees and subrecursive degrees of computational complexity, i.e. honest degrees, is elaborated in Kristiansen [6].) There is also a close relationship between the honest elementary degrees and the subrecursive hierarchies described in the book of Rose [16]. Let $\mathcal{E}^0, \mathcal{E}^1, \mathcal{E}^2, \ldots$ denote the classes in the Grzegorczyk hierarchy. Then the class $\mathcal{E}^3$ is exactly the functions elementary in an honest function of degree $\mathbf{0}$, the class $\mathcal{E}^4$ is exactly the functions elementary in an honest function of degree $\mathbf{0}'$, the class $\mathcal{E}^5$ corresponds to the degree $\mathbf{0}''$ and so on. By introducing an ω-jump in an obvious way, we will be able to climb beyond the Grzegorczyk classes and generate honest elementary degrees that correspond to higher levels in the transfinite hierarchies.

4 Intermediate and High$_n$ degrees

Definition 6. We define the n'th jump $f^{[n]}$ ($\mathbf{a}^{[n]}$) of an honest function f (degree a) by $f^{[0]} = f$ ($\mathbf{a}^{[0]} = \mathbf{a}$) and $f^{[n+1]} = f^{[n]'}$ ($\mathbf{a}^{[n+1]} = \mathbf{a}^{[n]'}$). We define the ω-jump f^ω of the honest function f by $f^\omega(x) = f^{[x]}(x)$.

We define low$_n$ (high$_n$) honest degrees analogously to low$_n$ (high$_n$) Turing degrees: An honest degree $\mathbf{a} \leq \mathbf{0}'$ is low$_n$ (high$_n$) when $\mathbf{a}^{[n]} = \mathbf{0}^{[n]}$ ($\mathbf{a}^{[n]} = \mathbf{0}^{[n+1]}$). Let $\mathbf{L}_n$ ($\mathbf{H}_n$) denote the set of low$_n$ (high$_n$) degrees. We say that an honest function f is low$_n$ (high$_n$) if $\deg(f) = \mathbf{a}$ for some low$_n$ (high$_n$) degree $\mathbf{a}$.

Note that $\mathbf{L}_0 = \{\mathbf{0}\}$ and $\mathbf{H}_0 = \{\mathbf{0}'\}$. Furthermore note that $f^{[n]}$ denotes the n'th jump of f whereas f^n denotes the n'th iterate of f. We have for example $f^{[3]7}(8) = f^{[3]6}f^{[3]}(8) = f^{[3]6}f^{[2]9}(8)$.

Lemma 7. *Let f, g, h be honest functions such that $h(x) \geq f'(x)$, and let y and z be any natural numbers. If*

$$g(x) = f(x) \text{ for any } x \text{ such that } y \leq x \leq h^{m+1}(z) \qquad (*)$$

then

$$g'(x) = f'(x) \text{ for any } x \text{ such that } y \leq x \leq h^m(z).$$

Proof. Fix x, y, z such that $y \leq x \leq h^m(z)$ and assume (*). We prove that $g'(x) = f'(x)$. The function f' is monotone. Thus $f'(x) \leq f'(h^m(z))$. Further, we have $h(v) \geq f'(v)$ for any v. Thus $f'(h^m(z)) \leq h^{m+1}(z)$. So we have $f'(x) \leq h^{m+1}(z)$, and thus

$$g(v) = f(v) \text{ for any } v \text{ such that } x \leq v \leq f'(x) \qquad (**)$$

follows by our assumption (*). By $x+1$ applications of (**) we have $g'(x) = g^{x+1}(x) = f^{x+1}(x) = f'(x)$. □

Lemma 8. *Let f and g be honest functions. Suppose there exists infinitely many y such that $y \leq x \leq f'(y) \Rightarrow g(x) = f(x)$. Then $f' \not\leq_E g$.*

Proof. First we prove that there for any k exist infinitely many z such that $g^k(z) = f^k(z)$ (*). Let k be any number. By assumption we can pick $y > k$ such that $g(x) = f(x)$ for any x such that $y \leq x \leq f'(y)$. For any $n \leq k$ we have $g^n(y) \leq f'(y)$ and $f^n(y) \leq f'(y)$. Thus $g^k(y) = f^k(y)$. This proves (*). Now $f'(x) > f^k(x)$ holds for any $x \geq k$. Hence there cannot exist k such that $f'(x) \leq g^k(x)$, and $f' \not\leq_E g$ follows by the Growth Theorem. □

Lemma 9. *Let f and g be honest functions. Suppose there exists infinitely many x such that $g(x) \geq f'(x)$. Then $g \not\leq_E f$.*

Proof. If $g(x) = f'(x)$ holds for infinitely many x, then there cannot exist a fixed k such that $g(x) \leq f^k(x)$ holds for every x. Hence we have $g \not\leq_E f$ by the Growth Theorem. □

Theorem 10. *There exists an honest degree $\mathbf{a}$ such that $\mathbf{0}^{[n]} < \mathbf{a}^{[n]} < \mathbf{0}^{[n+1]}$ for every natural number n.*

Proof. Let $f(x) = 2^x$. So $\deg(f) = 0$. Let $d_0 = 0$ and $d_{i+1} = f^\omega(d_i)$. Further, let $G(0) = 0$ and

$$G(x+1) = \begin{cases} f'(x+1) & \text{if } d_{3i} \leq x+1 < d_{3i+1} \text{ for some } i \\ G(x) & \text{otherwise} \end{cases}$$

Let $g(x) = \max(G(x), f(x))$.

We must prove that g is an honest function. It is easy to see that g is monotone and that $g(x) \geq 2^x$. It is a little bit harder to see that g has elementary graph, i.e. that the relation $g(x) = y$ is elementary. We shall just argue briefly for this and leave the details to the reader. The function f^ω is honest whenever f is honest. Hence it is possible to decide elementarily in j and x whether $d_j \leq x$. But then it is also possible to compute the j such that $d_j \leq x < d_{j+1}$ elementary in x. Furthermore, the relation $f'(x) = y$ is also elementary, and thus the relation $G(x) = y$ is elementary. So both f and G have elementary graphs, and thus g has elementary graph since $g(x) = \max(G(x), f(x))$.

We have to prove that $f^{[n]} <_E g^{[n]} <_E f^{[n+1]}$ holds for any n. It follows straight away from the Growth Theorem that $f^{[0]} \leq_E g^{[0]} \leq_E f^{[1]}$, and then the monotonicity of the jump operator yields $f^{[n]} \leq_E g^{[n]} \leq_E f^{[n+1]}$ for every n. We are left to prove that also $f^{[n+1]} \not\leq_E g^{[n]}$ and $g^{[n]} \not\leq_E f^{[n]}$ hold for every n.

First we prove $g^{[n]} \not\leq_E f^{[n]}$. By Lemma 9 it is sufficient to prove that there exist infinitely many x such that $g^{[n]}(x) = f^{[n+1]}(x)$. Fix an arbitrary n. Chose any j such that $j = 3i$ and $d_j > n + 2$. Then

$$d_{j+1} = f^\omega(d_j) = f^{[d_j]}(d_j) > f^{[n+2]}(d_j) = f^{[n+1]d_j+1}(d_j) > f^{[n+1]n+1}(d_j).$$

So by the definition of $g^{[0]}$ and choice of j we have

$$g^{[0]}(x) = f^{[1]}(x) \text{ for any } x \text{ such that } d_j \leq x \leq f^{[n+1]n+1}(d_j) \qquad (0)$$

Then, from (0), Lemma 7, and $f^{[n+1]}(x) \geq f^{[2]}(x)$ we have

$$g^{[1]}(x) = f^{[2]}(x) \text{ for any } x \text{ such that } d_j \leq x \leq f^{[n+1]n}(d_j) \qquad (1)$$

Then, from (1), Lemma 7, and $f^{[n+1]}(x) \geq f^{[3]}(x)$ we have

$$g^{[2]}(x) = f^{[3]}(x) \text{ for any } x \text{ such that } d_j \leq x \leq f^{[n+1]n-1}(d_j) \qquad (2)$$

...and so we proceed to get

$$g^{[n]}(x) = f^{[n+1]}(x) \text{ for any } x \text{ such that } d_j \leq x \leq f^{[n+1]}(d_j) \qquad (n)$$

This proves that there exist infinitely many x such that $g^{[n]}(x) = f^{[n+1]}(x)$.

Now we turn to the proof of $f^{[n+1]} \not\leq_E g^{[n]}$. Let $d_{3i+2} \leq x < d_{3i+3}$. We have $G(x) = G(d_{3i+1} \dot{-} 1) = f'(d_{3i+1} \dot{-} 1) \leq f'(d_{3i+1}) \leq f^\omega(d_{3i+1}) = d_{3i+2} \leq x$. Hence we have $g(x) = \max(f(x), G(x)) = f(x)$ for any x such that $d_{3i+2} \leq x < d_{3i+3}$. Now we can use Lemma 7 to prove that there exist infinitely many y such that

$$g^{[n]}(x) = f^{[n]}(x) \text{ for any } x \text{ such that } y \leq x \leq f'(y).$$

and then $f^{[n+1]} \not\leq_E g^{[n]}$ follows by Lemma 8. □

Now we can solve a problem stated as open by Meyer and Ritchie [13]. The function 2^x is a *backbone* function for the Grzegorczyk hierarchy, i.e. if we let $f(x) = 2^x$ and $\mathcal{E}^{n+3} = \{\psi \mid \psi \leq_E f^{[n]}\}$, then we have the usual Grzegorczyk hierarchy $\mathcal{E}^3 \subset \mathcal{E}^4 \subset \mathcal{E}^5 \subset \ldots$ from stage three and upwards. If we use the intermediate function g from the previous proof as the backbone function in place of 2^x we will get a slightly different hierarchy $\mathcal{C}^3 \subset \mathcal{C}^4 \subset \mathcal{C}^5 \subset \ldots$. The classes in this hierarchy have all the significant closure properties of the Grzegorczyk classes, i.e. they will be closed under limited recursion, closed under the bounded μ-operator, etcetera, but for each $i \geq 3$ we have $\mathcal{E}^i \subset \mathcal{C}^i \subset \mathcal{E}^{i+1}$. Meyer and Ritchie ask if there exists backbone function which generates such an intermediate hierarchy. We see that the answer to their question is positive. Moreover, the high$_n$ function given in the proof of the next theorem will generate a hierarchy $\mathcal{D}^3 \subset \mathcal{D}^4 \subset \mathcal{D}^5 \subset \ldots$ such that $\mathcal{E}^i \subset \mathcal{D}^i \subset \mathcal{E}^{i+1}$ for i such that $3 \leq i < n$ and $\mathcal{E}^{i+1} = \mathcal{D}^i$ for $i \geq n$. Our proof technique suggests that we can extend these results to an extended Grzegorczyk hierarchy and ordinals above ω. Let β be an ordinal such that $\mathcal{E}^\beta$ is a class in an extended Grzegorczyk hierarchy. Then, by using our proof techniques, it should be possible to find a backbone function which generates a hierarchy $\mathcal{F}$ such that $\mathcal{E}^\alpha \subset \mathcal{F}^\alpha \subset \mathcal{E}^{\alpha+1}$ for $\alpha < \beta$, and $\mathcal{E}^{\alpha+1} = \mathcal{F}^\alpha$ for $\alpha \geq \beta$.

Theorem 11. *The set $\mathbf{H}_{n+1} \setminus \mathbf{H}_n$ is nonempty, i.e. for any number n there exists an honest degree $\mathbf{a}$ such that $\mathbf{a}^{[n]} = \mathbf{0}^{[n+1]}$, but $\mathbf{0}^{[k]} < \mathbf{a}^{[k]} < \mathbf{0}^{[k+1]}$ for any $k < n$.*

Proof. We prove that the set $\mathbf{H}_{18} \setminus \mathbf{H}_{17}$ is nonempty. Let $f(x) = 2^x$. So $\deg(f) = \mathbf{0}$. We define the sequence $\{d_i\}_{i\in\omega}$ by $d_{4i+1} = f^{[18]}(d_{4i})$, $d_{4i+2} = f^{[17]}(d_{4i+1})$, $d_{4i+3} = f^{[17]}(d_{4i+2})$, and $d_{4i+4} = f^{[17]}(d_{4i+3})$. Let $d_0 = 0$. Further, let $G(0) = 0$ and let

$$G(x+1) = \begin{cases} f'(x+1) & \text{if } d_{4i} \leq x+1 < d_{4i+1} \text{ for some } i \\ G(x) & \text{otherwise} \end{cases}$$

Let $g(x) = \max(G(x), f(x))$.

We must prove that g is an honest function. In the proof of Theorem 10 we argue that a function very similar to g is honest. This argument can easily be adapted to prove that g is honest. We skip the details. Further we obviously have $f^{[n]} \leq_E g^{[n]} \leq_E f^{[n+1]}$ for any n. (We have $f^{[0]}(x) \leq g^{[0]}(x) \leq f^{[1]}(x)$, and hence $f^{[n]} \leq_E g^{[n]} \leq_E f^{[n+1]}$ for $n \in \omega$ follows by the Growth Theorem and the monotonicity of the jump operator.) Thus we are left to prove that $f^{[18]} \leq_E g^{[17]}$ and $f^{[17]} \not\leq_E g^{[16]}$.

First we prove $f^{[18]} \leq_E g^{[17]}$. We need

$$g^{[17]}(d_{4i}) = f^{[18]}(d_{4i}) \text{ whenever } d_{4i} \geq 18. \tag{Claim}$$

Straightaway from the definitions above we have $g^{[0]}(x) = f^{[1]}(x)$ for any x and i such that $d_{4i} \leq x < d_{4i+1} = f^{[18]}(d_{4i})$. Chose any i such that $d_{4i} \geq 18$. Then $f^{[18]}(d_{4i}) = f^{[17]d_{4i}+1}(d_{4i}) > f^{[17]18}(d_{4i})$. Hence $g^{[0]}(x) = f^{[1]}(x)$ for any x and

i such that $d_{4i} \leq x \leq f^{[17]18}(d_{4i})$. Now we can use Lemma 7 seventeen times to establish $g^{[17]}(x) = f^{[18]}(x)$ for any x and i such that $d_{4i} \leq x \leq d_{4i}$. This completes the proof of (Claim). Note that $g^{[17]}(x) \geq f^{[17]}(x)$ holds for every x because $g(x) \geq f(x)$ holds for every x. Now pick any sufficiently large x. (Pick any x such that there exists j such that $x \geq d_j \geq 18$.) Then there exists unique i such that $d_{4i} \leq x < d_{4i+4}$. Fix that i. Further we have

$$
\begin{aligned}
g^{[17]5}(x) &\geq g^{[17]5}(d_{4i}) && g^{[17]5} \text{ is montone} \\
&= g^{[17]4} f^{[18]}(d_{4i}) && \text{(Claim)} \\
&= g^{[17]4}(d_{4i+1}) && \text{def. of } d_j \\
&\geq g^{[17]} f^{[17]3}(d_{4i+1}) && \text{because } g^{[17]}(x) \geq f^{[17]}(x) \\
&= g^{[17]}(d_{4i+4}) && \text{def. of } d_j \\
&= f^{[18]}(d_{4i+4}) && \text{(Claim)} \\
&\geq f^{[18]}(x) . && f^{[18]} \text{ is monotone}
\end{aligned}
$$

Hence we see that $f^{[18]}(x) \leq g^{[17]5}(x)$ for every sufficiently large x. Hence $f^{[18]}(x) \leq g^{[17]5+k}(x)$ for some fixed k. Hence $f^{[18]} \leq_E g^{[17]}$ by the Growth Theorem.

We turn to the proof of $f^{[17]} \not\leq_E g^{[16]}$. First we need to establish

$$g^{[0]}(x) = f^{[0]}(x) \text{ for any } x, i \text{ such that } d_{4i+2} \leq x < d_{4i+4}. \quad (**)$$

Fix x, i such that $d_{4i+2} \leq x < d_{4i+4}$. Then we have $G(x) = G(d_{4i+1} \dot{-} 1) = f'(d_{4i+1} \dot{-} 1) \leq f^{[17]}(d_{4i+1}) = d_{4i+2} \leq x$. So $G(x) \leq x$, and thus $g^{[0]}(x) = g(x) = \max(f(x), G(x)) = f(x) = f^{[0]}(x)$. This proves (**). The definition of d_{4i+4} says that $d_{4i+4} = f^{[17]} f^{[17]}(d_{4i+2})$. Hence we have

$$g^{[0]}(x) = f^{[0]}(x) \text{ for any } x, i \text{ such that } d_{4i+2} \leq x \leq f^{[16]16} f^{[17]}(d_{4i+2}). \quad (0)$$

Now, apply Lemma 7 with $y = d_{4i+2}$ and $z = f^{[17]}(d_{4i+2})$. Then, by (0) and the lemma we have

$$g^{[1]}(x) = f^{[1]}(x) \text{ for any } x, i \text{ such that } d_{4i+2} \leq x \leq f^{[16]15} f^{[17]}(d_{4i+2}). \quad (1)$$

One more application of Lemma 7, again with $y = d_{4i+2}$ and $z = f^{[17]}(d_{4i+2})$, gives

$$g^{[2]}(x) = f^{[2]}(x) \text{ for any } x, i \text{ such that } d_{4i+2} \leq x \leq f^{[16]14} f^{[17]}(d_{4i+2}). \quad (2)$$

Now go on and apply Lemma 7 another fourteen times to establish

$$g^{[16]}(x) = f^{[16]}(x) \text{ for any } x, i \text{ such that } d_{4i+2} \leq x \leq f^{[17]}(d_{4i+2}). \quad (16)$$

From (16) and Lemma 8 it follows that $f^{[17]} \not\leq_E g^{[16]}$. □

In the previous proof we simply wrote down a neat definition of a function g and proved that g was high$_{n+1}$ and not high$_n$. Such a proof is more informative than a proof based on a Jump Inversion (Interpolation) theorem since it exhibit a transparent description of a nontrivial high$_{n+1}$ function. In the next section we will prove that nontrivial low$_{n+1}$ functions exist, but the proof will be based on a Jump Interpolation theorem. It is also possible to give a simple transparent definition of a nontrivial low$_{n+1}$ function, but our proof that the function indeed is low$_{n+1}$ (but not low$_n$) is complicated and contains many tedious technical details.

5 Jump interpolation and High$_n$ degrees.

Assume $\mathbf{a} \geq \mathbf{0}'$. It is proved in [5] that there exists $\mathbf{b}$ such that $\mathbf{b}' = \mathbf{a}$. (a Jump Inversion theorem). We are going to strengthen this result and prove that not only does there exist a $\mathbf{b}$ such that $\mathbf{b}' = \mathbf{a}$, but also one below $\mathbf{c}$ whenever $\mathbf{c} \leq \mathbf{a} \leq \mathbf{c}'$. This comes close to what we call a Jump Interpolation theorem in classical degree theory, and is considerably harder to prove than the Jump Inversion theorem in [5].

Lemma 12. *Let g and f be honest functions such that (i) $f(x+1) \geq 2^{f(x)}$, (ii) $g(x+1) \geq 2^{g(x)}$, and (iii) $f(x) \geq g(x)$. Then there exists an honest function h such that (I) $h \equiv_E g$, (II) $h(x) \leq f(x)$, (III) $h(x+1) \geq 2^{h(x)}$, and (IV) $h(x) \leq f(y) \Rightarrow h(x+1) \leq f(y+1)$ for all x, y.*

Proof. Assume that the functions f and g are given. We define $h(x)$ by recursion on x. Let $h(0) = g(0)$. Now suppose h is defined on x. Find the least y such that $h(x) \leq f(y) < f(y+1) < g(x+1)$. Let $h(x+1) = f(y+1)$ if such a y exists. Let $h(x+1) = g(x+1)$ if such a y does not exist.

We prove that (I), (II), (III), and (IV) hold when we assume (i), (ii), and (iii). It follows straightaway from the definition of h that (II) and (IV) holds. We turn to the proof of (III). Case: $h(x+1) \neq g(x+1)$. Then there exists y such that $h(x) \leq f(y) < f(y+1) = h(x+1)$. Thus we have $h(x+1) = f(y+1) \geq 2^{f(y)} \geq 2^{h(x)}$ by (i). Case: $h(x+1) = g(x+1)$. Then we have $h(x) \leq g(x) \leq g(x+1) = h(x+1)$ and $h(x+1) \geq 2^{h(x)}$ follows by (ii). This completes the proof of (III). It is easy to see that the graph of h is elementary, and hence h must be honest. Further, it is obvious that $g(x) \geq h(x)$, and hence $h \leq_E g$ by the Growth Theorem. Since we have $f(x) \geq g(x)$ (by assumption), we can by inspecting the definition of h see that $h(2x) \geq g(x)$. Hence we have $g \leq_E h$ by the Growth Theorem. This completes the proof of (I). □

Definition 13. *Let f be an honest function such that $f(x+1) \geq 2^{f(x)}$. Let $\mathcal{S}_f(0) = 0$.*

$$\mathcal{S}_f(x+1) = \begin{cases} f(i+1) \text{ if } x+1 = f(i) \text{ for some } i \\ \mathcal{S}_f(x) \quad \text{otherwise} \end{cases}$$

(The function $\mathcal{S}_f$ is well defined since f is an injection.) Further, let $\mathcal{I}_f(x) = \max(\mathcal{S}_f(x), 2^x)$.

Lemma 14. *Let f be an honest function. Then $\mathcal{I}_f$ is also honest.*

Proof. We have the equivalence $S_f(x+1) = y \Leftrightarrow$

$$(\exists i < y)[f(i) = x+1 \wedge f(i+1) = y] \vee (\forall i < y)[f(i) \neq x+1 \wedge S_f(x) = y].$$

All logical operations on the right hand side of the equivalence are elementary. The elementary functions are closed under limited recursion, and the graph of f is elementary by assumption. It follows that the graph of S_f is elementary. It is easy to that the graph of $\mathcal{I}_f$ will be elementary whenever the graph of S_f is. It is even easier to see that $\mathcal{I}_f$ is monotone and that $\mathcal{I}_f(x) \geq 2^x$. □

Lemma 15. *Let f be an honest function such that $f(x+1) \geq 2^{f(x)}$. Then we have (i) $\mathcal{I}_f(f(x)) = f(x+1)$ and*

$$(ii)\ f(z) \leq x < f(z+1) \Rightarrow \mathcal{I}_f(x) = \max(f(z+1), 2^x).$$

Proof. (i) holds since

$$\mathcal{I}_f(f(x)) = \max(S_f(f(x)), 2^{f(x)}) = \max(f(x+1), 2^{f(x)}) = f(x+1).$$

The first and the second equality are respectively due to the definitions of $\mathcal{I}_f$ and S_f, whereas the third equality holds by assumption. (ii) follows easily from (i) and the definition of $\mathcal{I}_f$. □

Lemma 16. *Let g be an honest function such that $g(x+1) \geq 2^{g(x)}$. Then $\mathcal{I}_g$ is honest and $\mathcal{I}'_g \equiv_E g$.*

Proof. Let k be the fixed number $g(0)$. It follows from Lemma 15 (i) that $g(x) = \mathcal{I}_g^x(k)$ (*). Furthermore, it follows from (*) that there exists a fixed number m such that $(\mathcal{I}'_g)^m(x) \geq g(x)$ and that there exists a fixed number n such that $\mathcal{I}'_g(x) \leq g^n(x)$. We have also proved that $\mathcal{I}_g$ is an honest function. Hence we have $\mathcal{I}'_g \equiv_E g$ by the Growth Theorem. □

Lemma 17. *We have $\mathcal{I}_{f'} \leq_E f$ for any honest f such that $f(x+1) \geq 2^{f(x)}$.*

Proof. For any z we have

$$\mathcal{I}_{f'}(f'(z)) = f'(z+1) = f^{z+2}(z+1) \leq f^2 f^{z+1}(z) = f^2(f'(z)).\ (*)$$

Lemma 15 guarantees that the first equality in (*) holds. The rest of the relations in (*) hold trivially. Now, fix any x and pick the unique z such that $f'(z) \leq x < f'(z+1)$. If $\mathcal{I}_{f'}(x) = 2^x$, we trivially have $\mathcal{I}_{f'}(x) \leq f^2(x)$. If $\mathcal{I}_{f'}(x) \neq 2^x$, we have $\mathcal{I}_{f'}(x) = f'(z+1)$ by Lemma 15 (ii). Further, by the monotonicity of f, Lemma 15 (i) and by (*) we have $\mathcal{I}_{f'}(x) = f'(z+1) = \mathcal{I}_{f'}(f'(z)) \leq f^2(f'(z)) \leq f^2(x)$. This proves that $\mathcal{I}_{f'}(x) \leq f^2(x)$ for any x. Hence $\mathcal{I}_{f'} \leq_E f$ follows by the Growth Theorem. □

Lemma 18. *Let g and f be honest functions such that (i) $g(x) \leq f(x)$, (ii) $g(x+1) \geq 2^{g(x)}$, (iii) $f(x+1) \geq 2^{f(x)}$, and (iv) $g(x) \leq f(y) \Rightarrow g(x+1) \leq f(y+1)$ for all x, y. Then we have $\mathcal{I}_g \leq_E \mathcal{I}_f$.*

Proof. We assume (i), (ii), (iii), and (iv). Lemma 14 says that $\mathcal{I}_f$ and $\mathcal{I}_g$ are honest functions. Thus, by the Growth Theorem it is sufficient to prove that $\mathcal{I}_g(x) \leq \mathcal{I}_f^2(x)$. Pick any x and fix the unique i such that $g(i) \leq x < g(i+1)$. Suppose $\mathcal{I}_g(x) = 2^x$. Then $\mathcal{I}_g(x) \leq \mathcal{I}_f^2(x)$ holds trivially. Suppose $\mathcal{I}_g(x) \neq 2^x$. Then we have $\mathcal{I}_g(x) = g(i+1)$ (†) by Lemma 15 (ii). Now, fix the unique j such that $f(j) \leq g(i) < f(j+1)$. Now we have

$$\begin{aligned}
\mathcal{I}_f^2(x) &\geq \mathcal{I}_f^2(f(j)) & &\mathcal{I}_f \text{ is montone, and } x \geq f(j) \\
&= f(j+2) & &\text{Lemma 15 (i)} \\
&\geq g(i+1) & &\text{(iv) and } g(i) \leq f(j+1) \\
&= \mathcal{I}_g(x) & &\text{(†)}
\end{aligned}$$

Hence we have $\mathcal{I}_g(x) \leq \mathcal{I}_f^2(x)$ for every x, and the proof is complete. □

Theorem 19. *Let* **a** *and* **b** *be honest degrees such that* $\mathbf{0}' \leq \mathbf{a} \leq \mathbf{b} \leq \mathbf{a}'$. *Then there exists an honest degree* **c** $\leq$ **a** *such that* $\mathbf{c}' = \mathbf{b}$.

Proof. Chose $f \in \mathbf{a}$ such that $f(x+1) \geq 2^{f(x)}$. (This is possible since $\mathbf{a} \geq \mathbf{0}'$.) Then we also have $f'(x+1) \geq 2^{f'(x)}$. Chose $g_0 \in \mathbf{b}$ such that $g_0(x+1) \geq 2^{g_0(x)}$ and $g_0(x) \leq f'(x)$. (This is possible since $\mathbf{a}' \geq \mathbf{b} \geq \mathbf{0}'$.) By Lemma 12 there exists honest $g \in \mathbf{b}$ such that $g(x+1) \geq 2^{g(x)}$ and $g(x) \leq f'(x)$ and $g(x) \leq f'(y) \Rightarrow g(x+1) \leq f'(y+1)$. Now $\mathcal{I}_g \leq_E \mathcal{I}_{f'}$ follows by Lemma 18. We have $\mathcal{I}_{f'} \leq_E f$ by Lemma 17, and hence $\mathcal{I}_g \leq_E f$. It follows from Lemma 16 that $\mathbf{b} = \deg(\mathcal{I}'_g)$. It also follows from lemmas above that $\mathcal{I}_g$ is honest. Hence let $\mathbf{c} = \deg(\mathcal{I}_g)$ and the theorem holds. □

Theorem 20. *For any honest degree* **a** *there exists an honest degree* **b** *such that* $\mathbf{a} < \mathbf{b}$ *and* $\mathbf{a}' = \mathbf{b}'$.

Proof. A similar theorem is proved in [5], and we just sketch a proof here. Pick any honest $f \in \mathbf{a}$. First we define a sequence of numbers $\{d_i\}_{i \in \omega}$ such that it is possible to decide elementary in x whether $d_i = x$ for some i. Further the distance between d_i and d_{i+1} should be sufficiently large. E.g. $d_{i+1} = f'f'(d_i)$ is sufficient, and any larger distance will do, e.g. $d_{i+1} = f^\omega(d_i)$. Further, let $G(0) = 0$ and

$$G(x+1) = \begin{cases} f'(x+1) & \text{if } d_i = x+1 \text{ for some } i \\ G(x) & \text{otherwise} \end{cases}$$

Finally, define g by $g(x) = \max(G(x), f(x))$. The theorem holds when $\mathbf{b} = \deg(g)$. The function g is honest. Since $g(x) \geq f(x)$ we have $f \leq_E g$ by the Growth Theorem. By Lemma 9 we have $g \not\leq_E f$. The proof that $f' \equiv_E g'$ relies on the Growth Theorem and that the distance from d_i to d_{i+1} is large. □

Corollary 21. *The set* $\mathbf{L}_{n+1} \setminus \mathbf{L}_n$ *is nonempty, i.e. for any number* n *there exists an honest degree* **a** *such that* $\mathbf{a}^{[n]} = \mathbf{0}^{[n]}$, *but* $\mathbf{0}^{[k]} < \mathbf{a}^{[k]} < \mathbf{0}^{[k+1]}$ *for any* $k < n$.

Proof. This follows straightaway from Theorem 19, Theorem 20, and the monotonicity of the jump operator. □

6 The $\ll$-relation and the "caps to"-relation

Definition 22. We define *the least upper bound* (l.u.b.) and the *greatest lower bound* (g.l.b.) of the partially ordered set of honest degrees in the usual way. A structure where each pair of elements has both a l.u.b. and a g.l.b. is called *a lattice*.

Let $\max[f,g](x) \stackrel{\text{def}}{=} \max(f(x), g(x))$ and $\min[f,g](x) \stackrel{\text{def}}{=} \min(f(x), g(x))$ for any honest functions f and g. The function $\max[f,g]$ is called *the join* of f and g, and $\max[\cdot,\cdot]$ is called *the join operator* (on functions). The function $\min[f,g]$ is called *the meet* of f and g, and $\min[\cdot,\cdot]$ is called *the meet operator* (on functions).

The join and the meet operator on the honest functions induce operators on the honest elementary degrees. A proof can be found in [5]. Thus the next definition makes sense, i.e. the definition is not dependent on the choices of honest $f \in \mathbf{a}$ and honest $g \in \mathbf{b}$.

Definition 23. Let $f \in \mathbf{a}$ and $g \in \mathbf{b}$ be honest functions. We define $\mathbf{a} \cup \mathbf{b} \stackrel{\text{def}}{=} \deg(\max[f,g])$, and we define $\mathbf{a} \cap \mathbf{b} \stackrel{\text{def}}{=} \deg(\min[f,g])$. The degree $\mathbf{a} \cup \mathbf{b}$ is called *the join* of $\mathbf{a}$ and $\mathbf{b}$, and $\cup$ is called the *the join operator* (on degrees). The degree $\mathbf{a} \cap \mathbf{b}$ is called *the meet* of $\mathbf{a}$ and $\mathbf{b}$, and $\cap$ is called the *the meet operator* (on degrees).

When we say that a degree $\mathbf{a}$ *cups to* a degree $\mathbf{b}$, we mean that there exists a degree $\mathbf{c}$ strictly below $\mathbf{b}$ such that $\mathbf{a} \cup \mathbf{c} = \mathbf{b}$. Analogously, that $\mathbf{a}$ *caps to* $\mathbf{b}$, means that there exists a degree $\mathbf{c}$ strictly above $\mathbf{b}$ such that $\mathbf{a} \cap \mathbf{c} = \mathbf{b}$.

The structure of elementary honest degrees is a lattice. This was first proved by Machtey [11]. He showed that $\deg(\max[f,g])$ is the l.u.b. of $\deg(f)$ and $\deg(g)$, and that $\deg(\min[f,g])$ is the g.l.b. of $\deg(f)$ and $\deg(g)$. (His proof is based on Turing machines, tape constructible functions, and computability-theoretic constructions. A simpler proof based solely on the Growth Theorem can be found in Kristiansen [5].) Later, in [12], Machtey proves a very strong density theorem for the lattice of elementary honest degrees. He proves that for any degrees $\mathbf{a}, \mathbf{b}$ such that $\mathbf{a} < \mathbf{b}$ there exists degrees $\mathbf{c}_0, \mathbf{c}_1$ strictly between $\mathbf{a}$ and $\mathbf{b}$ such that $\mathbf{a} = \mathbf{c}_0 \cap \mathbf{c}_1$ and $\mathbf{b} = \mathbf{c}_0 \cup \mathbf{c}_1$. Recently, in [7], the author have proved that if $\mathbf{0} \ll \mathbf{a} < \mathbf{b}$, then $\mathbf{a}$ cups to $\mathbf{b}$. (The $\ll$-relation is defined below.) Roughly speaking, this means that almost any degree cups up to any degree above it. (We just have possible exceptions for degrees which in some sense lie close to $\mathbf{0}$.) In this sense the "cups to"-relation is (almost) trivial. In contrast the "caps to"-relation is not trivial. In the sequel we shall see that some degrees just slightly above an arbitrary degree $\mathbf{a}$ do not cap to $\mathbf{a}$.

Definition 24. A binary function ρ is a *universal function* for an honest degree $\mathbf{a} = \deg(f)$ if for every unary function $\xi \leq_E f$ there exists an n such that $\xi(x) = \rho(n,x)$. Let f and g be honest functions. We write $f \ll g$ when there is a universal function ρ for the degree $\deg(f)$ such that $\rho \leq_E g$. We also write $\ll$ for the corresponding relation on the degrees.

A unary function ϕ *majorizes* a unary function ψ *almost everywhere* (a.e.) when there is a number k such that for all $x > k$ we have $\psi(x) < \phi(x)$. Notation: $\psi(x) \stackrel{(a.e.)}{<} \phi(x)$.

That the relation $\mathbf{a} \ll \mathbf{b}$ holds means that in some sense $\mathbf{b}$ lies far above $\mathbf{a}$. The situation $\mathbf{a} \ll \mathbf{b}$ implies that $\mathbf{a} < \mathbf{b}$, but there exist degrees $\mathbf{a}, \mathbf{b}$ such that $\mathbf{a} < \mathbf{b}$ and $\mathbf{a} \not\ll \mathbf{b}$. We know that there exists a whole $\ll$-dense set of honest degrees between $\mathbf{a}$ and $\mathbf{a}'$. (See Meyer and Ritchie [13] and Kristiansen [5].) Yet it is not known if the structure of honest degrees is $\ll$-dense, i.e. if there for any $\mathbf{a}, \mathbf{b}$ such that $\mathbf{a} \ll \mathbf{b}$ does exists $\mathbf{c}$ such that $\mathbf{a} \ll \mathbf{c} \ll \mathbf{b}$. (This is stated as an open problem in Meyer and Ritchie [13].) The next theorem gives a characterisation of the $\ll$-relation. Meyer and Ritchie [13] prove related results.

Lemma 25. *The equivalence* $g \ll f \Leftrightarrow (\exists m)(\forall k)[g^k(x) \stackrel{(a.e)}{<} f^m(x)]$ *holds for any honest functions f and g.*

Proof. Let $\mathcal{T}_n$ be the usual Kleene T-predicate for n-ary functions, and let $\mathcal{U}$ be a function which picks the result of a computation from a computation tree. ($\mathcal{U}$ is an elementary function and $\mathcal{T}_n$ is an elementary predicate.) We need the following version of the Kleene Normal Form Theorem.

(KNF) An n-ary function ψ is elementary in an honest function f (in symbols $\psi \leq_E f$) iff there exist a computable (recursive) index e for ψ and a fixed number k such that

$$\{e\}(x_1,\ldots,x_n) = \mathcal{U}(\mu y \leq f^k(\max(x_1,\ldots,x_n))[\mathcal{T}_n(e,x_1,\ldots,x_n,y)]) \ .$$

The proof of (KNF) is rather straight forward. Details can be found in [8]. Now, let us turn to the proof of the lemma. Assume the right hand side of the equivalence. Then we can pick a number m such that $g^k(x) \stackrel{(a.e.)}{<} f^m(x)$ holds for every choice of k. (Fix this m throughout the proof.) This implies that for every fixed k there exists n such that $g^k(x) \leq n + f^m(x)$ (*). Let ξ be any unary function elementary in g. By (KNF) we have a computable index e for ξ and a fixed number k such that

$$\xi(x) = \mathcal{U}((\mu t \leq g^k(x))[\mathcal{T}_1(e,x,t)]) \stackrel{(*)}{=} \mathcal{U}((\mu t \leq n + f^m(x))[\mathcal{T}_1(e,x,t)]) \ .$$

Let $\rho(\langle e,n \rangle, x) \stackrel{def}{=} \mathcal{U}((\mu t < n + f^m(x))[\mathcal{T}_1(e,x,t)])$. Then $\rho \leq_E f$, and for every unary function $\xi \in \deg(g)$ there exists a number n such that $\xi(x) = \rho(n,x)$. Hence $g \ll f$.

Now assume $g \ll f$. Let $\psi(x) = (\max_{i \leq x} \max_{j \leq x} \rho(i,j)) + 1$, where ρ is a universal function for $\deg(g)$ and $\rho \leq_E f$. Now $\psi \leq_E f$, so the Growth Theorem yields a fixed n such that $\psi(x) \leq f^n(x)$. It is easy to verify that ψ majorizes (a.e.) every function which is elementary in g. Since $g^k \leq_E g$ for every fixed k, we have $g^k(x) \stackrel{(a.e)}{<} \psi(x) \leq f^n(x)$ for every fixed k. This completes the proof. □

Lemma 26. *We have* $\min(f^m(x), g^n(x)) \leq \min[f,g]^{m+n}(x)$ *for any honest functions f and g.*

Proof. Suppose the lemma does not hold, i.e. there exists a number k such that $\min(f^m(k), g^n(k)) > \min[f,g]^{m+n}(k)$. Let us say that $\min(f^m(k), g^n(k)) = f^m(k)$. (Do a symmetrical argument if it is the case that $\min(f^m(k), g^n(k)) = g^n(k)$.) Since $f^m(k) > \min[f,g]^{m+n}(k)$, we must have $\min[f,g]^{m+n}(k) \geq g^n(k)$. Now the following contradiction emerges: $g^n(k) \geq \min(g^n(k), f^m(k)) = f^m(k) > \min[f,g]^{m+n}(k) \geq g^n(k)$. □

Theorem 27. *If* $\mathbf{a} \ll \mathbf{b}$, *then* $\mathbf{b}$ *does not cap to* $\mathbf{a}$.

Proof. Assume $\mathbf{a} \ll \mathbf{b}$ and pick honest $f \in \mathbf{a}$ and honest $g \in \mathbf{b}$. Then we have $(\exists m)(\forall k)[f^k(x) \stackrel{(a.e)}{<} g^m(x)]$ (i) by Lemma 25. Now, assume for the sake of a contradiction that $\mathbf{b}$ caps to $\mathbf{a}$. So there exists an honest h such that $\min[g,h] \leq_E f$ and $h \not\leq_E g$. Hence we have a fixed number n such that $\min[g,h](x) < f^n(x)$ (ii) and $(\forall k)(\exists^\infty x)[h(x) > g^k(x)]$ (iii). (Both (ii) and (iii) follow by the Growth Theorem.) We shall see that for any m and x there exist n and $x_0 > x$ such that $f^{(m+1)n}(x_0) > g^m(x_0)$. This contradicts (i) and thus the proof is complete.

Let m and x be any numbers. By (iii) we can pick $x_0 > x$ such that $h(x_0) > g^m(x_0)$ (iv). Hence

$$g^m(x_0) = \min(g^m(x_0), h(x_0)) \qquad \text{(iv)}$$
$$\leq \min[g,h]^{m+1}(x_0) \qquad \text{Lemma 26}$$
$$< f^{(m+1)n}(x_0) \qquad \text{(ii)}$$

□

Corollary 28. *If* $\mathbf{0}'$ *caps to* $\mathbf{a}$, *then* $\mathbf{a}$ *is not a* low_1 *degree.*

Proof. Assume that $\mathbf{a}$ is low_1. Pick honest $g \in \mathbf{a}$ and honest $f \in \mathbf{0}'$. Since $\mathbf{a}$ is low_1, we have $g' \leq_E f$. Thus we get $g'(x) \leq f^m(x)$ for some fixed m by the Growth Theorem. Thus $g^k(x) \stackrel{(a.e)}{<} g^{x+1}(x) = g'(x) \leq f^m(x)$ for any fixed k. Thus $g \ll f$ by Lemma 25. Thus $\mathbf{a} \ll \mathbf{0}'$. Thus $\mathbf{0}'$ does not cap to $\mathbf{a}$ by Theorem 27. □

That $\mathbf{b}$ caps to $\mathbf{a}$ is trivially first-order definable from the $\leq$-relation in the structure of honest degrees. Thus it would have been nice if the converse of Theorem 27 also held, i.e. that $\mathbf{a} \ll \mathbf{b}$ if and only if $\mathbf{b}$ does not cap to $\mathbf{a}$. If so were, we could define the $\ll$-relation from the $\leq$-relation. Unfortunately we have not been able to prove that the converse of Theorem 27 holds, but neither have we been able to construct a counterexample. However, Theorem 27 yields an interesting definable subset of the honest degrees which generates the whole structure under join. Let $A = \{\mathbf{a} \mid \mathbf{a} \text{ caps to } \mathbf{0}\} \cup \{\mathbf{0}\}$. Then A is first-order definable from the $\leq$-relation. Further we have $\mathbf{0} \not\ll \mathbf{a}$ for any $\mathbf{a} \in A$. Let $\mathbf{b}$ be any honest degree strictly above $\mathbf{0}$. By a theorem of Machtey there exist $\mathbf{c}_0, \mathbf{c}_1$ such that $\mathbf{0} = \mathbf{c}_0 \cap \mathbf{c}_1$ and $\mathbf{b} = \mathbf{c}_0 \cup \mathbf{c}_1$. So $\mathbf{c}_0, \mathbf{c}_1 \in A$ and hence each honest degree equals the join of two degrees in A.

References

1. Ambos-Spies, K.: On the structure of polynomial time degrees of recursive sets. (Habilitationsschrift) Forschungsbericht Nr. 206/1985. Universität Dortmund, Dortmund, Germany 1985 (P.O. Box 500500, D-4600 Dortmund 50).
2. Beklemishev, L.D.: A proof-theoretic analysis of collection. *Archive for Mathematical Logic* **37** (1998) 275–296.
3. Beklemishev, L.D.: Induction rules, reflection principles, and provably recursive functions. *Annals of Pure and Applied Logic* **85** (1997) 193–242.
4. Basu, S.K.: On the structure of subrecursive degrees. *Journal of Computer and System Sciences* **4** (1970) 452–464.
5. Kristiansen, L.: A jump operator on honest subrecursive degrees. *Archive for Mathematical Logic* **37** (1998) 105–125.
6. Kristiansen, L.: Information content and computational complexity of recursive sets. *Logical Foundations of Mathematics, Computer Science and Physics – Kurt Gödel Legacy* (eds. Hájek) 235–246. Springer Lecture Notes in Logic, Springer-Verlag 1996.
7. Kristiansen, L.: Fragments of Peano arithmetic and subrecursive degrees. (Submitted.)
8. Kristiansen, L.: Papers on subrecursion theory. Dr Scient Thesis, ISSN 0806-3036, ISBN 82-7368-130-0, Research report 217, Department of Informatics, University of Oslo 1996. (This report contains [5] and [6].)
9. Ladner, R.E.: On the structure of polynomial time reducibility. *Journal of the Association for Computing Machinery* **22** (1975) 155–171.
10. Machtey, M.: Augmented loop languages and classes of computable functions. *Journal of Computer and System Sciences* **6** (1972) 603–624.
11. Machtey, M.: The honest subrecursive classes are a lattice. *Information and Control* **24** (1974) 247–263.
12. Machtey, M.: On the density of honest subrecursive classes. *Journal of Computer and System Sciences* **10** (1975) 183–199.
13. Meyer A.R., Ritchie D.M.: A classification of the recursive functions. *Zeitschr. f. math. Logik und Grundlagen d. Math. Bd.* **18** (1972) 71–82.
14. Odifreddi, P.: *Classical recursion theory.* Amsterdam London New York Tokyo: North-Holland 1989.
15. Péter, R.: *Rekursive Funktionen.* Budapest: Verlag der Ungarischen Akademie der Wissenschaften 1957 [English translation: New York: Academic Press 1967].
16. Rose, H.E.: *Subrecursion. Functions and hierarchies.* Oxford: Clarendon Press 1984.
17. Soare, R. I.: *Recursive enumerable sets and degrees.* Berlin Heidelberg New York London Paris Tokyo: Springer-Verlag 1987.

Star-extremal Circulant Graphs

Ko-Wei Lih[1], Daphne D.-F. Liu[2], and Xuding Zhu[3]

[1] Institute of Mathematics, Academia Sinica, Nankang, Taipei, Taiwan 115.
makwlih@sinica.edu.tw
[2] Department of Mathematics and Computer Science, California State University, Los Angeles, Los Angeles, CA 90032, USA. dliu@calstatela.edu
[3] Department of Applied Mathematics, National Sun Yat-sen University, Kaoshiung, Taiwan 804. zhu@math.nsysu.edu.tw

Abstract. A graph is called star-extremal if its fractional chromatic number is equal to its circular chromatic number (also known as the star chromatic number). We prove that members of a certain family of circulant graphs are star-extremal. The result generalizes some known theorems of Sidorenko [7] and Gao and Zhu [5]. Then we show relations between circulant graphs and distance graphs and discuss their star-extremality. Furthermore, we give counter-examples to two conjectures of Collins [3] on the asymptotic independence ratios of circulant graphs. This article is an extended abstract of a paper with the same title which will appear elsewhere.

1 Introduction

Given a positive integer n and a set $S \subseteq \{0, 1, 2, \cdots, \lfloor n/2 \rfloor\}$, let $G(n, S)$ denote the graph with vertex set $V(G) = \{0, 1, 2, \cdots, n-1\}$ and edge set $E(G) = \{uv : |u - v|_n \in S\}$, where $|x|_n = \min\{|x|, n - |x|\}$ is the *circular distance modulo* n. Then $G(n, S)$ is called the *circulant graph* of order n with the *generating set* S.

In this article, we explore the star-extremality of circulant graphs. A graph is called *star-extremal* if its fractional chromatic number and circular chromatic number, defined below, are equal.

A *fractional coloring* of a graph G is a mapping c from $\mathcal{I}(G)$, the set of all independent sets of G, to the interval $[0, 1]$ of real numbers such that $\sum \{c(I) : x \in I \text{ and } I \in \mathcal{I}(G)\} \geq 1$ for any vertex x in G. The *fractional chromatic number* $\chi_f(G)$ of G is the infimum of the *weight*, $w(c) = \sum \{c(I) : I \in \mathcal{I}(G)\}$, of a fractional coloring c of G.

Suppose k and d are positive integers such that $k \geq d$. A (k,d)-*coloring* of a graph $G = (V, E)$ is a mapping c from V to $\{0, 1, \cdots, k-1\}$ such that $|c(x) - c(y)|_k \geq d$ for any edge xy in G. The *circular chromatic number* $\chi_c(G)$ of G is the infimum of k/d among all (k, d)-colorings of G. The circular chromatic number is also known as the *star-chromatic number* in the literature [8].

For any graph G, it is well-known that

$$\max\left\{\omega(G), \frac{|V(G)|}{\alpha(G)}\right\} \leq \chi_f(G) \leq \chi_c(G) \leq \chi(G) \text{ and } \lceil \chi_c(G) \rceil = \chi(G), \quad (1)$$

where $\omega(G)$ is the *clique number* (i.e., the maximum number of vertices of a complete subgraph in G); $\alpha(G)$ is the *independence number* (i.e., the maximum number of vertices of an independent set in G.) Hence, a graph G is star-extremal if the equality holds in the second inequality in (1).

The star-extremality for circulant graphs was first discussed by Gao and Zhu [5]. They proved that if all the vertices of $G(n, S)$ have degree ≤ 3, then $G(n, S)$ is star-extremal. On the other hand, there exist non star-extremal circulant graphs (see [5] or Section 3.) In general, it seems a difficult problem to determine whether or not an arbitrary circulant graph is star-extremal.

As circulant graphs are vertex-transitive, we know that

$$\chi_f(n, S) = n/\alpha(n, S) , \qquad (2)$$

where $\chi_f(n, S)$ and $\alpha(n, S)$ denote respectively, the fractional chromatic number and independence number for $G(n, S)$. Therefore determining the independence number of a circulant graph is equivalent to determining its fractional chromatic number. So far, to the best of our knowledge, the complexity of determining the independence number or the clique number of circulant graphs is unknown, i.e., it is unknown whether or not this is an NP-complete problem.

In Section 2, we focus on the family of circulant graphs with the generating set S being consecutive integers. Given integers $k < k' \leq n/2$, let $S_{k,k'}$ denote the set $\{k, (k+1), \cdots, k'\}$. We will determine the exact value of $\alpha(n, S_{k,k'})$ for any $n \geq 2k'$ and $k' \geq (5/4)k$. This result is used to prove that the circulant graphs $G(n, S_{k,k'})$ are star-extremal for all $n \geq 2k'$ and $k' \geq (5/4)k$.

Given a finite set S of positive integers, the *distance graph*, denoted as $G(Z, S)$, has as vertex set all integers Z and edges connecting u and v if $|u - v| \in S$. Thus, given S, the distance graph $G(Z, S)$ can be viewed as the limit of the circulant graphs $G(n, S)$ as n approaches infinity. In Section 3, we explore a relationship between circulant graphs and distance graphs for general sets S based on the property of star-extremality.

The *independence ratio* of a graph G is defined as $\alpha(G)/|V(G)|$. In Section 4, we show that for a given S, the fractional chromatic number of the distance graph $G(Z, S)$ is equal to the reciprocal of the asymptotic independence ratio of circulant graphs $G(n, S)$ as $n \to \infty$. Applying this fact, we present counter-examples to two conjectures of Collins [3] on the asymptotic independence ratio of circulant graphs.

2 Circulant graphs with interval generating sets

Given integers $k < k' \leq n/2$, recall that $S_{k,k'} = \{k, (k+1), \cdots, k'\}$. We determine the exact values of $\alpha(n, S_{k,k'})$ for all n and $k' \geq (5/4)k$. Then we use this result to show that such circulant graphs are star-extremal.

One of the tools we shall use is the following *multiplier method*, which was first used in [5]. Given a circulant graph $G(n, S)$ and a positive integer t, let

$$\lambda_t(n, S) = \min\{|ti|_n : i \in S\} ,$$

and let
$$\lambda(n,S) = \max\{\lambda_t(n,S) : t = 1,2,3,\cdots\},$$
where the multiplications ti are carried out modulo n and $|x|_n$ is the circular distance modulo n. For any positive integer t, the mapping c on $\{0,1,2,\cdots,n-1\}$ defined by $c(i) = ti$ is an $(n, \lambda_t(n,S))$-coloring for $G(n,S)$ (multiplications are carried out modulo n.) Hence, $\chi_c(n,S) \leq n/\lambda(n,S)$. Combining this with (1) and (2), the following result can be obtained:

Lemma 1. *([5]) For any circulant graph $G(n,S)$, $\lambda(n,S) \leq \alpha(n,S)$. Moreover, if $\lambda(n,S) = \alpha(n,S)$, then $\chi_f(n,S) = \chi_c(n,S) = n/\alpha(n,S)$, that is, $G(n,S)$ is star-extremal.*

The value of $\lambda(n,S)$ can be obtained in a finite number of calculation steps. To be precise, we have the following:

Lemma 2. *Given a circulant graph $G(n,S)$, $\lambda(n,S) = \lambda_t(n,S) = |ts|_n = |t(-s)|_n$ for some $1 \leq t \leq \lceil n/2 \rceil$ and $s \in S$.*

For the sub-family of $G(n, S_{k,k'})$ with $k' = 2k - 1$, Sidorenko [7] has shown that $\alpha(n, S_{k,2k-1}) = 2k$ for $6k - 2 \leq n \leq 8k - 3$. Gao and Zhu [5] proved that $\lambda(n, S_{k,2k-1}) = 2k$ for $6k - 2 \leq n \leq 8k - 3$. Hence, by Lemma 1, the following result is obtained.

Theorem 3. *([5]) If $k' = 2k - 1$, the circulant graphs $G(n, S_{k,k'})$ are star-extremal for all n where $6k - 2 \leq n \leq 8k - 3$.*

Other special sub-families of the circulant graphs $G(n, S_{k,k'})$ that have been studied include the following two:

Theorem 4. *([5]) Given $n, k' \in Z^+$, $k' \leq n/2$, $G(n, S_{1,k'})$ is star-extremal and $\chi_f(n, S_{1,k'}) = \chi_c(n, S_{1,k'}) = n/\lfloor \frac{n}{k'+1} \rfloor$.*

Theorem 5. *([5]) Suppose $k' = k + l \leq n/2$. If $n - 2k' < \min\{k, l\}$, then $G(n, S_{k,k'})$ is star-extremal and $\chi_f(n, S_{k,k'}) = \chi_c(n, S_{k,k'}) = n/k$.*

The proofs of Theorems 4 and 5 are obtained, respectively, by showing $\alpha(n, S_{1,k'}) = \lambda(n, S_{1,k'}) = \lfloor \frac{n}{k'+1} \rfloor$, and $\alpha(n, S_{k,k'}) = \lambda(n, S_{k,k'}) = k$ under the assumptions on k and k'.

We note here that the circulant graphs $G(n, S_{1,k'})$ in Theorem 4 are indeed powers of the cycle C_n on n vertices. Given n and r, the r-th *power* of C_n, denoted by C_n^r, has the same vertex set as C_n and u, v are adjacent if their distance on the cycle C_n is not greater than r. Therefore, by definition, $G(n, S_{1,k'}) = C_n^{k'}$.

In their study of the circular chromatic number of planar graphs Gao, Wang and Zhou [4] defined a family of planar graphs Q_n called *triangular prisms*, which have the vertex set $V = \{u_0, u_1, u_2, \cdots, u_{n-1}\} \cup \{v_0, v_1, v_2, \cdots, v_{n-1}\}$, and the edge set E consisting of two n-cycles $(u_0, u_1, \cdots, u_{n-1})$ and $(v_0, v_1, \cdots, v_{n-1})$ and $2n$ edges (u_i, v_i), (u_{i+1}, v_i) for every $0 \leq i \leq n - 1$ ($u_0 = u_n$). In [4], the argument to compute the values of $\chi_f(Q_n)$ is long. The proof can be shortened considerably by applying known results in circulant graphs. The family of planar graphs Q_n are precisely the second powers of even cycles. Indeed, $(v_0, u_1, v_1, u_2, v_2, \cdots, u_{n-1}, v_{n-1}, u_0)$ is a cycle of length $2n$, and $Q_n \cong C_{2n}^2$. Hence, by Theorem 4, $\chi_f(Q_n) = \chi_c(Q_n) = 2n/\lfloor \frac{2n}{k+1} \rfloor$.

Now we consider the general family of circulant graphs $G(n, S_{k,k'})$. We view the vertices of $G(n, S_{k,k'})$ as circularly ordered, and denote by $[a, b]$ the set of integers $\{a, a+1, a+2, \cdots, b\}$, where the addition is taken under modulo n. For example, $[2, 5] = \{2, 3, 4, 5\}$ and $[5, 2] = \{5, 6, \cdots, n-1, 0, 1, 2\}$.

Lemma 6. *Suppose I is an independent set of $G(n, S_{k,k'})$. Then for any i, the intersection $I \cap [i, i+k+k'-1]$ has cardinality at most k.*

Lemma 7. *Suppose $G = G(n, S_{k,k'})$ with $n = q(k+k') + r$, $0 \leq r \leq k+k'-1$, then*

$$\lambda(G) \geq \begin{cases} \lambda_q(G) = qk, & 0 \leq r \leq k' ; \\ \lambda_{q+1}(G) = qk + r - k', & k'+1 \leq r \leq k'+k-1 . \end{cases}$$

Theorem 8. *Suppose $G = G(n, S_{k,k'})$, $k' \geq (5/4)k$. Let $n = q(k+k') + r$, $0 \leq r \leq k+k'-1$, then*

$$\alpha(G) = \lambda(G) = \begin{cases} qk, & 0 \leq r \leq k' ; \\ qk + r - k', & k'+1 \leq r \leq k'+k-1 . \end{cases}$$

Equivalently, $\alpha(G) = \lambda(G) = qk + \max\{0, r - k'\}$.

Proof. Let $n = q(k+k') + r$, $0 \leq r \leq k+k'-1$. By Lemmas 1 and 7, it suffices to show that $\alpha(G) \leq qk + \max\{0, r - k'\}$.

Assume to the contrary that $\alpha(G) > qk + \max\{0, r-k'\}$. Let I be a maximum independent set of G. Regard I as a disjoint union of I-intervals, where an I-interval is a maximal interval $[a, b]$ consisting of vertices in I. By Lemma 6 the length (number of vertices) of any I-interval is between 1 and k. Assume that the independent set I chosen has the minimum number of I-intervals among all maximum independent sets of G.

Two I-intervals $[a, b]$ and $[c, d]$ are called *consecutive* if $[b+1, c-1] \cap I = \emptyset$. Note that the consecutive "relation" is not symmetric, i.e., $[a, b]$ and $[c, d]$ being consecutive does not imply that $[c, d]$ and $[a, b]$ are consecutive. (Indeed, $[c, d]$ and $[a, b]$ are not consecutive if $[a, b]$ and $[c, d]$ are consecutive, and I contains more than two I-intervals.) For two consecutive I-intervals $[a, b]$ and $[c, d]$, the cardinality of the set $[b+1, c-1]$ is called the *gap* between them.

Note that if $[a, b]$ and $[c, d]$ are two consecutive I-intervals, then $b + k' + 1 \leq c + k - 1$ (and symmetrically, $b - k + 1 \leq c - k' - 1$). For otherwise, for any point $d \in [b+1, c-1]$, the set $I' = I \cup \{d\}$ is independent with $|I'| > |I|$, a contradiction.

Now we show that the gap between any two consecutive I-intervals is at most $k-2$. Assume to the contrary that $[a, b]$ and $[c, d]$ are consecutive I-intervals such that $|[b+1, c-1]| \geq k-1$. Since $b \in I$, it follows that $[b+k, b+k'] \cap I = \emptyset$. Because

$|[b+1, c-1]| \geq k-1$, it follows that $[b+1, b+k-1] \cap I = \emptyset$. Hence $[b+1, b+k'] \cap I = \emptyset$. Then, we partition the interval $[b + k' + 1, b]$ $(= [0, n-1] - [b+1, b+k'])$ into sub-intervals of length $k + k'$, except the last sub-interval which may have size less than $k + k'$ (when $r \geq k' + 1$.) If $0 \leq r \leq k'$, then the number of such sub-intervals is equal to q. By Lemma 6, $|I| \leq qk$, contrary to our assumption. If $k' + 1 \leq r \leq k + k' - 1$, then the number of such sub-intervals is equal to $q+1$, and the last interval has size $r - k'$. Again, it follows from Lemma 6 that $|I| \leq qk + r - k'$, contrary to our assumption. Therefore, the gap between any two consecutive I-intervals is at most $k - 2$.

Next we show that the gap between any two consecutive I-intervals is greater than $2(k' - k)$. Assume to the contrary that $[a, b]$ and $[c, d]$ are consecutive I-intervals with gap t, $t \leq 2(k' - k)$. Let

$$I' = (I \cup [b+1, c-1]) - ([b+k'+1, c+k-1] \cup [b-k+1, c-k'-1]) \ .$$

It is clear that I' is an independent set of G with $|I'| \geq |I| + t - 2(t - (k' - k)) \geq |I|$. Hence, I' is a maximum independent set with less intervals than I, which is contrary to our assumption.

We conclude that the gap between any two consecutive I-intervals is between $2(k' - k) + 1$ and $k - 2$. By the assumption $k' \geq \frac{5k}{4}$, the gap between any two consecutive I-intervals is between $\frac{k}{2} + 1$ and $k - 2$. This implies that any set of consecutive k vertices in G intersects at most two I-intervals.

Choose two consecutive I-intervals $[a, b]$ and $[c, d]$ such that $|[a, b]| + |[c, d]|$ is the largest among all pairs of consecutive I-intervals. Since the gap between $[a, b]$ and $[c, d]$ is at most $k - 2$, it follows that $|[a, d]| \leq k$, for otherwise $a + k \in [c, d]$, contrary to the assumption of I being an independent set. Therefore $|[a, b]| + |[c, d]| \leq k - 2(k' - k) - 1$.

Let $[u, v]$ be the I-interval preceding $[a, b]$ (i.e., $[u, v]$ and $[a, b]$ are consecutive I-intervals), and let $[x, y]$ be the I-interval following $[c, d]$. By the facts that the gap between any two consecutive I-intervals is between $\frac{k}{2} + 1$ and $k - 2$, and that I is a maximum independent set, we have $I \cap [b - k + 1, c - k' - 1] = [u, v]$ and $I \cap [b + k' + 1, c + k - 1] = [x, y]$. Moreover, according to the choice of $[a, b]$ and $[c, d]$, we have $|[u, v]| \leq |[c, d]|$ and $|[x, y]| \leq |[a, b]|$. Therefore $|[u, v]| + |[x, y]| \leq k - 2(k' - k) - 1 \leq \frac{k}{2} - 1$ (since $k' \geq \frac{5k}{4}$). Let

$$I' = (I \cup [b+1, c-1]) - ([u, v] \cup [x, y]) \ .$$

It is clear that I' is an independent set with

$$|I'| \geq |I| + \frac{k}{2} + 1 - (\frac{k}{2} - 1) > |I| \ ,$$

contrary to our assumption that I is a maximum independent set. □

Corollary 9. *If $k' \geq (5/4)k$, then $G(n, S_{k,k'})$ is star-extremal.*

Theorem 10. *Suppose $G = G(n, S_{k,k'})$ with $n = q(k+k') + r$, $0 \leq r \leq k+k'-1$ and*

$$q \geq \frac{1}{k - k'} - \frac{k'}{k + k'} - \frac{kk'}{(k - k')(k + k')} \ ,$$

then G is star-extremal. Moreover, the values of $\alpha(G)$ and $\lambda(G)$ are the same as in Theorem 8.

Proof. First we consider the case that $0 \leq r \leq k'$. By Lemmas 1 and 7, it suffices to show that $\alpha(G) \leq qk$. Assume to the contrary, $\alpha(G) \geq qk + 1$, and let I be a maximum independent set of G. We define the I-intervals to be the same as in the proof of Theorem 8 and assume that the set I chosen has the minimum number of I-intervals among all maximum independent sets of G. Then the gap between any two consecutive I-intervals was shown to be between $2(k' - k) + 1$ and $k - 2$ in the proof of Theorem 8.

Now we show that for any i, $|I \cap [i, i + k + k' - 1]| \leq 2k - k'$. Let a be the least element of $I \cap [i, i + k + k' - 1]$, and let $[a, b]$ be the first non-empty intersection of an I-interval with $[i, i + k + k' - 1]$. Note that when $a = i$, $[a, b]$ may only be a part of an I-interval. Similarly, let $[c, d]$ be the last non-empty intersection of an I-interval with $[a, a + k]$. It is clear that $[c, d] \neq [a, b]$, for otherwise $[b, b + k] \cap I = \emptyset$, contradicting to the fact that the gap between any two consecutive I-intervals is at most $k - 2$.

It is easy to see that $I \cap [c - (k' - k), c - 1] = \emptyset$ (since the gap between any two consecutive I-intervals is at least $2(k' - k) > k' - k$). Also, we have $I \cap [c + k, d + k'] = \emptyset$ and $I \cap [a + k, a + k'] = \emptyset$, because all the elements of $[c + k, d + k']$ are neighbors of elements in $[c, d]$, and all the elements of $[a + k, a + k']$ are neighbors of a. Let $A = [a, c - (k' - k) - 1] \cup [d + 1, a + k - 1]$ and let $B = [a + k', c + k - 1] \cup [d + k' + 1, a + k + k' - 1]$. Then the following property is clear:

$$I \cap [i, i + k + k' - 1] \subset I \cap [a, a + k + k' - 1] = [c, d] \cup (I \cap (A \cup B)) .$$

For each vertex $x \in A$, we have $x + k' \in B$. This one-to-one correspondence implies $|I \cap (A \cup B)| \leq |A| = |B|$. Because $|A| + |[c, d]| = k - (k' - k)$, we conclude that $|I \cap [i, i + k + k' - 1]| \leq k - (k' - k) = 2k - k'$.

For each $0 \leq i \leq n - 1$, let $n_i = |I \cap [i, i + k + k' - 1]|$. Then

$$(k + k')|I| = \sum_{i=0}^{n-1} n_i \leq n(2k - k') .$$

Since $|I| \geq qk + 1$ and $n \leq q(k + k') + k'$, so

$$(k + k')(qk + 1) \leq (q(k + k') + k')(2k - k') .$$

The result of the theorem can be obtained by simplifying the inequality above. □

3 Circulant graphs and distance graphs

Circulant graphs and distance graphs are closely related. Given a finite set S of positive integers, the distance graph $G(Z, S)$ can be viewed as the limit of the sequence of circulant graphs $G(n, S)$ as n approaches infinity. Therefore, for given S, if $G(n, S)$ is star-extremal for all n, then $G(Z, S)$ is star-extremal. However, the reverse of this fact is not always true. Take $S = \{1, 3, 4, 5\}$, it is known [1] that $G(Z, S)$ is star-extremal, while it was proved [5] that $\chi_f(10, S) = 5 < \chi_c(10, S) = 6$, so $G(10, S)$ is not star-extremal.

A *homomorphism* (or edge-preserving map) from a graph G to another graph H is a mapping, $f : V(G) \to V(H)$, such that if $uv \in E(G)$ then $f(u)f(v) \in E(H)$. If such a homomorphism exists, we say G admits a homomorphism to H and denote this by $G \to H$. If $G \to H$, then by composition of functions, we have $\chi_f(G) \leq \chi_f(H)$ and $\chi_c(G) \leq \chi_c(H)$. Let max S denote the largest number in the set S.

Lemma 11. *Given S, then $G(Z, S) \to G(n, S)$ for all $n \geq 2 \max S$.*

Corollary 12. *Given S, then $\chi_c(Z, S) \leq \chi_c(n, S)$ and $\chi_f(Z, S) \leq \chi_f(n, S)$ for all $n \geq 2 \max S$.*

Theorem 13. *Given S, if $G(Z, S)$ is star-extremal, then there exists $m \in Z^+$, such that $G(km, S)$ is star-extremal for any $k \in Z^+$.*

Proof. Suppose $\chi_c(Z, S) = \chi_f(Z, S) = p/q$ (We may assume this is a rational number by a result in [2]). Let $d = \max S$. According to Corollary 12, it is enough to show that there exists some $m \geq 2d$ such that there is a (p, q)-coloring for any $G(km, S)$, because we would then have

$$p/q = \chi_f(Z, S) \leq \chi_f(mk, S) \leq \chi_c(mk, S) \leq p/q \ .$$

Because $\chi_c(Z, S) = p/q$, by a result in [5], there exists a (p, q)-coloring f of $G(Z, S)$, $f : Z \to [0, p-1]$. Partition $Z^+ \cup \{0\}$ into blocks such that each block consists of p^d consecutive vertices. Consider the restriction of f to these blocks. By the pigeonhole principle, there exist two blocks with the same color sequence. Let x and y be the leading vertices of these two blocks with $x < y$, i.e., $f(x+i) = f(y+i)$, $0 \leq i \leq p^d - 1$. Let $m = y - x$. Then it is straightforward to verify that the restriction of f to the set of vertices $[0, m-1]$ is a (p, q)-coloring for $G(m, S)$.

For $k \geq 2$, define a mapping $f' : [0, km - 1] \to [0, p-1]$ by $f'(v) = f(v \bmod m)$. It is clear that f' is a (p, q)-coloring for $G(km, S)$. The proof is complete. □

4 Independence ratio

Let $S = \{a_1, a_2, \cdots, a_l\}$ be a set of integers with $a_1 < a_2 < \cdots < a_l$. In the study of the asymptotic independence ratio of the circulant graphs $G(n, S)$, Collins introduced the S-graph, denoted by $G(S)$, which has vertex set $V = \{0, 1, 2, \cdots, a_1 + a_l - 1\}$ and edge set $E = \{uv : |u - v| \in S\}$. Note that $G(S)$ is not necessarily a circulant graph.

Given n and S, let $\mu(n,S)$ and $\mu(S)$ denote the independence ratio of the circulant graph $G(n,S)$ and the S-graph $G(S)$, respectively. That is,

$$\mu(n,S) = \frac{\alpha(n,S)}{n} \quad \text{and} \quad \mu(S) = \frac{\alpha(G(S))}{a_1 + a_l} .$$

For a given set S, the *asymptotic independence ratio* $L(S)$ of S is defined in [3] by

$$L(S) = \lim_{n \to \infty} \mu(n,S) .$$

According to (2), we have $\mu(n,S) = 1/\chi_f(n,S)$. Combining this with the fact that $\chi_f(Z,S) = \lim_{n \to \infty} \chi_f(n,S)$, the following result is obtained.

Theorem 14. *For any given S, $L(S) = 1/\chi_f(Z,S)$.*

A set $S = \{a_1, a_2, \cdots, a_l\}$, $a_1 < a_2 < \cdots < a_l$, $l \geq 2$, is called *reversible* if $a_1 + a_l = a_2 + a_{l-1} = \cdots = a_{\lfloor \frac{l}{2} \rfloor} + a_{\lceil \frac{l}{2} \rceil}$. Collins [3] proved that $L(S) = \mu(S)$ if S is reversible, and made the following conjecture.

Conjecture 15. *([3]) Suppose S is a reversible set, $S = \{a_1, a_2, \cdots, a_l\}$, $a_1 < a_2 < \cdots < a_l$. If n is an integer, $n > a_1 + 2a_l$, then $\alpha(n,S) = \lfloor n\mu(S) \rfloor$.*

We now give a counter-example to Conjecture 15. The interval set $S_{k,k'}$ studied in Section 2 is reversible. However, by Lemma 6 and Theorem 8, we have $\mu(S_{k,k'}) = \frac{k}{k+k'}$, and $\alpha(n, S_{k,k'}) \neq \lfloor n\mu(S_{k,k'}) \rfloor$ when $k' \geq (5/4)k$, $n = q(k+k') + r$ and $r \geq k' + 1$.

For a non-reversible set S, Collins [3] gave two methods for constructing reversible sets from S. Let $S = \{a_1, a_2, \cdots, a_l\}$ and let $x = a_{l-1} + a_l$, $y = a_1 + a_l$. Define $\hat{S} = S \cup (x - S)$ (here $x - S$ is the set formed by taking x minus any element in S) and $\tilde{S} = S \cup (y - S)$. Collins [3] showed that $L(S) \geq \max\{\mu(\hat{S}), \mu(\tilde{S})\}$ and made the following conjecture.

Conjecture 16. *([3]) $L(S) = \max\{\mu(\hat{S}), \mu(\tilde{S})\}$.*

For a counter-example to Conjecture 16, take $S = \{1, 2, 3, 6\}$. It is known [6] and easy to see that $\omega(Z,S) = \chi(Z,S) = 4$, so $\chi_f(Z,S) = \chi_c(Z,S) = 4$. Hence $L(S) = 1/4$. But $\hat{S} = \{1, 2, 3, 6, 7, 8\}$, $\tilde{S} = \{1, 2, 3, 4, 5, 6\}$ and $\mu(\hat{S}) = 2/9$, $\mu(\tilde{S}) = 1/7$.

References

1. G. J. Chang, L. Huang and X. Zhu, *Circular chromatic numbers and fractional chromatic numbers of distance graphs*, European J. Comb., to appear.
2. G. J. Chang, D. D.-F. Liu and X. Zhu, *Distance graphs and T-coloring*, manuscript.
3. K. L. Collins, *Circulants and sequences*, SIAM J. Disc. Math. 11 (1998) 330-339.
4. G. Gao, Y. Wang and H. Zhou, *Star chromatic numbers of some planar graphs*, J. Graph Theory 27 (1998) 33-42.

5. G. Gao and X. Zhu, *Star-extremal graphs and the lexicographic product*, Disc. Math. 152 (1996) 147-156.
6. D. D.-F. Liu, *T-colorings of graphs*, Disc. Math. 101 (1992) 203-212.
7. A. F. Sidorenko, *Triangle-free regular graphs*, Disc. Math. 91 (1991) 215-217.
8. A. Vince, *Star chromatic number*, J. Graph Theory 12 (1988) 551-559.

On the Approximability of Physical Map Problems using Single Molecule Methods

(Extended Abstract)

Laxmi Parida

Department of Computer Science, Courant Institute of Mathematical Sciences,
New York University, NY10012, USA. parida@cs.nyu.edu

Abstract. Single molecule approaches such as optical mapping are capable of constructing physical maps (ordered restriction maps) of DNA molecules, but can only do so by combining a population of data in the presence of errors from various sources. Various statistical and other heuristic approaches have been proposed to estimate the physical map in the presence of orientation errors, along with false positive and false negative errors [AMS97,GP98,MP97,KS98,LDW98,Ree97]. We improve the hardness result of a combinatorial model (EBFC problem identified in [GP96,MP97]) that was shown to be NP-hard in [DHM97]; we show that the EBFC problem is MAX SNP hard, give bounds on the approximation achievable and give a polynomial time 0.878-approximation algorithm. A variant of a related problem called the BFC problem [GP96,MP97], was shown to be NP-hard in [AMS97]; we show that both the problems are MAX SNP hard and give bounds on the approximation achievable. We also show that the combinatorial problem underlying the other approach, identified in [Par97] as the Consistency Graph (CG) problem, is MAX SNP hard, give bounds on the approximation achievable and give a polynomial time 0.817-approximation algorithm.
We also explore the complexity of the physical map problem that models the other errors in the single molecule approach, such as presence of spurious molecules in the data, missing fragments of molecules and sizing errors. We show that the first can actually be solved in polynomial time (contrast this with the result in [AMS97] that this problem is NP-complete under added restrictions), while the other two problems are MAX SNP hard. For the latter two problems, we also give bounds on the approximability factors.

1 Introduction

One of the primary goals of many efforts in molecular biology, including the Human Genome Project, is to determine the entire sequence of the human genome and to establish important links to genetics. In this context an important step is to construct *restriction maps* of portions of the DNA [Coo94]. A restriction enzyme cleaves DNA at some fixed sites called the *restriction sites* consisting of well defined sequences. An ordered restriction map specifies the location of these identifiable markers or restriction sites in a DNA molecule. A single molecule approach called optical mapping [WHS95,Men+95,Sch+93] is a new approach for the rapid production of ordered restriction maps. Optical mapping fixes elongated DNA molecules onto surfaces by electrostatic interactions. The fixation conditions are carefully controlled in order to allow enzymatic reactions on the surface. The restriction sites are detected by flourescence microscopy. The size of the resulting individual restriction fragments is determined by relative fluorescence intensity and apparent molecular contour length measurements and other parameters. The fragment size information is used to obtain the ordered restriction maps.

Consider the following idealized version of the problem: assume that the output of the single molecule approach can be processed to generate a discretized binary string of the molecule indicating the presence of restriction sites along it. (This resolution is not necessarily at the level of base pairs.) If the technology were perfect, that would suffice as a physical map (modulo the resolution). However, because of imperfect digestion rates (the rate or the probability of a site in a molecule being cleaved by the enzyme), not all sites are represented in that string (*false negatives*) and some sites might get wrongly represented, due to experimental errors (*false positives*). It is not unusual to assume a lower bound on the digestion rate, say at least 10% of the molecules are correctly cleaved by the enzyme. In the absence of any such assumption, a false cut would be indistinguishable from a true one. The other possible errors are as follows. The different samples are not necessarily laid down along the same direction on the slide. Specifically, each sample is laid down along one of two anti-parallel directions (*orientation uncertainty error*). Furthermore, a few bad or spurious molecules might accidentally enter the data population (*spurious molecule error*). Also, it is possible, particularly in large molecules, that some of the fragments might be lost (*missing fragment error*). Due to the different measurement schemes used, the sizes of the fragments may also vary for each of the sample molecule in the population (*sizing error*). Thus there are basic technological problems getting the discretized strings with reasonably consistent alignment of each. In order to get all the sites, a population of the data must be used, and the restriction sites are those obtained by consensus from the population. The task is to isolate the consensus restriction sites (called the *consensus map*) from the data population. Each cut site in the consensus map is termed a *consensus cut site*. (A description of the problem is presented in [Par97].)

The most widely studied problem is the one that incorporates the false positive and false negative errors along with orientation uncertainties [AMS97,GP98,KS98,LDW98,MP97,Ree97]. The approaches used here can be broadly categorized [Par97] as (1) using an explicit map and (2) using mutual agreement amongst the data in the population. Broadly speaking, in the first approach, the true positives (which have to be estimated from the data) are maximized while in the second approach all the molecules are assigned the correct orientation by increasing the coherence in the data. The approaches used in [AMS97,GP98,MP97] can be identified with the first approach of using an explicit map. The underlying combinatorial problem (called the EBFC problem) for this approach presented in [GP96,MP97], was shown to be NP-hard in [DHM97]. A variant of a related problem called the BFC problem was shown to be NP-hard in [AMS97]; we show that both the problems are MAX SNP hard. We also show that the EBFC problem is MAX SNP hard [1], give bounds on the approximation achievable and give a polynomial time 0.878-approximation algorithm. The approaches in [KS98,LDW98,Ree97] and the weighted-clique heuristic in [AMS97] can be identified with the second approach of exploiting mutual agreement in the data to arrive at a consensus map. The underlying combinatorial problem in this approach has been identified in [Par97] and formalized using a cost function (called the CG problem). We show a relation between these problems and the graph theoretic maximum cut problem. We show that the problems are MAX SNP hard by using appropriate reductions of the maximum cut problem to these problems. We show that achieving an approximation ratio of $1-\Upsilon/3$ for the BFC, EBFC and the CG problems is NP-hard where $1-\Upsilon$ denotes the upper bound on the polynomial time approximation factor of the maximum cut problem. We also show that our problems could be reduced to the maximum cut problem and based on these reductions we provide a polynomial time 0.878-approximation algorithm for the EBFC problem and 0.817-approximation algorithm for the CG problem. Both the algorithms are based on semi-definite programming using the algorithm in [GW94].

Next we study the complexity of the physical map problem using data that has some unknown spurious molecules thrown into the population. This problem has been shown to be NP-complete [AMS97] under the condition that the *total number of spurious molecules is known exactly*. Relaxing this condition (that is, the exact number of spurious molecules is not known), we show that this has a polynomial time algorithm based on optimizing a submodular function.

We also study the physical map problem incorporating missing fragments. One can envisage two cost functions for the general physical map problem: one that counts the number of cuts in each molecule corresponding to a cut of the consensus physical map and the other that just counts the number of cuts in the consensus map [Par97]. The physical map problem with missing fragments using the second cost had been shown to be NP-hard [AMS97]; we show that both the approaches to the problem to be MAX SNP-hard. We do this by showing that the maximum cut problem can be reduced to a very special case of this problem. We give a gap preserving reduction and based on this reduction prove that achieving an approximation ratio $1-\Upsilon/6$ for the problem using the first cost function and a ratio $(1-\Upsilon/6)\frac{p_{max}}{p_{min}}$ using the second cost function is NP-hard where p_{max} and p_{min} are the (bounds on) the maximum and minimum digestion rates in the data. Usually the (bounds on) digestion rates are uniform. Lastly, we study the physical map problem incorporating the sizing errors (the same fragment having different sizes in the data). A version of this problem (using the

---

[1] A follow-up work to this, on the computational complexity of the k-populations problem appears in [PM98].

second cost function) had been shown to be NP-hard [AMS97]; we show that both the approaches to the problem are MAX SNP hard, and give bounds on the approximability of the problem, based on our results on the missing fragments problem. However, it is important to bear in mind that the hardness proofs are for worst case scenario of the problems and in real life, the data arise from a well-controlled (benevolent) process.

The paper is organized as follows. In Section 2, we deal with the problem of incorporating the orientation errors: we first discuss the complexity of the problems using and explicit map (EBFC and other problems) and then discuss the problem using mutual agreement of data (CG and other problems). Section 3 deals with the spurious molecules, and Section 4 and 5 deal with missing fragments and sizing error respectively. We conclude the paper in Section 6.

2 Modeling orientation errors

In this section we consider the physical map problem that takes the false positive, false negative and the orientation uncertainty error into account. Various approaches have been suggested to deal with this problem [AMS97,GP98,KS98,LDW98,MP97,Ree97]. The approaches used here can be broadly categorized [Par97] as (1) using an explicit map and (2) using mutual agreement amongst the data in the population. In [AMS97,MP97], the authors use the first approach, and, in [KS98,LDW98,Ree97] and the weighted-clique heuristic in [AMS97] the authors use the second approach of exploiting mutual agreement in the data to arrive at a consensus ordered restriction map.

2.1 Using an explicit map (The EBFC Problem)

In this approach, in some sense we guess a map that best "fits" the input data. In [Par97] this notion has been formalized to give rise to appropriate optimization problems [GP96,MP97].

We formalize the problem as follows. Given m molecules with n sites each, and, p_j as the digestion rate for column j, obtain an alignment of the molecules such that the total number of 1's in the consensus cut columns, J, which is at least mp_J in each, is maximized (p_J is the digestion rate for site J). This is called the Binary Flip Cut (BFC) problem [GP96,MP97]. We show that BFC is MAX SNP hard and give an upper bound on the polynomial time approximation factor of the problem. In [AMS97], the authors showed that the problem is NP-hard but the cost function was different, we call this altered cost function BFC_{max}, for uniformity of notation. In BFC_{max}, the total number of consensus cut columns, K, which is at least mp_J in each consensus cut column, is maximized. We show at the end of this section that even BFC_{max} is MAX SNP-hard and give an upper bound on the polynomial time approximation factor of the problem.

We associate indicator variables X_i, $i = 1, 2, \ldots, m$, with every row which takes a value 1 if the molecule is flipped and 0 otherwise. Let Y_j, $j = 1, 2, \ldots, n$, be an indicator variable associated with every column that takes on a value of

1 if it is a consensus cut and 0 otherwise. Define the *conjugate* of column j as $\bar{j} = n - j + 1$. BFC can be modeled as the following optimization problem:

$$\max \left\{ \sum_{j=1}^{n} Y_j \left(\sum_{i=1}^{m} \left(M_{ij}(1 - X_i) + M_{i\bar{j}} X_i \right) - mp_j \right) \right\}. \quad (1)$$

M is the input binary matrix. Note that the term mp_j is used to ensure that the number of 1's along a consensus cut site j (with the rows flipped, if required) is at least mp_j. In other words, for a given alignment (which is an assignment of boolean values to X_i, $i = 1, 2, \ldots, m$, and Y_j, $j = 1, 2, \ldots, n$) we count the number of 1's in every column j, that has $Y_j = 1$, less mp_j.

Assume n is even[2]. Suppose, for every pair of columns, j and $\bar{j}$, we know whether both are consensus cut or neither are consensus cuts, then the remaining columns are such that exactly one of j and $\bar{j}$ is a consensus cut. This problem is called the exclusive BFC (EBFC) problem ([GP96,MP97]). This problem was shown to be NP-hard in [DHM97] using a similar reduction as in [AMS97] to show the hardness of the BFC$_{\max}$ problem.

Formally, the EBFC problem ([GP96,MP97]) is as follows. Given m binary molecules of length n each, determine the flip for each molecule and an assignment of either j or $\bar{j}$ as a cut (but not both) for j, $1, \le j \le n/2$, such that the total number of 1's in the cut sites is maximized.

We prove the following lemma about the EBFC problem.

Lemma 1. *EBFC is a special case of the BFC problem.*

Proof. Let $S_j = |\{i \mid M_{ij} = 1 \text{ AND } M_{i\bar{j}} = 1\}|$, and where $\bar{j} = n - j + 1$ let $\bar{S}_j = |\{i \mid M_{ij} = 1 \text{ XOR } M_{i\bar{j}} = 1\}|$. Further, let

$$p_j = p_{\bar{j}} = \frac{\bar{S}_j + 2S_j}{2m}. \quad (2)$$

Note that S_j is the count of the number of symmetric cuts and $\bar{S}_j$ is the total number of non-symmetric cuts in columns j and $\bar{j}$. Irrespective of the assignment of orientations to the molecules/rows, j and $\bar{j}$ always has at least S_j many 1's. The 1's corresponding to $\bar{S}_j$, gets distributed between j and $\bar{j}$ depending on the alignment. We claim that under this definition of p_j for the BFC problem, it is the same as the EBFC problem. It can be verified that under these conditions that $Y_j + Y_{\bar{j}} = 1$ holds for all j, since the definition of p_j ensures that only one of j or $\bar{j}$ is a consensus cut in the optimal alignment (and that is the one with the higher number of 1's). If the number of 1's is equal in both, we can arbitrarily pick only one without changing the cost. □

For the sake of completeness we give the following definitions.

---

[2] If n is odd, we simply remove the middle site, that is, the site $(n+1)/2$, and the problem remains unchanged.

Max Cut (MC) problem: Given a graph, find a partition of the vertices into disjoint sets, S_1 and S_2, such that the number of edges with one vertex in S_1 and the other in S_2 is maximized.

Bipartite Max Cut (BMC) problem: Given a bi-partite weighted graph with edge weights in $\{+1, -1\}$, find a partition of the vertices into disjoint sets, S_1 and S_2, such that the sum of the weights of edges with one vertex in S_1 and the other in S_2 is maximized.

Theorem 2. *EBFC is NP-hard. Further, there exists a constant $\epsilon > 0$ such that approximating EBFC within a factor of $1 - \epsilon$ is NP-hard.*

Proof. We reduce an instance of an MC problem to an instance of an EBFC problem: we show this reduction in two steps (steps 1 and 2) for the sake of clarity. Showing merely the relationship between the optimal solutions for the two problems would show that the EBFC problem is NP-hard; however, we also show how a solution to the Max Cut problem can be constructed given *any* solution (not necessarily optimal) to the EBFC problem. Further, we show that if the cost of the former is close to the optimal, so is the cost of the latter.

The proof proceeds in the following three steps. Let C_X^* denote the cost of the optimal solution and C_X denote the cost of any solution of the problem X.

Step 1. We show the reduction of an instance of the MC problem with e edges to an instance of the BMC problem with

(**1.1**) correspondence between the two solutions,
(**1.2**) $C_{MC}^* = C_{BMC}^*/2$, $C_{MC} \geq C_{BMC}/2$, and,
(**1.3**) the number of negative edges in the BMC is $2e$.

Step 2. We show the reduction of an instance of the BMC problem to an instance of the EBFC problem with

(**2.1**) correspondence between the two solutions, and,
(**2.2**) $C_{EBFC} - e^- = C_{BMC}$, where e^- is the number of edges with negative weights in BMC.

Step 3. Finally, we show that the reduction is *gap-preserving*.

For some $\epsilon > 0$, let C^* denote the optimal solution and $\tilde{C}$ denote an approximate solution with $\tilde{C}_{EBFC} \geq (1-\epsilon)C_{EBFC}^*$.

$$\begin{aligned}
\tilde{C}_{MC} &\geq \frac{\tilde{C}_{BMC}}{2} & \text{(using Step 1.2)} \\
&= \frac{\tilde{C}_{EBFC} - 2e}{2} & \text{(using steps 1.3 \& 2.2)} \\
&\geq \frac{(1-\epsilon)C_{EBFC}^* - 2e}{2} & \text{(by definition of } \tilde{C}_{EBFC}) \\
&= \frac{(1-\epsilon)(C_{BMC}^* + 2e) - 2e}{2} & \text{(using Step 2.2)} \\
&= \frac{(1-\epsilon)C_{BMC}^* - 2e\epsilon}{2} \\
&= (1-\epsilon)\frac{C_{BMC}^*}{2} - e\epsilon \\
&\geq (1-\epsilon)C_{MC}^* - (\epsilon)2C_{MC}^* & \text{(since } C_{MC}^* \geq e/2) \\
&= (1-3\epsilon)C_{MC}^*
\end{aligned} \quad (3)$$

This shows that given a PTAS for EBFC, we can construct a PTAS for MC, which is a contradiction, hence EBFC does not have a PTAS.

Now, we prove each of the steps from 1 to 2. We give the construction here and the details will appear elsewhere.

Step 1. MC to BMC reduction. Consider an MC problem with vertices and edges $(V, E), n = |V|, e = |E|$. Let a solution be of size K, and, the partition of the vertices induced by this solution be S_1 and S_2.

<u>Reduction</u>: Construct an instance of BMC with $(\tilde{V}, \tilde{E})$ as follows: For each $v_i \in V$, with degree d_i, construct $2(d_i + 1)$ vertices, $V_{gadget_i} = \{v'_{i0}, v'_{i1}, \ldots, v'_{id_i}, v''_{i0}, v''_{i1}, \ldots, v''_{id_i}\}$. Further, $wt(v'_{ij}, v''_{ij}) = wt(v'_{i0}, v''_{ij}) = wt(v'_{ij}, v''_{i0}) = -1, j = 1, 2, \ldots, d_i$. Thus, v_i gives rise to $3d_i$ edges with negative weight. Also if $v_1 v_2 \in E$ then $wt(v'_{10} v''_{20}) = wt(v'_{20} v''_{10}) = +1$. It can be seen that this construction gives a bipartite graph with $\tilde{V} = V' \cup V''$ where $v'_x \in V', v''_x \in V''$.

Thus the BMC has $2n + 2e$ vertices, and, $2e$ edges with weights $+1$, and, $2e$ edges with negative weights. Recall for any graph $\sum_i d_i = 2e$.

Step 2. BMC to EBFC reduction. Consider a BMC $((V_1, V_2), E), V_1 = \{v_1^1, v_2^1, \ldots, v_m^1\}, V_2 = \{v_1^2, v_2^2, \ldots, v_n^2\}$, and, number of edges with negative weights be e^-. Let a solution be of size K and partition of vertices, $V_1 \cup V_2$, induced by this solution be S_1 and S_2.

<u>Reduction</u>: Construct an instance of EBFC $[M_{ij}]$ with m rows and $2n$ columns as follows. If $wt(v_i^1 v_j^2) = 1$, then $M_{ij} = 1, M_{i\bar{j}} = 0$. If $wt(v_i^1 v_j^2) = -1$, then $M_{ij} = 0, M_{i\bar{j}} = 1$. If $v_i^1 v_j^2$ is not an edge in the BMC, then $M_{ij} = M_{i\bar{j}} = 0$.

This concludes the proof of the inapproximability of the EBFC problem. □

Corollary 3. *Achieving an approximation ratio $1 - \Upsilon/3$ for EBFC is NP-hard.*

Theorem 4. *BFC is NP-hard. Further, there exists a constant $\epsilon > 0$ such that approximating BFC within a factor of $1 - \epsilon$ is NP-hard. Also achieving an approximation ratio $1 - \Upsilon/3$ for BFC is NP-hard.*

We also show the following results for the BFC_{max} problem.

Theorem 5. *BFC_{max} is NP-hard. Further, there exists a constant $\epsilon > 0$ such that approximating BFC_{max} within a factor of $1-\epsilon$ is NP-hard. Further, achieving an approximation ratio $(1 - \Upsilon/3)\frac{p_{max}}{p_{min}}$ for BFC_{max} is NP-hard.*

Proof. Under the definition of p_j's as in equation (2) the BFC_{max} is the same as the EBFC problem (the number of consensus cuts is always $n/2$, when the molecules have n sites), hence BFC_{max} is NP-hard.

Next, we show that if we have a PTAS for BFC_{max}, we have a PTAS for BFC, which would be a contradiction. Given a BFC let $p_{min} = \min_j p_j$, and $p_{max} = \max_j p_j$. Let $\tilde{X}$ denote an approximate solution and X^* denote the optimal solution. Recall that BFC_{max} optimizes the number of consensus columns. Let N^* denote the number of consensus cut columns when the solution is optimal, with C^* as the BFC cost, and let $\tilde{N}$ denote the number of consensus columns in a solution that is not necessarily optimal, with $\tilde{C}$ as the corresponding BFC cost. Thus if BFC$_{\max}$ has a PTAS let $\frac{\tilde{N}}{N^*} \geq \epsilon$ for some $0 < \epsilon \leq 1$. Note that N^* is the number of consensus cuts. Since $\tilde{C} \geq \tilde{N} p_{\min}$ and $C^* \leq N^* p_{\max}$, we have the following:

$$\frac{\tilde{C}}{C^*} \geq \frac{\tilde{N} p_{\min}}{N^* p_{\max}} \geq \epsilon \frac{p_{\min}}{p_{\max}}. \tag{4}$$

□

Approximability of the EBFC problem. We first give a reduction of an instance of the EBFC problem to an instance of the MC problem. This reduction is best described in two steps: first reducing an instance of the EBFC problem to an instance of the BMC problem and then reducing this instance of the BMC problem to that of an MC problem. Using this reduction we give a polynomial time 0.878-approximation algorithm for a general instance of the EBFC problem.

EBFC to BMC reduction. Given an $m \times 2n$ binary matrix $[a_{ij}]$ for the EBFC problem, we first pre-process the matrix as follows: if $a_{ij} = a_{i\bar{j}} = 1$, then we make the assignment $a_{ij} = a_{i\bar{j}} = 0$. This does not affect the algorithm since one of j or $\bar{j}$ is a cut and in all the configurations there will be contribution of 1 towards the solution due to these two values. Notice that this can only *improve* the approximation factor of the solution.

After the pre-processing, we generate a bipartite graph with vertices $m + n$ with the first partition having m vertices and the second n vertices. The weights are assigned as follows: if $a_{ij} = 1$, then there is an edge between vertex v_i of the first partition and v_j, a vertex in the second partition with a weight of 1; if $a_{i\bar{j}} = 1$, then there is an edge between vertex v_i of the first partition and v_j, a vertex in the second partition with a weight of -1; if $a_{ij} = a_{i\bar{j}} = 0$, then there no edge between the vertices v_i of partition 1 and v_j of partition 2.

There exists a correspondence between a solution to the BMC problem and a solution to the EBFC problem with the corresponding costs defined as $\tilde{C}_{BMC}$ and $\tilde{C}_{EBFC}$ respectively. If the solution to the BMC problem is optimal, its cost is denoted by C^*_{BMC} and the corresponding solution to the EBFC problem is also optimal with the cost denoted as C^*_{EBFC}. Then, the following hold:

$$C^*_{EBFC} = C^*_{BMC} + L, \quad \tilde{C}_{EBFC} \geq \tilde{C}_{BMC} + L. \tag{5}$$

L is the number of 1's in the right half of the input matrix ($L = \sum_{i=1}^{m} \sum_{j=n+1}^{2n} a_{ij}$). Notice that all the rows of the matrix can be flipped to interchange the left and the right half of the matrix.

BMC to MC reduction. Consider an instance of BMC $(V, E^+ \cup E^-)$ where the edges with weight 1 are in E^+ and the ones with negative weight are in E^-. Further, without loss of generality, the number of vertices on the left partition of the BMC problem is m and on the right partition is n.

Construct an instance of the MC problem with $|V|+m'$ vertices and $|E^+|+m'+|E^-|$ edges as follows: if a vertex v has an edge of weight -1 incident on it, then replace it by a pair of vertices u, w connected by an edge $[u,w]$; all the edges of weight $+1$ incident on v are now incident on u. All the edges of weight -1 incident on v are now incident on w with the weight changed to $+1$.

Let the solution to the MC problem include l_1 edges which correspond to the original edges with weight -1, m_1 edges that correspond to the new single edges introduced and p of the original edges (which had a weight of 1 in the BMC problem). Then

$$\tilde{C}_{MC} = p + m_1 + l_2, \tag{6}$$

and the cost of the BMC problem by the construction is,

$$\tilde{C}_{BMC} = p - l_1. \tag{7}$$

Let $L = |E^-|$. Since $l_1 + l_2 = L$, we have from equations (6) and (7),

$$m_1 + L \leq C^*_{BMC}. \tag{8}$$

$m_1 + L$ is the trivial solution obtained by having all the new vertices (with edges having weight -1 incident to them) in one partition and the rest of the vertices in the other partition.

Assuming we can obtain a solution for the MC problem with cost $\tilde{C}_{MC} \geq \epsilon C^*_{MC}$, for some $0 < \epsilon \leq 1$, we obtain the following.

$$\begin{aligned}
\tilde{C}_{BMC} &= \tilde{C}_{MC} - (m_1 + L) && \text{(from eqn (8))} \\
&\geq \epsilon C^*_{MC} - (m_1 + L) \\
&\geq \epsilon(C^*_{BMC} + (m_1 + L)) - (m_1 + L) && \text{(from eqn (8))} \\
&= \epsilon C^*_{BMC} - (1-\epsilon)(m_1 + L) \\
&\geq \epsilon C^*_{BMC} - (1-\epsilon)(C^*_{BMC}) && \text{(from eqn (8))} \\
&\geq (2\epsilon - 1)C^*_{BMC}.
\end{aligned} \tag{9}$$

A 0.878-approximation algorithm. In this section, using the constructions presented in the last section, we present an algorithm that achieves an approximation factor of 0.878 for the EBFC problem.

Let $[a_{ij}]$ be the matrix for the EBFC problem and L be the total number of 1's in the right half of this matrix, then the following holds:

$$2L \geq \sum_i \sum_j a_{ij} \geq C^*_{EBFC}, \tag{10}$$

Recall that all the rows (molecules) can be flipped without altering the problem so that the above holds. Further, let $\tilde{C}_{BMC} \geq \delta C^*_{BMC}$.

$$\begin{aligned}
\tilde{C}_{EBFC} &= \tilde{C}_{BMC} + L && \text{(using equation (5))} \\
&\geq \delta C^*_{BMC} + L \\
&\geq \delta(C^*_{EBFC} - L) + L && \text{(using equation (5))} \\
&= \delta C^*_{EBFC} + (1-\delta)L \\
&\geq \frac{(\delta+1)}{2} C^*_{EBFC} && \text{(using equation (10))}
\end{aligned} \tag{11}$$

When $\delta = 2\epsilon - 1$, from derivation (9), then we obtain $\tilde{C}_{EBFC} \geq \epsilon C^*_{EBFC}$, if we have $\tilde{C}_{MC} \geq \epsilon C^*_{MC}$. Using the algorithm presented in [GW94] to obtain 0.878-approximation algorithm for the MC problem, we obtain a 0.878-approximation algorithm for the EBFC problem.

2.2 Using mutual agreement of data (The CG, WCG Problems)

The second approach can be broadly described as guessing the correct alignment of molecules by studying a few molecules (say $d \geq 2$) at a time and building the entire solution from this (possibly with some back-tracking). We assume a somewhat idealized version of the problem and formalize the problem as the d-wise Match (dM) problem. Let the number of molecules be m, each having n sites. For a fixed d ($d \geq 2$ and $d << m$), we assume that we can orient the d molecules so that they have maximum agreement between them. This is done by enumerating all the 2^{d-1} possible configurations where each molecule can have left-to-right or right-to-left orientation with respect to a reference molecule whose orientation is fixed. This assigns an orientation to each molecule of the sample size of d molecules. We associate a cost with each configuration of the d molecules, X, as $A^X(i_1, i_2, \ldots, i_d)$, and define the cost as follows, for some fixed $\delta > 0$:

$$A^X(i_1, i_2, \ldots, i_d) = \begin{cases} \text{\# of cut sites that are within } \delta \text{ of each other in } all \\ \text{the } d \text{ molecules, given } X. \end{cases} \quad (12)$$

There could be other (more sophisticated) ways of defining the cost and in principle, this alignment could model other errors as well. Given the sample of d molecules, we assign the configuration $X_{\min}$ that minimizes the the cost (defined by equation (12)). This configuration $X_{\min}$ implicitly assigns an orientation to each of the d molecules. Thus, orientation can be assigned to each molecule of every possible d-sized sample of the m molecules. If the orientation of any one of the molecules is changed, the cost associated with the molecules increases by say δ. Also, a molecule belongs to $\binom{m}{d-1}$ samples and could have different orientations assigned to it in the different samples. The aim is to assign an orientation to $every$ molecule, so that the sum of the deviation δ from the optimal in each of the $\binom{m}{d}$ samples is minimized, or, the cost of alignment due to each of the samples is maximized. This optimization problem is termed the d-match (dM) problem. It is assumed that once the orientation of each molecule is known the positions of the consensus cut sites can be estimated quite simply.

For the sake of simplicity, we study the $pairwise$ or 2-wise match problem which is as follows. Given m molecules with n sites each, with false positive and negative errors and orientation uncertainties, and a fixed $\delta > 0$, find an alignment to the molecules so that it has the maximum 2-$wise$ $match$ where $A^X(i_1, i_2)$, is defined as

$$A^X(i_1, i_2) = \begin{cases} \text{\# of cut sites that are within } \delta \text{ of each other in} \\ \text{both the molecules, given } X. \end{cases} \quad (13)$$

and X denotes an alignment, i.e., (1) both are in the same orientation, say left-to-right, or (2) they have an opposite orientation, say one is left-to-right and the other is right-to-left. Thus it is the following optimization problem:

$$\max_{\text{(over all alignments)}} \left\{ \sum_{i_1=1}^{m} \sum_{i_2=i_1}^{m} A^X(i_1, i_2) \right\}. \quad (14)$$

Informally, the task is to *maximize the sum of the pairwise match cost*.

Graph-theoretic Formulation. For the $d = 2$ case, we map the problem to a graph problem. Notice that if there are only two molecules, there can be only one alignment (either both are in the same orientation or one of them is in the opposite orientation) and there is no conflict. If there are 3 molecules, it is possible that considering two of them assigns an orientation to each of the molecule and the third molecule may or may not support this decision. In general for n molecules we capture this in a graph structure as described below.

Graph construction. Given a 2-wise match problem, a complete graph $\mathcal{G}$ is constructed with every vertex v_i corresponding to a molecule i.

Edge Labels. Let $X = S$ denote an alignment where both the molecules i and j have the same orientation, and $X = O$ denote the alignment where one of them is is left-to-right while the other is right-to-left. Every edge $e_{ij} = v_i v_j$ with $A^S(i,j) \neq A^O(i,j)$ is labeled by label $L(v_i v_j)$ as follows:

$$L(e_{ij}) = \begin{cases} \text{Same} & A^S(i,j) > A^O(i,j), \\ \text{Opposite} & A^S(i,j) < A^O(i,j). \end{cases}$$

Recall that $A^X(i,j)$ is the cost of the alignment X using equation (13). We remove those edges that have $A^S(i,j) = A^O(i,j)$, thus the graph $\mathcal{G}$ is not necessarily a complete graph.

Edge Weights. The edge weight $Wt(e_{ij})$ is defined as $Wt(e_{ij}) = \max(A^S(i,j), A^O(i,j))$.

Consistency Condition. We define a predicate on any set of three vertices (of the complete graph) as follows: v_i, v_j, v_k are *consistent* if either all the three or exactly one edge is labeled *Same*.

It can be verified that the three molecules can be assigned unique orientations only if all of the three pairwise labels are *Same* or exactly two of the pairwise labels are *Opposite*. Thus a consistent set of vertices give a unique alignment to the corresponding set of molecules. Further, a labeled graph $\mathcal{G}$ is said to be *consistent* if every three vertices v_i, v_j, v_k is consistent.

The Constrained Optimization Problem: Given the labeled and edge weighted graph $\mathcal{G}$, the problem is to obtain a set of edges $\mathcal{S}$, such that for all edges $e \notin \mathcal{S}$ the labels of the edges is changed from *Same* to *Opposite* or *vice-versa*, so that (1) the graph with these new labels is consistent, and, (2) the sum of the weights of the edges $e \in \mathcal{S}$ is maximized. This is called the **Weighted Consistency Graph (WCG)** problem.

We define a special case of the WCG problem, the Consistency Graph (CG) problem: Given a labeled graph $\mathcal{G}$, find the maximum number of edges that retain the labels to get a consistent graph. This is the WCG problem under the assumption that all the edges are of equal weight.

Theorem 6. *The CG problem is NP-hard. Further, there exists a constant $\epsilon > 0$ such that approximating CG within a factor of $1 - \epsilon$ is NP-hard.*

Proof. We give the proof in two steps. In step 1 we show a reduction of an instance of a maximum cut (MC) problem to an instance of the CG problem and show that $C_{MC} = 2C_{CG} - e$ where C_X is a solution to the problem X and e is the number of edges in the MC problem. In step 2 we show that the reduction is *gap-preserving*.

Step 1. Given an instance of the MC problem with n vertices and e edges, we construct an instance of CG by simply labeling every edge as *Opposite*. A consistent graph is such that the vertices can be partitioned into two sets S_1 and S_2 such that $\forall v_i, v_j \in S_1$ (or S_2), $L(e_{ij}) = Same$, and, $\forall v_i \in S_1, v_j \in S_2, L(e_{ij}) = Opposite$. There are only two kinds of consistent triangles: (1) all labels are *Same* or (2) exactly one label is *Same*. It can be verified that only these two kinds of triangles (and no other) exist for the consistent graph whose vertices are given by $S_1 \cup S_2$.

Given a solution of size e' to the CG, which is the number of edges with label *Opposite* (since there was no edge with label *Same*), we can show that the solution to the MC problem is of size $2e' - e$. It can also be verified that increasing the solution to the CG problem by $x > 0$, increases the solution to the MC problem by x.

Step 2. Let $\tilde{C}_{CG}$ denote an approximate solution and C_{CG}^* denote the optimal solution. Then $\tilde{C}_{CG} \leq (1+\epsilon) C_{CG}^*$.

$$\begin{aligned}
\tilde{C}_{MC} &= 2\tilde{C}_{CG} - e & \text{(using step 1)} \\
&\geq 2(1-\epsilon)C_{CG}^* - e & \text{(by defn of } \tilde{C}_{MC}) \\
&\geq (1-\epsilon)(C_{MC}^* + e) - e & \text{(by step 1)} \\
&= (1-\epsilon)C_{MC}^* - e\epsilon \\
&\geq (1-3\epsilon)C_{MC}^* & \text{(since } C_{MC}^* \geq e/2).
\end{aligned} \quad (15)$$

This shows that given a polynomial time approximation scheme (PTAS) for the CG problem, we can construct a PTAS for the MC problem (using step 1 gives the correspondence between the two solutions), which is a contradiction; hence the CG problem does not have a PTAS. This concludes the proof. □

Corollary 7. *Achieving an approximation ratio $1 - \Upsilon/3$ for the CG problem is NP-hard.*

This directly follows the theorem and the fact that the approximation factor can be no more than $1 - \Upsilon$, unless P=NP [Has97].

Corollary 8. *The WCG and dM problems are MAX SNP hard. Further, achieving an approximation ratio $1 - \Upsilon/3$ for the WCG and dM problems is NP-hard.*

The same proof goes through for the WCG problem as well, by simply assigning weights along with the labels to the edges. Since WCG is a special case of the d-wise match problem (with $d = 2$), all the results for the WCG problem also hold for the dM problem.

A 0.817-approximation algorithm. Let the maximum cut problem on a graph with weights $+1$ be called the Positive Max Cut (PMC) problem and the one with weights $+1$ or -1 be called Negative Max Cut (NMC). We show that given an arbitrary instance of the CG problem, we can construct an instance of the NMC problem, and then construct an instance of the PMC problem. As all the weights in the PMC problem are positive, we can use the Goemans and Williamsons' semi-definite programming based algorithm to get a 0.878-approximation of the PMC problem. Next we use this solution to obtain an approximate solution for the NMC problem, and using that solution we obtain a solution for the CG problem.

We describe this in three steps. In Step 1, we describe the reduction of an instance of the CG problem to an instance of an NMC problem and also describe the correspondence between a solution to the NMC problem and that of the CG problem. In Step 2, we describe the reduction of an instance of an NMC problem to an instance of the PMC problem and describe the correspondence between a solution to the PMC problem and that of the NMC problem. These two steps give the algorithm and finally in Step 3, we argue that the algorithm gives an approximation factor of 0.817.

Step 1. Given an instance of the CG problem given by graph $\mathcal{G}_{CG}$, with m vertices and n edges labeled as *Same* or *Opposite*, we construct an instance of NMC on the graph $\mathcal{G}_{NMC}$, by assigning a weight of -1 to all those edges labeled *Same* and assigning a weight of $+1$ to the *Opposite* labeled edges. Thus the number of vertices of $\mathcal{G}_{NMC}$ is m and the number of edges is n.

Next, we claim that an optimal solution to the NMC instance gives an optimal solution to the CG problem, and an approximate solution to the CG problem can be constructed from the approximate solution to the NMC instance.

Let L be the total number of edges labeled *Same* in the CG problem or labeled -1 in the NMC problem. Given a solution of the form $p_1 - l_2$ of the NMC instance, where p_1 is the number of edges with weight $+1$ and l_1 is the number of edges with weight -1 in the cut, the solution C_{CG} to the CG problem is

$$C_{CG} = p_1 + l_1, \qquad (16)$$

where l_1 is the number of edges with weight -1 not in the cut. The p_1 edges corresponding to the $+1$ labels are the ones in the CG instance which are labeled *Opposite*, and, the l_2 edges corresponding to the -1 labels are the ones in the CG instance which are labeled *Same*, and these edges *do not switch labels* in the solution. Thus $L = l_1 + l_2$. Also, it can be verified that an optimal solution in the NMC instance gives an optimal solution in the CG problem.

Step 2. Given an instance of the NMC problem, we construct an instance of the PMC problem (with weight on the edges as +1) by replacing every edge with a negative weight by two edges and a vertex, each edge having a weight of 1. If L is the number of edges with weight -1, then the PMC instance has $m + n/2 + L$ vertices.

Next, we claim that that an optimal solution to the PMC instance, gives an optimal solution to the NMC problem, and, an approximate solution to the NMC problem can be constructed from the approximate solution to the PMC instance.

Now, we give the correspondence between the solutions in each of the problem. Notice that the edges introduced in the reduction come in pairs. Let the solution to the PMC problem include l_1 edges which are *not* paired, $2l_2$ paired edges and p of the original edges (which had a weight of 1 in the NMC problem). Then

$$C_{PMC} = p + 2l_1 + l_1. \tag{17}$$

The edges that come in pairs correspond to the l_2 edges which are not in the cut in the NMC instance and the edges corresponding to l_1 which are not in pairs correspond to the edges in the cut of the NMC instance. Thus the cost of the NMC problem by this construction is $C_{NMC} = p_1 - l_1$. It can be verified that the optimal solution in one corresponds to the optimal one in the other.

Step 3. Thus from equations (16) and (17), we have

$$C_{PMC} = L + C_{CG}. \tag{18}$$

Now, we make the following observation:

$$2L \leq p_1 + l_1. \tag{19}$$

This holds since $2L$ is the cost of the trivial solution to the constructed PMC problem which corresponds to the zero solution in the NMC instance, (where all the vertices belong to just one partition!), hence this must be smaller than the cost of any other non-trivial solution (*viz.*, $p_1 + l_2$). Also, if C_{CG}^* is the optimal solution then,

$$p_1 + l_1 \leq C_{CG}^*. \tag{20}$$

Finally, we use the algorithm presented in [GW94] to obtain a 0.878-approximation algorithm for the PMC problem. Note that we could not directly use it on the NMC instance due to the -1 weights. Let $\tilde{C}_X$ denote an approximate solution and C_X^* denote the optimal solution to problem X.

$$\begin{aligned}
\tilde{C}_{CG} &= \tilde{C}_{PMC} - L & \text{(from eqn (18))} \\
&\geq 0.878 C_{PMC}^* - L & \text{(from [GW94])} \\
&\geq 0.878(C_{CG}^* + L) - L & \text{(from eqn (18))} \\
&= 0.878 C_{CG}^* - 0.122L \\
&\geq 0.878 C_{CG}^* - 0.122(C_{CG}^*/2) & \text{(from eqns (19) and (20))} \\
&\geq 0.817 C_{CG}^*.
\end{aligned}$$

This concludes the argument.

0.817-approximation algorithm for the WCG problem. Assigning the appropriate weights to the graph, *i.e.*, positive weights to edges labeled *Opposite* and negative weights to edges labeled *Same*, and, using the same steps as in the CG problem, we get similar results for the WCG problem. Also note that in the NMC to PMC reduction if an edge has weight $-w$ it is replaced by two edges with weight w each incident on a vertex as for the CG problem.

3 Modeling spurious molecules

The Binary Partition Cut (BPC) problem takes into account the presence of spurious or bad molecules (along with false positive and negative errors). We can define two kinds of cost functions for the problem: one that maximizes the number of cuts in each molecule corresponding to a cut of the consensus physical map (BPC) and the other that maximizes the number of cuts in the consensus map ($\text{BPC}_{\max}$). Recall that each consensus cut site must satisfy the digestion rate criterion. A variation of the latter where the information that *the total number of bad molecules is known exactly*, is shown to be NP-complete in [AMS97]. However, we show that the two original problems have efficient polynomial time solutions.

Let the input binary $m \times n$ matrix be $[M_{ij}]$. Associate indicator variables X_i, $i = 1, 2, \ldots, m$, with every row which takes a value 1 if the molecule is good and 0 if it is spurious. Let Y_j, $j = 1, 2, \ldots, n$, be an indicator variable associated with every column that takes on a value of 1 if it is a consensus cut and 0 otherwise. Let the digestion rate be p_j for column j, $j = 1, 2, \ldots, n$. Modeling the BPC problem exactly along the lines of the BFC problem, as in equation (1), we obtain the following optimization problem:

$$\max \left\{ \sum_{j=1}^{n} Y_j \left(\sum_{i=1}^{m} X_i M_{ij} - p_j m \right) \right\}. \tag{21}$$

Consider the corresponding minimization problem $\min \left\{ \sum_{i=1}^{m} \sum_{j=1}^{n} -M_{ij} X_i Y_j + \sum_{j=1}^{n} m p_j Y_j \right\}$, which is a submodular function [NW88], hence BPC has a polynomial time solution.

Corollary 9. *The $\text{BPC}_{\max}$ problem has a polynomial time solution.*

Proof. If n_o is the number of consensus cuts in the optimal solution then there exists no sub-optimal solution with $n > n_o$. This is because the new optimal alignment can be obtained from this sub-optimal giving a larger n_o, which is a contradiction. Hence a solution to the BPC problem gives a solution to the $\text{BPC}_{\max}$ problem. □

4 Modeling missing fragments

The Binary Shift Cut (BSC) problem takes into account missing fragments (along with false positive and negative errors). As in the BPC problem formulation, we can define two kinds of cost functions for the problem: one that maximizes the number of cuts in each molecule corresponding to a cut of the consensus physical map (BSC) and the other that maximizes the number of cuts in the consensus map (BSC_{max}). Recall that each consensus cut site must satisfy the digestion rate criterion. The BSC_{max} is shown to be NP-complete in [AMS97]. We show that both the approaches to the problem are MAX SNP hard and also give bounds on the approximation factors achievable.

We show that BSC is MAX SNP-hard in the following theorem. The proof is similar to the proof of theorem 2, however the details vary since, in the context of the BSC problem, the following holds : $j \neq \bar{\bar{j}}$, where $\bar{j}$ is the the *conjugate* of j. In the construction used in the theorem, $\bar{\bar{j}} = j + 1$. Recall that for the EBFC problem $j = \bar{\bar{j}}$.

Theorem 10. *The BSC problem is NP-hard. Further, there exists a constant $\epsilon > 0$ such that approximating this problem within a factor of $1 - \epsilon$ is NP-hard.*

Proof. We prove the result for a special case of the BSC problem where every molecule is such that either the left or the right fragment (not both) is missing; the missing fragment is exactly one unit in all the molecules. In the aligned configuration, a column j is in a cut only if the number of cuts is at least $p_j m$, which is defined in the proof of step 2. The proof has three steps. Let C_X^* denote the cost of the optimal solution and C_X denote the cost of any solution of the problem X.

Step 1. We show a reduction of an instance of the maximum cut (MC) problem with n vertices and e edges to an instance of the bipartite maximum cut (BMC) problem (which is the maximum cut problem on a bipartite graph with weights $+1$ or -1 on the edges) with

  (**1.1**) correspondence between the two solutions,

  (**1.2**) $4C_{MC}^* = C_{BMC}^* - 4e - 3n$, $4C_{MC} \geq C_{BMC} - 4e - 3n$, and,

  (**1.3**) the number of edges with positive weights in the BMC is $8e + 2n$.

Step 2. We show the reduction of an instance of the BMC problem to an instance of the BSC problem with

  (**2.1**) correspondence between the two solutions, and,

  (**2.2**) $2C_{BSC} - c = C_{BMC}$, where c is the number of 1's in the BSC matrix.

Step 3. For some $\epsilon > 0$, let C^* denote the optimal solution and $\tilde{C}$ denote an approximate solution with $\tilde{C}_{BSC} \geq (1-\epsilon)C^*_{BSC}$.

$$\begin{aligned}
\tilde{C}_{MC} &\geq \tfrac{\tilde{C}_{BMC}-4e-3n}{4} & &\text{(using Step 1.2)}\\
&= \tfrac{2\tilde{C}_{BSC}-12e-5n}{4} & &\text{(using Steps 1.3 \& 2.2)}\\
&\geq \tfrac{(1-\epsilon)2\tilde{C}^*_{BSC}-12e-5n}{4} & &\text{(by defn of } \tilde{C}_{BSC})\\
&= \tfrac{(1-\epsilon)(C^*_{BMC}+8e+2n)-12e-5n}{4} & &\text{(using Step 2.2)} & &(22)\\
&= \tfrac{(1-\epsilon)C^*_{BMC}-4e-3n}{4} - \tfrac{(8e+2n)\epsilon}{4}\\
&\geq (1-\epsilon)C^*_{MC} - 2.5e\epsilon & &\text{(using Step 1.2)}\\
&\geq (1-\epsilon)C^*_{MC} - (2.5\epsilon)2C^*_{MC} & &\text{(since } C^*_{MC} \geq e/2)\\
&= (1-6\epsilon)C^*_{MC}.
\end{aligned}$$

This shows that given a polynomial time approximation scheme (PTAS) for BSC, we can construct a PTAS for MC (using steps 1 and 2 that give the correspondence between the two solutions), which is a contradiction; hence BSC does not have a PTAS.

We skip the details of steps 1 and 2 here. □

Corollary 11. *Achieving an approximation ratio $1 - \Upsilon/6$ for BSC is NP-hard.*

Theorem 12. *The $BSC_{\max}$ problem is NP-hard. Further, there exists a constant $\epsilon > 0$ such that approximating this problem within a factor of $1 - \epsilon$ is NP-hard and achieving an approximation ratio $(1 - \Upsilon/6)\frac{p_{max}}{p_{min}}$ for $BSC_{\max}$ is NP-hard.*

5 Modeling sizing errors of the fragments

The Binary Sizing error Cut (BSeC) problem takes into account varying size of fragments in the molecules (along with false positive and negative errors). As in the BPC problem formulation, we can define two kinds of cost functions for the problem: one that maximizes the number of cuts in each molecule corresponding to a cut of the consensus physical map (BSeC) and the other that maximizes the number of cuts in the consensus map ($BSeC_{\max}$). Recall that each consensus cut site must satisfy the digestion rate criterion. The $BSeC_{\max}$ is shown to be NP-complete in [AMS97]. We show that both the approaches to the problem are MAX SNP hard and also give bounds on the approximation factors achievable.

Note that a special class of this problem can be viewed as a restriction of the BSC problem where the smaller fragments are considered to have missing (end) fragments. Hence we directly have the following theorems as a consequence of the results on the BSC and $BSC_{\max}$ problems.

Theorem 13. *The BSeC problem is NP-hard. Further, there exists a constant $\epsilon > 0$ such that approximating it within a factor of $1-\epsilon$ is NP-hard and achieving an approximation ratio $1 - \Upsilon/6$ for BSeC is NP-hard.*

Theorem 14. *$BSeC_{\max}$ problem is NP-hard. Further, there exists a constant $\epsilon > 0$ such that approximating this problem within a factor of $1 - \epsilon$ is NP-hard and achieving an approximation ratio $(1 - \Upsilon/6)\frac{p_{max}}{p_{min}}$ for $BSeC_{\max}$ is NP-hard.*

6 Conclusion

In this paper we have studied the computational complexity of physcial mapping problems using single molecule methods. We have resolved the open problem regarding the hardness of the CG and the BFC problems; we have shown that both the problems are MAX SNP hard and given bounds on the approximation factors achievable. We have also given a polynomial time 0.817 approximation algorithm for the CG problem. We have improved the known result on the EBFC problem: we have shown that the problem is MAX SNP hard and presented a polynomial time 0.878 approximation algorithm. We have also improved the known results on the variations of the physical map problem that take other errors into account such as presence of spurious molecules in the data, missing fragments of molecules and sizing errors. We have shown that the first can actually be solved in polynomial time (contrast this with the result in [AMS97] that this problem is NP-complete under added restrictions), while the other two problems are MAX SNP hard. For the latter two problems, we have also given bounds on the approximability factors. The inapproximability results provide justification and guidelines for the design of heuristics for the problems in practice.

References

[AMS97] T.S. ANANTHARAMAN, B. MISHRA AND D.C. SCHWARTZ, "Genomics via Optical Mapping II: Ordered Restriction Maps," *Journal of Computational Biology*, 4(2):91–118, 1997.

[AKK96] S. ARORA, D. KARGER, AND M. KARPINSKI, "Polynomial time approximation schemes for dense instances of NP-hard problems," In *Proc. STOC*, 1996.

[Cai+95] W. CAI ET AL., "Ordered Restriction Endonuclease Maps of Yeast Artificial Chromosomes Created by Optical Mapping on Surfaces," *Proc. Natl. Acad. Sci., USA*, **92**:5164–5168, 1995.

[Coo94] N. G. COOPER (editor), *The Human Genome Project - Deciphering the Blueprint of Heredity*, University Science Books, Mill Valley, California, 1994.

[DHM97] V. Dančík, S. Hannehalli, and S. Muthukrishnan. Hardness of flip-cut problems for optical mapping. *J. Computational Biology*, 4(2), 1997.

[GJ79] M.R. GAREY AND D.S. JOHNSON, *Computer and Intractability: A Guide to the Theory of NP-Completeness*, W.H. Freeman and Co., San Francisco 1979.

[GP96] D. Geiger and L. Parida. A model and solution to the DNA flipping string problem. Technical Report TR1996-720, Courant Inst. of Math. Sciences, New York University, May 1996.

[GP98] D. GEIGER, L. PARIDA, "Mass Estimation of DNA Molecules & Extraction of Ordered Restriction Maps in Optical Mapping Imagery", to appear in *Algorithmica*.

[GW94] M. X. GOEMANS, D. P. WILLIAMSON, ".878-approximation algorithms for MAX CUT and MAX 2SAT", *Proceedings of the Twenty-Sixth Annual ACM Symposium on Theory of Computing*, pp 422-431, Montreal, Quebec, Canada, 23-25 May 1994.

[Has97] J. HÅSTAD, "Some optimal inapproximability results", *Proceedings of the Twenty-Ninth Annual ACM Symposium on Theory of Computing*, pp 1-10, El Paso, Texas, 4-6 May 1997.

[KS98] R. M. KARP, R. SHAMIR, "Algorithms for Optical Mapping", In *Proceedings of the Annual Conference on Computational Molecular Biology*, (RECOMB98), ACM Press, 1998.

[LDW98] J. K. LEE, V. DANCIK, M. S. WATERMAN, "Estimation for restriction sites observed by optical mapping using reversible-jump markov chain monte carlo", In *Proceedings of the Annual Conference on Computational Molecular Biology*, (RECOMB98), ACM Press, 1998.

[Men+95] X. MENG ET AL., "Optical Mapping of Lambda Bacteriophage Clones Using Restriction Endonuclease," *Nature Genetics*, **9**:432–438, 1995.

[MP97] S. MUTHUKRISHNAN AND L. PARIDA, "Towards Constructing Physical Maps by Optical Mapping: An Effective Simple Combinatorial Approach," In *Proceedings First Annual Conference on Computational Molecular Biology*, (RECOMB97), ACM Press, 209–215, 1997.

[NW88] G. NEMHAUSER, L. WOLSEY, *Integer and Combinatorial Optimization*, Wiley Interscience Series in Discrete Math and Optimization, 1988.

[PY91] C. PAPADIMITRIOU, M. YANNAKAKIS, "Optimization, approximation and complexity classes", *Journal of Computer and System Sciences*, **43**: 425–440, 1991.

[Par97] L. PARIDA, "A Uniform Framework for Ordered Restriction Map Problems", to appear in *Journal of Computational Biology*.

[PM98] L. PARIDA AND B. MISHRA, Partitioning k clones: Hardness results and practical algorithms for the k-populations problem. In *Proceedings of the Second Annual Conference on Computational Molecular Biology (RECOMB98)*, pages 192–201. ACM Press, 1998.

[Ree97] J. REED, *Optical Mapping*, Ph. D. Thesis, Dept of Chemistry, New York University, June 1997.

[Sam+95] A. SAMAD ET AL., "Mapping the Genome One Molecule At a Time—Optical Mapping," *Nature*, **378**:516–517, 1995.

[Sch+93] D.C. SCHWARTZ ET AL., "Ordered Restriction Maps of *Saccharomyces cerevisiae* Chromosomes Constructed by Optical Mapping," *Science*, **262**:110–114, 1993.

[WHS95] Y.K. WANG, E.J. HUFF AND D.C. SCHWARTZ, "Optical Mapping of the Site-directed Cleavages on Single DNA Molecules by the RecA-assisted Restriction Endonuclease Technique," In *Proc. Natl. Acad. Sci. USA*, **92**:165–169, 1995.

[WSK84] M.S. WATERMAN, T.F. SMITH AND H. KATCHER, "Algorithms for Restriction Map Comparisons," *Nucleic Acids Research*, **12**: 237–242, 1984.

[Wat95] M.S. WATERMAN, *An Introduction to Computational Biology: Maps, Sequences and Genomes*, Chapman Hall, 1995.

On the Weakness of Conditional Equations in Algebraic Specification

Arno Schönegge[1] and David Kempe[2]

[1] Xcc Software AG, D-76137 Karlsruhe, Germany.
Arno.Schoenegge@xcc.de
[2] Department of Computer Science, Cornell University, Ithaca, NY, USA.
kempe@cs.cornell.edu

Abstract. In the initial semantics approach to algebraic specification, one usually restricts oneself to conditional equations. It is well-known that this restriction does not affect expressiveness, at least if the introduction of auxiliary (hidden) functions is allowed: all the computable data types can be specified using conditional equations and initial algebra semantics [BT82].
However, the situation is quite different concerning the specification *without* hidden functions. In this case, the completeness for computable data types is lost for the initial specification methods (cf. [Maj79,TWW82,BT87]) as well as for the loose ones (cf. [Ber93,KS98]). Instead of considering these two approaches separately, it is natural to ask how their expressive powers compare to each other. It is already known that the loose specification methods (even allowing full first-order logic) are not more powerful than the initial methods (even using equations only) [Sch97]. In this paper we give the answer to the opposite question. More precisely, we provide an example of a computable data type which has a monomorphic quantifier-free specification under loose semantics and prove that it fails to possess a conditional equational specification under initial semantics (both without hidden functions). Thus, our main result is that, without hidden functions, the restriction to conditional equations *does* affect expressiveness—even if compared to quantifier-free axioms.

1 Introduction

Most algebraic specification languages can be roughly classified as belonging to either the initial or the loose approach (see [Wir95] or [SW98] for an overview).[1]

In the initial approach, the semantics of a specification is defined to be the initial[2] model of the axioms, i.e., it is characterized by "no junk" (each data item can be constructed using only the constants and functions in the signature) and "no confusion" (two data items are equivalent iff they can be proved equal from

---
[1] Several alternative approaches have been proposed (e.g. final algebra semantics [Wan81]), but are, of less practical relevance.
[2] We give precise definitions in Section 2.

the axioms) [GTWW75,EM85]. In order to guarantee executability and the existence of initial models, one usually restricts the axioms to conditional equations or even equations only. Some examples of specification languages following the initial approach are Clear, OBJ, and ACT-ONE.

In the loose approach the semantics of a specification is defined to be the set of all term-generated models of the axioms [GGM76,WPP+83]. Usually, full first-order logic or its quantifier-free subset can be used for the axioms. Examples of specification languages following the loose approach are ASL and the Larch shared language.

A problem that immediately arises is to investigate the expressive power of the various approaches. In fact, this question has been studied over several years. One fundamental result is that, if equipped with hiding mechanisms, all common algebraic specification methods are adequate for computable data types [BT82,BT87]. Here, hiding means that local definitions of auxiliary functions can be added for reasons of specification only; these functions are hidden from the user of the specified data type.

However, if we do not allow hidden functions—as we suppose throughout this paper—the situation changes substantially. Majster [Maj79] and Bergstra & Tucker [BT87, Theorem 4.8] gave examples of computable data types which cannot be specified using conditional equations and initial algebra semantics. Orejas [Ore79], Thatcher et al. [TWW82, Theorem 5], and also Bergstra & Meyer [BM84] proved that conditional equations are strictly more powerful than equations.

Concerning the loose approach, the corresponding questions have been answered only recently. Berghammer [Ber93] showed that the computable data type $(\mathbb{Z}; 0, x+1, x-1)$ of integers cannot be monomorphically specified with first-order formulas under loose semantics (cf. also [Sch97]). An example of a computable data type which demonstrates that full first-order axioms are strictly more powerful than quantifier-free formulas was provided in [KS98].

Thus, the relevant questions concerning expressiveness without hidden functions are answered for both the initial and the loose approach. However, instead of considering these methods separately, it is natural to oppose the initial methods to the loose ones. This results in two questions:

- Is the loose approach more powerful than the initial approach?
- Is the initial approach more powerful than the loose approach?

The answer to the first question is already known to be "no": the data type of integers mentioned above does have an equational specification under initial semantics. However, to our knowledge, there has been no satisfactory answer to the second question. On the one hand, it is known that quantifiers increase expressiveness: the data type (of completeness of finite graphs) given in [KS98] can be specified under loose semantics using first-order axioms but not without quantifiers. In particular, it cannot be specified under initial semantics using conditional equations. The remaining problem deals with the case that quantifiers are dispensed with even in the loose approach. This paper is intended to close the gap. We are going to prove the following:

Theorem 1. *There is a computable data type that can be (monomorphically) specified under loose semantics using quantifier-free axioms, but not under initial semantics using conditional equations (both without hidden functions).*

This result is, although very much in the style of the results mentioned above, less obvious. At first sight, one might suspect the opposite—as we will discuss in Section 3. The proof of the theorem is done constructively by providing a suitable data type (Section 4) and proving its specifiability with loose methods (Section 5) and its non-specifiability with initial methods (Section 6). First, we recall some basic definitions.

2 Preliminaries

We assume the reader to be familiar with the very basic notions of algebraic specification (cf. e.g. [EM85,Wir90,LEW96]) like those of (many-sorted) *signature* $\Sigma = (S, F)$, (total) Σ-*algebra* $A = ((s^A)_{s \in S}, (f^A)_{f \in F})$, and Σ-*homomorphism*.

The set of *terms* $T(\Sigma, X)$ with variables taken from X is defined as usual. Terms without variables are called *ground terms*. A Σ-algebra A is called *term-generated* if for each of its carrier elements $a \in s^A$ $(s \in S)$, there is a denotation, i.e. a ground term $t \in T(\Sigma, \emptyset)$ with $t^A = a$. The class of all term-generated Σ-algebras is denoted by $Gen(\Sigma)$.

Two Σ-algebras A, B are called *isomorphic* if there is a bijective Σ-homomorphism $h : A \to B$. The isomorphism class of a Σ-algebra A, also called the "abstract" *data type*, is denoted by $[A]$. Sometimes, we simply speak of the *data type* A. For a class C of Σ-algebras, $A \in C$ is called *initial in C* if for all $B \in C$ there exists a unique Σ-homomorphism $h : A \to B$.

A Σ-algebra is called *computable* if it is isomorphic to a *computable number algebra*, an algebra in which all carrier sets are decidable subsets of $\mathbb{N}$ and all functions are computable.

Atomic Σ-formulas are Σ-equations $t_1 = t_2$ (where $t_1, t_2 \in T(\Sigma, X)$ are terms of the same sort) and the boolean constant **false**. *(First-order) Σ-formulas* are built from the atomic ones with the logical connectives $\neg$ and $\wedge$, and the quantifier $\exists$. Further logical operators such as **true**, $\neq$, $\vee$, $\to$, $\leftrightarrow$, and $\forall$ are regarded as abbreviations. *Conditional equations* are formulas of the form:

$$t_1 = t'_1 \wedge \cdots \wedge t_k = t'_k \to t = t'.$$

The usual *satisfaction* of a Σ-formula φ by a Σ-algebra A w.r.t. a valuation $v : X \to A$ is denoted by $A, v \models \varphi$. We write $A \models \varphi$, and say that φ is *valid in A*, if $A, v \models \varphi$ holds for all valuations v. For a set Φ of Σ-formulas, we write $A \models \Phi$ if $A \models \varphi$ for all $\varphi \in \Phi$.

A *(first-order algebraic) specification* $SP = (\Sigma, \Phi)$ consists of a signature Σ and a finite set Φ of Σ-formulas, called *axioms*. Two semantic functions are defined corresponding to the two main approaches:

- *loose semantics*: $Mod(SP) := \{A \in Gen(\Sigma) \mid A \models \Phi\}$
- *initial algebra semantics*:[3] $I(SP) := \{A \in Mod(SP) \mid A \text{ initial in } Mod(SP)\}$.

The elements of $Mod(SP)$ and $I(SP)$ are called the *models* and *initial models*, respectively. A specification $SP = (\Sigma, \Phi)$ is said to *specify* a Σ-algebra A (and also the corresponding abstract data type $[A]$)

- *under loose semantics* if $Mod(SP) = [A]$
- *under initial algebra semantics* if $I(SP) = [A]$.

A specification SP is called *monomorphic* if any two of its models are isomorphic, i.e., if there is—up to isomorphism—at most one element in $Mod(SP)$.[4] In particular, a specification SP specifies a Σ-algebra A under loose semantics if and only if SP is monomorphic and $A \in Mod(SP)$.

While in the initial algebra approach, one usually restricts oneself to conditional equations (in order to guarantee executability and the existence of initial models), in the loose semantics approach quantifier-free or arbitrary first-order formulas are used as axioms.

A Σ-algebra A with $\Sigma = (S, F)$ can be specified *with hidden functions* (under loose or initial algebra semantics) if there is a super-signature $\Sigma' = (S, F \cup HF)$ (where HF denotes the additional hidden function symbols) and a Σ'-algebra A' with A as its Σ-reduct, i.e. $A'|_\Sigma = A$, such that A' possesses a specification (under loose or initial algebra semantics, respectively).

3 Discussion of the problem

The aim is to prove that there is a computable data type that can be specified using quantifier-free loose methods but not using the conditional initial approach (both without hidden functions).

This claim is very much in the style of the results mentioned in the introduction (e.g. [TWW82, Theorem 2 and 5], [Maj79, Lemma 6], [BM84], and [BT87, Theorem 4.8]). Nevertheless, it is somewhat surprising—as we are going to discuss in this section.

First, remember that any finite set of quantifier-free formulas can be rewritten in conjunctive normal-form preserving equivalence and thus into a finite set of formulas of the form:

$$t_1 = t'_1 \wedge \cdots \wedge t_n = t'_n \rightarrow t_{n+1} = t'_{n+1} \vee \cdots \vee t_m = t'_m.$$

Therefore, the main difference between quantifier-free axioms and conditional equations is whether there are disjunctions allowed on the right hand side of implications.

---

[3] The results of the paper remain valid if in the definition of initial algebra semantics $Alg(SP) := \{A \in Alg(\Sigma) \mid A \models \Phi\}$ is used instead of $Mod(SP)$.

[4] Our definition of monomorphicity differs from the one given in [Wir90] where *'exactly one'* is required instead of *'at most one'*.

While these disjunctions express a kind of indeterminism, on the other hand, assuming that the quantifier-free specification is monomorphic, everything is determined by the axioms. Consequently, one might be tempted to suppose that a "deterministic" specification can always be written without those disjunctions, i.e. with conditional equations and (disjunctions of) negated equations (which correspond to implications with empty right hand side) only. The latter are covered by the initial constraint and can thus be omitted. Hence, this informal argument suggests that any data type specifiable with quantifier-free axioms and loose semantics can also be specified with conditional equations under initial semantics, i.e. just the opposite of our claim.

Another reason for our result being somewhat surprising is that there are several classical characterizations for theories to be equivalent to a set of conditional equations (e.g. closure of the models against direct products), see e.g. [CK90, Section 6.2], [Mak87, Theorem 3.5], [Fag80], and [Mal71]. Since we start from a quite restrictive condition, namely the monomorphic specifiability without quantifiers, one might hope that one of these characterizations could be applied.

Finally, many data types which at first sight seem to be promising candidates for being specifiable using loose methods but not using initial methods turn out to possess a (tricky) specification with conditional equations. For instance, Bergstra & Meyer [BM84] demonstrated this for the data type of finite sets of integers equipped with a cardinality function, although one would at first suppose that (the statement of) the axiom

$$\mathtt{insert}(x, s) \neq s \;\to\; \mathtt{card}(\mathtt{insert}(x, s)) = \mathtt{succ}(\mathtt{card}(s))$$

cannot be expressed using conditional equations.[5]

4 The data type of segments

In this section we provide an example of a data type, named **SEGS**, that can be monomorphically specified using quantifier-free axioms but fails to possess a conditional equational specification under initial semantics. Roughly speaking, **SEGS** (for *segments*) is the data type of finite initial segments of the natural numbers where the set *Seg* of segments is defined as follows:

$$Seg \;:=\; \Big\{ \{0\},\; \{0,1\},\; \{0,1,2\},\; \{0,1,2,3\},\; \ldots \Big\}.$$

---

[5] The trick is to code inequalities in equational language, e.g. the inequality $x \neq y$ by $\mathtt{card}(\mathtt{insert}(x, \mathtt{insert}(y, \mathtt{empty}))) = \mathtt{succ}(\mathtt{succ}(\mathtt{zero}))$. Similar techniques can be applied for other data types; for further examples see e.g. [Sch98, Section 10.1].

To formally define the data type **SEGS**, we fix the signature:

```
signature Σ_SEGS
    sorts       Nat, Set, Seg
    functions   zero   :                    → Nat
                succ   : Nat                → Nat
                empty  :                    → Set
                insert : Nat × Set          → Set
                red    : Set                → Seg
end signature
```

and declare the interpretation as follows:

$$\text{Nat}^{\text{SEGS}} := \mathbb{N} = \{0, 1, 2, \ldots\}$$
$$\text{Set}^{\text{SEGS}} := \{M \subset \mathbb{N} \mid M \text{ finite}\}$$
$$\text{Seg}^{\text{SEGS}} := \{\emptyset\} \cup Seg$$
$$\text{zero}^{\text{SEGS}} := 0$$
$$\text{succ}^{\text{SEGS}}(n) := n + 1$$
$$\text{empty}^{\text{SEGS}} := \emptyset$$
$$\text{insert}^{\text{SEGS}}(n, M) := M \cup \{n\}$$
$$\text{red}^{\text{SEGS}}(M) := \begin{cases} M & \text{, if } M \in Seg \\ \emptyset & \text{, otherwise.} \end{cases}$$

For the construction of the data type, we borrowed from two ideas:

- First, in finite model theory (cf. e.g. [EF95]), one mostly considers *finite graphs* as models because these are in some sense the most general structures. Furthermore, *connectedness* of graphs is a property of specific importance (cf. e.g. [Gai82, p. 124], [FSV95, Section 5], or [EF95]). Our example is similar: the segments can be regarded as connected, one-dimensional graphs.
- Second, we adopted a technique which has been applied by Thatcher *et al.* [TWW82, Section 3] in order to prove that conditional equations are strictly more powerful than equations. It uses *"reduction functions"* (here the function **red**) to map the problem of defining a predicate onto the problem of defining equality (which is the crucial point in algebraic specification).

Our claim is that the data type **SEGS** has a quantifier-free specification under loose semantics but no conditional specification under initial semantics. The following two sections are devoted to its proof.

5 Specifiability using loose methods

We start with the simpler task, namely the proof of specifiability, which is done by constructing a suitable specification.

Theorem 2. *There is a quantifier-free specification SP which specifies* **SEGS** *under loose semantics, i.e. Mod(SP) = [***SEGS***].*

Before providing an appropriate specification, let us agree on two abbreviations in order to improve readability:[6]

$$member(x, u) :\equiv \left(insert(x, u) = u\right)$$
$$segp(u) \quad :\equiv red(u) \neq red(empty).$$

Using these abbreviations, we can specify **SEGS** as follows:

---
specification SP_{SEGS}
  signature Σ_{SEGS}
  variables x, y : Nat
          u, v : Set
  axioms   (1) zero $\neq$ succ(x)
          (2) succ(x) = succ(y) $\rightarrow$ x = y
          (3) $\neg$ member(x, empty)
          (4) member(x, insert(y, u)) $\leftrightarrow$ (x = y $\vee$ member(x, u))
          (5) insert(x, insert(y, u)) = insert(y, insert(x, u))
          (6) segp(insert(zero, empty))
          (7) segp(u) $\wedge$ member(x, u) $\rightarrow$ segp(insert(succ(x), u))
          (8) member(succ(x), u) $\wedge$ $\neg$ member(x, u) $\rightarrow$ $\neg$ segp(u)
          (9) segp(u) $\wedge$ u $\neq$ v $\rightarrow$ red(u) $\neq$ red(v)

end specification

Lemma 3. *The quantifier-free specification* SP_{SEGS} *specifies* **SEGS** *under loose semantics, i.e.* $Mod(SP_{SEGS}) =$ **SEGS**.

Proof. We can easily verify that all axioms are valid in **SEGS**. Therefore, **SEGS** is a model of SP_{SEGS} and it remains to prove the monomorphicity of SP_{SEGS}.

For that purpose, we use the fact that a specification SP is monomorphic iff SP fixes the ground equations, i.e. iff for all ground equations $t_1 = t_2$, either $SP \models t_1 = t_2$ or $SP \models t_1 \neq t_2$ (cf. [Wir90, Fact 2.3.2]). We check separately every kind of ground equation:

(a) ground equations of the sort Nat:
  Due to axioms (1) and (2), we get $SP \models t_1 \neq t_2$ whenever t_1 and t_2 differ syntactically; otherwise, $SP \models t_1 = t_2$ holds trivially.

---
[6] Note, that we are not adding any new symbols to the signature.

(b) ground equations of the sort Set:
We first consider equations of the form $member(t, t')$. These are fixed due to the inductive definition of $member$ in axioms (3) and (4). Now, let $t_1 = t_2$ be an arbitrary ground equation of the sort Set. In the case that there is a ground term t with $SP_{SEGS} \models member(t, t_1)$ but $SP_{SEGS} \models \neg member(t, t_2)$ (or vice versa) we immediately obtain $SP_{SEGS} \models t_1 \neq t_2$. Otherwise,

$$SP_{SEGS} \models member(t, t_1) \iff SP_{SEGS} \models member(t, t_2)$$

holds for all ground terms t of the sort Nat. A proof by induction using axiom (5) yields $SP_{SEGS} \models t_1 = t_2$.

(c) ground equations of the sort Seg:
We start with ground equations of the form $segp(t)$. If $t^{\mathbf{SEGS}} \in Seg$, we obtain $SP_{SEGS} \models segp(t)$ using axioms (6) and (7). Otherwise, axiom (8) implies $SP_{SEGS} \models \neg segp(t)$.
Now let $\mathtt{red}(t_1) = \mathtt{red}(t_2)$ be an arbitrary ground equation. In case that $SP_{SEGS} \models t_1 = t_2$, we immediately obtain $SP_{SEGS} \models \mathtt{red}(t_1) = \mathtt{red}(t_2)$. The same holds in case that $SP_{SEGS} \models \neg segp(t_1)$ and $SP_{SEGS} \models \neg segp(t_2)$. In all remaining cases, we get $SP_{SEGS} \models \mathtt{red}(t_1) \neq \mathtt{red}(t_2)$ by applying axiom (9).

□

6 Non-Specifiability using initial methods

Having proved that the data type **SEGS** can be monomorphically specified without quantifiers, in order to establish theorem 1, it remains to show that **SEGS** fails to possess a conditional equational specification under initial semantics. This is formalized in the following theorem.

Theorem 4. *There is no conditional equational specification SP which specifies* **SEGS** *under initial semantics, i.e.* $I(SP) = [\mathbf{SEGS}]$.

The idea for the proof is to construct non-desired models for possible conditional equational axiomatizations. As non-desired models, we consider the data types $\mathbf{SEGS}_m$ which differ from **SEGS** in that they also treat some non-segments like segments, namely those sets $M \subseteq \mathbb{N}$ which have exactly one gap of length one sufficiently far away from 0. More precisely, we define $\mathbf{SEGS}_m$ just as **SEGS**, except for:

$$\mathrm{Seg}^{\mathbf{SEGS}_m} := Seg \cup \left\{ M \setminus \{n\} \mid M \in Seg, n \geq m \right\}$$

$$\mathtt{red}^{\mathbf{SEGS}_m}(M) := \begin{cases} M, & \text{if } M \in Seg \\ M, & \text{if } M \cup \{n\} \in Seg \text{ for some } n \geq m \\ \emptyset, & \text{otherwise.} \end{cases}$$

For any conditional equation φ, we determine a number m such that

$$\mathbf{SEGS} \models \varphi \implies \mathbf{SEGS}_m \models \varphi. \qquad (\star)$$

In order to choose m, it suffices to consider the number $inserts(\varphi)$ of occurrences of the symbol insert in φ. We are going to prove that $(\star)$ holds whenever $m > inserts(\varphi)$. This is done by constructing modifications v' of valuations v such that $\mathbf{SEGS}_m, v \models \neg\varphi$ implies $\mathbf{SEGS}, v' \models \neg\varphi$.

The proof of theorem 4 needs some preparations. First, we show that we can restrict ourselves to conditional equations of a certain "simple" form. More precisely, we call a conditional Σ_{SEGS}-equation

$$t_1 = t'_1 \wedge \cdots \wedge t_k = t'_k \rightarrow t = t'$$

simple if it has the following properties:

- if $t_i = t'_i$ (with $i \in \{1, \ldots, k\}$) is a Seg-equation, then it is of the form $\mathbf{red}(x) = \mathbf{red}(\mathbf{empty})$ for some variable x of the sort Set, and
- if $t = t'$ is a Seg-equation, then it is of the form $\mathbf{red}(x) = \mathbf{red}(y)$ for some variables x, y of the sort the Set.

Lemma 5. *For a finite set Φ of conditional equations over Σ_{SEGS}, there is a finite set Φ' of simple conditional equations over Σ_{SEGS} such that:*[7]

- $A \models \Phi' \iff A \models \Phi$ *for all* $A \in \{\mathbf{SEGS}\} \cup \{\mathbf{SEGS}_m \mid m \in \mathbb{N}\}$
- $inserts(\Phi') = inserts(\Phi)$.

Proof. First, we eliminate variables of the sort Seg: if y is such a variable, we can replace all occurrences of y by the term $\mathbf{red}(x)$, where x is a new variable of the sort Set. This does not affect validity in *term-generated* models.

Therefore, we can assume Seg-equations to be of the form $\mathbf{red}(t_1) = \mathbf{red}(t_2)$. Moreover, we can assume the terms t_1, t_2 of the sort Set to be variables, because the following transformations of conditional equations preserve equivalence:

$$(\cdots \wedge \mathbf{red}(t_1) = \mathbf{red}(t_2) \wedge \cdots \rightarrow \cdots)$$
$$\rightsquigarrow (\cdots \wedge x_1 = t_1 \wedge x_2 = t_2 \wedge \mathbf{red}(x_1) = \mathbf{red}(x_2) \wedge \cdots \rightarrow \cdots)$$

$$(\cdots \rightarrow \mathbf{red}(t_1) = \mathbf{red}(t_2))$$
$$\rightsquigarrow (\cdots \wedge x_1 = t_1 \wedge x_2 = t_2 \rightarrow \mathbf{red}(x_1) = \mathbf{red}(x_2)).$$

(x_1, x_2 are new variables of the sort Set.)

It remains to prove that the Seg-equations on the left hand side of implications can even be transformed to the form $\mathbf{red}(x) = \mathbf{red}(\mathbf{empty})$. To show this, we make use of the fact that for all $A \in \{\mathbf{SEGS}\} \cup \{\mathbf{SEGS}_m \mid m \in \mathbb{N}\}$ and all valuations v:

$A, v \models \mathbf{red}(x_1) = \mathbf{red}(x_2)$
$\iff A, v \models x_1 = x_2 \vee (\mathbf{red}(x_1) = \mathbf{red}(\mathbf{empty}) \wedge \mathbf{red}(x_2) = \mathbf{red}(\mathbf{empty}))$.

Hence, a conditional equation of the form

$$\cdots \wedge \mathbf{red}(x_1) = \mathbf{red}(x_2) \wedge \cdots \rightarrow \cdots$$

---
[7] $inserts(\Phi) := \max\{inserts(\varphi) \mid \varphi \in \Phi\}$.

can be replaced by the following two conditional equations:

$$\cdots \wedge x_1 = x_2 \wedge \cdots \to \cdots$$
$$\cdots \wedge \mathtt{red}(x_1) = \mathtt{red}(\mathtt{empty}) \wedge \mathtt{red}(x_2) = \mathtt{red}(\mathtt{empty}) \wedge \cdots \to \cdots$$

It is easily seen that none of the applied transformations affects the number of insert symbols. □

Most easy cases in the final proof can be dealt with using the following fact.

Lemma 6. *Let $v : X \to \mathbf{SEGS}$ be a valuation and $m \in \mathbb{N}$. Then*

(a) *for any equation $t_1 = t_2$ of the sort* Nat *or* Set:

$$\mathbf{SEGS}_m, v \models t_1 = t_2 \iff \mathbf{SEGS}, v \models t_1 = t_2.$$

(b) *for variables x of the sort* Set:

$$\mathbf{SEGS}_m, v \models \mathtt{red}(x) = \mathtt{red}(\mathtt{empty}) \implies \mathbf{SEGS}, v \models \mathtt{red}(x) = \mathtt{red}(\mathtt{empty}).$$

Proof. Statement (a) is obvious, since $\mathbf{SEGS}_m$ and $\mathbf{SEGS}$ do not differ in their evaluation of terms of the sort Nat or Set.

To prove (b) we simply observe that $\mathbf{SEGS}_m, v \models \mathtt{red}(x) = \mathtt{red}(\mathtt{empty})$ implies $v(x) \notin \mathit{Seg}$ and thus $\mathbf{SEGS}, v \models \mathtt{red}(x) = \mathtt{red}(\mathtt{empty})$. □

Since the modification of the valuation should preserve Set-equalities, we have to change Set-variables simultaneously if they are "connected" through equations. Let us define this relationship formally. Two variables x_1, x_2 of the sort Set are said to be *adjacent* in a set Φ of Set-equations if there is an equation of the form $t_1(x_1) = t_2(x_2)$ in Φ. Adjacency between Set-variables and the symbol empty is defined similarly.

The reflexive, symmetric, transitive closure of adjacency (in Φ) is obviously an equivalence relation on Set-variables and the symbol empty. We denote it by $\sim_\Phi$. The equivalence class $[\mathtt{empty}]_{\sim_\Phi} = \{x \mid x \sim_\Phi \mathtt{empty}\}$ will be of particular interest, allowing to formulate the following estimation.

Lemma 7. *Let Φ be a set of equations of the sort* Set *and v a valuation such that $\mathbf{SEGS}_m, v \models \Phi$. Then, $|v(x)| \leq \mathit{inserts}(\Phi)$ for all* Set-*variables $x \in [\mathtt{empty}]_{\sim_\Phi}$.*

Proof. Let $x \in [\mathtt{empty}]_{\sim_\Phi}$. Then there is a sequence $(x_1, x_2, \ldots, x_k)$ of distinct Set-variables such that $x_k :\equiv x$ and Φ contains distinct equations of the form:

$$t_1(\mathtt{empty}) = t_1'(x_1), \quad t_2(x_1) = t_2'(x_2), \quad \ldots, \quad t_k(x_{k-1}) = t_k'(x_k).$$

For any $i \in \{1, \ldots, k\}$ (with $x_0 :\equiv \mathtt{empty}$ and $v(\mathtt{empty}) := \emptyset$), one obtains $|v(x_i)| \leq \mathit{inserts}(t_i) + |v(x_{i-1})|$ as well as $|v(x_{i-1})| \leq \mathit{inserts}(t_i') + |v(x_i)|$, and therefore

$$\Big||v(x_i)| - |v(x_{i-1})|\Big| \leq \mathit{inserts}\Big(t_i(x_{i-1}) = t_i'(x_i)\Big).$$

Using the above inequation, the desired estimation is derived as follows:

$$\begin{aligned}
|v(x)| &= |v(x_k)| - |v(x_0)| \\
&= \sum_{i \in \{1,\ldots,k\}} |v(x_i)| - |v(x_{i-1})| \\
&\leq \sum_{i \in \{1,\ldots,k\}} \Big||v(x_i)| - |v(x_{i-1})|\Big| \\
&\leq \sum_{i \in \{1,\ldots,k\}} inserts\Big(t_i(x_{i-1}) = t'_i(x_i)\Big) \\
&\leq inserts(\Phi).
\end{aligned}$$

$\square$

These preparations allow us to prove the following key lemma.

Lemma 8. *Let φ be a simple conditional equation over the signature Σ_{SEGS} and $m > inserts(\varphi)$. Then*

$$\mathbf{SEGS} \models \varphi \implies \mathbf{SEGS}_m \models \varphi.$$

Proof. Let φ be a *simple* conditional equation, i.e. of the form:

$$\varphi \equiv \Big(t_1 = t'_1 \wedge \cdots \wedge t_k = t'_k \rightarrow t = t'\Big).$$

We will show the implication

$$\mathbf{SEGS} \models \varphi \implies \mathbf{SEGS}_m \models \varphi$$

for $m > inserts(\varphi)$ by contraposition and therefore assume $\mathbf{SEGS}_m \not\models \varphi$. Hence, there is a valuation $v : X \to \mathbf{SEGS}_m$ such that $\mathbf{SEGS}_m, v \models \neg\varphi$, i.e. $\mathbf{SEGS}_m, v \models t_1 = t'_1 \wedge \cdots \wedge t_k = t'_k$ and $\mathbf{SEGS}_m, v \models t \neq t'$.

If $t = t'$ is an equation of the sort Nat or Set, we use lemma 6(a) to obtain $\mathbf{SEGS}, v \models t \neq t'$, and lemma 6(a) and (b) for $\mathbf{SEGS}, v \models t_1 = t'_1 \wedge \cdots \wedge t_k = t'_k$. Therefore, $\mathbf{SEGS}, v \models \neg\varphi$, i.e. $\mathbf{SEGS} \not\models \varphi$.

If $\mathbf{SEGS}, v \models t \neq t'$, we can again use lemma 6 to obtain $\mathbf{SEGS}, v \models \neg\varphi$, i.e. $\mathbf{SEGS} \not\models \varphi$. Therefore, we can focus on the (interesting) case that $t = t'$ is a Seg-equation with $\mathbf{SEGS}, v \models t = t'$. Since φ is simple, the equation $t = t'$ has to be of the form $\mathbf{red}(x_1) = \mathbf{red}(x_2)$ for appropriate Set-variables x_1, x_2. Together with $\mathbf{SEGS}_m, v \models \mathbf{red}(x_1) \neq \mathbf{red}(x_2)$ and $\mathbf{SEGS}, v \models \mathbf{red}(x_1) = \mathbf{red}(x_2)$, the definitions of $\mathbf{SEGS}$ and $\mathbf{SEGS}_m$ yield:

(a) $v(x_1) \neq v(x_2)$,
because otherwise, $\mathbf{SEGS}_m, v \models \mathbf{red}(x_1) = \mathbf{red}(x_2)$.
(b) $v(x_1), v(x_2) \notin Seg$,
otherwise, the set $v(x_1)$ or the set $v(x_2)$ would be a segment, and (a) would imply $\mathbf{SEGS}, v \models \mathbf{red}(x_1) \neq \mathbf{red}(x_2)$.

(c) there is some $n \geq m$ such that $v(x_1) \cup \{n\} \in Seg$ or $v(x_2) \cup \{n\} \in Seg$;
w.l.o.g. (if necessary, exchange x_1 and x_2) let $v(x_1) \cup \{n\} \in Seg$,
because otherwise, (b) would yield $\mathbf{SEGS}_m, v \models \mathtt{red}(x_1) = \mathtt{red}(\mathtt{empty})$
as well as $\mathbf{SEGS}_m, v \models \mathtt{red}(x_2) = \mathtt{red}(\mathtt{empty})$. Obviously this implies $\mathbf{SEGS}_m, v \models \mathtt{red}(x_1) = \mathtt{red}(x_2)$.

(d) $|v(x_1)| > m$,
because, due to statements (b) and (c), the set $v(x_1)$ has to be of the form $\{0, 1, \ldots, n-1, n+1, \ldots, l\}$ for some $l \geq n+1 \geq m+1$.

Now, let Φ denote the set of Set-equations in $t_1 = t'_1 \wedge \cdots \wedge t_k = t'_k$. We define a modified valuation v' as follows:

$$v'(x) := \begin{cases} v(x) \cup \{n\} & \text{, if } x \text{ is a variable of sort Set such that } x \notin [\mathtt{empty}]_{\sim_\Phi} \\ v(x) & \text{, otherwise} \end{cases}$$

This modified valuation has the following properties:

(e) $v'(x_1) \in Seg$,
because $x_1 \notin [\mathtt{empty}]_{\sim_\Phi}$ by (d) and lemma 7, and therefore, by (c), even $v'(x_1) = v(x_1) \cup \{n\} \in Seg$.

(f) $v'(x_1) \neq v'(x_2)$,
otherwise, $v(x_1) \cup \{n\} = v(x_2) \cup \{n\}$ (even if $x_2 \in [\mathtt{empty}]_{\sim_\Phi}$ because then, $v'(x_2) = v(x_2) = v(x_1) \cup \{n\}$ would imply $n \in v(x_2)$). Hence, we obtain $v(x_2) = v(x_1) \cup \{n\}$ or $v(x_2) = v(x_1)$, contradicting (b) or (a), respectively.

(g) $\mathbf{SEGS}, v' \models \mathtt{red}(x_1) \neq \mathtt{red}(x_2)$,
which follows from (e) and (f) and the definition of $\mathtt{red}^{\mathbf{SEGS}}$.

It remains to prove that

(h) $\mathbf{SEGS}, v' \models t_1 = t'_1 \wedge \cdots \wedge t_k = t'_k$.
We consider each of the equations $t_i = t'_i$ separately and distinguish several cases using the precondition $\mathbf{SEGS}_m, v \models t_i = t'_i$ in each of them:

  ○ If $t_i = t'_i$ is a Nat-equation, lemma 6(a) yields $\mathbf{SEGS}, v \models t_i = t'_i$. Hence, we obtain $\mathbf{SEGS}, v' \models t_i = t'_i$, because the evaluations of Nat-variables under v and v' are identical.

  ○ If $t_i = t'_i$ is an equation of the sort Set, we again distinguish two cases. If $t_i = t'_i$ contains the symbol empty or a variable from $[\mathtt{empty}]_{\sim_\Phi}$, then all Set-variables occurring in $t_i = t'_i$ belong to $[\mathtt{empty}]_{\sim_\Phi}$ (due to the definition of $\sim_\Phi$). As v' coincides with v on these variables, we can use the same argument as for Nat-equations.
  If $t_i = t'_i$ contains neither the symbol empty nor a Set-variable from $[\mathtt{empty}]_{\sim_\Phi}$, then $t_i = t'_i$ is of the form $t_i(x_i) = t'_i(x'_i)$ for appropriate variables x_i, x'_i with $x_i, x'_i \notin [\mathtt{empty}]_{\sim_\Phi}$. Thus, $v'(x_i) = v(x_i) \cup \{n\}$ and also $v'(x'_i) = v(x'_i) \cup \{n\}$. Together with $\mathbf{SEGS}, v \models t_i(x_i) = t'_i(x'_i)$, we get $\mathbf{SEGS}, v' \models t_i(x_i) = t'_i(x'_i)$.

◦ If $t_i = t'_i$ is an equation of the sort Seg, since φ is simple, it must be of the form $\texttt{red}(x) = \texttt{red}(\texttt{empty})$.
In case that $x \in [\texttt{empty}]_{\sim_\Phi}$, one obtains $v'(x) = v(x)$, and it follows from the precondition that $\textbf{SEGS}_m, v' \models \texttt{red}(x) = \texttt{red}(\texttt{empty})$. Applying lemma 6(b) shows $\textbf{SEGS}, v' \models \texttt{red}(x) = \texttt{red}(\texttt{empty})$.
In case that $x \notin [\texttt{empty}]_{\sim_\Phi}$, the definition of v' yields $v'(x) = v(x) \cup \{n\}$. For $v(x) = \emptyset$, we obtain $v'(x) \notin Seg$ because of $n \geq m > 0$. Otherwise, if $v(x) \neq \emptyset$, the precondition $\textbf{SEGS}_m, v \models \texttt{red}(x) = \texttt{red}(\texttt{empty})$ implies $v'(x) \notin Seg$. Thus, $v'(x) \notin Seg$ holds in both cases, and therefore $\textbf{SEGS}, v' \models \texttt{red}(x) = \texttt{red}(\texttt{empty})$.

Statements (g) and (h) imply $\textbf{SEGS}, v' \models \neg\varphi$, and thus $\textbf{SEGS} \not\models \varphi$. □

Now, we can turn to the proof of the main theorem.

Proof of theorem 4. For contradiction, we assume that there is a conditional equational specification $SP = (\Sigma_{SEGS}, \Phi)$ with $I(SP) = [\textbf{SEGS}]$. Moreover, let Φ' denote the set of *simple* conditional equations constructed from Φ as in lemma 5, and $m > inserts(\Phi) = inserts(\Phi')$.

By lemma 8, it follows that $\textbf{SEGS}_m \models \Phi'$, and thus $\textbf{SEGS}_m \models \Phi$ (see lemma 5).

On the other hand, it can be easily verified that there is no homomorphism $h: \textbf{SEGS} \to \textbf{SEGS}_m$. For $M_m := \{0, 1, \ldots, m-1, m+1\}$ one would obtain

$$\begin{aligned} M_m &= \texttt{red}^{\textbf{SEGS}_m}(M_m) \\ &= \texttt{red}^{\textbf{SEGS}_m}(h(M_m)) \\ &= h(\texttt{red}^{\textbf{SEGS}}(M_m)) \\ &= h(\emptyset) \\ &= \emptyset. \end{aligned}$$

Consequently, the data type $\textbf{SEGS}$ is not initial in $Mod(SP)$—contradicting the assumption. □

7 Conclusions

We have provided an example of a computable data type which possesses a quantifier-free monomorphic specification, but fails to have a conditional equational specification under initial semantics (both without hidden functions). Our result is in the line of the work done by Majster [Maj79], Bergstra & Tucker [BT87], Thatcher et al. [TWW82] and others (e.g. [Ore79,BM84,Ber93]) and answers an important question which has been left open. Roughly speaking, it shows that the restriction to conditional equations—which is standard in the initial approach to algebraic specification—means a restriction in expressiveness.

However, the result is of relevance not only in the scope of algebraic specification. For instance, it can be read as a result in proof theory. This is due to the fact that both the model of a monomorphic quantifier-free specification and the initial model of conditional equations can be characterized by the ground equations derivable from the axioms. Thus, in proof-theoretic terms, the question we dealt with is: Given a finite, quantifier-free set Φ of formulas (over a functional signature which allows to construct a ground term for any sort) such that for any two ground terms t_1, t_2 of the same sort either $t_1 = t_2$ or $t_1 \neq t_2$ is derivable. Is there a finite set Φ' of conditional equations (over the same signature) such that exactly the same ground equations are derivable from Φ'? Theorem 1 answers this question negatively.

Furthermore, our result might be of some interest in the context of logic programming and rewriting: it provides further justification for the generalizations to *disjunctive* logic programming (e.g. [LMR92,TZ92,NL95]) and to *quasi*-Horn clause rewriting (e.g. [Kap88,BG94]).

References

[Ber93] R. Berghammer. On the characterization of the integers: the hidden function problem revisited. *Acta Cybernetica*, 11(1-2):85–96, 1993.

[BG94] L. Bachmair and H. Ganzinger. Rewrite-based equational theorem proving with selection and simplification. *Journal of Logic and Computation*, 4(3):217–247, 1994.

[BM84] J.A. Bergstra and J.J. Meyer. On specifying sets of integers. *Elektron. Informationsverarb. u. Kybernet.*, 20:532–541, 1984.

[BT82] J.A. Bergstra and J.V. Tucker. The completeness of the algebraic specification methods for computable data types. *Information and Control*, 54:186–200, 1982.

[BT87] J.A. Bergstra and J.V. Tucker. Algebraic specifications of computable and semicomputable data types. *Theoret. Comput. Sci.*, 50:137–181, 1987.

[CK90] C.C. Chang and H.J. Keisler. *Model Theory*, volume 73 of *Studies in Logic and the Found. of Math.* Elsevier Science Publishers B. V., 1990.

[EF95] H.-D. Ebbinghaus and J. Flum. *Finite Model Theory.* Perspectives in Mathematical Logic. Springer, Berlin, 1995.

[EM85] H. Ehrig and B. Mahr. *Fundamentals of Algebraic Specification 1, Equations and Initial Semantics*, volume 6 of *EATCS Monographs on Theoretical Computer Science.* Springer, Berlin, 1985.

[Fag80] R. Fagin. Horn clauses and database dependencies (extended abstract). In *Conference Proceedings of the Twelfth Annual ACM Symposium on Theory of Computing*, pages 123–134, 1980.

[FSV95] R. Fagin, L. Stockmeyer, and M. Vardi. On monadic NP vs. monadic co-NP. *Information and Computation*, 120:78–92, 1995.

[Gai82] H. Gaifman. On local and non-local properties. In J. Stern, editor, *Proc. of the Herbrand Symposium, Logic Colloquium'81*, pages 105–135. North-Holland, 1982.

[GGM76] V. Giarratana, F. Gimona, and U. Montanari. Observability concepts in abstract data type specifications. In *5th Symposium on Mathematical Foundations of Computer Science*, volume 45 of *LNCS*, pages 576–587. Springer, Berlin, 1976.

[GTWW75] J.A. Goguen, J.W. Thatcher, E.G. Wagner, and J.B. Wright. Abstract data types as initial algebras and the correctness of data representations. In *Proc. Conf. on Computer Graphics, Pattern Recognition and Data Structures*, pages 89–93, 1975.

[Kap88] S. Kaplan. Positive/negative conditional rewriting. In M.P. Chytil, L. Janiga, and V. Koubek, editors, *Mathematical Foundations of Computer Science*, volume 324 of *LNCS*, pages 381–395. Springer, Berlin, 1988.

[KS98] D. Kempe and A. Schönegge. On the power of quantifiers in first-order algebraic specification. In *Annual Conference of the European Association for Computer Science Logic*, 1998.

[LEW96] J. Loeckx, H.-D. Ehrich, and M. Wolf. *Specification of abstract data types*. Verlag Wiley-Teubner, 1996.

[LMR92] J. Lobo, J. Minker, and A. Rajasekar. *Foundations of Disjunctive Logic Programming*. MIT-Press, 1992.

[Maj79] M.E. Majster. Data types, abstract data types and their specification problem. *Theoret. Comput. Sci.*, 8:89–127, 1979.

[Mak87] J.A. Makowsky. Why Horn formulas matter in computer science: Initial structures and generic examples. *J. Comput. Sys. Sci.*, 34:266–292, 1987.

[Mal71] A.I. Malćev. *The Metamathematics of Algebraic Systems: Collected Papers 1936-1967*. Translated and edited by B.F. Wells III. North-Holland, Amsterdam, 1971.

[NL95] G. Nadathur and D.W. Loveland. Uniform proofs and disjunctive logic programming. In D. Kozen, editor, *Proc. of the Tenth Annual Symp. on Logic in Comp. Sci.*, pages 148–155. IEEE Computer Society Press, 1995.

[Ore79] F. Orejas. On the power of conditional specifications. *ACM SIGPLAN Notices*, 14(7):78–81, 1979.

[Sch97] A. Schönegge. The hidden function question revisited. In *Proceedings of the 6th International Conference on Algebraic Methodology and Software Technology*, volume 1349 of *LNCS*, pages 451–464. Springer, Berlin, 1997.

[Sch98] A. Schönegge. *Specifiability of computable data types (in german)*. PhD thesis, Universität Karlsruhe, Fakultät für Informatik, Shaker Verlag, 1998.

[SW98] D.T. Sannella and M. Wirsing. Specification languages. In E. Astesiano, H.-J. Kreowski, and B. Krieg-Brückner, editors, *Algebraic Foundations of Systems Specifications*. Chapman and Hall, 1998. To appear.

[TWW82] J.W. Thatcher, E.G. Wagner, and J.B. Wright. Data type specification: Parameterization and the power of specification techniques. In *ACM Trans. on Prog. Languages and Systems*, volume 4, pages 711–732, 1982.

[TZ92] J.V. Tucker and J.I. Zucker. Deterministic and nondeterministic computation and Horn programs, on abstract data types. *Journal of Logic Programming*, 13(1):23–55, 1992.

[Wan81] M. Wand. Final algebra semantics and data type extensions. *J. Comput. System Sci.*, 19:27–44, 1981.

[Wir90] M. Wirsing. *Algebraic Specification*, volume B of *Handbook of Theoretical Computer Science*, chapter 13. Elsevier Science Publishers B. V., 1990.

[Wir95] M. Wirsing. Algebraic specification languages: an overview. In *Proc. 10th Workshop on Specification of Abstract Data Types*, volume 906 of *LNCS*, pages 81–115. Springer, Berlin, 1995.

[WPP+83] M. Wirsing, P. Pepper, H. Partsch, W. Dosch, and M. Broy. On hierarchies of abstract data types. *Acta Informatica*, 20:1–33, 1983.

A Decision Algorithm for Prenex Normal Form Rational Presburger Sentences Based on Combinatorial Geometry

Naoki Shibata, Kozo Okano, Teruo Higashino, and Kenichi Taniguchi

Department of Informatics and Mathematical Science, Osaka University
Machikaneyama 1-3, Toyonaka, Osaka 560-8531, Japan.
{n-sibata,okano,higashino,taniguchi}@ics.es.osaka-u.ac.jp

Abstract. In this paper, we propose a decision algorithm for theory of rationals with addition (the theory consisting of rational variables, rational constants, $+, -, =, <, \wedge, \vee, \neg, \forall$ and $\exists$) which runs in $r\alpha^d n^{\beta d(b+1)^a}$ time and it uses some techniques in combinatorial geometry, where α and β are proper constants, and r, n, d, a and b denote the maximum bit length of coefficients, the number of inequalities, the number of variable, the number of quantifier alternations and the maximum length of the consecutive same quantifiers in the input sentence respectively. The known fastest decision algorithm was Ferrante and Rackoff's algorithm which runs in $r\gamma^d n^{\delta d(2b+1)^a}$ time, where γ and δ are proper constants.

1 Preface

Decision algorithms for prenex normal form sentences of the theory of rationals with addition, which we call PRP sentences, (the theory consisting of rational variables, rational constants, $+, -, =, <, \wedge, \vee, \neg, \forall, \exists$) are used in the tests of protocols, timing verification of hardware, and so on[1, 2]. There has been a lot of work on the problem of finding precise time and space complexity of this algorithm[3, 4, 5, 6, 7]. Ferrante and Rackoff have proposed a decision algorithm for PRP sentences which runs in $r\gamma^d n^{\delta d(2b+1)^a}$ time, where γ and δ are constants, and r, n, d, a, b denote the maximum bit length of coefficients, the number of inequalities, the number of variables, the number of quantifier alternations and the maximum length of the consecutive same quantifiers in the input sentence, respectively[8, 9]. This is the fastest decision algorithm until now as long as authors know. In this paper, we propose a faster algorithm which runs in $r\alpha^d n^{\beta d(b+1)^a}$ time (α and β are proper constants) and it uses some techniques in combinatorial geometry. The dominant part in the time complexity of Ferrante and Rackoff's is the term with double exponential, where our algorithm has an advantage.

The overview of our algorithm is as follows. From an input PRP sentence, we construct an arrangement (a set of all subspaces obtained by dividing a whole d-dimensional space with hyperplanes corresponding to all inequalities in the

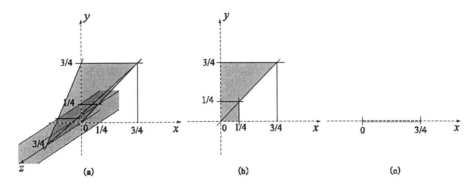

Fig.1 Transformation of an arrangement in execution of the algorithm.

sentence, where d is the number of variables in the sentence). In each of subspaces, any point has the same truth value when we evaluate the point in the matrix[1] of the input PRP sentence. Therefore, we assign each truth value of an adequate point in the matrix to the corresponding subspace. Next, we make a projection of those subspaces into $(d-s)$-dimensional subspace where s is the number of the innermost series of the same quantifiers. The value of the input PRP sentence is decided by the 0-dimensional arrangement obtained at last.

2 Algorithm

2.1 Outline

We give a brief look of a decision process for PRP sentence $F = \forall x \exists y \exists z \{x \geq 0 \land y \geq 0 \land [(y - x \geq 0 \land y + z \leq 3/4 \land z \geq 0) \lor (x \leq 1/4 \land y \leq 1/4)]\}$, giving some definitions relating to combinatorial geometry.

First, we divide a whole d-dimensional space into subspaces to make an arrangement from F where d denotes the number of variables in a given sentence. For F, we use a 3-dimensional space, because F has three variables. Here, we briefly introduce the notion of the arrangement. A plane divides a space into three subspaces, *i.e.* a subspace above the plane, a subspace below the plane and the plane itself. In general, a set of planes divides a space into points, segments without their ends, polygons without their periphery, and polyhedra without their surfaces. Each of these subspaces is called a face. Also, a point, a segment without ends, a polygon without periphery, and a polyhedron without surfaces are called a 0-face, a 1-face, a 2-face and a 3-face, respectively. A set of subspaces made by dividing a space with a set of planes is called an arrangement.

---

[1] A matrix of PRP sentence is a logical expression which is obtained by taking all quantifiers off from the sentence.

A point, a line, a plane and a whole 3-dimensional space is called 0-flat, 1-flat, 2-flat and 3-flat, respectively.

For F, we make an arrangement using seven planes corresponding to inequalities in E. Some part of the arrangement and the domain in which E is true is shown in Fig.1 (a), where E is an expression $x \geq 0 \wedge y \geq 0 \wedge [(y - x \geq 0 \wedge y + z \leq 3/4 \wedge z \geq 0) \vee (x \leq 1/4 \wedge y \leq 1/4)]$, the matrix of F. This domain is a subset of the arrangement made by dividing the space with planes corresponding to inequalities in E.

We will extend these definitions mentioned above to general dimension ones. These definitions follows the literature [10].

Definition 1. In a Cartesian coordinate system $\{v_1, ..., v_d\}$, if a flat fl contains a line perpendicular to an axis v_i, fl is *perpendicular to an axis v_i*. If a flat fl contains a line parallel to an axis v_i, fl is *parallel to the axis v_i*. □

Definition 2. Without loss of generality, we introduce a hyperplane h which is not perpenticular to the axis v_d. As with an optional point $x = (x_1, x_2, ..., x_d)$ on h, there exists a unique set of rational numbers which meet $x_d = \eta_d + \sum_{i=1}^{d-1} \eta_i x_i$. For a point $p = (\pi_1, \cdots, \pi_d)$, we call that p is above, on and below h if $\eta_d + \sum_{i=1}^{d-1} \eta_i \pi_i$ is greater than π_d, equal to π_d, and less than π_d, respectively. Notations h^+ and h^- represent a set of points above h and that below h, respectively. Now, we assume no element of a given set of planes $H = \{h_1, ..., h_n\}$ is perpendicular to any axis v_i. For a given hyperplane h and a point p, we define $v_i(p)$ as follows.

$$v_i(p) = \begin{cases} +1 \ (p \in h_i^+) \\ 0 \ (p \in h_i) \\ -1 \ (p \in h_i^-) \end{cases}$$

A *face* is a set of points such that the values $v_1(p), ..., v_n(p)$ are all the same for any point p in the set. A face which can be contained in a k-flat and cannot be contained in a $(k-1)$-flat is called a *k-face*. Especially a 0-face is also called a vertex. □

Definition 3. A finite set H of hyperplanes in a d-dimensional space divides the space into faces of various dimensions. We call this set of faces an *arrangement* which was made by dividing a whole d-dimensional Euclidian space. It is also simply called the arrangement of H. The elements of the arrangement A of H are called *faces contained in A*. If a k-flat fl contains k-face f which is contained in A, fl is called a *flat contained in A*. □

Next, we assign a boolean value u to each face f contained in the arrangement A obtained from E, where u is the value obtained by substituting the coordinate of some point in f to E. By this process, the value obtained by substituting the coordinate of any point p in the space to E should be equal to the value assigned to the face which contains p. The faces painted in grey in Fig. 1(a) are the faces to which true is assigned.

Then, we eliminate the quantifier $\exists z$. This is a process to make a projection of grey area in Fig.1(a) onto a (x,y)-plane. By this process, we obtain Fig 1(b).

To make this projection, we extract all 1-flats (lines) from Fig 1(a), then make their projections, then make a 2-dimensional arrangement from these projections, and finally assign true values to faces of the arrangement iff the faces are contained in the shadow of grey-painted domain of Fig 1(a).

Definition 4. A *projection of a point* $p = (V_1, V_2, ..., V_d)$ to $(v_1, v_2, ..., v_{d-s})$-space is $(V_1, V_2, ..., V_{d-s})$. □

Definition 5. A *projection of a face* f to $(v_1, v_2, ..., v_{d-s})$-space is a set of all projections of points contained in f. □

Definition 6. A *projection of a flat* fl to $(v_1, v_2, ..., v_{d-s})$-space is a set of all projections of points contained in fl. □

Definition 7. A *projection of an arrangement* B to $(v_1, v_2, ..., v_{d-s})$-space is an arrangement of a set of all projections of flats such that they are contained in B and their projections are $(d - 1 - s)$-flats. □

Generally, for any face f contained in an arrangement A, there exists a set $\mathcal{F}$ corresponding to f such that $\mathcal{F}$ satisfies the following two conditions: (1)$\mathcal{F}$ is a subset of $\mathbf{pr}A$ (projection of A) and (2) a set of points contained in the projection of f is the same as that contained in all elements of $\mathcal{F}$(refer Lemma 1 in section 3.3).

Definition 8. Let A' be a *projection of an arrangement* A (*cf.* Definition 7). A *projection of an assigned arrangement* w.r.t. quantifier Q to $(v_1, v_2, ..., v_{d-s})$-space is an arrangement A' which is assigned boolean values as follows:

If $Q = `\exists$': True is assigned to every f' contained in A' iff there exists a face f such that true is assigned to f and the projection of f contains f'.

If $Q = `\forall$': False is assigned to every f' contained in A' iff there exists a face f such that false is assigned to f and the projection of f contains f'. □

Because quantifiers of the input formula are $\forall x \exists y \exists z$, we eliminate $\exists y$ in the same way, and obtain a 1-dimensional assigned arrangement in the next step. This arrangement is represented in Fig.1(c). In fact, our algorithm eliminates a series of the same quantifiers simultaneously. In other words, we obtain Fig.1(c) from Fig.1(a) directly.

As mentioned above, existential quantifier elimination is to obtain a shadow of a true domain by casting ray parallel to the eliminating axis. On the other hand, eliminating a universal quantifier is to obtain a domain which is not in a shadow of false domain by casting ray parallel to the eliminating axis. Finally we eliminate the remaining $\forall x$ and obtain a 0-dimensional assigned arrangement. This arrangement is composed of a point assigned false. Therefore, the input formula is decided to be false.

2.2 Details of the algorithm.

First, we describe the subroutines ARRANGE, ASSIGN, PROJECT, and then describe the main routine MAIN. ARRANGE converts a set of hyperplanes into the data structure of its arrangement (we call this operation "making an arrangement"). ARRANGE is used in MAIN and PROJECT. ASSIGN assigns

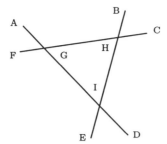
Fig.2 An example of an Arrangement.

truth values of PRP expression to an arrangement. PROJECT makes a projection of an assinged arrangement.

We use a set of coordinates of d affine independent points [2] (Let $P = \{p_0, \cdots, p_k\}$ s.t. k is finite. If $x = \Sigma_{i=0}^{k} \lambda_i p_i$ and $\Sigma_{i=0}^{k} \lambda_i = 1$ hold, x is called an affine combination of P. If there is no $p_i \in P$ s.t. p_i is an affine combination of $P - \{p_i\}$, we call that P is affine independent).

Fig.3 shows the data structure of the arrangement in Fig.2. Here, each rectangle represents a face. Each face f has a set of edges connected to all superfaces of f, and a set of edges to all subfaces. A face also has additional information which contains the following information:

- The dimension of the face.
- The truth value assigned to the face.
- The coordinate of the point, if the face is 0-face.

If a face is unbounded, we use enough large values as its coordinate(e.g. We use the coordinate of a point which is on a ray AG and sufficiently far from a point G).

Algorithm ARRANGE
    ◇ INPUT: A set H of hyperplanes.
    ◇ OUTPUT: An arrangement of H.
    See [10] for its algorithm.

The subroutine ASSIGN assignes the truth value of the input PRP expression to the corresponding face in the input arrangement. The function INNER returns a coordinate in the input face, such as a center of gravity of the face.

Algorithm ASSIGN
    ◇ INPUT: The matrix E of the input sentence, an arrangement A of a set of hyperplanes containing a set of hyperplanes corresponding to all inequalities in E.
    ◇ OUTPUT: An assigned arrangement which is assigned truth value of E.
        ▷ *Assign the truth value of E to faces contained in A.*
1   for each $f \in A$
2   Assign the truth value of $E(\text{INNER}(f))$ to f;

---

[2] According to circumstances, we use more than d coordinates.

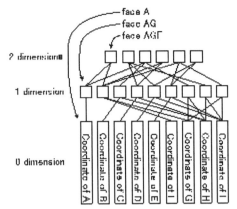

Fig.3 The Data structure of Fig. 2.

```
3       next
4       return A;
5    end
```

The subroutine PROJECT is a subroutine which makes a projection of an arrangement. The function EXT used in this algorithm is a function making a flat which contains a face given as its argument and whose dimension is the same as the face.

Algorithm PROJECT

◇ INPUT: An integer value s, an quantifier q, an assigned arrangement A which meets the following condition: for all axes, there exists a hyperplane hp which is contained in A and perpendicular to the axis.
◇ OUTPUT: A projection of A.

```
1       S := {EXT(f)|f ∈ A, f is (d − 1 − s)-face};
2       A' := ARRANGE(S);
3       if q = '∃' then u := true else u := false;
        ▷ Assign ¬u to all f' ∈ A'.
4       for each f' ∈ A' do assign ¬u to f' next
        ▷ Assign truth value to faces contained in A'.
5       for f ∈ A s.t. u is assigned to f
6           for each f' ∈ A'
7               Let x₁, ..., xₖ be projections of vertices of f;
                ▷ Check whether an optional point INNER(f') in f' is contained in sh f.
8               if (INNER(f') = Σₘ₌₁ᵏ aₘxₘ ∧ 0 < aₘ∧ There exists a₁, ..., aₖ which
                    satisfy aₘ < 1 ∧ Σₘ₌₁ᵏ aₘ = 1)
9                   then assign u to f'; endif
10          next
11      next
12      return A';
13   end
```

The routine MAIN of our algorithm decides the truth value of the input PRP sentence. The function AFFINE used in MAIN returns a set of points such that they are on a hyperplane h corresponding to the inequality in and the set is composed of enough many points so that they can represent h.

Algorithm MAIN
⋄ INPUT: A PRP sentence F.
⋄ OUTPUT: The truth value of F.

1. Let $Q_1, ..., Q_d$ be quantifiers of input PRP sentence. Let $v_1, ..., v_d$ be variables quantified by each quantifier.
2. $S :=$ (the set of all inequalities contained in F) $\bigcup$ ($\bigcup_{i=1}^{d}\{v_i = 0\}$);
3. $H := \emptyset$;
4. **for** each $h \in S$
5. $t :=$ AFFINE(h);
6. $H := H \bigcup \{t\}$;
7. **next**
8. $A :=$ ARRANGE(H);
9. $A :=$ ASSIGN(A,matrix of F);
   ▷ i is the number of remaining quantifiers.
10. $i := d$;
11. **while** $i \geq 1$ **do**
12.
13.   **if** $Q_i =$ '$\forall$' **then** $q :=$ '$\exists$' **else** $q :=$ '$\forall$'
14.
    ▷ Assign the number of the innermost sequence of same Q_i to s.
15.   **if** $q \in \{Q_1, ..., Q_i\}$ **then**
16.   determine s, s.t. $Q_{i-s} = q \land$ for each $i - s + 1 \leq k \leq i$, $Q_k \neq q$;
17.   **else**
18. $s := i$;
19.   **endif**
20.
21. $A :=$ PROJECT(A,s,Q_i);
22.
23. $i := i - s$;
24. **endwhile**
25.
    ▷ A is composed of one point.
26. **return** Truth value of the point of A;
27. **end**

3 Proof of correctness and analysis of time complexity of each routine

In this section, we describe each specification of every routine, and give the proofs of their correctness and analysis of time complexity of each routine.

Hereafter, we will use these notations.

n : the number of inequalities occurring in the input sentence
d : the number of variables occurring in the input sentence
l : the length of the input sentence
a : the number of quantifier alternation of the input sentence
b : the maximum length of the consecutive same quantifiers

3.1 ARRANGE

Specification:
   INPUT: A set H of hyperplanes
   OUTPUT: An arrangement of H

Proof of correctness. Refer [10].

Time complexity: Let g be the maximum bit length of the denominator and the numerator of every coordinate points which represents hyperplanes in H. Thus the time complexity is $O(g \cdot \gamma^d |H|^d)$ and the size of output is also the same order, where γ is some constant[10].

Analysis of the time complexity: The number of faces contained in the arrangement is $O(|H|^d)$ (see [10]). The coordinates of the vertices in the output arrangement are solutions of simultaneous linear equations with d unknowns. Therfore, the bit lengths of the coordinates are $g \cdot \gamma^d$. Since the time complexity is a product of the number of faces and the maximum bit length of vertices in the output arrangement, the time complexity is $O(g \cdot \gamma^d |H|^d)$ and the size of the output is also the same order.

3.2 ASSIGN

Specification:
   INPUT: A PRP expression E and an arrangement A of the set of
           hyperplanes which represents a set of hyperplane which is
           correspond to inequalities in E.
   OUTPUT: An arrangement A' which is assigned truth values of E.

Proof of correctoness. It is clear by Definition 3, and definitions of PRP expression and an assigned arrangement. □

Time complexity: The time complexity of ASSIGN is polynomial order of l, and the size of output arrangement is $O(l)$, where l is the size of A.

Analysis of time complexity: The loop from line 1 to line 3 runs in time of polynomial order of l. Therefore, the time complexity of this algorithm is polynomial order of l. The size of output arrangement is equal to that of the input arrangement.

3.3 PROJECT

Specification:
   INPUT: An assigned arrangement A which contains
   hyperplanes perpendicular to every axis,
   the number s of variables quantified by the innermost
   series of same quantifiers, quantifier Q.
   OUTPUT: A projection of A to $(d' - s)$-dimensional space
   by quantifier Q, where d' is dimension of A.

Proof of correctness. By Lemma 1 which is described later, the following condition holds: for any face $f \in A$ and $f' \in \mathbf{pr}\ A$, f' is contained in $\mathbf{sh}\ f$, iff any point in f' is contained in $\mathbf{sh}\ f$.

We now describe the correctness of line 8.

Let $\{x_1, ..., x_k\}$ be a set of vectors of coordinates of vertices of a convex polyhedron C. If and only if any vector p of coordinates is contained in C, there exist $a_1, ..., a_k$ such that they satisfy the following expression (1). We can say the same thing if we substitute $\{x_1, ..., x_k, y_1, ..., y_l\}$ for $\{x_1, ..., x_k\}$ ($y_1, ..., y_m$ are all contained in C) (see Lemma 4 in appendix). Therefore, the correctness of line 8 is proved.

$$p = \sum_{m=1}^{k} a_m x_m \wedge 0 < a_m < 1 \wedge \sum_{m=1}^{k} a_m = 1 \qquad (1)$$

□

Time complexity: Let m and r be the number of faces in A and the maximum bit length of coordinates in A, respectively. Let α, η and θ be constants. The data size of the output A' is bounded by $r\alpha^{b+1}m^{b+1}$ and the total time complexity is bounded by $r\alpha^{\eta(b+1)}m^{\theta(b+1)}$.

Analysis of time complexity: Let A be the arrangement of H. Then the number of faces in A is $O(|H|^d)$. Let m and r be the number of faces in A and the maximum bit length of coordinates in A, respectively. The number of $(d-s-1)$-flats contained in A is less than or equal to $r\alpha^{s+1}m^{s+1}$. Since $s \le b$ holds, the number of $(d-s-1)$-flats contained in A is less than $r\alpha^{\eta(b+1)}m^{\theta(b+1)}$. The projection $\mathbf{pr}\ A$ is an arrangement of a set of hyperplanes whose number is the same as the number of hyperplanes contained in A. Thus, $\mathbf{pr}\ A$ contains $O(|H|^d)^{b+1}$ faces. Thus, the number of faces in the output arrangement is less than m^{b+1}.

Let r be the maximum bit length of coordinates in A. The maximum bit length of coordinates in $\mathbf{pr}\ A$ is less than $r\alpha^{b+1}$.

We now describe the time complexity of line 8. Since the number of vertices is less than m^{b+1}, the number of inequalities is less than $d + 2m^{b+1} + 1$ and the number of variables is less than m^{b+1}. Since this is a linear programing, the time complexity of this decision is $O((m^{b+1})(d + 2m^{b+1} + 1)^3 r)$[11].

The time complexity between line 5 and line 11 is as follows: since the number of faces in A is m and that of faces in A' is m^{b+1}, it is polynomial order of m^{b+2}. Therefore, this subroutine runs in $r\alpha^{\eta(b+1)}m^{\theta(b+1)}$ time.

Lemma 1. Let S be a set of faces in **pr** A. If there exist hyperplanes such that one of the hyperplanes is not parallel to each of all axes [3], there exists a subset S' in S for all face f in A such that set of all points in S' is equal to the projection of f to $(v_1, ..., v_{d-s})$ space.
Proof. See Appendix.

3.4 MAIN

Specification:
    INPUT: A PRP sentence F.
    OUTPUT: The truth value of F.
Proof of correctness. The proof of the correctness of MAIN is achieved if all four propositions below are proved.

Definition 9. The arguments of a function $\mathcal{VAL}$ are an assigned arrangement A and a coordinate p. $\mathcal{VAL}$ returns the truth value assigned to the face which contains p. □

Definition 10. Let A be an assigned arrangement and $Q_1, ..., Q_i$ be quantifiers. We introduce a new function $\mathcal{FML}$ as follows:

$$\mathcal{FML}(A, (Q_1, ..., Q_i)) = Q_1 v_1, ..., Q_i v_i \mathcal{VAL}(A, (v_1, ..., v_i))$$

□

Proposition 1. When the control reaches line 12 of MAIN for the first time, the truth value of $\mathcal{FML}(A, (Q_1, ..., Q_i))$ is equal to the truth value of F and A contains hyperplanes perpendicular to each of all axes.

Proposition 2. On line 12 of MAIN, the truth value of $\mathcal{FML}(A, (Q_1, ..., Q_i))$ is equal to the truth value of F and A contains hyperplanes perpendicular to each of that axes.

Proposition 3. When the control reaches line 25 of MAIN for the first time, the following three conditions hold: (1)A is a 0-dimensional arrangement, (2) $i = 0$ and (3) t is the truth value of the input PRP sentence.

Proposition 4. The control goes out of the while loop at some time.
Proof of Proposition 1. It is clear, because in each face the truth value of E doesn't change, and A contains hyperplane perpendicular to all axes line 2. □
Proof of Proposition 2: Let a_1 and a_2 be the value of $\mathcal{FML}(A, (Q_1, ..., Q_i))$ on line 12 and line 22, respectively. From the correctness of subroutines PROJECT and Lemma 2, we can say $a_1 = a_2$.

From Lemma 1, a projection of an arrangement which contains hyperplanes perpendicular to all axes contains hyperplanes perpendicular to all axes. Thus, the correctness of Proposition 2 is proved. □

---
[3] Here we describe the reason why this condition is needed. There is an arrangement of plane $x = 0$ in a 3 dimensional space. We want to make projection of this arrangement to (x, y)- plane which contains the line $x = 0$. We can make it if we add a plane $z = 0$ which is not parallel to z axis to the 3 dimensional arrangement.

Lemma 2. Let A be an arrangemet which contains hyperplanes perpendicular to all axes, and **pr** A projection of A to $(v_1, ..., v_{i-s})$-space by quantifier Q. For all $V_1, ..., V_{i-s}$, we can say:

$$\mathcal{VAL}(\mathbf{pr}\ A, (V_1, ..., V_{i-s}))$$
$$= Q\ v_{i-s+1}, ..., Q\ v_d \mathcal{VAL}(A, (V_1, ..., V_{i-s}, v_{i-s+1}, ..., v_d))$$

Proof. Let f be a face contained in A. Let $\mathcal{F}$ be a set of all faces in **pr** A such that **sh** f contains them. By Lemma 1, the set of points contained in elements of $\mathcal{F}$ is equal to **sh** f. We describe the case $Q = `\exists$'. By the algorithm, if true is assigned to f, true is assigned to all faces in $\mathcal{F}$. Thus, for each point p in the face f to which true is assigned, **sh** p is contained in a face to which true is assigned. If false is assigned to f, the assigned values of faces in $\mathcal{F}$ can be either true or false. If false is assigned to a face f' in $\mathcal{F}$, false is assigned to every face f'' such that $f' \subseteq$ **sh** f''. Therefore, if **sh** p' is contained in a face to which false is assigned, all points p'' such that **sh** $p' =$ **sh** p'' are contained in face to which false is assigned. We can prove the case $Q = `\forall$' in the similar way. Thus, the correctness of this lemma is proved. □

Proof of Proposition 3. It is clear. □

Proof of Proposition 4. On line 22, s is more than zero. Thus, i decreases in each loop. Therefore, the correctness of this lemma is proved. □

Time complexity: The time complexity of the MAIN routine, which is the time complexity of whole algorithm, is $r \cdot \alpha^{\beta d} n^{\gamma d(b+1)^a}$, where α, β and γ are proper constants.

Analysis of time complexity The time complexity of line 5 is polynomial order of l. The time complexity is $O(r \cdot \gamma^d n^d)$ and the size of A is the same order. On line 9, the size of the returned value A from ASSIGN is $O(l \cdot \gamma^d n^d)$. Here we describe the time complexity of the loop between line 11 and line 24. On line 12, let m be the number of faces which is contained in A and r' be the maximum bit length of coordinates of A. The time complexity of line 21 is $O(mr)$, where α, η and θ are constants. The size of A is $r'\alpha^{b+1} m^{b+1}$. Let m_s be the number of faces in A before the control reaches the loop and r_s the maximum bit length of coordinates of A before the control reaches the loop. Since the number of iteration is a, the time complexity of line 21 of a-th loop is $r_s \alpha^{\eta ab} m_s^{\gamma d(b+1)^a}$. Since $d = O(ab)$ holds, the total time complexity is $r\alpha^d n^{\beta d(b+1)^a}$, where α and β are constants. □

4 A brief explanation of Ferrante and Rackoff's algorithm

In this section, we briefly explain the time complexity of algorithm of Ferrante and Rackoff.

Let F be the PRP sentence to be decided, and E be the matrix of F. Inequalities contained in E can be represented in the form v_d **op** t_i, where **op** is $<, =$, or $>$, and t_i is of the form $t_i = \sum_{j=1}^{d-1} c_j v_j$, where c_i's are rationals.

The following lemma holds.

Lemma 3.

$$\exists v_m E(v_1, ..., v_m) = \bigvee_{\substack{v_m = t_i \text{ or} \\ v_m = (1/2)(t_i + t_j) \\ \text{or } v_m = +\infty \text{ or } v_m = -\infty}} E(v_1, ..., v_m)$$

$$\forall v_m E(v_1, ..., v_m) = \bigwedge_{\substack{v_m = t_i \text{ or} \\ v_m = (1/2)(t_i + t_j) \\ \text{or } v_m = +\infty \text{ or } v_m = -\infty}} E(v_1, ..., v_m)$$

□

The principle of Ferrante and Rackoff's algorithm is to eliminate all quantifiers in the input sentence in turn using Lemma 3 and then decide truth from the obtained expression. The righthand side of both equation in Lemma 3 has $(n^2 + 2) = O(n^2)$ subexpressions where n is number of inequalities in $E(v_1, ..., v_m)$. Thus, If one quantifier is eliminated using Lemma 3, the number of inequation increases to the third power. If the same type of this quantifier is succeeded, the second quantifier can be distributed to each term. Thus eliminating two successing quantifiers makes the number of inequation to the fifth power.

Generally, we can obtain $n^{d(2b+1)^a}$ as the upperbound of the number of inequalities contained in the expession obtained by eliminating all quantifiers by the method above. The maximum bit length of coefficients increases by five times whenever one quantifier is eliminated. Therefore, the upperbound of the maximum bit length of coefficients in the expression is $r5^d$. The time complexity of their algorithm is polynomial of $r5^d \cdot n^{d(2b+1)^a}$. Thus, the time complexity of their algorithm is $r\gamma^d n^{\delta d(2b+1)^a}$.

5 Conclusion

In this paper, we proposed a decision algorithm for PRP sentences which runs in $r\alpha^d n^{\beta d(b+1)^a}$ and it uses some techniques in combinatorial geometry, where α and β are proper constants, and r, n, d, a and b denote the maximum bit length of coefficients, the number of inequalities, the number of variable, the number of quantifier alternations and the maximum length of the consecutive same quantifiers in the input sentence respectively.

In the future we plan to improve the average time complexity and make it applicable to larger and practical problems.

References

[1] T. Higashino, J. Kitamichi and K. Taniguchi : "Presburger Arithmetic and its Application to Program Developments," Journal of Japan Society for Software Science and Technology, Vol.9, 6, pp.31-39, 1992. (In Japanese).

[2] A. Nakata, T. Higashino and K. Taniguchi : "Time-Action Alternating Model for Timed LOTOS and its Symbolic Verification of Bisimulation Equivalence", Proceedings of Joint International Conference on 9th Formal Description Techniques and 16th Protocol Specification, Testing, and Verification (FORTE/PSTV'96), pp.279-294, 1996.
[3] A.R.Bruss and A.Meyer : "On time-space classes and their relation to the theory of real addition," *Thoret. Comput. Sci.* **11** pp.59-69, 1980.
[4] L. Berman : "The complexity of logical theories," *Theoret. Comput. Sci.* **11** pp.71-77, 1980.
[5] V. Weispfenning : "The complexity of linear problems in fields," *J. Symbolic Computation* **5** pp.3-27, 1988.
[6] C.Hosono and Y.Ikeda : "A formal derivation of the decidability of the theory SA," *Thoret. Comput. Sci.* **127** pp.1-23, 1994.
[7] M.J.Fischer and M.O.Rabin : "Super exponential complexity of Presburger Arithmetic," SIAM-AMS Proc. **VII**(AMS,Providence,RI), 1974.
[8] J.Ferrante and C.Rackoff : "A decision procedure for the first order theory of real addition with order," *SIAM J. Comput.* **4**, pp.69-76, 1975.
[9] J.E. Hopcroft and J.D. Ullmann : *Introduction to automata theory, languages and computation*, Addison-Wesley, pp.355-357, 1979.
[10] H. Edelsbrunner : *Algorithms in Combinatorial Geometry*, Springer-Verlag, 1987.
[11] L.G. Khachiyan : "Polynomial algorithms for linear programming," *Dokl. Akad. Nauk SSSR* **244**, pp.1093-1096, 1979.

Appendix

In this paper, we assume the properties below without proofs. Here, we also give Lemma 4 and a proof of Lemma 1.

Property 1. In a d-dimensional space, iff a crossing of $m(\leq d)$ hyperplanes is a $(d-m)$-flat, the normal vectors of these hyperplanes are linearly independent.
□

Property 2. Let fl be a k-flat in $(v_1, ..., v_d)$ space. Let S be a set of axes which are parallel to fl. If a projection of fl to $(v_1, ..., v_{d-s})$ is k'-flat, $k - k' = |S|$ holds.
□

Property 3. In an arrangement A, for every $k(\leq d-1)$-flat fl in A, there exists $d - k$ hyperplanes whose normal vectors are linearly independent, and fl is contained in all of them.
□

Property 4. If the same dimensional flats fl and fl' meet $fl \subseteq fl'$, then $fl = fl'$ holds.
□

Property 5. Let fl be a flat in $(v_1, ..., v_d)$ space. Let S be a set of axes such that each of which is in $v_{d-s+1}, ..., v_d$ and parallel to fl. Let fl' be a crossing of fl and a hyperplane hp such that hp is perpendicular to an optional axis v_i in S. fl' is equal to a projection of fl to $(v_1, ..., v_{d-s})$ space.
□

Property 6. Let fl a $k(< d)$-flat in A. If and only if fl is parallel to an axis j, all hyperplanes which contain fl and contained in A are parallel to the axis j. □

Property 7. Let fl_1 and fl_2 be the same dimensional flat in $(v_1, ..., v_d)$ space. Let sh fl_1 and sh fl_2 be projections of fl_1 and fl_2 to $(v_1, ..., v_{d-s})$ space, respectively. If $fl_1 \neq fl_2$ and sh $fl_1 =$ sh fl_2 hold, a flat which contains both fl_1 and fl_2 is parallel to one or more axes in v_{d-s+1}axis, ..., $v_d$axis. □

Property 8. If and only if an arrangement A is equal to an arrangement B, a set of hyperplanes contained in A is equal to that of B. □

Lemma 4. Let S be a set of hyperplanes which meets $|S| \leq d$. The following statements (A) and (B) are equivalent.
 (A) Normal vectors of hyperplanes in S are linearly independent.
 (B) Let S_1 and S_2 be sets of hyperplanes such that $S_1 \subseteq S, S_2 \subseteq S$ and $S_1 \neq S_2$ hold. The crossing of all hyperplanes in S_1 is not equivalent to the crossing of all hyperplanes in S_2.

Proof.
 (A) $\Rightarrow$ (B): If the number of hyperplanes in S_1 and S_2 dIf and only ifers, the dimensions of their crossings are different and it contradicts (B). Thus, we only think the case that the number of hyperplanes in S_1 and that of S_2 are the same. We use contradiction. We assume that the crossings are the same. Let fl be the crossing. The crossing of hyperplanes in $S_1 \bigcup S_2$ is also the same as fl. From $S_1 \neq S_2$, S_1 contains a hyperplane which is not contained in S_2. Thus, $S_1 \bigcup S_2$ contains more hyperplanes than S_1, and the crossings of hyperplanes in $S_1 \bigcup S_2$ should be flat in less dimension than that of S_1. This contradicts the assumption.

 (A) $\Leftarrow$ (B): We prove (B) $\wedge \neg$ (A) doesn't hold. Let S be $\{h_1, h_2, ..., h_{|S|}\}$. We substitute h_1 for S_1 and $h_1 \bigcap h_2$ for S_2, and $h_1 \neq h_1 \bigcap h_2$ holds. Otherwise, let fl_i be $h_1 \bigcap ... \bigcap h_i$ and we can say $fl_i \neq fl_{i+1}$ for $1 \leq i \leq |S|-1$. From Property 1, the dimension of fl_i is not equal to that of fl_{i+1}. From the definition of fl_i and the assumption $fl_{i+1} \neq fl_i$, fl_{i+1} is a subset of fl_i. Thus, the dimension of fl_{i+1} is less than that of fl_i. Therefore, fl_k is a $(d-k)$-flat where $1 \leq k \leq |S|$. From Property 1, the normal vectors of hyperplanes in S should be linearly independent and that contradicts $\neg$ (A). □

Proof of Lemma 1.
 The proof consists of five steps, (S1) to (S5).

 (S1) Let fl be an optional flat contained in A. Let $i(\leq d-1-s)$ be the dimension of sh fl. We prove that there exists an i-flat c such that c is contained in A and meets sh $fl =$ sh c.

 Let S be a set of axes which are parallel to fl and S is a subset of $\{v_{d-s+1}$axis, ..., $v_d$axis$\}$. From Property 2, fl is an $i+|S|$-flat. Because A contains hyperplanes each of which is perpendicular to every axis, fl is crossing to each hyperplane $j = 0$ for every element j of S. Let P be a set of such hyperplanes. From Property 6, any $(d-1)$-flat α such that α is contained in fl and contained in A is parallel to each element in S. Thus, the normal vectors of $(d-1)$-flat contained in P or α are linearly independent. There exists a crossing c of all hyperplanes in P and fl. By Property 1, c is an i-flat. Let α' be a set of $(d-1)$-flats which are contained in A and contain c. $\alpha' = \alpha \bigcup P$ holds. There is no axis which is parallel to c in v_{d-s+1}axis, ..., $v_d$axis. By Property 5, **sh** fl is equal to **sh** c.

(S2) We prove that there exist s $(d-1)$-flats such that they are contained in A and contain c and their crossing is not parallel to any of v_{d-s+1}axis, ..., $v_d$axis.

Because c is not parallel to any of v_{d-s+1}axis, ..., $v_d$axis, we can say the following statements from Property 6. There exists a hyperplane hp for each axis in $\{v_{d-s+1}$axis, ..., $v_d$axis$\}$ such that hp is contained in A and hp contains c and hp is not parallel to j. Let β' be a set of such hyperplanes. Because the number of v_{d-s+1}axis, ..., $v_d$axis is s, the number of elements of β' is less than $s+1$. Because the flat c contains i-flats and Property 3 holds, there are $(d-i)$ $(d-1)$-flats which contain c. By assumption, $d-i > s$ holds. Therefore, there exists a combination of s $(d-1)$-flats which contain c and any of their crossing c' is not parallel to any of v_{d-s+1}axis, ..., $v_d$axis. Let β be a set of these s $(d-1)$-flats. c' is the flat we were trying to prove existence.

(S3) We prove that there exists a set of $(d-1-s)$-flats such that every $(d-1-s)$-flat contains c and each $(d-1-s)$-flat is contained in A and the set is equal to the set of $(d-i-s)$ $(d-1-s)$-flats such that their normal vectors are linearly independent and these are contained in pr A

We can find $(d-i-s)$ $(d-1-s)$-flats by the following way. By definition, $\beta \subset \alpha'$ holds. Let γ be a set $\alpha' - \beta$. Because α' and β are sets of $d-i$ and s hyperplanes, respectively. γ contains $(d-i-s)$ hyperplanes.

Every crossing of each hyperplane in γ and all hyperplanes in β is a $(d-1-s)$-flat which is not parallel to any of v_{d-s+1}axis, ..., $v_d$axis. Let δ be a set of these $(d-1-s)$-flat. Because the crossing of all hyperplanes in β is a $(d-s)$-flat and the normal vectors of all hyperplanes in γ and β are linearly independent, δ contains $d-i-s$ flats by Property 2. Because the normal vectors of hyperplanes in α' are linearly independent, the elements in δ are all different. We can say the normal vectors of hyperplanes in δ are linearly independent as follows: because the normal vectors of hyperplanes in α' are linearly independent, crossings of all two combinations of flats in δ are different. All projections of crossings of flats in δ are also different. The reasons are the following. If and only if we let fl_1 and fl_2 be crossings of some flats in δ and assume **sh** $fl_1 =$ **sh** fl_2 holds, fl_1 and fl_2 are the same dimensional flat. From $fl_1 \neq fl_2$ and Property 7, all hyperplanes in β, which contain both fl_1 and fl_2, should be parallel to one or more of v_{d-s+1}axis, ..., $v_d$axis. This is a contradiction. Therefore, from Lemma 4, the normal vectors of projections of flats in δ are linearly independent.

(S4) We prove that there exists a flat in pr A which is equal to projection sh fl of each flat in A.

All elements in H' are $(d-1-s)$-flats. Because each element of δ is a $(d-1-s)$-flat contained in A, and all, projections of the elements in δ are

$(d-1-s)$-flats, projections of all elements in δ are elements of H'. Because a set of all hyperplanes in β and γ, used in definition of δ, is α', and the crossings of all hyperplanes in α' is c, a crossing of $(d-1-s)$-flats in δ is the i-flat c. Because the normal vectors of projections of flats in δ are linearly independent, a crossing c' of projections of flats in δ is an i-flat. Because **sh** c is contained in projections of flats in δ, **sh** $c \subseteq c'$ holds, as well as Property 4 holds, $c = c'$ holds. Thus, **sh** c is equal to a crossing of projections of all flats in δ. Therefore, we can say **sh** c is contained in **pr** A. Thus (S4) is proved.

(S5) **We prove that there exist sets of faces which are contained in pr A and equal to projections sh f of each face f in A.**

Let B be an arrangement which has the minimal number of faces and holds the following condition: there exist sets of faces which are equal to a projection of each face f in A. Note that B uniquely corresponds to A. There exist sets of faces which is equal to a projection of each flat in A. From Property 8, $B = $ **pr** A holds. Therefore, (S5) is proved. □

The intuitive meaning of some symbols used in proof of Lemma 1 is described in the following table.

c : An i-flat which meets **sh** $fl = $ **sh** c
α' : A set of $(d-1)$-flats which contain c and contained in A
β : A set of $(d-1)$-flats which contain c' and contained in A
δ : Projections of $(d-1-s)$-flat which is contained in δ and contain **sh** c and contained in **pr** A

Lower Bounds on Negation-Limited Inverters

Shao Chin Sung[1] and Keisuke Tanaka[2]

[1] School of Information Science,
Japan Advanced Institute of Science and Technology,
1-1 Asahidai Tatsunokuchi Ishikawa, 923-1292, Japan.
son@jaist.ac.jp

[2] NTT Information and Communication Systems Laboratories,
1-1 Hikarinooka Yokosuka-shi, Kanagawa 239-0847, Japan.
keisuke@isl.ntt.co.jp

Abstract. In this paper, we consider the complexity of *negation-limited inverters*, circuits each of which inverts n input variables with only $\lceil \log(n+1) \rceil$ negation gates. We improve the upper bound on depth and the lower bound on size of negation-limited inverters. Moreover, we give a superlinear lower bound $\Omega(n \log n)$ on size of minimum-depth negation-limited inverters. As far as we know, this is the first superlinear lower bound on size of non-monotone circuits with a weak depth restriction $O(\log n)$.

1 Introduction

We do not know much about the complexity of combinational circuits (i.e., circuits over basis $\{\wedge, \vee, \neg\}$) for explicitly defined Boolean functions. Even a superlinear lower bound for size of combinational circuits is not known. The complexity of monotone circuits (i.e., combinational circuits *without* negation gates ($\neg$)) for many explicitly defined Boolean functions are well understood. For example, exponential lower bounds for size of monotone circuits are known [2, 7, 8]. Exponential gaps between monotone and combinational circuit complexity are also shown [7, 12]. Thus, we cannot generally derive strong lower bounds for the combinational circuit complexity using those bounds for the monotone circuit complexity.

In this situation, there is no doubt that it is necessary to understand the effect of negation gates in order to obtain strong lower bounds for the combinational circuit complexity. This motivates the study of *negation-limited circuit complexity*, the complexity for circuits in which the number of negation gates is limited.

Markov [6] gave the number of negation gates which are necessary and sufficient to compute an arbitrary Boolean function with arbitrarily fixed number of outputs. Especially, he showed that $\lceil \log(n+1) \rceil$ negation gates are sufficient to compute *any* n-input Boolean function with arbitrarily fixed number of outputs (all logarithms in this paper are base two). In [4, 5], Fischer constructed a polynomial size circuit that contains only $\lceil \log(n+1) \rceil$ negation gates, and inverts n

input variables. Owing to this result, for *any* n-input Boolean function F^n, if there exists a polynomial size combinational circuit for F^n, there also exists a polynomial size circuit with only $\lceil \log(n+1) \rceil$ negation gates for F^n.

As Fischer's results show, the circuit which inverts n input variables with only $\lceil \log(n+1) \rceil$ negation gates, which we call a *negation-limited inverter*, plays an important role in the study of negation-limited circuit complexity. Tanaka and Nishino [11], and Beals, Nishino and Tanaka [3] improved the Fischer's construction of a negation-limited inverter, and gave a non-trivial linear lower bound on size of it.

In this paper, we continue to study on the complexity of negation-limited inverters. We show a $D + 3\lceil \log(n+1) \rceil$ upper bound on depth, where D is the minimum depth of monotone sorting networks. If the minimum depth of monotone circuits computing the *majority* function is not less than D, our upper bound matches the lower bound shown in [11]. We also show a $(7 + 1/3)n$ lower bound on size of negation-limited inverters. Furthermore, we also show a superlinear lower bound $\Omega(n \log n)$ on size of minimum-depth negation-limited inverters under a natural assumption (see Section 6). As far as we know, this is the first superlinear lower bound on size of non-monotone circuits with a weak depth restriction $O(\log n)$.

2 Definitions and preliminaries

In this paper, we mainly consider circuits over basis $\{\wedge, \vee, \neg\}$. Let x be the vector of the input variables $x_1, \ldots, x_n$, i.e., $x = (x_1, \ldots, x_n)$. Let $F^n = (f_1^n, \ldots, f_m^n)$ be an n-input m-output *Boolean function* $\{0,1\}^n \to \{0,1\}^m$. In a circuit, we denote by Υ_G the Boolean function computed at the output of a gate G. A circuit with output gates $Y_1, \ldots, Y_m$ is said to *compute* $F^n = (f_1^n, \ldots, f_m^n)$ if $\Upsilon_{Y_j}(x) = f_j^n(x)$ for all $1 \le j \le m$ and for any input x.

The size and depth of a circuit are respectively the number of gates and the length of the longest directed path in it. We call a circuit including at most s negation gates ($\neg$) an *s-circuit*. When $s = 0$, an s-circuit is said to be *monotone*. We denote by $C^s(F^n)$ and $D^s(F^n)$ respectively the minimum size and the minimum depth of s-circuits computing F^n. $C^s(F^n)$ and $D^s(F^n)$ are undefined if F^n cannot be computed with s negation gates.

A theorem of Markov [6] precisely determines the number of negation gates which are necessary and sufficient to compute an arbitrary Boolean function F^n. A *chain* $\alpha = (\alpha_1, \ldots, \alpha_k)$ in the Boolean lattice $\{0,1\}^n$ is an sequence $\alpha^1 \le \ldots \le \alpha^k \in \{0,1\}^n$ ($\le$ is defined componentwise). The *decrease* of F^n on α is the number of $i \le k$ such that $f_j^n(\alpha^{i-1}) > f_j^n(\alpha^i)$ for some $1 \le j \le m$. We define $d(F^n)$ to be the maximum decrease of F^n on all chains α. Markov has shown that $\lceil \log(d(F^n) + 1) \rceil$ negation gates are necessary and sufficient to compute an arbitrary F^n. Therefore, $C^s(F^n)$ and $D^s(F^n)$ are defined only for $s \ge \lceil \log(d(F^n) + 1) \rceil$.

Let $\|x\|$ denote the number of ones in x, i.e., $\|x\| = \sum_{i=1}^n x_i$. We define the n-input parity function, denoted by $PARITY^n$, as a Boolean function such that

$PARITY^n(x) = 1$ if and only if $\|x\|$ is odd, and for $0 \le k \le n$ the n-input k-th threshold function, denote by T_k^n, as a Boolean function such that $T_k^n(x) = 1$ if and only if $\|x\| \ge k$.

Inverter, denoted by $INV^n = (INV_1^n, \ldots, INV_n^n)$, is an n-input n-output Boolean function, where

$$INV_i^n(x) = \neg x_i \quad \text{for each } 1 \le i \le n.$$

¿From the theorem of Markov, $r = \lceil \log(n+1) \rceil$ negation gates are necessary and sufficient to compute INV^n. An r-circuit computing INV^n is called a *negation-limited inverter*. In this paper, we concentrate only on the case $n = 2^r - 1$ for some $r \ge 2$.

In an arbitrary negation-limited inverter, we label the negation gates as $N_1, \ldots, N_r$ such that there is no path from N_j to N_k if $j > k$. Such a labeling can be uniquely found, since there exists a path pass through all r negation gates in any negation-limited inverters [11]. We also label the i-th output gate of negation-limited inverters as Y_i, i.e., $\Upsilon_{Y_i}(x) = INV_i^n(x) = \neg x_i$.

3 Previous works

The complexity of negation-limited inverters had been investigated in [1, 3, 5, 9, 11]. For the upper bounds for negation-limited inverters, the following propositions have been shown.

Proposition 1 (Beals, Nishino, and Tanaka [3]).

$$C^r(INV^n) \le O(n \log n).$$

Proposition 2 (Fischer [5]).

$$D^r(INV^n) \le D^0(T_1^n, \ldots, T_n^n) + 5r - O(1).$$

On the other hand, for the lower bounds for negation-limited inverters, the following proposition has been shown.

Proposition 3 (Tanaka and Nishino [11]).

$$C^r(INV^n) \ge 5n + 3r - O(1) \text{ and } D^r(INV^n) \ge D^0(T_{\lceil n/2 \rceil}^n) + 3r.$$

4 Upper bound on depth

In this section, we consider the upper bound on depth of negation-limited inverters. Recall that $n = 2^r - 1$.

Theorem 4.

$$D^r(INV^n) \le D^0(T_1^n, \ldots, T_n^n) + 3r.$$

Tanaka and Nishino [11] show that $(\neg \Upsilon_{N_1}(x), \ldots, \neg \Upsilon_{N_r}(x))$ is the binary representation of $\|x\|$, i.e.,

$$\sum_{j=1}^{r}(1 - \Upsilon_{N_j}(x))2^{r-j} = \|x\|. \tag{1}$$

Then, $\neg x_i$ for each $1 \leq i \leq n$ can be represented as follows:

$$\neg x_i = 1 \text{ if and only if } \|x\| - x_i + \sum_{j=1}^{r}\Upsilon_{N_j}(x)2^{r-j} \geq 2^r - 1. \tag{2}$$

Fischer [5] shows that we can construct a negation-limited inverter such that N_j for $1 \leq j \leq r$ is in depth $D^0(T_1^n, \ldots, T_n^n) + 3j - 2$.

Proof (of Theorem 4). Let us define Boolean functions $g_{j,k}$ with $n+r$ variables for $0 \leq j \leq r$ and $2^r - 2^{r-j} \leq k < 2^r$ as follows:

$$g_{j,k}(a,b) = 1 \text{ if and only if } \|a\| + b_1 2^{r-1} + \cdots + b_j 2^{r-j} \geq k,$$

where $a \in \{0,1\}^n$ and $b \in \{0,1\}^r$.

Note that $g_{0,k}(a,b) = T_k^n(a)$ for $0 \leq k < 2^r$. Therefore, $g_{0,k}$ for $0 \leq k < 2^r$ can be computed by depth $D^0(T_1^n, \ldots, T_n^n)$. Also note that we have for $0 \leq j < r$ and $2^r - 2^{r-j-1} \leq k < 2^r$,

$$g_{j+1,k} = g_{j,k} \vee (b_{j+1} \wedge g_{j,k-2^{r-j-1}}),$$

and therefore $g_{j+1,k}$ for $2^r - 2^{r-j-1} \leq k < 2^r$ can be computed by depth two if $g_{j,k}$ for $2^r - 2^{r-j} \leq k < 2^r$ and b_{j+1} are given.

For each $1 \leq i \leq n$, if $a = (x_1, \ldots, x_{i-1}, 0, x_{i+1}, \ldots, x_n)$ and $b = (\Upsilon_{N_1}(x), \ldots, \Upsilon_{N_r}(x))$, from (2) we have $g_{r,2^r-1}(a,b) = INV_i^n(x)$. Since $\Upsilon_{N_j}(x)$ can be computed by depth $D^0(T_1^n, \ldots, T_n^n) + 3j - 2$ for each $1 \leq j \leq r$ (see [5]), $g_{j,k}$ for $0 \leq j \leq r$ and $2^r - 2^{r-j} \leq k < 2^r$ can be computed by depth $D^0(T_1^n, \ldots, T_n^n) + 3j$, i.e., $D^r(INV_i^n) \leq D^0(T_1^n, \ldots, T_n^n) + 3r$ for $1 \leq i \leq n$. □

The size of the circuit constructed in this theorem is $O(n^2 \log n)$. Our upper bound matches the lower bound in Proposition 3 if $D^0(T_{\lceil n/2 \rceil}^n) \geq D^0(T_i^n)$ for all $1 \leq i \leq n$.

5 Lower bound on size

In this section we consider the lower bound on size of negation-limited inverters. We improve the lower bound in Proposition 3 by more than $2n$.

Theorem 5.

$$C^r(INV^n) \geq (7 + 1/3)n + r/3 - O(1).$$

Theorem 5 is obtained from Proposition 6 shown by Sung [10] and Lemma 7.

Proposition 6 (Sung [10]).

$$C^{r-1}(PARITY^n) \geq (5 + 1/3)n + r/3 - O(1).$$

Lemma 7.

$$C^r(INV^n) \geq C^{r-1}(PARITY^n) + 2n + 1.$$

Proof. From (1), we have $\Upsilon_{N_r} = \neg PARITY^n$. It implies that any negation-limited inverter contains an $(r-1)$-subcircuit computing $PARITY^n$ whose output gate is predecessor of N_r. In the following, we show that there are at least n $\wedge$ gates each of which is on some paths from N_r. Similarly, by a dual argument of $\wedge$ and $\vee$, we can find at least n $\vee$ gates each of which is on some paths from N_r. Then, the lemma is satisfied.

Let $a^1, \ldots, a^n \in \{0,1\}^n$ such that $a^i = (a_1^i, \ldots, a_n^i)$ and $a_j^i = 1$ if and only if $j \neq i$, and let $\mathbf{1} = (1, \ldots, 1)$. Note that $\Upsilon_{Y_i}(\mathbf{1}) = 0$ and $\Upsilon_{Y_i}(a^i) = 1$, and from (1) $\Upsilon_{N_r}(\mathbf{1}) = 0$, $\Upsilon_{N_r}(a^i) = 1$, and $\Upsilon_{N_j}(\mathbf{1}) = \Upsilon_{N_j}(a^i) = 0$ for all $1 \leq j < r$.

First, we show that for each $1 \leq i \leq n$ there exists a path from N_r to Y_i such that each gate G on it satisfies $\Upsilon_G(a^i) = 1$. Suppose such a path does not exists, i.e., each path from N_r to Y_i contains a gate G such that $\Upsilon_G(a^i) = 0$. Then, $\Upsilon_{Y_i}(a^i) = 1$ even if we fix Υ_{N_r} to 0. Furthermore, $\Upsilon_{Y_i}(a^i) = 1$ even if we fix Υ_{N_j} to 0 for all $1 \leq j \leq r$, since $\Upsilon_{N_j}(a^i) = 0$ for $1 \leq j < r$. Note that the circuit becomes monotone, since we fix Υ_{N_j} to 0 for all $1 \leq j \leq r$. From the monotonicity, $a^i \leq \mathbf{1}$ implies that $\Upsilon_{Y_i}(a^i) \leq \Upsilon_{Y_i}(\mathbf{1}) = 0$. It contradicts to $\Upsilon_{Y_i}(a^i) = 1$. Therefore, there exists a path from N_r to Y_i on which all gates output 1. We denote such a path by P_i.

Then, we show that each P_i contains an $\wedge$ gate such that it does not contained in any $P_{i'}$ for $i' \neq i$. Since $\Upsilon_{N_r}(a^{i'}) = 1$ and $\Upsilon_{Y_i}(a^{i'}) = 0$ for any $i' \neq i$, there exists at least one $\wedge$ gate on P_i. By G_i we denote the last $\wedge$ gate on P_i. Then $\Upsilon_{G_i}(a) = 1$ implies that $\Upsilon_{Y_i}(a) = 1$ for all $a \in \{0,1\}^n$. Since $\Upsilon_{G_i}(a^i) = 1$ and $\Upsilon_{G_{i'}}(a^i) = 0$, we have $G_i \neq G_{i'}$ if $i \neq i'$. Thus, there are at least n $\wedge$ gates each of which is on some paths from N_r. □

6 Superlinear lower bound on size of minimum-depth negation-limited inverters

In this section, we ignore size and depth of negation gates. This is a natural formulation, since it can be considered as circuits over basis $\{\wedge, \vee, \overline{\wedge}, \overline{\vee}\}$, where we limit the total number of $\overline{\wedge}$ (NAND) and $\overline{\vee}$ (NOR) gates to r. We consider the lower bound on size of negation-limited inverters whose depth are minimum (i.e., $D^r(INV^n)$) under the following natural assumption: For all $1 \leq i \leq n$,

$$D^0(T^n_{\lceil n/2 \rceil}) \geq D^0(T^n_i).$$

This relation is widely believed to be true even it is still open.

Then, from [11] and Theorem 4, we have

$$D^r(INV^n) = D^0(T^n_{\lceil n/2 \rceil}) + 2r.$$

Therefore, in any minimum-depth negation-limited inverter, N_j is in depth $D^0(T^n_{\lceil n/2 \rceil}) + 2j - 2$ for each $1 \leq j \leq r$.

Theorem 8. *Any minimum-depth negation-limited inverter has size at least*

$$2n \log(n+1) + 3n - O(1).$$

Let g and h be any Boolean functions, and let $A \subseteq \{0,1\}^n$ be a set of assignments. We say $g = h$ over A if $g(a) = h(a)$ for all $a \in A$. We define sets A_k^j of assignments as follows:

$$A_k^j = \{a \in \{0,1\}^n \mid \|a\| \bmod 2^{r-j} = k\},$$

where $1 \leq j \leq r$ and $0 \leq k < 2^{r-j}$.

Let G be a gate such that $\Upsilon_G = \neg x_i$ over A_k^j. It is clear that G is not a negation gate. Note that each $\wedge$ gate and each $\vee$ gate in a negation-limited inverter can be represented by a monotone function of x and $N_1, \ldots, N_r$. Especially, each $\wedge$ gate and each $\vee$ gate in depth $D^0(T^n_{\lceil n/2 \rceil}) + 2j - 1$ or in depth $D^0(T^n_{\lceil n/2 \rceil}) + 2j$ can be represented by a monotone function of x and $N_1, \ldots, N_j$, since N_j for $1 \leq j \leq r$ is in depth $D^0(T^n_{\lceil n/2 \rceil}) + 2(j-1)$.

Lemma 9. *Let g_1 and g_2 be monotone functions of x and $N_1, \ldots, N_j$. If $g_1 = \neg x_{i_1}$ over $A_{k_1}^j$ and $g_2 = \neg x_{i_2}$ over $A_{k_2}^j$ for some $i_1 \neq i_2$, then $g_1 \neq g_2$.*

Proof. It is clear that $g_1 \neq g_2$ if $k_1 = k_2$, since there exists an assignment $a \in A_{k_1}^j$ such that $a_{i_1} \neq a_{i_2}$, i.e., $g_1(a) \neq g_2(a)$. Assume without loss of generality that $k_1 < k_2$. Let $a^1 \in A_{k_1}^j$ and $a^2 \in A_{k_2}^j$ such that $a^1 \leq a^2$, $\|a^1\| = k_1$ with $a_{i_1}^1 = 0$ and $a_{i_2}^1 = 1$, and $\|a^2\| = k_2$ with $a_{i_1}^2 = 1$ and $a_{i_2}^2 = 1$. Then, we have $g_1(a^1) = \neg a_{i_1}^1 = 1$. Note that from (1) and $k_1 < k_2 < 2^{r-j}$, $\Upsilon_{N_{j'}}(a^1) = \Upsilon_{N_{j'}}(a^2) = 1$ for $j' < j$. From the monotonicity of g_1, $a^2 \geq a^1$ and $\Upsilon_{N_{j'}}(a^1) = \Upsilon_{N_{j'}}(a^2)$ for $j' < j$ imply that $g_1(a^2) \geq g_1(a^1) = 1$. It contradicts to $g_2(a^2) = \neg a_{i_2}^2 = 0$. □

Lemma 10. *Suppose G is a gate such that $\Upsilon_G = \neg x_i$ over A_k^j for $2 \leq j \leq r$. The following statements are satisfied.*

(i) *There exists a path from N_j to G, which does not contain any negation gate except N_j.*
(ii) *Any path of (i) contains at least one $\wedge$ gate and at least one $\vee$ gate.*
(iii) *There is a unique path from N_j to G if G is in depth $D^0(T^n_{\lceil n/2 \rceil}) + 2j$.*

Proof. Let $a^1, a^2, a^3 \in A_k^j$ such that $\|a^1\| = k$ with $a_i^1 = 0$, $\|a^2\| = k + 2^{r-j}$ with $a^2 \geq a^1$ and $a_i^2 = 1$, $\|a^3\| = k + 2 \cdot 2^{r-j}$ with $a_i^3 = 1$, and $\|a^4\| = k + 3 \cdot 2^{r-j}$ with $a_i^4 = 0$. Then we have $\Upsilon_G(a^1) = \Upsilon_G(a^4) = 1$ and $\Upsilon_G(a^2) = \Upsilon_G(a^3) = 0$. From (1), we have $\Upsilon_{N_j}(a^1) = \Upsilon_{N_j}(a^3) = 1$ and $\Upsilon_{N_j}(a^2) = \Upsilon_{N_j}(a^4) = 0$, and $\Upsilon_{N_{j'}}(a^1) = \Upsilon_{N_{j'}}(a^2)$ and $\Upsilon_{N_{j'}}(a^3) = \Upsilon_{N_{j'}}(a^4)$ for all $j' \neq j$.

(i) Suppose there does not exist a path which satisfies (i). Then, Υ_G is a monotone function of x and all $N_{j'}$ for $j' \neq j$. From the monotonicity of Υ_G, $a^2 \geq a^1$ and $\Upsilon_{N_{j'}}(a^2) = \Upsilon_{N_{j'}}(a^1)$ imply that $\Upsilon_G(a^2) \geq \Upsilon_G(a^1) = 1$. It contradicts to $\Upsilon_G(a^2) = 0$.

(ii) Suppose there exists a path from N_j to G on which all gates except N_j are $\wedge$ gates. It implies that $\Upsilon_G(a) = 0$ if $\Upsilon_{N_j}(a) = 0$ for any $a \in \{0, 1\}^n$. It contradicts to $\Upsilon_{N_j}(a^4) = 0$ and $\Upsilon_G(a^4) = 1$. Similarly, suppose there exists a path from N_j to G on which all gates except N_j are $\vee$ gates. It contradicts to $\Upsilon_{N_j}(a^3) = 1$ and $\Upsilon_G(a^3) = 0$.

(iii) Since N_j is in depth $D^0(T^n_{\lceil n/2 \rceil}) + 2j - 2$ and G is in depth $D^0(T^n_{\lceil n/2 \rceil}) + 2j$, all paths from N_j to G contain no negation gate except N_j. From (ii), any path from N_j to G contain three gates such that one negation gate N_j, (including G) one $\wedge$ gate and one $\vee$ gate. Thus, there are at most two such paths from N_j to G.

Suppose there are two paths from N_j to G. Then, $\Upsilon_G = (\Upsilon_{N_j} \vee \Upsilon_{H_1}) \wedge (\Upsilon_{N_j} \vee \Upsilon_{H_2})$, or $\Upsilon_G = (\Upsilon_{N_j} \wedge \Upsilon_{H_1}) \vee (\Upsilon_{N_j} \wedge \Upsilon_{H_2})$ for some gates H_1 and H_2. However, it is impossible, since $\Upsilon_G(a^3) = 0$ and $\Upsilon_{N_j}(a^3) = 1$ imply that $\Upsilon_G \neq (\Upsilon_{N_j} \vee \Upsilon_{H_1}) \wedge (\Upsilon_{N_j} \vee \Upsilon_{H_2})$, and $\Upsilon_G(a^4) = 1$ and $\Upsilon_{N_j}(a^4) = 0$ imply that $\Upsilon_G \neq (\Upsilon_{N_j} \wedge \Upsilon_{H_1}) \vee (\Upsilon_{N_j} \wedge \Upsilon_{H_2})$. □

Lemma 11. *Suppose there exists a gate G in depth $D^0(T^n_{\lceil n/2 \rceil}) + 2j$ such that $\Upsilon_G = \neg x_i$ over A_k^j. Then, there exists a gate H in depth $D^0(T^n_{\lceil n/2 \rceil}) + 2j - 2$ such that $\Upsilon_H = \neg x_i$ over $A_{k'}^{j-1}$ for some $0 \leq k' \leq 2^{r-j+1}$.*

Proof. We prove only for the case that G is an $\vee$ gate. The case that G is an $\wedge$ gate can be proved by a similar argument.

Suppose G is an $\vee$ gate. From (iii) of Lemma 10, there exist two gates H_1 and H_2 such that Υ_{H_1} and Υ_{H_2} are monotone functions of x and $N_1, \ldots, N_{j-1}$, and $\Upsilon_G = \Upsilon_{H_1} \vee (\Upsilon_{H_2} \wedge \Upsilon_{N_j})$. We will show that $\Upsilon_{H_1} \wedge \Upsilon_{N_j} = 0$ over A_k^j. Then, we have $\Upsilon_G = \Upsilon_{H_2} = \neg x_i$ for all $a \in A_k^j$ such that $\Upsilon_{N_j}(a) = 1$, Since $A_k^{j-1} = \{a \in A_k^j \mid \Upsilon_{N_j}(a) = 1\}$, $\Upsilon_{H_2} = \neg x_i$ over A_k^{j-1}. Since G is in depth $D^0(T^n_{\lceil n/2 \rceil}) + 2j$, H is in depth $\leq D^0(T^n_{\lceil n/2 \rceil}) + 2j - 2$. From Lemma 10, there exists a path from N_{j-1} to H with length at least 3. Thus H is in depth $D^0(T^n_{\lceil n/2 \rceil}) + 2j - 2$.

Now we show that $\Upsilon_{H_1} \wedge \Upsilon_{N_j} = 0$ over A_k^j. Suppose there exists an assignment $a \in A_k^j$ such that $\Upsilon_{H_1}(a) = \Upsilon_{N_j}(a) = 1$. It implies that $\Upsilon_G(a) = \neg a_i = 1$. There exists an assignment $a' \in A_k^j$ such that $a' \geq a$, $\|a'\| = k + 2^{r-j}$, and $a_i' = 1$. From (1) $\Upsilon_{N_j}(a') = 0$ and $\Upsilon_{N_{j'}}(a') = \Upsilon_{N_{j'}}(a)$ for $j' < j$. From the monotonicity, $\Upsilon_{H_1}(a') \geq \Upsilon_{H_1}(a) = 1$. It implies that $\Upsilon_G(a') = \Upsilon_{H_1}(a') = 1$. It contradicts to $a_i' = 1$. □

Let H_3 be the gate which computes $\Upsilon_{H_2} \wedge \Upsilon_{N_j}$. Note that $\Upsilon_{H_3}(a) = \Upsilon_{H_2}(a)$ if $\Upsilon_{N_j}(a) = 1$ for all $a \in A_k^j$, i.e., $\Upsilon_{H_3} = \neg x_i$ over A_k^{j-1}. Since H_2 is in depth $D^0(T_{\lceil n/2 \rceil}^n) + 2j - 2$ and G is in depth $D^0(T_{\lceil n/2 \rceil}^n) + 2j$, H_3 is in depth $D^0(T_{\lceil n/2 \rceil}^n) + 2j - 1$.

Proof (of Theorem 8). For each $1 \leq i \leq n$, the i-th output gate Y_i is in depth $D^0(T_{\lceil n/2 \rceil}^n) + 2r$, and $\Upsilon_{Y_i} = \neg x_i$, i.e., $\Upsilon_{Y_i} = \neg x_i$ over A_0^r. From Lemma 9, $Y_i, \ldots, Y_n$ are n different gates. By applying Lemma 11 to Y_i for each $1 \leq i \leq n$, we will find $2n$ gates, $G_1, \ldots, G_n$ in depth $D^0(T_{\lceil n/2 \rceil}^n) + 2r - 1$, and $H_1, \ldots, H_n$ in $D^0(T_{\lceil n/2 \rceil}^n) + 2r - 2$, such that $\Upsilon_{G_i} = \Upsilon_{H_i} \neg x_i$ over A_k^{r-1} for some $0 \leq k < 2$. From Lemma 9, $G_1, \ldots, G_n$ and $H_1, \ldots, H_n$ are $2n$ different gates. Again applying Lemma 11 to H_i for each $1 \leq i \leq n$, we will find n different gates in depth $D^0(T_{\lceil n/2 \rceil}^n) + 2r - 3$ and n different gates in depth $D^0(T_{\lceil n/2 \rceil}^n) + 2r - 4$. After $r - 1$ applications of Lemma 11, we find n different gates in depth $D^0(T_{\lceil n/2 \rceil}^n) + j$ for each $3 \leq j \leq 2r$, and n different gates, $H_1', \ldots, H_n'$, in depth $D^0(T_{\lceil n/2 \rceil}^n) + 1$ or $D^0(T_{\lceil n/2 \rceil}^n) + 2$ such that $H_i' = \neg x_i$ over A_k^1 for some $0 \leq k < 2^{r-1}$ and for each $1 \leq i \leq n$. Thus, we find $(2r - 1)n$ gates in depth $> D^0(T_{\lceil n/2 \rceil}^n)$.

Finally, there are at least $4n - O(1)$ gates in depth $\leq D^0(T_{\lceil n/2 \rceil}^n)$ [11]. □

7 Conclusion

In this paper, we have presented a superlinear lower bound on size of minimum-depth negation-limited inverters under a natural assumption. We have also shown a new upper bound on depth and a new lower bound on size of negation-limited inverters.

On the complexity of negation-limited inverters, G. Turán posed the following question: *Is the size of any $c \log n$ depth inverter using $c \log n$ negation gates superlinear?* Though Turán's question remains open, we hope that our work represents a step towards its resolution.

References

1. Akers, S.B. Jr.: On maximum inversion with minimum inverters. *IEEE Transactions on Computers* **C-17(2)** (1972) 134–135
2. Alon, N., Boppana, R.B.: The monotone circuit complexity of Boolean functions. *Combinatorica* **7(1)** (1987) 1–22
3. Beals, R., Nishino, T., Tanaka, K.: More on the complexity of negation-limited circuits. In Proc. of the 27th ACM Symp. on Theory of Computing (1995) 585–595
4. Fischer, M.J.: The complexity of negation-limited networks—a brief survey. In Lecture Notes in Computer Science **33** (1974) 71–82, Springer-Verlag
5. Fischer, M.J.: Lectures on network complexity. Technical Report TR-1104, Department of Computer Science, Yale University, June 1974, Revised 1977, 1996. Revised April 1977, April 1996.

6. Markov, A.A.: On the inversion complexity of a system of functions. *J. ACM* **5(4)** (1958) 331–334
7. Raz, R., Wigderson, A.: Monotone circuits for matching require linear depth. *J. ACM* **39(3)** (1992) 736–744
8. Razborov, A.A.: Lower bounds on monotone complexity of the logical permanent. *Mathematical Notes of the Academy of Sciences of the USSR*, **37** (1985) 485–493 (Originaly in Russian).
9. Santha, M., Wilson, C.: Limiting negations in constant depth circuits. *SIAM J. Comput.* **22(2)** (1993) 294–302
10. Sung, S.C.: Negation-limited circuit complexity of symmetric functions. Technical report (1997) JAIST
11. Tanaka, K., Nishino, T.: On the complexity of negation-limited Boolean networks (preliminary version). In Proc. of the 26th ACM Symp. on Theory of Computing (1994) 38–47
12. Tardos, É.: The gap between monotone and non-monotone circuit complexity is exponential. *Combinatorica* **7(4)** (1987) 141–142

Springer Series in
Discrete Mathematics and Theoretical Computer Science

Editors

Douglas Bridges
Department of Mathematics
University of Waikato
Private Bag 3105
Hamilton
New Zealand
e-mail: douglas@waikato.ac.nz

Cristian S. Calude
Department of Computer Science
The University of Auckland
Private Bag 92019
Auckland
New Zealand
e-mail: cristian@cs.auckland.ac.nz

Advisory Editorial Board

John Casti
Santa Fe Institute

R.L. Graham
AT&T Research
Murray Hill

Helmut Jürgensen
University of Western Ontario
University of Potsdam

Anil Nerode
Cornell University

Arto Salomaa
Turku University

Gregory J. Chaitin
IBM Research
Division

Joseph Goguen
Oxford University

Edsger W. Dijkstra
University of Texas at Austin

Juris Hartmanis
Cornell University

Grzegorz Rozenberg
Leiden University

Springer-Verlag Singapore's series in *Discrete Mathematics and Theoretical Computer Science* is produced in cooperation with the Centre for Discrete Mathematics and Theoretical Computer Science of the University of Auckland, New Zealand. This series brings to the research community information about the latest developments on the interface between mathematics and computing, especially in the areas of artificial intelligence, combinatorial optimization, computability and complexity, and theoretical computer vision. It focuses on research monographs and proceedings of workshops and conferences aimed at graduate students and professional researchers, and on textbooks primarily for the advanced undergraduate or lower graduate level.

For details of forthcoming titles, please contact the publisher at:

Springer-Verlag Singapore Pte. Ltd.
#04-01 Cencon I
1 Tannery Road
Singapore 347719
Tel: (65) 842 0112
Fax: (65) 842 0107
e-mail: rebec@cyberway.com.sg
http://www. springer.com.sg

Springer Series in
Discrete Mathematics and Theoretical Computer Science

D.S. Bridges, C.S. Calude, J. Gibbons, S. Reeves, I.H. Witten (Eds.), *Combinatorics, Complexity and Logic.* Proceedings, 1996. viii, 422 pages.

L. Groves, S. Reeves (Eds.), *Formal Methods Pacific '97.* Proceedings, 1997. Viii, 320 pages.

G.J. Chaitin, *The Limits of Mathematics: A Course on Information Theory and the Limits of Formal Reasoning.* 1998. xii, 148 pages.

C.S. Calude, J. Casti, M.J. Dinneen (Eds.), *Unconventional Models of Computation.* Proceedings, 1998. viii, 426 pages.

K. Svozil, *Quantum Logic.* 1998. xx, 214 pages.

J. Grundy, M. Schwenke, T. Vickers (Eds.), *International Refinement Workshop and Formal Methods Pacific '98.* Proceedings, 1998. viii, 381 pages

G. Păun (Ed.), *Computing with Bio-Molecules: Theory and Experiments.* 1998. x, 352 pages

C.S. Calude (Ed.), *People and Ideas in Theoretical Computer Science.* 1998.vii, 341 pages

C.S. Calude, M.J. Dinneen (Eds.), *Combinatorics, Computation and Logic'99.* Proceedings, 1999. viii, 370 pages